The Correspondence of John Tyndall

VOLUME 3

THE CORRESPONDENCE OF JOHN TYNDALL

VOLUME 3

The Correspondence, January 1850–December 1852

Edited by
Ruth Barton, Jeremiah Rankin, and Michael S. Reidy

General Editors
James Elwick, Roland Jackson, Bernard Lightman,
and Michael S. Reidy

UNIVERSITY OF PITTSBURGH PRESS

Published by the University of Pittsburgh Press, Pittsburgh, Pa., 15260
Copyright © 2017, University of Pittsburgh Press
All rights reserved
Manufactured in the United States of America
Printed on acid-free paper
10 9 8 7 6 5 4 3 2 1

Cataloging-in-Publication is available from the British Library

ISBN 13: 978-0-8229-4509-3
ISBN 10: 0-8229-4509-56

CONTENTS

ACKNOWLEDGMENTS

This volume has been a highly collaborative affair. At Montana State University (MSU), a large team of undergraduate and graduate students participated in the initial transcription of the letters. These include Michael Barton, Michael Conrad, Jake Rubow, Shannon Zera, Alexandre Manigault, Estella Terrazas, Jerry Jessee, Paige Madison, Clinton Colgrove, Carla Hoopes, Dane Hoskins, Erin Dockins, and Joey Morrison. Special credit needs to be extended to Daniel Zizzamia, who coordinated much of this work and oversaw the computer server which housed the transcriptions. Joshua Howe served as the project's highly skilled postdoctoral researcher during the early stages of the volume. At the University of Auckland, Louis Gerdelan, Jeremiah Rankin, and Stephen Sequeira helped transcribe letters. Jeremiah's participation steadily increased throughout the project, and his contributions especially stand out. We (two) have been grateful for his close knowledge of the letters and the careful arguments that contributed to reliable dating. Thus, when his research assistantship ran out and large amounts of work remained, we asked him to continue as a fellow editor. For the last sixteen months of the project we have been three editors.

Other researchers, beyond our home institutions, also helped make the project successful. The tireless work of Roland Jackson has been especially important for this volume, as it was through his detective work that the Tyndall-Francis letters were discovered. William Brock was always helpful with esoteric details, Michael Wright assisted with classical allusions, Hsiang-Fu Huang and Geoffrey Belknap provided careful photography of damaged manuscripts, and James Braund and Hanna Roman helped with the translations of German and French letters.

Federal funding and institutional support was crucial. This included a Social Sciences and Humanities Research Council of Canada Grant and a National Science Foundation Collaborative Research Grant (award #0924426). At MSU, we received support from Renee Pera, vice president for research and economic development; Martha Potvin, provost and vice president of academic affairs; Karlene Hoo, dean of the graduate college; Nicol Rae, dean of the college of letters and science; and David Cherry, associate dean of the college of letters and science. At the University of Auckland, the Department of History and School of Humanities provided administrative support. And our dogs, Arlo and Gummi Gigi, helped keep us sane.

LIST OF ABBREVIATIONS

Annal. Chim. et Phys.	*Annales de Chimie et de Physique*
BAAS	British Association for the Advancement of Science
Brit. Assoc. Rep.	*Report of the British Association for the Advancement of Science*
Brock, 'Queenwood'	William H. Brock, 'Queenwood College Revisited', Article XVII in *Science for All: Studies in the History of Victorian Science and Education* (Aldershot: Ashgate Variorum, 1996).
Brock and MacLeod, *Hirst Journals*	William H. Brock and Roy M. MacLeod, eds, *Natural Knowledge in Social Context: The Journals of Thomas Archer Hirst FRS.* (London: Mansell microfiche, 1980).
Brock and Meadows, *Lamp of Learning*	W. H. Brock and A. J. Meadows, *The Lamp of Learning: Taylor & Francis and the Development of Scientific Publishing* (London: Taylor & Francis, 1984)
Carlyle, *Latter-Day Pamphlets*	Thomas Carlyle, *Latter-Day Pamphlets* (London: John Chapman, 1850).
Carlyle, *Past and Present*	Thomas Carlyle, *Past and Present* (London: Chapman and Hall, 1843)
DM	Deutsche Museum, Munich
Emerson, *Representative Men*	Ralph Waldo Emerson, *Representative Men: Seven Lectures* (London: John Chapman, 1850).
Emerson, *Poems*	Ralph Waldo Emerson, *Poems* (London: Chapman Brothers, 1847).
ERE	Michael Faraday, *Experimental Researches in Electricity*, 3 vols. (London: R. & J. E. Taylor: 1839–55).
Faraday Correspondence	Frank A. J. L. James, *The Correspondence of Michael Faraday, Volume 4, 1849–1855* (London: Institution of Electrical Engineers, 1999).
Faraday, 'ERE [N]'	Michael Faraday, 'Experimental Researches in Electricity [N]', a series of numbered articles with consecutively numbered paragraphs: 'ERE 22 continued', *Phil. Trans.* 139 (1849), pp. 19–41; 'ERE 26', *Phil. Trans.* 141 (1851), pp. 29–84; 'ERE 27', *Phil. Trans.* 141 (1851), pp. 85–122; 'ERE 28', *Phil. Trans.* 142 (1852), pp. 25–56.

FRS	Fellow of the Royal Society
Hirst, 'Journals'	Thomas Archer Hirst Journals, Bound Typescript, Volume 2, 1850–1855 (RI MS JT/2/32b).
IC HP	Imperial College of Science, Technology and Medicine, College Archives, Huxley Papers
Jackson, 'Tyndall and Diamagnetism'	Roland Jackson, 'John Tyndall and the Early History of Diamagnetism', *Annals of Science* 72 (2015), pp. 435–89.
Journal	John Tyndall's Journal, 1848–55, typescript (RI MS JT/2/13b)
JRL Frankland	John Rylands Library, University of Manchester, Frankland Letters
Jungnickel & McCormmach	Christa Jungnickel and Russell McCormmach, *Intellectual Mastery of Nature*, 2 vols. (Chicago: Chicago University Press, 1986; paperback 1990)
LT	Louisa Tyndall, John Tyndall's widow
LT, 'Biography'	Louisa Tyndall's typescript biography of her husband (RI MS JT/5/9–11)
NRCC Plücker	National Research Council of Canada Archive, Plücker letters fonds: 1814–1963
ODNB	*Oxford Dictionary of National Biography*
OED	*Oxford English Dictionary*
Phil. Mag.	*Philosophical Magazine*
Phil. Trans.	*Philosophical Transactions of the Royal Society of London*
Poggend. Annal.	*Poggendorff's Annalen der Physik und Chemie*
RDS	Royal Dublin Society
RGO MS.RGO 6	Royal Greenwich Observatory Archives, Papers of George Airy
RI	Royal Institution
RI MS JT	Royal Institution, London, John Tyndall Papers
Roy. Soc. Proc.	*Proceedings of the Royal Society of London*
RS	Royal Society
Russell, *Edward Frankland*	Colin A. Russell, *Edward Frankland: Chemistry, Controversy, and Conspiracy in Victorian England* (Cambridge: Cambridge University Press, 1996)
StA JDF	St Andrews, Papers of James David Forbes
StBPL T&F	St Bride Printing Library, Taylor and Francis papers
StJCL JJS	St. John's College Library, Cambridge, Papers of James Joseph Sylvester

Frontispiece: Portrait of John Tyndall by Henderson of Halifax.
Reproduced by permission of the Royal Society of London (RS.9261).

INTRODUCTION TO VOLUME 3

The 305 letters in this volume span three full calendar years, from January 1850 to December 1852. As the volume begins, Tyndall was a PhD student living in Marburg. He was unknown, almost broke, and working himself to the brink of mental and physical exhaustion in his determination to forge a reputation in science. In the period covered by this volume, he completed his degree, published his first scientific papers, became a regular participant in the British Association meetings, established friendships with leading men of science in Berlin and London, was elected FRS, and applied for, but failed to obtain, various scientific positions. As the volume ends, he was preparing for his first lecture at the Royal Institution of Great Britain, the catalyst for a profound transition in his life. Taken together, the letters offer a behind-the-scenes view of nineteenth-century publishing processes, the practices and challenges of diamagnetic research, the application procedures for university positions, the use of patronage in establishing a scientific career, and the often anxious and weary-worn personality of our ambitious protagonist.

A Life Viewed Through Letters

The letters in this volume reveal Tyndall's preoccupations in his early career. Geographically, he moved often. He spent the first five months of 1850 in Marburg before returning to England in mid-June to investigate employment opportunities. While there, he spent time in London, Lancashire, Yorkshire, and Hampshire, plus two days in Edinburgh at a British Association meeting. After making the acquaintance of Britain's leading natural philosophers and visiting old friends, he returned to Marburg to continue research on diamagnetism. He worked hard in Marburg for the next six months, both at his experiments and literary work. In mid-April 1851, he made a strategic move to Berlin, where he spent two months carrying out further research and, more importantly, making friendly connections with leading German researchers. In June 1851, the mid-point of this volume, he returned to England and the teaching position at Queenwood College that he had left almost three years previously. He was still at Queenwood, seeking scientific employment, eighteen months later when this volume ends.

In Marburg in early 1850, in collaboration with Hermann Knoblauch, the newly appointed "extraordinary professor" of physics,[1] Tyndall was preoccupied with an investigation into diamagnetism. Only fifteen letters are extant from this five-month period. Twelve, between Tyndall and his protégés, Jemmy Craven and Thomas Hirst in Halifax, are lengthy reflections on life's opportunities and obstacles, religious beliefs, and the meaning of life. Tyndall also sought advice and assistance on employment. Surveying was one option but a "literary connexion" became increasingly appealing. He submitted various pieces to magazines and weekly papers, and although a few were published, when he left Marburg for England in mid-June they had yet to bring him any money. Friends offered loans, but Tyndall preferred financial independence.

Over forty letters date from the following four months in England. Introspective reflections on life and discussions with friends and mentors over his career direction and financial prospects continue. Tyndall hoped to find some way to fund a longer stay in Germany in order to continue his magnetic investigations, which he felt confident were leading him in reputation-making directions. A few letters show significant new strategic connections. Early in his visit Tyndall called on Michael Faraday, doyen of British natural philosophers and the pre-eminent researcher on magnetism in Britain, and on Richard Taylor and William Francis, publisher and editors of the *Philosophical Magazine*, to which he had submitted reports of his research with Knoblauch. In August, he gave a successful, but controversial, presentation on his diamagnetic research at the Edinburgh British Association meeting; this initiated ongoing discussion with William Thomson, one of the most brilliant natural philosophers of his generation who, although close to Tyndall in age, was already a professor at Glasgow. He also made useful connections through the Carlylean literary friends of Hirst, which offered further hope of the kind of literary work he sought.

In September, Tyndall deferred an offer from George Edmondson, principal of Queenwood College, to return to his previous teaching position. Instead, he decided to return to Germany to continue his research, even if this meant borrowing money. A last minute offer from Francis to translate German and French articles into English for the *Philosophical Magazine* enabled him to limit his burden of debt. He wrote almost weekly to Francis on translation matters and, often, about his own publications. (Francis probably wrote almost as often, but Tyndall did not save those letters.) From these six months in Marburg there are thirty-two extant letters, twenty of them from Tyndall to Francis. Tyndall also maintained contact with Faraday and Thomson, informing them of the progress of his research. There are very few personal letters from this period. He was sharing lodgings with Hirst, who had accompanied him to Germany, and, when not translating, he was absorbed in his research.

In mid-April 1851 Tyndall accepted Edmondson's renewed offer of a position at Queenwood; he left Marburg to spend two months in Berlin. The nineteen letters from these two months are more personal, for Tyndall wrote regularly to Hirst in Marburg, telling him about his successes in Berlin and asking him, often abruptly, to perform tasks on his behalf. There are no letters discussing translation matters, suggesting that Tyndall had already done all for which Francis could afford to pay him.

Tyndall's return to England and Queenwood in late June 1851 was a major life transition. He settled down to teaching and paying off his debts; at the same time he resumed translation work for Francis and endeavored to maintain his experimental investigations and publishing output. His first preoccupation was his appearance at the Ipswich British Association meeting, at which both Faraday and Thomson were present and where Tyndall made further useful acquaintances. The most important was Edward Sabine, General Secretary of the British Association and Treasurer of the Royal Society. At Queenwood, Tyndall hoped that a better employment opportunity would turn up. Over the next eighteen months he applied for positions in Canada, New South Wales, and Ireland.

The letters from Queenwood reveal much about Tyndall's personal insecurities, the process of collecting testimonials, which was crucial to any application for a scientific position, and the ways in which patrons, especially Sabine, assisted his advance in scientific circles. Previous themes continue: introspective letters to Hirst reflecting on life's opportunities and hurdles; letters to Francis about the selection of articles for translation and the politics of publishing; and carefully worded letters maintaining contact with Faraday and Thomson.

Tyndall's Correspondents

Two correspondents, Thomas Hirst and William Francis, dominate this volume. Over half the letters in the volume, and well over half its length, are taken up by these two correspondents. Hirst was a young surveyor whom Tyndall had mentored in Halifax. In the previous volume of correspondence, Tyndall addressed the young men he was encouraging to pursue self-improvement as "dear boys." During the years covered by this volume, Hirst transitioned from "boy" to Tyndall's "other self."[2] Hirst came to share Tyndall's Carlylean philosophy and, also, his scientific ambitions. In October 1850, they traveled together to Marburg where Hirst began his own doctorate. Their letters are religious, philosophical, moralistic and deeply introspective. They analyze their own motives and ambitions, and comment critically on many colleagues. We have forty-seven letters from Tyndall to Hirst and thirty-nine from Hirst to Tyndall.

The second outstanding correspondent is William Francis, who managed and then edited the *Philosophical Magazine* (*Phil. Mag.*). They probably met when Tyndall visited Richard Taylor's publishing firm in late June 1850. Francis had been with the company since 1837, he became official editor of the *Phil. Mag.* in 1851, and was made a partner in the firm in 1852, renamed Taylor and Francis.[3] He was important as a source of employment and advice. Most of the letters concern the articles which Tyndall was translating for the *Phil. Mag.*, others show Tyndall using his contact with Francis to get his own articles published, and a small number show that Francis, who had extensive scientific contacts, was a significant counselor and confidant as Tyndall made his way, somewhat awkwardly, onto the stage of Victorian science. We have seventy-four letters from Tyndall to Francis, though only four from Francis to Tyndall.

Other correspondents can be thought of in four groups. Tyndall maintained contact with many old friends from his surveying and school teaching days, including the Steuarts and the Wynnes, mentors who used their influence on Tyndall's behalf. A second group are the German scientists, both Marburgers and Berliners. Tyndall was effusive in his journal about the importance of the time he spent in Germany, both for the research done in Knoblauch's and Gustav Magnus's laboratories and for the warm welcome and generous assistance given by professors, young and old. Tyndall felt an outsider in Britain, but at home among the Germans. Heinrich Debus was especially important. He was a Marburg friend who took a position teaching chemistry at Queenwood early in 1851. When Tyndall returned to Queenwood a few months later he worked alongside Debus in the Queenwood laboratory. (As with volume two, these letters demonstrate the close connection between Queenwood and Marburg, with Edward Frankland, Debus, Tyndall and Hirst all doing stints at both.)

Thirdly, British men of science, well-known in Tyndall's later career, appear among his correspondents for the first time in this volume—as mentors, patrons, friends, and rivals. Tyndall's admiration for Faraday (twelve letters) is evident from their first meeting in mid-1850. In effusive language he called Faraday his "unconscious benefactor," a "preacher of toil, courage, and humility."[4] Sabine (fourteen letters) was a more important patron in formal matters: he took charge of garnering signatures for Tyndall's nomination certificate for the FRS in 1851–52 and invited Tyndall to serve as a section secretary at the British Association meeting in 1852. Thomson appears as a competitor. They first clashed at the 1850 British Association meeting and their relationship remained tense, from Tyndall's side at least, for decades. We also have the beginnings of an important friendship, when Tyndall wrote T. H. Huxley a thank-you letter, after Huxley signed his Royal Society nomination certificate.[5]

Lastly, a large set of letters (about forty) in August-October 1851 concern the testimonials which Tyndall requested from eminent scientific figures in Germany, Britain, France, and Switzerland. This group is far wider than those few Germans and British with whom he was already in correspondence.

The collection of letters for these years is far from complete. Even for the Tyndall-Hirst correspondence there are letters for which we do not have replies, or replies for which we do not have the originating letter. It is clear that Tyndall did not systematically save his correspondence. Hirst saved almost everything he received from Tyndall; Francis saved most of the business letters he received from Tyndall but not the personal letters; Tyndall saved the personal letters he received from Francis but not the business letters. We do not know how much is missing, but the letters here and records of correspondence in Tyndall's journal indicate that we have less than half in these years, perhaps less than a third.

Those letters that we do have shine light on how Tyndall navigated his way through the scientific world of nineteenth-century Germany and Britain. There is considerable overlap between Tyndall's journals, available in both manuscript and typescript transcription in the Royal Institution (RI) archives, and these letters. The letters and personal journals from the RI have previously been used to a small extent in biographical writing and in studies of Tyndall's philosophical, religious, and poetical interests.[6] The previously unknown Tyndall-Francis letters are the most significant contribution made by this volume, for they have not been used by any previous scholars and overlap little with the content of Tyndall's journals. They reveal the extent and nature of Tyndall's contributions to the *Philosophical Magazine* and the ways in which he used it to advance his own career interests.

The Puzzling Science of Diamagnetism

Tyndall was deeply engaged in original, experimental work on magnetism during the years covered by this volume. He worked first with Knoblauch in Marburg, then for two months in Magnus's laboratory in Berlin, and finally alongside Debus when he returned to Queenwood. Shortly after beginning intense and systematic investigations with Knoblauch in late 1849, he had a significant breakthrough: "A few days ago a light dawned in upon me," he wrote in his first journal entry of 1850.[7] Tyndall's initial researches into magnetism and diamagnetism, as outlined in both his letters and journals, illustrate the manner in which he pursued subjects, and the combination of experimentation, instrumentation, and theoretical insight that characterized his approach in all his subsequent research in the physical sciences.[8]

Faraday had shown that all bodies were affected by either a magnetic or a diamagnetic force. A magnetic substance such as iron was attracted by the poles of a magnet, setting itself axially between them; a diamagnetic substance such as bismuth was repelled, setting itself at right angles (or equatorial) to the poles. Julius Plücker, a young German researcher who had taken up the study of diamagnetism, focused specifically on the optic axes of crystals. He argued that the poles of a magnet repelled or attracted the optic axes with a force that was independent of both the magnetic or diamagnetic force. He thought he had found an entirely new and distinct force in nature, which he termed the "optic axis" force. Faraday further refined his own and Plücker's experiments, arguing that a "magne-crystallic" force within a crystal interacted with the known forces of magnetism and diamagnetism. He suggested, like Plücker, that a new force was at work, believing that it arose from the symmetrical and ordered manner in which the physical structure of the crystals cohered.

Knoblauch, who was repeating Faraday's experiments, asked Tyndall to join him. While Knoblauch was largely consumed by other work, Tyndall followed an exhaustive, systematic research strategy. Knoblauch's laboratory housed a powerful electromagnet, which enabled Tyndall to extend and perfect the experiments of both Plücker and Faraday. The variation of diamagnetic force with the orientation of the plane of polarization of crystals, discovered by Plücker and Faraday, was puzzling. Tyndall hit on the idea of examining the cleavage plane of the crystals rather than simply their optic axis. His meticulous experiments were carefully planned to control different variables, and made use of the most sensitive instruments available. Whereas Plücker had used only a limited number of polished crystals bought from a mineralogist's shop, Tyndall experimented with almost a hundred different types of crystals, both natural and artificial, many secured from Robert Bunsen's laboratory in Marburg. He manipulated the crystals in many different ways. He cut them into cubes and disks, boiled them in muriatic acid (hydrochloric acid), and polished their surfaces. In other experiments, he ground the crystals into powder, and then reconstituted them in powerful vices to form thick bars and thin wafers. In this way, he could both ensure the purity of his crystals and manipulate their cleavage planes and optic axes, placing them between the poles of a magnet with their cleavage plane either parallel or perpendicular to their optic axis. Through these experiments, Tyndall believed he could interpret the behavior of diamagnetic crystals by peering into their molecular structure. As he told Hirst, "this cutting and cleaving has almost rendered the very constitution of crystals transparent and I seem almost to see into their very molecular arrangement."[9] A focus on the underlying mechanical structure of matter became characteristic of Tyndall's explanatory theories.

Tyndall and Knoblauch published the results of these experiments jointly, in both Poggendorff's *Annalen* and the *Philosophical Magazine* in 1850; these were Tyndall's first scientific publications. The first paper (February and March 1850) was an unequivocal refutation of Plücker's earlier work; the second (July and December 1850) was much longer and outlined Tyndall's and Knoblauch's own findings, which contradicted the earlier work of both Plücker and Faraday. It was strongly critical of Plücker's research. Plücker, Tyndall argued, had erred in his experiments partly because his crystals were not entirely pure, but also because he had not tested a sufficient number of crystals or exercised the extreme care needed to measure the exceedingly small effects of the diamagnetic force. Said simply, he had not been sufficiently methodical. Faraday was treated more respectfully; Tyndall expressed "unqualified admiration" for the "perfect exactitude" of his experiments.[10] He concluded that no new force was required to explain the experimental results of both Plücker and Faraday, but that the known forces of magnetism and diamagnetism were modified by the physical arrangement of the crystal's molecules. Perhaps equally importantly, he also felt that he had learned scientific and life lessons from the investigation: "on what an insecure foundation the most brilliant discoveries of physical science rest" and "the power of resolute patience to break down difficulty."[11]

In his subsequent experimental work and publications, Tyndall expanded his research first to show the similarities between the laws of diamagnetism and magnetism (published in April 1851 and September 1851), then to questions concerning the polarity of bismuth (published in November 1851), and finally moving beyond crystals and magnetism to analyze the role of organic structure in the absorption of heat (the first paper he sent to the Royal Society, and read to members at a meeting in January 1853). In this latter research, Tyndall again demonstrated his patient, systematic approach to experimentation. By cutting pieces of wood into different shapes and sizes, he manipulated the physical arrangement of their fibers, enabling him to measure specifically how the organic structure of wood affected the transfer of heat.

Tyndall's research required the use of sophisticated experimental apparatus. Several letters reveal the importance and difficulty of obtaining sufficiently sensitive instruments. In April 1852, when J. D. Forbes of Edinburgh inquired about the rheostat and tangent galvanometer which Tyndall used in his research, he replied at length. He described his rheostat, drew a careful diagram, sent a sample of the wire used, and suggested that Forbes visit Queenwood to see the instrument. "I purchased mine in Berlin," he advised Forbes, "the price was somewhat under three pounds—I hardly think you will get any thing of the kind at present in England."[12] The superiority of German equipment, which was the best available at the time, is a recurring theme in Tyndall's letters. When seeking a new thermopile (thermosaüle in German) in mid-1852, Tyndall engaged a Berlin instrument-maker recommended by Knoblauch. Several letters between Tyndall and Hirst, covering a six-month period, discuss the purchase, design, manufacture, and safe delivery of the instrument. The process was complicated by the apparent unreliability of the instrument-maker, Tyndall's reliance on Knoblauch and Hirst as intermediaries, and the length of time required to convey queries and instructions between Berlin, Marburg, and Queenwood. Tyndall was prepared to endure such frustrations in order to obtain a first-class instrument.

Tyndall did not rely solely on instrument-makers; he exercised his own ingenuity to devise the sophisticated apparatus his research required. He invented an extremely sensitive and precise torsion balance which, when combined with Knoblauch's very large electromagnet, enabled him to measure the small variations in diamagnetic force with the orientation of the crystal. He later devised an instrument for his research on the transmission of heat through organic material. "I have invented a <u>ducky</u> little instrument for experiments on conduction," he wrote to Hirst. "I look upon it with somewhat of the feelings of a father."[13] Tyndall discovered how to keep the current of a battery constant and thus pass a consistent amount of heat through pieces of wood at a measurable rate. The combination of superior German equipment and Tyndall's ingenuity in designing delicate instruments meant that he obtained more precise experimental data than other researchers.

Tyndall presented his results to the annual meetings of the British Association each fall. His initial work on diamagnetism, critiquing the work of Plücker and Faraday, formed the subject of the paper he read at Edinburgh in September 1850. It was his first appearance before an audience of British men of science. By then, his article (with Knoblauch) in the July issue of the *Philosophical Magazine* had been read by other researchers on diamagnetism, including Faraday, Thomson, and Forbes. His presentation and the published reports of the session—including a short abstract in the *Athenaeum* and a full report in the *Literary Gazette*—ensured that a much wider group knew that he had dared to critique the great Faraday.

It was during this meeting that Tyndall first had an open confrontation with Thomson, which he recounted in letters to his friends in dramatic terms. He described how, during the discussion period after his paper, he outduelled Thomson in a lengthy argument about the significance of Tyndall's results. "[S]uffice it to say it came to a regular hand to hand contest between us," he wrote smugly and a bit misleadingly to Mrs. Steuart, "it was in point of fact the keenest piece of amusement I ever enjoyed."[14] He was far more humble in his journal, recording his doubts and fears, and how, before the meeting, he forced himself to read Emerson to control his mounting "terror."[15]

In contrast to this triumph, the Ipswich meeting in July 1851 was beset by difficulties. Of four papers Tyndall submitted, one was sent to the Chemical rather than the Physical Section, another was omitted from the program entirely, and a third was given so late in the day that he was forced to rush through it. Tyndall did not even note his time spent at the Ipswich meeting in his journal.

In 1852, he was prepared to miss the Belfast meeting, deciding to attend only when Sabine, President-elect for the 1852–53 year, offered him one of the secretarial posts of the Physical Section. According to Tyndall, Thomson again seemed to do everything in his power to hamper Tyndall's success. Yet, despite the roadblocks set by his "scientific antagonist," Tyndall thoroughly enjoyed the meeting, partly because it enabled him to return to Ireland. "Nearly 5 years have now elapsed since my foot rested on Irish ground," he wrote in his journal. "I was then unknown, I now return with a certain small celebrity attached to my name. Little did I imagine 5 years ago that I should mingle among the magnates of the British Association, and mingle among them by a fairly won right, at my next visit."[16] As his journal makes clear, Tyndall also enjoyed the meeting because he felt he had, in a sense, scientifically arrived.

Within two months of the Belfast meeting, Tyndall's focused dedication to science paid off. Henry Bence Jones, an active member of the Royal Institution, heard of Tyndall when visiting Germany and approached him about giving one of the RI's prestigious Friday evening lectures.[17] His lecture in February 1853, entitled "On the Influence of Material Aggregation upon the Manifestation of Force," was highly successful, providing Tyndall with new and hard-earned professional opportunities. It was a succinct summary of his experimental work undertaken during the three years covered by this volume.

"Literary Connexions"

For two years in Marburg, Tyndall lived on the savings he had made while working as a surveyor during the railway boom, but in early 1850, after submitting his thesis and making early advances in diamagnetism, his savings were running out. "I see a rich, unexplored field before me," he explained to Hirst in May 1850, "and my object now is to fix myself in a position to cultivate it."[18] A major theme of these letters, from first to last, pertains to this "object" of placing himself in a financial position to continue his scientific researches. He considered returning to his former teaching position at Queenwood, seeking a position on the English Survey, or returning to work under the surveyor Richard Carter of Halifax, but none of these options would sufficiently enable him to maintain research momentum and build a scientific reputation. Through 1850 we find him, instead, putting persistent effort into securing what he called a "literary connexion"—that is, some "periodical, scientific or literary" willing to pay him for regular contributions, which he could send from Germany while continuing his magnetic researches.[19]

Tyndall was already a published writer. During the 1840s, he had sent a steady flow of poems and short articles under various pseudonyms to Irish, Lancashire and Yorkshire newspapers. In early 1850, encouraged by Hirst and using him as an intermediary, he attempted to publish a series of "Letters from Germany" in the *People's Journal* and in *Howitt's Journal*. He considered collecting the "Letters" together into a book to be funded by subscriptions. It was not until mid-year that any success came from these efforts. Tyndall received his first-ever "literary earnings"[20]—two pounds—for "Propensities and Its Equivalents" in the *Leader*.

While Tyndall's connection with the *Leader* did not provide reliable income, his connection with Francis proved a crucial source of further literary work. In early October 1850, just before Tyndall left England to return to Germany, Francis offered Tyndall regular employment as translator of German and French scientific papers for inclusion in the *Phil. Mag.* Tyndall received two pounds per sheet (that is sixteen pages), amounting to about twenty-five to thirty pounds per year, which enabled him to continue his researches in Germany for eight months rather than the six he had planned.

Tyndall's contribution to the *Phil. Mag.* developed over time. In early 1851, Francis proposed he write a regular column, "Reports on the progress of the physical sciences," informing English readers of important research published in European (most often French and German) scientific journals. Each "Report," the first of which appeared in March, contained translations of two or more related papers, which were often abridged and usually framed by comments by Tyndall elucidating their significance. From late 1851, Tyndall took on a broader, editorial role, writing book reviews and advising on the merit of papers submitted by British scientific men.

His relationship to Francis and the magazine also led to other translation and publishing opportunities. He translated all of the physical science sections of a new edition of Justus Liebig's *Annual Report of the Progress of Chemistry and the Allied Sciences, Physics, Mineralogy, and Geology*, work he undertook for Bence Jones. Francis managed to get Tyndall appointed as the Berlin correspondent of the *Literary Gazette,* although he turned this down, perhaps because the offer came as he was about to leave Berlin. Francis also convinced John Van Voorst, a publisher of natural history books, to commit to a treatise on physics, although one never appeared. In late 1852,

Tyndall also became one of four new editors of the *Scientific Memoirs*, a compilation of translations of the most important papers published in foreign journals.

Tyndall's friendship with Francis and his relationship with the *Phil. Mag.* both proved of immense value in his pursuit of a scientific career. Much more was at stake than just money. Francis was a valuable source of scientific papers and volumes, and was unstinting in meeting Tyndall's frequent requests. The numerous letters between Tyndall and Francis highlight how the former manipulated his relationship with the *Phil. Mag.* to position himself and his work within the scientific community. Tyndall advised Francis—at first as a translator and, later, in an editorial capacity—on which authors, papers, and subjects should be included and which excluded. He often proposed translations or endorsed papers that familiarized readers with his field of inquiry—especially with previous investigations that he intended to either challenge or cite in his own future publications—so that they could better appreciate the significance of his research.

Rather than hiding this fact from Francis, he often gave it as the main reason a translation was needed. His first "Report," subtitled "Electromagnetism," for instance, focused squarely on his own topic. "I have just commenced a review of the matter as it at present stands," he wrote to Francis, "one inducement being that in another month or so I shall send you an investigation of my own on the subject and [it] may be thus made clear beforehand."[21] His own work—his second single-authored paper—was published in the next number. "You have given me a discretionary power and I must use it," he explained, in justification of his self-serving choice of topics.[22] While Francis sometimes disagreed with Tyndall's advice—whether for practical, personal, or financial reasons (see letter 0589, for example)—it was usually followed, enabling Tyndall to influence the contents of the *Phil. Mag.* to his advantage.

Tyndall also facilitated the publication of papers by his German friends and benefactors, repaying their kindness by introducing their work to English readers. He proposed and translated papers by Magnus, Knoblauch, Dove, Riess, Bunsen, and Clausius, among others, for publication in the *Phil. Mag.* and *Scientific Memoirs*. His "Report" in the July 1851 number of the *Phil. Mag.* contained translations of papers by Dove and Knoblauch, both friends from his time in Germany who would soon write testimonials on his behalf. At Tyndall's suggestion, the authors' names appeared at the top of the page, a departure from the practice in previous "Reports" in which Tyndall's name alone occupied such a prominent position (as explained in letter 0498). Tyndall went to great lengths to fulfill Bunsen's request for the publication of his lengthy paper on the volcanic phenomena of Iceland. After approaches to

Francis to include it in the *Phil. Mag.* and Playfair to arrange its publication by the Cavendish Society were rebuffed, Tyndall eventually translated the paper himself for inclusion in the *Scientific Memoirs*. Tyndall's efforts on behalf of his friends were appreciated—Bunsen wrote expressing his thanks[23]—and helped strengthen valuable relationships.

Tyndall's self-promotion through the pages of the *Phil. Mag.* was not limited to his translations. He also made use of his connection with Francis to gain rapid publication of his own research. He badgered Francis in 1851 to reserve space for his intended "memoir on magnetism," writing in mid-January, late January, and early February reminding him to reserve space in the March number.[24] Sometimes he was not forthcoming as to why he wished to rush things into print, while at other times he was explicit. As he was finishing his paper on the polarity of bismuth later in the year, he wrote to Francis: "With regard to my own little paper my object was to pack it off to Toronto along with the others and thus render the packet weightier."[25] Tyndall felt the modern pressure to publish, and he used the magazine and close relationship to Francis to bolster his résumé.

Testimonials: the Search for a Scientific Position

During his second stay in Marburg, Tyndall was once again confronted with the prospect of either staying in Germany or returning to England. Tyndall heeded Francis's advice that time spent near London would help his chances, so he returned to his teaching position in Queenwood. While not the scientific position he desired, Tyndall's time at Queenwood did indeed enhance his career prospects. He was able to continue his original research in the well-equipped student laboratory, which Tyndall fitted out more extensively with instruments he purchased prior to leaving Germany.[26] He was also able to hone his expertise as a lecturer and demonstrator. As his reputation spread, he was invited to deliver popular lectures in local towns.

Tyndall had been at Queenwood less than two weeks when, in late July, he heard about a position in natural philosophy open at the University of Toronto. He immediately discussed his strategy with Francis and wrote to Faraday for counsel, emphasizing his commitment to science and his desire to stay in England rather than go abroad. The delicate and time-consuming process of applying for a scientific position began.

A major aspect of applying for university positions in the mid nineteenth century entailed amassing testimonials. Acquiring them for Toronto dominates Tyndall's correspondence from late July to mid-October in 1851.[27] It was customary to present testimonials from as many influential people as possible, so Tyndall searched far and wide. Beginning with Magnus and Poggendorff, both of whom he knew from his time in Berlin, he approached the most eminent men of science in Germany and Britain, plus a few from Switzerland and France. For those he did not know personally, he developed a standard request format, to George Biddell Airy, Plücker, and Augustus De la Rive, for instance.[28] He sent copies of his published work with many requests, but often noted that expense had prevented him sending all the materials that he had intended.[29]

Supplying testimonials was an onerous task for leading scientific men and many had conventions for avoiding it, as some replies received by Tyndall illustrate. Airy appealed to his position as "a servant of the government" who was unable to offer advice on a matter "which is not related to my own profession."[30] He did, however, offer to respond in a positive manner if anyone approached him. Forbes suggested that he made it a "strict rule to confine myself in such cases to the limits of my personal knowledge of the facts."[31] Faraday replied that it was his custom to express opinion only when approached confidentially. His letter, though, was very warm and he gave Tyndall permission to use it.[32]

Tyndall was grateful to those who wrote fulsome testimonials and critical of those who were brief, as if he did not fully appreciate the extent of the demands that he, and others like him, made on the goodwill of others. Thomson admitted his was "short,"[33] but Tyndall told Francis it was nothing more than a "closely clipped statement of facts."[34] In his journal he was even harsher with his criticisms. "Received an admirable testimonial from Sylvester, frank and generous. I could not help contrasting it with that of Thomson which is almost niggardly. Time no doubt will place me and Mr Thomson in our exact and true relations."[35] The perceived slight seems to have reinforced his already low opinion of Thomson, though it did not preclude further friendly relations between them.[36] Other testimonials reflected growing respect between Tyndall and his mentors. Not only did Sabine enclose a testimonial based on a close reading of Tyndall's publications, but Sabine also wrote a warm letter, which Tyndall then copied in his journal and recopied in several letters to his friends.[37] Sabine then also offered to nominate Tyndall as FRS, with the idea of cementing his connection to English science even if he were to end up in a colony abroad.

Tyndall found it much easier to gather testimonials from his colleagues in Germany, where he was well known, than from those in England. But by the beginning of October, he told Hirst, he had "strong testimony from the principle men" in both countries, as well as France and Switzerland.[38] He used these testimonials in applying for other positions as they became vacant: a position in experimental philosophy and chemistry at the University of Sydney, which opened in November 1851 as he was sending off the Toronto application, and, in mid-1852, a chair of natural philosophy in Queen's College at

Galway. In preparing his application for the latter position, he collected new testimonials to bolster his already large collection.

The process of seeking testimonials and applying for scientific positions demonstrates the importance of patronage in pursuing a career in science. Tyndall worked assiduously to enroll eminent scientific men—most notably Faraday and Sabine—as patrons. He sought counsel from Faraday both because he respected his opinion immensely and for more strategic reasons; he hoped to secure him as a mentor and guide. Requests for testimonials strengthened existing relationships and initiated new ones. Moreover, Tyndall sent printed copies of his testimonials to mentors and patrons, both past and prospective. Those collected for the Toronto application, were sent to the Wynnes, the Steuarts, Faraday, Thomson, and Sabine, among others.[39] By demonstrating his wide connections and rising status in the scientific world, the testimonials showed him to be worthy of their support.

Tyndall received none of these positions, but the process itself helped raise his profile within the scientific community. It led to his election as FRS and earned him the support of influential scientists who worked to keep him from moving abroad. One has a feeling that he would have accepted any of these positions if one had been offered to him, but also that he would have waited to the very last minute for the chance to stay in Britain. Indeed, the process seemed to harden his resolve and heighten his confidence that something suitable would arise. He was willing to stay at Queenwood for a year or two, he related to Sabine, "if at the end of that time a reasonable prospect awaited me of obtaining in these kingdoms the means of carrying out my investigations and of supplying my material wants."[40] In the meantime, Tyndall was drawn increasingly into the vibrant scientific world of Victorian London.

An Examined Life

While Tyndall reserved his more intimate self-reflections to his journals, many of his letters in this volume are deeply personal and introspective. Letters between Tyndall and Hirst are particularly revealing, as they analyze their own motives and ambitions, and weigh their own merits and deficiencies. These letters highlight their shared high-minded ambitions and idealism, and their admiration—or hero-worship—for Carlyle. Carlyle's ideals—hard work, independence, self-discipline, patience, perseverance, and self-control—were the qualities Tyndall and Hirst believed they needed to succeed in their chosen paths. Following Carlyle, they had only contempt for idleness, dissimulation, hypocrisy, frivolity, and mere convention. They were vigilant in guarding against such faults in themselves and critical of those who failed to meet their high standards.

Tyndall's Carlylean spirituality and morality was a secularising transmutation of Christian belief.[41] When George Wynne enquired about his religious beliefs in 1850, Tyndall responded with care—and copied the core of his reply into his journal. He began by emphasizing what he had in common with traditional Christian belief: "the writings of John in the New Testament" were the "epitome" of his religion. "The religion there taught is not the religion of the understanding but a deeper one." He emphasized that his beliefs had not been altered by the two years he had been in Germany. His guides were still Carlyle and Emerson, the writers who "have exercised and still continue to exercise a great influence upon me."[42] Tyndall also offered Carlylean advice to friends and protégés. Idleness, he told Margaret Ginty, is "the greatest sin to which humanity is subject."[43] When this volume opens, Tyndall was something of a spiritual mentor to Craven and Hirst. Hirst was torn between a Carlylean vision of life and the orthodox Christianity of his family and many friends, but within a few months he came to share Tyndall's ideals. Hirst moved from protégé to "soul friend"; as Tyndall reflected in his journal, "he thinks my thoughts and shares my creed."[44] They resolved to be friends in the "ideal" Platonic sense and devoted lengthy letters to reflections on friendship, love, intimacy, and the meaning of life, often through commentary on poetry.[45] When lodging together in Marburg, they reflected together on essays from Montaigne or Sophocles' *Lives* (of Greek philosophers), rather as Christians might meditate on passages from the *Bible*. Hard work, in whatever task to which they had been "called," was their mantra; they gave new meanings to the language of the religion that they had rejected.

Carlyle's social and moral vision resonated with, and reinforced, Tyndall's understanding of life. Tyndall experienced life as struggle. His ambitions, like those of many upward-aspiring members of the lower orders, required self-discipline: a struggle against one's own inclinations to rest, drink, make merry with friends, and marry, rather than to work hard, save money, study for long hours, and practice Malthusian restraint. It was also a struggle to learn the social codes of a different class. As Tyndall revealingly wrote to Sabine, "I have endeavoured, with what success I know not, to render myself fit for the companionship of cultured men."[46]

Tyndall suffered deep-seated angst about coming to science as an outsider. It was a struggle to make one's way in science without the privileges and inside knowledge that came with a cultured background. The process of his election as FRS was particularly fraught. When Francis first raised the topic, Tyndall admitted that he was entirely ignorant "as to how matters are transacted in the Royal Society and [such] places."[47] When Sabine put him forward for nomination, anxiety dominated. He worried about the cost until reassured by Sabine. In writing to Francis regarding obtaining signatures for his application, Tyndall asserted his aversion to patronage or honors that were offered for reasons other than merit—out of charity, for instance, or because of effective lobbying. "One word in your last letter I must take exception to, and that is the word 'prevail'. —I would not prevail on anybody. I would sooner wait for 7 years and gain the matter on the grounds of having worked up to it than accept it tomorrow through interest."[48] Once elected, a chance remark from Sabine led Tyndall to feel that he may have misrepresented himself; he felt he should confess his lowly social origins in case Sabine would wish to withdraw his patronage. These letters point to a recurring theme: Tyndall's struggle to reconcile his Carlylean ideal of self-reliance—on which his self-respect and, he believed, the respect of his peers, relied—with the practicalities of scientific advancement.

Tyndall construed his experimental research as a struggle—against nature, rivals, and his own physical and mental limits. During his periods of intensely-focused research, Tyndall worked himself to the brink of exhaustion, demonstrating a remarkable work-ethic, strict self-discipline, and debilitating self-denial. He woke before six a.m. most mornings, and after sponging himself with cold water, put in productive hours in the laboratory. When at Queenwood, he would attend to his teaching duties during the day, but then return to his experiments at night, working often until ten p.m. He became exceedingly abstemious, believing that he did some of his best thinking when he was starving himself. The letters are filled with references to Tyndall being overworked, underfed, and physically and mentally exhausted. He began to have headaches, which could keep him in bed for an entire day. Others mentioned his degraded physical appearance. Du Bois, Hirst told Tyndall, "tells me you look pale and care worn."[49] Mrs. Edmondson, Tyndall related to Hirst, was "concerned for that young man—he looks like an old one—he will most certainly break down—in fact one needs only to look at him to see that he is broken down already."[50] Characteristically, Tyndall quoted Mrs. Edmondson's concerns proudly, as evidence of his toughness.

As well as scrutinizing his own motives and conduct, Tyndall was extremely sensitive to how others treated him. His insecurities and his deep sense of struggle led him to misinterpret minor, unintended slights as evidence of disapproval or antipathy, and to expect—often mistakenly—that others shared his sense of competition and rivalry. He persuaded himself that he did not care that David Brewster, who had not replied to a letter, was "cold and indifferent" towards him, but then Brewster disproved Tyndall's interpretation by shaking his hands warmly and apologizing for the unanswered letter.[51] He expected that Plücker would be "dreadfully angry" when he and Knoblauch disproved his theories and was later surprised that Plücker wrote a generous testimonial.[52] This expectation of confrontation was in conflict with other ideals he sometimes espoused. Faraday drew out his best aspirations.

When Faraday told Tyndall that science benefitted when each person shared ideas and accepted corrections, Tyndall agreed: "When Science is a Republic as you say it gains."[53] In practice, though, he conceived science and his larger life in terms of conflict rather than cooperation.

Another unattractive analogue of his constant self-examination was analysis of the character of colleagues, which often turned into self-righteous criticism. Tyndall and Hirst reinforced one another's condescending criticisms of friends. They looked down on Noll, the Marburg physician who gave them medical advice, because he lacked ambition: "The summit of Noll's wishes is to be respectable country doctor, with a comfortable fire-side and a wife."[54] Knoblauch, to whose collaboration and equipment Tyndall owed his first publications, and Edmondson, who twice hired him, were both denigrated as "a reed shaken by the wind."[55] The former, Tyndall and Hirst seemed to imply, was not fixed in his ambitions. The latter—a regular target of Tyndall's exasperation and contempt—was seen to lack decisiveness and force of character; he was (quoting Carlyle) "a well cultivated cabbage plant" rather than a strong tree.[56]

The dark side of Tyndall's insecurity and ambition is revealed most clearly in his accounts of the 1852 British Association meeting at Belfast. In recounting the meeting to Hirst, he summed up Thomson as his "old scientific antagonist," overlooking his fellow-physicist's past generosity.[57] As at previous meetings, Tyndall interpreted Thomson's ebullience and self-confidence in debate as combative, and his sometimes thoughtless interruptions and criticisms as intentionally undermining. Tyndall thought Thomson unnecessarily cut short his first presentation. The following day, Tyndall complained that Thomson had "made a long communication, so spun out as to throw me quite to the end of the day and leave me a very small audience." Although Thomson "delivered some very eulogistic remarks regarding my 'admirable discoveries' afterwards"—evidence that he "is a decent soul at bottom"—Tyndall attributed his sabotaging actions to insecurity and jealousy: "he is greatly afraid of his fame. I think it will never be extraordinary."[58]

Even the admiring Hirst occasionally questioned Tyndall's defensiveness and egotism. When Tyndall wrote, about the Brewster incident mentioned above, that "I contrived to feel indifferent about Sir David's indifference whispering to myself that a day might come when he perhaps could not afford to be so great as he was,"[59] Hirst replied with an unusually admonitory letter, one he obviously struggled to write to his mentor: "Science may be demanding too much of thee . . . tending to swamp the Man in the Philosopher."[60] Tyndall spent the following Sunday in rare self-criticism: "Tom's letter has I think been of service to me . . . and induced me to withdraw for a time to higher and holier contemplation."[61] He copied the letter into his journal, but then turned his Sunday contemplation to a portrait of the just-deceased Duke of Wellington. He felt "instructed in fortitude and courage" and resolved to care less about priority and fame: "When I see great men squabbling about questions of priority," he told Hirst, "I know that they have not the spirit; that their object is fame and the admiration of men."[62] The struggles revealed in these letters, between his high ideals on the one side, and his deep-rooted anxiety, driving ambition, and overweening but brittle pride on the other, define Tyndall's character during this period.

Conclusion

The letters in this volume chart Tyndall's social maneuvering as he made his way in the Victorian scientific world. The period witnessed remarkable political, social, and economic developments. Prussian troops arrived in Marburg as he finished his dissertation; he visited the Great Exhibition during one of his returns to London. Yet, these momentous events are increasingly relegated to the background, as Tyndall's life and his letters begin to focus more narrowly on efforts to earn a scientific reputation.

That these efforts relied so heavily on self-promotion through the pages of the *Philosophical Magazine* came as a surprise. All of his work for the magazine—his translations, book reviews, and "Reports"—enabled him to get his name in print, a known quantity who was doing yeoman's work for the good of science. It also gave him an inside, working knowledge of the process and politics of scientific journal publication. As Francis confided to him, "the Magazine has a commercial as well as a scientific aspect and in order to print the good papers it is sometimes necessary to give one of little or no value."[63] We, too, gain inside knowledge into the publishing process. The Tyndall to Francis correspondence was found very late in this project. Without it we would be ignorant of the extent of Tyndall's involvement in the *Philosophical Magazine* and oblivious to the extent to which he used Francis's position as editor and publisher to his own advantage.

The letters also highlight how Tyndall approached his own research published in the *Philosophical Magazine*. The extreme care, patience and exactitude he took in devising his experiments, building and setting up his apparatus, and then conducting his experiments formed the foundation of his success. Protracted investigations demanded intense periods of focused attention. As he explained it to Hirst, "hard work resembles teetotalism—it is easier for a drinker to abstain altogether than to half abstain—and the more I work the easier and pleasanter I find my work."[64] When he was engaged in these intense periods of work, it eclipsed everything else. Such hard work satisfied Tyndall's Carlylean ethics, where deep immersion was necessary to win the struggle against nature, colleagues, and himself.

The letters offer an intimate portrait of Tyndall as he established his research trajectory, published his work, and presented it to his peers. The boundaries of physics were being formed during this period, although differently in different countries. Tyndall's training in Germany was quite different from that available in England or Scotland. The Germans were establishing chairs in "experimental physics" when positions in Britain were still called "natural philosophy." While Faraday called himself an "experimental philosopher" and Thomson preferred the more general term "naturalist,"[65] Tyndall showed his German education by calling himself a "physicist." Because the letters follow Tyndall as he established his research, searched for a scientific position, and forged a name for himself in science, they offer a valuable perspective on the creation of a "physicist" in the mid-Victorian era.

Part of Tyndall's success required that he endear himself to others. He worried incessantly about his social position and the opportunities that would or would not come his way. What is most striking in these letters is how viscerally Tyndall reacted to how other people thought of him. He was extremely despondent when he was received coldly or even coolly; he was overjoyed when he was greeted with warmth. A small slight could last for months in Tyndall's mind. Even Louisa, who rarely allowed negative attributes to sneak into her unpublished biography of her husband, noted this tendency. "We find him exhibiting a considerable degree of susceptibility to the nature of his reception at the hands of other men," she perceptively wrote, "which in some cases almost verges upon suspicious watchfulness."[66] Many of these suspicions were created entirely in Tyndall's mind, a fraught and troubled space full of friendship and envy, kindness and egotism, high ambition and deep insecurity, which the letters in this volume enable us to explore.

Notes

1. "Extraordinary professors" were able to take fees from students, but did not receive a university salary.
2. Journal, 2 March 1851, JT/2/13b/523.
3. Brock and Meadows, *The Lamp of Learning*, chapter 5.
4. Letter 0575.
5. Letter 0582.
6. See, for example, W. H. Brock, N. D. McMillan and R. C. Mollan (eds.), *John Tyndall: Essays on a Natural Philosopher* (Dublin: Royal Dublin Society, 1981); Ruth Barton, "John Tyndall, Pantheist: A Rereading of the Belfast Address," *Osiris* 3 (1987), pp. 111–34; S. Kim, *John Tyndall's Transcendental Materialism and the Conflict between Religion and Science in Victorian England* (Lewiston, Queenston, Lampeter: Mellen University Press, 1996); Bernard Lightman and Michael S. Reidy (eds.), *The Age of Scientific Naturalism: Tyndall and His Contemporaries* (London: Pickering & Chatto, 2014).
7. Journal, 1 January 1850, JT/2/13b/473.

8. For a discussion of Tyndall's research during these years, see Jackson, "Tyndall and Dia-magnetism"; A. J. Meadows, "Tyndall as a Physicist" and Irena M. McCabe, "Tyndall, the Chemical Physicist" (both in Brock et. al., *John Tyndall*, pp. 81–92 and 93–102).

9. Letter 0403.

10. Letter 0418.

11. Letter 0403.

12. Letter 0620.

13. Letter 0638.

14. Letter 0420; compare letter 0418.

15. Journal, 7 August 1850, JT/2/13b/503.

16. Journal, 29 August 1852, JT/2/13b/581.

17. See Sophie Forgan, "Tyndall at the Royal Institution," in Brock et. al., *John Tyndall*, pp. 49–59 (on p. 49). Tyndall told Hirst about Bence Jones's approach in letter 0673; the formal invitation came from the RI Secretary, John Barlow (letter 0675) when the details had already been discussed with Bence Jones (see letter 0674).

18. See letters 0398 (n. 16) and 0400.

19. See, for instance, letters 0401 and 0408.

20. Journal, 2 October 1850, JT/2/13b/513. The article was published in late June but Tyndall did not know of the payment until October.

21. Letter 0458.

22. Letter 0459.

23. Letter 0680.

24. Letters 0461, 0462, and 0464.

25. Letter 0530.

26. Edmondson provided the needed funds (see letters 0495 and 0497). On the Queenwood laboratory in relation to science education, see Brock, "Queenwood" and D. Thompson, "Queenwood College, Hampshire," *Annals of Science* 11, no. 3 (1955), pp. 246–54.

27. The testimonials are not included in the correspondence, as they are open letters to potential committees rather than strictly confidential letters to Tyndall.

28. Letters 0518, 0522, and 0523.

29. Enclosures are mentioned in letter 0518 to Airy in England, for example, while the prohibitive cost of postage is mentioned in letter 0522 to de la Rive in Germany.

30. Letter 0527.

31. Letter 0529.

32. Letter 0506. William Whewell (letter 0542) made a similar offer.

33. Letter 0528.

34. Letter 0530.

35. Journal, 2 July 1852, JT/2/13b/573.

36. See, for example, letters 0560 and 0635.

37. See letter 0559 from Sabine, Tyndall's Journal (JT/2/13b/550–51), and for example, letters 0563, 0568, and 0577.

38. Letter 0540

39. Letters 0573, 0577, 0586, 0555, 0560, and 0559.

40. Tyndall quoted these words he had written to Sabine in letter 0540 to Hirst.

41. See Barton, "John Tyndall, Pantheist."

42. Letter 0397.

43. Letter 0517. See also Tyndall's advice to Hirst in letter 0604.
44. Journal, 2 March 1851, JT/2/13b/523.
45. For "ideal" friendship, see letters 0421 and 0424. For intimacy, see letters 0426–0428. For reflections on poetry, see, for example, the protracted discussions of Emerson in 1850 and Schiller in the latter half of 1851.
46. Letter 0639.
47. Letter 0475.
48. Letter 0603.
49. Letter 0689.
50. Letter 0540.
51. Letter 0661.
52. Letters 0392 and 0549.
53. Letters 0453 and 0465.
54. Letter 0480.
55. Letter 0481.
56. Letter 0504.
57. Letter 0661. Thomson had written a lengthy letter (0425), for instance, when Tyndall requested only a reference.
58. Journal, 4 September 1852, JT/2/13b/583.
59. Letter 0661.
60. Letter 0667.
61. Journal, 24 October 1852, JT/2/13b/588.
62. Letter 0673.
63. Letter 0589.
64. Letter 0484.
65. Jed Buchwald and Sungook Hong review the literature on the emergence of physics in "Physics," in David Cahan (ed.), *From Natural Philosophy to the Sciences* (Chicago: University of Chicago Press, 2003), esp. pp. 166–69. Ursula DeYoung notes the different terms used in *A Vision of Modern Science: John Tyndall and the Role of the Scientist in Victorian Culture* (New York: Palgrave Macmillan, 2011), pp. 8–9, although without elaborating on what Tyndall meant by "physics."
66. LT, "Biography," JT/2/10/515.

EDITORIAL PRINCIPLES

Our aim is to include all extant letters to and from John Tyndall. Some of the letters are from published sources, such as letters written by Tyndall to newspapers or journals, but the majority are personal letters that are transcribed either from the originals or from typescripts produced by Louisa Tyndall from originals, which may now be missing. One central editorial principle has been to transcribe, wherever possible, from original letters rather than Louisa's typescripts. This is because of Louisa's conscious and unconscious editorial interventions, a point discussed in more detail below. We have also aimed to reproduce, as accurately as possible, the text of the letters as they were written. Idiosyncrasies of spelling and punctuation in the handwritten letters have therefore been preserved, but, because Louisa made many inadvertent errors, mistakes appearing in typescript sources have been silently corrected. But, in general, non standard spellings (e.g., the use of dialect) have been retained even when in a typescript-only letter (though there is the possibility that the spelling is Louisa's, not her husband's). Where the transcription is based on a typescript transcription by Louisa (or one of her assistants), this will be indicated by "LT Transcript Only." Some of the letters in this volume are transcripts made by Tyndall, indicated by "JT Transcript Only," while others are transcripts of letters copied from the originals by Tyndall indicated by "LT Transcript of JT Transcript Only."

We have included illustrations that are part of the letter. In cases where it was possible to reprint the original handwritten image, we have done so. In other cases, volume editors have reconstructed drawings because it was not possible to reprint an image from the handwritten letter (due, for example, to deterioration of the original document).

The letters are presented in chronological order of writing. Where more than one letter has the same date, letters from Tyndall (in alphabetical order of addressee) come first, followed by letters to Tyndall (in alphabetical order of writer), with the exception that if the order of writing can be determined, that order takes precedence. Where letters cannot be dated precisely, they are placed at the latest likely date of writing. (In some cases, if it is possible, but unlikely, that they were written later, notes indicate such possibilities.) To indicate the beginning of a new year, the date of that year has been inserted in a large font before the first letter from that year.

For non-English language letters, both the transcription of the original and the English translation are presented. The transcription of the non-English letter appears first, followed by the English translation in a smaller font. Editorial notes are inserted in the English translation. Some foreign letters are in an original hand, while others exist only as transcriptions by Louisa or others who often did not have a full command of the language. In the latter case, these letters contain frequent spelling, grammatical, and interpretation errors that do not seem to have been committed by the original author. In these typescript-only cases, and only in these cases, spelling and grammatical errors have been corrected, as have poor interpretations where a better word choice was obvious from the context.

There is a standard format for each letter. The first line lists the author, if it is to Tyndall, or the recipient, if it is from Tyndall; the date; and the letter number. The second line lists any information supplied by the writer in their salutation, such as the place, date, or time of day when the letter was written. (In some cases, though, where the writer put the date at the foot of the letter we have left it there in order to reproduce the text accurately.) The opening salutation, the text of the letter, and closing then follow. Then come the source and editorial notes, the latter indicated in the text by Arabic numerals.

Editorial notes are intended to provide the contextualization necessary to understand the contents of each letter. There are basically six types of editorial notes. Many of them are informative notes on persons mentioned in the letters. If a person referred to in a letter is not described in a note, they have been mentioned more than twice in the volume and they will have an entry in the Biographical Register at the back of the volume. For the rest—those mentioned once or twice—biographical information will appear in the editorial note appended to the letter in which he or she is first mentioned. When there is a second reference to that person, a note will refer back to the letter in which he or she was first mentioned. Biographical notes on well-known figures such as Faraday or Huxley stress what that person was doing at the time when the letter was written rather than providing a full overview of the person's life.

A second type of editorial note identifies allusions and quotations. In the case of quotations, where possible, the first edition of the work is cited unless a different edition is mentioned in the letter. If a quote is not in English, a translation is given. Glosses on obscure places, abstruse words, and words used in an unfamiliar sense comprise the third type of editorial note.

The fourth type provides information about the context of the letter, drawing on such sources as contemporary publications, scholarly sources,

Tyndall's personal journals and field notes, and other letters in the Tyndall Correspondence. The fifth type of editorial note is cross-references to other letters in the Correspondence, referring to the letter number. The last type is information on significant textual changes that Tyndall made to the original that is thought to be relevant to the meaning of the letter.

There are a series of conventions for indicating when editors have given additional information about the text of a letter:

When Louisa Tyndall has made an annotation or marginal insertion that the editors have deemed important enough to retain, we have included her annotation in an editorial footnote.

When editors other than LT have made insertions, they are indicated by unitalicized square brackets, for example, [word].

When certain words are ambiguous and we have had to conjecture their meaning, they are indicated by italicized square brackets, for example, *[words]*; when we have considered such words illegible, the number of illegible words are indicated in full italicization and square brackets, for example, *[3 words illeg]*.

If a word or series of words are not present or have been destroyed—by inkblots, holes, or mold, for example—they are indicated as missing in italicized angular brackets, for example, *<3 words missing>*.

In extremely rare cases where it is clear that Louisa herself has intentionally destroyed a few words, they are indicated as excised in italicized angular brackets, for example, *<3 words excised>*.

Finally, a brief word about Louisa. She was responsible for collecting together many of Tyndall's letters in order to write a biography of her late husband. This biography was not completed. It is important to recognize that Louisa was not always reliable as a transcriber. She often threw out the manuscript letter when she had made her transcription, but some original manuscripts remain, which indicate her editing processes. Like any transcriber, she was sometimes inaccurate, but corrections show that she proofread her transcriptions. However, she also edited Tyndall's writing to make his letters seem more consistent or grammatically correct. For example, she made punctuation more formal, corrected grammar, and, occasionally, corrected quotations. She consistently corrected Tyndall's usual (deliberate) lowercase for "god" to "God." Her typewriter was unable to insert the superscripts that Tyndall used in abbreviations, or the "&c" symbols that stood for an address at the foot of the letter. (Modern typefaces are unable to deal with all the variant scrawls that the "etc." symbol took, and they are all formalized here to "&c.") Louisa also left out details that she considered too private or too salacious for public consumption. She equated postmark and date of writing and overlooked the time taken to write some letters. When she inferred dates from internal evidence, her reasoning was sometimes faulty. Similarly, her identifications of people are sometimes in error. Independent evidence has always been sought for her dates and identifications. Lastly, the patterns of punctuation differ between LT sources and manuscript sources. Tyndall himself used dashes more often than commas, semicolons, and stops, and he used dashes of many different lengths that are all standardized here to em-dashes. It is often difficult to distinguish between very short dashes and commas or very short dashes and stops. Similarly, capitalization is often ambiguous; distinguishing S/s for "science," for example, is a particular problem. Therefore, scholars for whom punctuation and capitalization are significant will need to consult the manuscripts. Users should also be aware that throughout this project the transcribers and editors have worked from scans and photographs rather than the original manuscripts.

NOTE ON MONEY

In the nineteenth century the currency used in England was the pound sterling, usually abbreviated to £ (or, occasionally, to L). A pound consisted of 20 shillings (abbreviation, s) and each shilling contained 12 pennies (d). A guinea was 21 shillings, that is £1 1s. A crown was 5 shillings (5s), and half a crown equalled two shillings and sixpence (2s 6d). The smallest coin was a farthing, one quarter of a penny (¼d).

In the notes to this volume the editors have used the following format: £3 4s. 5d. Several forms of representation were in contemporary use. Sometimes the numbers for pounds, shillings and pence were separated by dots, forward slashes or spaces; thus £3 4s. 5d. could be written as £3.4.5, £3/4/5 or £3 4 5. A zero could be represented by a dash; thus 7 shillings was often written as 7/- or just 7/. A half crown was often written as 2/6.

German currency is occasionally mentioned in this volume. The currency of the northern German states was the thaler (Thaler in German, usually abbreviated Th), the groschen or silbergroschen (Silbergroschen, abbreviated SG or Sg) and pfennig (Pfennig, abbreviated Pf). A thaler consisted of 30 groschen, a groschen of 12 pfennig.

This currency was introduced in Prussia in 1821 and spread to other northern German states following the currency union of 1837. Groschen were called Silbergroschen, especially in Prussia, to distinguish them from the pre-1821 coins of the same name. This system was in place until German unification in 1871.

In mid-century the exchange rate between British and German currencies was approximately 7 Th per £ (see letter 0496 for a specific example).

Canadian currency also appears in this volume. In British North America exchange rates between the pound sterling and local coinages were set locally. The Halifax pound, which became a widely used standard, set the rate of exchange between pounds sterling and the common local coinage, the Spanish silver dollar, at one pound: four dollars. In 1841 when the Halifax pound was set as the official rate across the Province of Canada (comprising the southern parts of modern-day Quebec and Ontario) the rate was modified slightly to one pound: 4 dollars 1 cent. (James Powell, *A History of the Bank of Canada* (Bank of Canada, 2005).)

NOTE TO READERS

A large number of the letters in this volume were not dated by their writers, and in some the first page is missing, leaving them, effectively, undated. Tyndall is the chief culprit. He sometimes gave only the day of the week; often he gave no indication at all. We have been able to date almost all letters sufficiently closely to present them in order of writing and here we outline the chief procedures we used in attributing dates. Between any pair of correspondents it has usually been possible to place the letters in order of writing; a few of these letters are dated which gives an initial time frame. References to external events and recent publications narrow down the dates further; we consider every allusion in a letter rather than rely on any one alone. Postmarks can be used, but cannot be assumed to be the same as date of writing. Day of starting, day of finishing (sometimes indicated in a postscript), and day of posting are distinguished. The journals of Tyndall and Hirst, and many of their letters, also refer to the sending and receiving of letters which sometimes gives a very close date.

Caution is needed at many points. When transcribing letters Louisa Tyndall used the postmark as if it was the date of writing; her dates are therefore often out by one or two days. She sometimes guessed, or based her dating on allusions in the letters, which often turned out to be incorrect. When using the journals of Tyndall and Hirst allowance must be made for the gaps in their journals. They often wrote them up only every few days, or once a week, so entries often cover more than their specified date or were written some days after their specified date. Hirst, who complained to Tyndall about his undated letters (see letter 0404), wrote the date on which he received letters on the back or at the top of Tyndall's letters; occasionally, though, when he could determine it, he put the date of writing at the top. Handwriting needs to be examined: sometimes we have been puzzled by dates, only to discover that they were inserted in another hand by some later collector or archivist. Sometimes the writer himself put the wrong date on the letter (for example, Hirst, in letters 0414 and 0488). Where day of the week and date contradict one another (and in the absence of other evidence) we take day of the week as more reliable.

Allusions to articles in *Poggendorff's Annalen* and the *Philosophical Magazine*, both of which were published monthly, provide supplementary information. The *Phil. Mag.* is especially useful because it seems to have been published on the first of the month; the exact date that *Poggendorff* was published seems to have been variable and is therefore less useful for dating.

From those letters for which we have full information, we have worked out standard postage times: four days between London and Marburg; one day between Berlin and Marburg post offices, but usually an extra day to get to the addressee; five days between Marburg and Halifax post offices; one day between Queenwood and London; one day between a Yorkshire town and a Lancashire town. Hence if we have the date received, for example, from a journal entry, we can infer the date sent. Finally, if the day of the week is given by the writer and we have achieved a narrow date range by the above procedures, then we can determine the precise date. For letters undated by the writer, the first editorial note describes how the attributed date was reached, but does not explain the prior process by which the sequence of letters between two correspondents was determined. In cases where our dating is less certain (for example, October and December 1851), we have explained our reasoning more fully.

There are some periods when Tyndall and Hirst (July–September 1850) and Tyndall and Francis (September 1851) were writing almost daily to each other. Letters crossed in the mail; they allude to missing letters; some are undated or were incorrectly dated. They also occasionally met each other in person, making it difficult for us to determine if allusions in letters pertained to these discussions or perhaps to missing letters. We think that we have sorted these letters into order and that our dates are reliable, but we must warn users of the volume that the task was difficult and that we could have overlooked some germane point.

The Tyndall letters in the Taylor and Francis archives present exceptional problems. Not only are many undated, but individual letters were often split up in the archive (probably because some sheets were used in the editing and printing processes); sometimes we could join together first halves and second halves, but in many cases we have partial letters only. Others parts may exist in the archive but have yet to be identified as written by Tyndall. In one case (letter 0550) the two halves of a letter came from two different archives—the Taylor and Francis Archive at the St Brides Printing Library and the Royal Dublin Society (where, for unknown reasons, a small number of Tyndall-Francis letters ended up). Moreover, many of the sheets are in very bad condition (because they were used by editor and typesetters and were written on extremely thin paper meant for long-distant postage), hence we have many conjectured readings.

At various points in the process of editing these letters we received newly discovered letters which changed our previous date determinations. New letters sometimes clarified what was previously an indeterminate allusion and we were pushed to reconsider the interpretations that led us to a date. We found that every allusion must be taken into account. We explain our process of attribution in order that users who find further relevant letters have the tools to assess our dates. Given this array of problems we would be grateful to be informed if any user finds that any of our dates are incorrect, or if any of our date ranges can be improved.

TIMELINE OF JOHN TYNDALL'S LIFE

Year	Event
c. 1822	John Tyndall born at Leighlin Bridge, County Carlow, Ireland
c. 1836	Attends John Conwill's National School in Ballinabranna
1839	Begins employment as civil assistant with Ordnance Survey of Ireland in Carlow
1840	Becomes civil assistant in the Ordnance Survey Office at Youghal, Cork County
1841	First appearance in print (poem in *Carlow Sentinel* under pseudonym 'W[alter] S[nooks]')
1842	Joins English Ordnance Survey in Preston
1843	Leads written criticism of Ordnance Survey in *Liverpool Mercury*; dismissed November
1844	Home in Ireland, mostly unemployed, until obtaining a position with the firm of Nevins and Lawton, Surveyors, of Manchester
1845–47	Works for Richard Carter, Surveyor, of Halifax
1845	Meets Thomas Archer Hirst in Halifax
1847	Begins teaching mathematics and surveying at Queenwood College, Hampshire
1848–50	At University of Marburg, working for PhD with Robert Bunsen and Friedrich Stegmann
1850	Returns to England: meets Michael Faraday and William Francis and attends his first British Association meeting, before returning to Marburg
1851	Moves from Marburg to Berlin (April–June); then returns to Queenwood
1852	Elected Fellow of the Royal Society
1852	Friendship developed with Thomas Henry Huxley
1853	First lecture at Royal Institution (RI)
1853	Meets Herbert Spencer
1853	Appointed Professor of Natural Philosophy at RI
1853	Awarded Royal Medal for magnetic work; declines when controversy arises over award
1857	Climbs Mont Blanc for first time with Hirst
1859	Demonstrates existence of greenhouse gases and climatic implications
1859	Joins Government School of Mines as Professor of Natural Philosophy
1860	*The Glaciers of the Alps* published
1860	Meets John Lubbock and his wife, Ellen Lubbock

Year	Event
1861	First ascent of the Weisshorn
1861	Delivers his first Christmas Lectures at RI
1862	*Mountaineering in 1861* published
1862	Begins to champion Mayer's priority in discovery of conservation of energy
1863	*Heat Considered as a Mode of Motion* published
1864	Awarded Royal Society's Rumford Medal
1864	First meeting of X Club
1865	*On Radiation* published
1865	Engages in public controversy over the efficacy of prayer
1866	Succeeds Michael Faraday as scientific adviser to Trinity House
1867	Michael Faraday dies; Tyndall becomes Superintendent of the RI
1867	*Sound: A Course of Eight Lectures* published
1868	Successfully climbs Matterhorn
1868	Discovers the cause of light scattering, to be known as 'Tyndall Effect'
1868	*Faraday as a Discoverer* published
1868	Lecture on 'Scientific Materialism' at the British Association
1868	Resigns from Royal School of Mines
1869	Joins Metaphysical Society
1870	'On the Scientific Use of the Imagination' lecture at British Association
1870	*Researches on Diamagnetism* published
1870	*Three Scientific Addresses* published
1870	*Researches on Diamagnetism and Magne-Crystallic Action* published
1871	Meets Louis Pasteur for first time while in Paris
1871	*Hours of Exercise in the Alps* published
1871	*Light and Electricity* published
1871	*Fragments of Science* published
1872	Debates over how to measure the efficacy of prayer, aka the 'Prayer-Gauge Debate'
1872	*Contributions to Molecular Physics in the Domain of Radiant Heat* published
1872	*The Forms of Water in Clouds and Rivers, Ice and Glaciers* published
1872–73	USA Lecture Tour
1873	*Six Lectures on Light* published
1874	As President of the BAAS, delivers the 'Belfast Address'
1876	Marries Louisa Charlotte Hamilton
1877	Develops 'Tyndallization' (lengthy sterilization to destroy heat-resistant spores)
1877	Tyndall and Louisa build summer cottage 'Alp Lusgen' at Bel Alp, northern side of Valais, above Brieg
1877	*Fermentation and its Bearings on Phenomena of Disease* published
1879	*Fragments of Science*, which had gradually expanded, first published in two volumes
1881	*Essays on the Floating Matter of the Air* published
1885	Tyndall and Louisa build Hindhead retreat, Surrey Downs

Year	Event
1887	Resigns from RI
1890	Fight with Gladstone in *The Times* over Irish Home Rule
1892	Hirst dies
1892	*New Fragments* published
1893	Dies Haslemere, Surrey, accidentally poisoned

TIMELINE OF EVENTS IN JOHN TYNDALL'S LIFE SPECIFIC TO VOLUME 3

Month and year	Biographical details	Notable social, political, and cultural events
early 1850s	pursued experimental research with Knoblauch on diamagnetic forces in crystalline bodies from late 1849 until late 1850	the Second Republic in France (est. 1848) became increasingly conservative and pro-Catholic
	sought income-making opportunities in journalism through 1850; protégés and patrons offered to loan JT money	political uncertainty in the German lands: Prussia, seeking a confederation with German states, was opposed by Austria, which was seeking to maintain the unity of the imperial Hapsburg dominions; both claimed leadership over German speaking lands
	exchanged many long introspective letters with Hirst	1849–1853 gradual expansion of the Zollverein (customs union) to include all non-Austrian Germany
		after "the hungry forties" and the threat of Chartist uprising, Britain had a period of stability; there was wide self-satisfaction when the Great Exhibition demonstrated the superiority of British manufactures
January 1850		the Anglican King's College in Toronto was reconstituted as the secular University of Toronto
February 1850	first, short, joint memoir with Knoblauch on diamagnetism published in Poggendorff's *Annalen* in February; English version published in *Phil Mag.* in March	

Month and year	Biographical details	Notable social, political, and cultural events
March 1850		Robert Stephenson's Menai Straits Bridge opened to rail traffic; Stephenson died only 4 months later
May 1850	finished writing up the second, longer joint memoir on diamagnetism and sent it to the *Phil. Mag*	the first live hippopotamus seen in Britain arrived at London Zoo from Egypt
June 1850	left Marburg (12th) and arrived in London (19th)	
	made journalistic and scientific connections—called on Faraday and at the *Philosophical Magazine* offices	
	went north at the end of the month; stayed at a friend's school in Lancashire, where, for 3 months, he earned his keep by teaching	
July 1850	second joint memoir with Knoblauch on diamagnetism published in *Phil. Mag.*	Robert Peel died (2nd)
	visited Hirst in Halifax (25th-29th); met Hirst's new circle of Carlylean literary friends, including Phillips	
August 1850	attended Edinburgh British Association meeting, but for only 2 days; gave presentation (2nd) on "magneto-optical" phenomena	the Australian Colonies Government Act (5th) gave the Australian states self-government
		Irish Franchise Act (14th) more than doubled the Irish electorate
		first undersea telegraph cable, laid between France and England, failed to work (27th)
September 1850	received his first journalistic earnings, £2 for an article in the *Leader* (published June)	300 years after Roman Catholic rites had been made illegal, the Roman Catholic hierarchy was restored in England and Wales; led to "no popery" riots
October 1850	William Francis offered employment translating scientific articles from German to English for publication in the *Phil. Mag.*	

Month and year	Biographical details	Notable social, political, and cultural events
October 1850	left London (7th) for Marburg, accompanied by Hirst arrived Marburg (10th) and began regular translations for Francis	
November 1850		the Prussians took over Marburg, in a conflict in which Bavaria was allied with Austria Tennyson appointed poet laureate (19th)
December 1850	German version of second joint memoir with Knoblauch on diamagnetism published in Poggendorff's *Annalen*	
January 1851	wrote (early Jan.) the first of his regular "Reports on the Progress of the Physical Sciences" for the *Phil. Mag.* (published March) worked intensively on diamagnetism (for 3 months)	introduction of the wet collodion process revolutionized photography: the glass plate negative enabled multiple high quality prints to be made
February 1851	short "waterjet" memoir published in the *Phil. Mag.* sent (4th) his single-authored "memoir on magnetism" to Francis of *Phil. Mag.* When publication was delayed he spent a month sending corrections and additions received (22nd) an invitation from Edmondson to return to his old position at Queenwood College	
March 1851		Owens College, Manchester, opened (12th) census of religious attendance in England (30th) showed that Church of England adherents were only just half of the observant population
April 1851	long memoir on magnetism published in the *Phil. Mag.* considered career options left Marburg (18th) for Berlin accepted the Queenwood position (20th)	

Month and year	Biographical details	Notable social, political, and cultural events
April 1851	arrived Berlin (21st) attended meeting of the Berlin Physical Society (25th)	
May 1851	set up his laboratory space and commenced work in the first week of May elected a member of the Berlin Physical Society (9th)	Great Exhibition opened (1st) in Hyde Park Prussia and Austria both sent troops to the Frankfurt area to enforce their visions of German unity. Prussia backed down
June 1851	left Berlin (22nd) for London; arrived London (26th) and travelled directly to Queenwood	
July 1851	called on Francis in London attended British Association in Ipswich (2nd-8th); shared lodgings with Francis began (late July) collecting testimonials in support of an application for the position of professor of physics at the new University of Toronto	
August 1851		gold was discovered in Ballarat, Victoria, a colony which had only one month previously been separated from New South Wales the Ecclesiastical Titles Act prevented Roman Catholic bishops adopting titles referring to British towns and counties
September 1851	important diamagnetism memoir published in *Phil. Mag.* (beginning of month)	
October 1851	application posted to Toronto at end of month	a second submarine telegraph cable laid between Dover and Calais performed slightly better than the cable laid in 1850
November 1851	memoir on the polarity of bismuth published in *Phil. Mag.* (beginning of month) Sabine offered to organise Tyndall's nomination for the FRS	

Month and year	Biographical details	Notable social, political, and cultural events
November 1851	submitted the same testimonials in application for the position of Professor of Experimental Philosophy at the new University of Sydney gave lecture in Romsey (11th)	
December 1851	process of collecting signatures for the FRS nomination certificate began	coup d'état by Louis Napoleon in France (2nd) to protect the Republic and his own position against monarchial and imperial aspirations
January 1852	heard (early in month) that he had not obtained the Sydney chair	France declared a new constitution (14th) which gave unprecedented powers to the president
February 1852	gave lectures in Warminster (9th) and Southampton (11th) first submission of memoir on "transmission of heat" to Sabine (Secretary of the RS) for *Phil. Trans.*; read to the Royal Society in January 1853	
March 1852	Sabine suggested that Tyndall apply for a grant from the Royal Society to assist with the costs of his experiments Hirst arrived on a two-month visit to England, during which he twice visited Tyndall at Queenwood gave another lecture in Romsey (30th)	
April 1852		the report of the Royal Commission on Oxford University recommended reforms, including opening Oxford to non-Anglicans
June 1852	elected FRS (3rd) introduced by his new friend Huxley, when formally admitted to the RS (17th) began application process for the chair of Natural Philosophy at the Queen's College of Galway, which he had heard was about to become vacant	

Month and year	Biographical details	Notable social, political, and cultural events
July 1852	awarded a grant of £50 by RS to assist with his researches	
August 1852	engaged in planning for the new series of *Scientific Memoirs*, established to publish translations of foreign language memoirs; he had been appointed an editor by William Francis	time was standardized at ports throughout Britain: time signals, based on a Time Ball set up on the Strand in London, were sent by railway telegraph to all the principal ports
	submitted application for Galway position	the report of the Royal Commission on Cambridge University made similar recommendations as those for Oxford
September 1852	first visit to Ireland in almost 5 years	the Duke of Wellington died (14th)
	Belfast (1st–7th), as one of the secretaries of Section A, at the British Association meeting	
	Leighlin (8th–9th), visited mother and old friends	
October 1852	translations well under way for first number of *Scientific Memoirs*	University of Sydney inaugurated (the Act which established the University had been passed 2 years previously)
	invited (in mid-October) to give a Friday evening lecture, early in 1853, at the Royal Institution	Great Exhibition closed (15th)
November 1852	first volume of new series of *Scientific Memoirs* published	State funeral for the Duke of Wellington (18th)
		in France, the Empire was re-established with Louis Napoleon as Napoleon III
December 1852	lectured at Andover (23rd)	

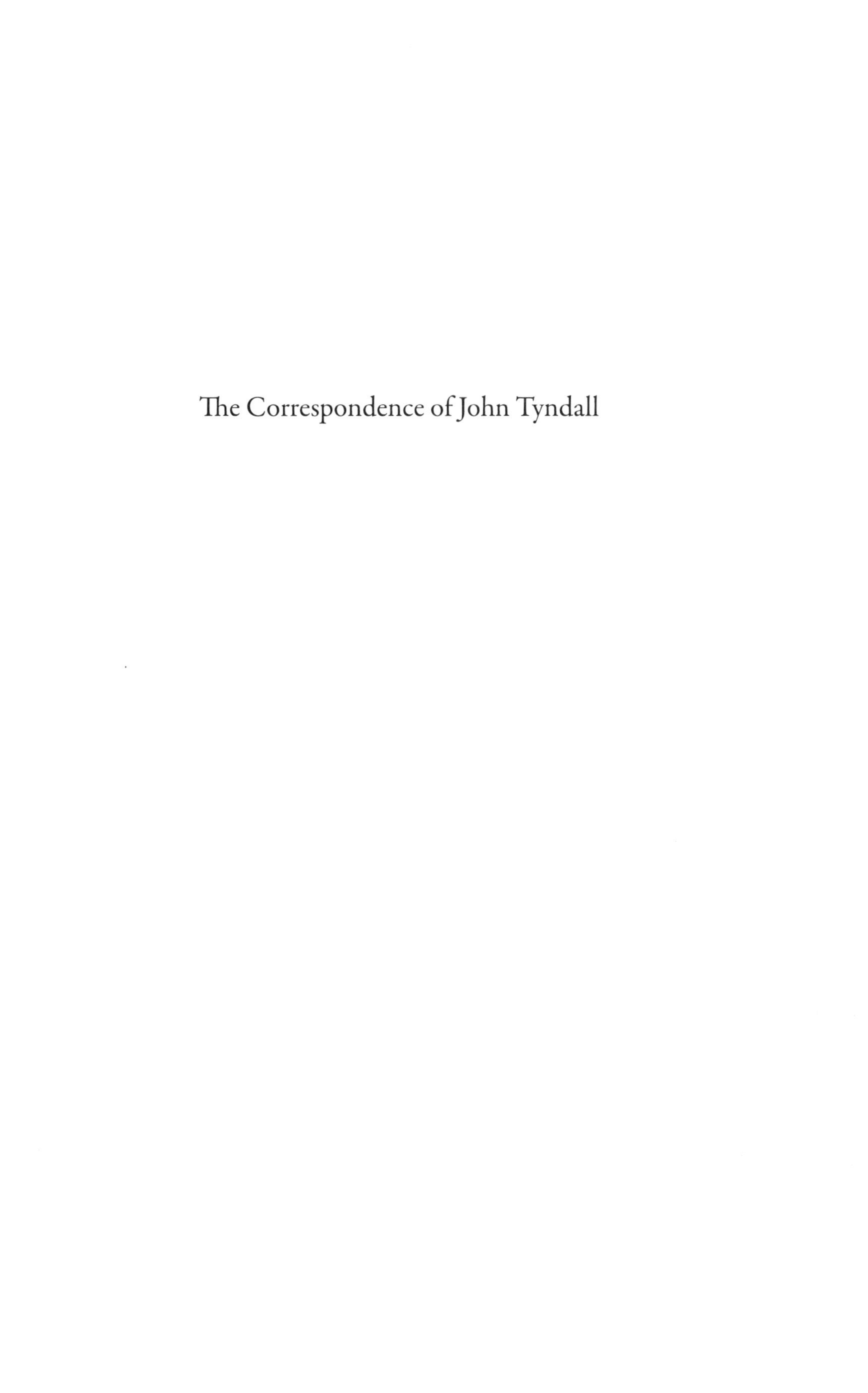

The Correspondence of John Tyndall

1850

<table>
<tr><td>To Thomas Archer Hirst</td><td>[8–14 January 1850][1]</td><td>0392</td></tr>
</table>

My Dear Tom,

Here's a bit of speculation—perhaps a few sketches such as the enclosed[2] can be disposed of somehow or other, will you enquire about it? Try the Peoples Journal, Howitts[3] or any other you think likely. a note somewhat in the following style may accompany the manuscript.

Dear Sir. I am requested by a friend at present studying in Germany to enquire whether a series of sketches of German University life somewhat similar in style to the enclosed would be acceptable to you—In case of negative, you perhaps be good enough to return this manuscript. <u>T.H.</u>

I dont know how it reads, it was written under disadvantages, being the product of bits & scraps of time snatched from my investigations. I am driven to this course because I see the probability of my remaining in Germany longer than I anticipated or else give up an investigation which promises to be of great interest. Professor Knoblauch and myself are experimenting on magnetism, in my last note to Jimmy I said we had as yet obtained no safe result— We have since obtained one, and this threatens a very beautiful theory which has obtained considerable celebrity for its propounder.[4] He will be dreadfully angry no doubt and will do all in his power to overturn us, but in this he will assuredly fail Tom, for Nature is on our side, and we shall limit ourselves strictly to declaring what she has confided to us. You see this will draw us into a discussion and it will be unpleasant to be obliged to run away before the thing is complete.

It will be necessary to preserve a copy of the manuscript beside you. Jimmy will help you to make it—I know he will.

Give me your opinion. Suppose I continue these sketches whether you dispose of them or not, how many subscribers do you think you and Jimmy could procure towards a book of 150 or 200 pages. I have a good many acquaintances here & there and by this means I may be enabled to accomplish something.

In case all else fails and you imagine that the thing is worth reading for really, I dont know, just write to Mr Edmonson Queenwood and obtain from him the address of Dr Charles Mackay. Send the manuscript to the latter with a note like the following:—

Dear Sir

The enclosed has reached me from my friend John Tyndall. He is engaged at present in a series of magnetic investigations which will detain him in Germany longer than he anticipated. Would you oblige me by a short answer to the following question? do you know of any periodical where a few sketches of German University life somewhat in the style of the enclosed would be received? I have tried the People's Journal and one or two more. Your acquaintance with literary matters will I dare say enable you at once to decide whether the proposition is likely to be entertained by any body or not.

I am Dr Sir | Very faithfully *[yrs]* | T.H.

Now boy, how are you getting on? Have you turned Turk, Jew, or Mahommedan,[5] since you last wrote me—you will be welcome to me in any guise. I should not be at all surprised if some fine day now lapped comfortably up in the carpet bag of futurity you should astonish the world by a volume entitled Travels of a Gentleman in Search of a Religion! and doubtless the matter will read very prettily, and you yourself will look back upon it with a feeling somewhat similar to mine sometimes when I call to mind a certain day on which I followed a hunt in Ireland and had to save myself from positive starvation by eating maggotty blackberries—Oh it is very pleasant to look back upon, and looks exceedingly well on paper—but the actual thing itself! As I have nothing better to say I will send you a little poem from Goethe, entitled Mason Lodge,[6] translated I believe by Carlyle. You have read it perhaps.

The Mason's ways are
A type of existence
And his persistence
Is as the days are
Of men in this world.

The future hides in it
Good hap and sorrow
We press still thorow
Naught that abides in it
Daunting us,—onward!

And solemn before us
Veiled the dark portal
Goal of all mortal—
Stars silent o'er us
Graves under us silent.

> But heard are the voices
> Voice of the sages
> The world and the ages
> Choose well your choice is
> Brief and yet endless;
>
> Here eyes do behold you
> In Eternity's stillness
> Here is all fullness
> Ye have to reward you
> Work and despair not'.

I'm making up my mind to be a thorough miser when I return to England. miser I wont say. but a downright money getter—I will rise up early and lie down late and eat the bread of carefulness till my purse is full to the throttle —I wish to heaven my friends were considerate and would pay me what they owe me[7]—I should then be able to push on a little longer, no one could have a better opportunity than I have at present, and here I must quit before half its advantages are reaped—no matter

We press still thorow | <u>Nought daunting us—onward</u>[8] | J. Tyndall

On second thought Tom dont send the thing to Charles Mackay till you hear again from me.[9]

The accompanying[10] has lain for some days beside me—The more I look into the matter the more firmly I find myself bound to this investigation, and therefore the more likely that my stay in Marburg will be prolonged. you have within a note to Mr Carter.[11] I must tell you what it is lest you should be angry with me—It is to ask him whether in case I should require it he would trust me with the loan of twenty pounds for 6 months. If he does not he is not a good fellow for nearly ten times the sum of my money lay for a long time in his hands and I gave him very little trouble about it. I should not however be disappointed even should he refuse me. —I think for popular reading the enclosed sketch might be altered with benefit but I'll let it go as it stands. Kiss Jimmy for me—and if you see Ada[12] soon tell her that I continue to love her very dearly.

J.

We[13] publish in the German Journals[14] of February, and I will send an English version of the matter to the Philosophical Transactions.[15]

RI MS JT/1/T/525

---

1. *[8–14 January 1850]*: dated from postmark, 14 January 1850. The letter was completed 'some days' (say between 3 and 6 days) after it was begun.

2. *the enclosed*: Tyndall sent a sketch on student life in Germany. He had published several previously, under the pseudonym Wat Ripton, in the *Preston Chronicle*, including 'The German Student', 24 February 1849, p. 3, and 'An Excursion in Germany', 19 May 1849, p. 3. According to LT, this sketch (or group of sketches) was not published. See LT, 'Biography', Vol. 1, p. 282.

3. *The People's Journal, Howitt's*: *The People's Journal* was a London weekly intended for the working classes. William Howitt worked as a contributor before establishing the rival *Howitt's Journal of Literature and Popular Progress* in 1847. This was financially unsuccessful and only three volumes were published. John Saunders then combined both into *the People's and Howitt's Journal. Of Literature, Art, and Popular Progress*.

4. *beautiful theory … its propounder*: J. Plücker's experiments and researches on diamagnetism.

5. *Have you turned … Mahommedan*: Hirst was in a period of religious confusion after an acquaintance and relatives had sought to persuade him that his doubts about Christian orthodoxy were dangerous (see letter 0388).

6. *Mason Lodge*: J. W. Goethe, who was a mason, wrote 'Masonic Lodge', translated by Carlyle as 'Mason's Lodge' or 'Mason-Lodge' (1827) and included at the end of Book III of *Past and Present*. A fourth verse is missing from the version here. Occasional variants in wording and punctuation may be due to different versions being available in 1850.

7. *my friends … owe me*: the unpaid debt owed to Tyndall by Thomas Foy serves as one example (see letters 0354, 0355, 0357).

8. *We press … onward*: lines 8 and 10 from the above poem.

9. *On second thought … from me*: written at the bottom left corner of the sheet, to the left of the signature.

10. *accompanying*: probably the first part of the letter, indicating that this letter was written over several days.

11. *note to Mᵣ Carter*: the note (missing) to Richard Carter was never sent.

12. *Kiss Jimmy … Ada*: perhaps children of friends, but see biographical entry for Ada Piercy. Numerous letters from Hirst to Tyndall in 1849 refer to Ada.

13. *We*: postscript written vertically across the left margin.

14. *German Journals*: published as H. Knoblauch and J. Tyndall, 'Ueber Das Verhalten Krystallisirter Körper Zwischen Den Polen Eines Magnetes', *Poggend. Annal.* 79, no. 2 (1850), pp. 233–41.

15. *Philosophical Transactions*: Tyndall probably meant the *Philosophical Magazine*, which is where the English version was published (cited letter 0395, n. 22).

From Thomas Archer Hirst   20[–28] January 1850[1]   0393

Halifax, Jan. 20ᵗʰ 1850

My dear Tyndall—

Judging from my long silence, few could credit the esteem in which I hold

your long dear letter:[2] or the number of times it was read & lingered on by syllables, nor yet the number of times that the walls of my room echoed my half
unconscious ejaculations of 'Bravo' 'God bless you' & the like; nor do I care
to press the point or excuse myself for I know I write to one who has already
felt & anticipated my reply—'Life is a string of Moods'[3] How humiliating yet
how true! I have attempted & partly executed one or two replies before this
Tyndall & destroyed them both, and why? Because I re-read them in a different mood. While fresh under the heart, emotions that your letter caused,
I was all abandonment, had you been within reach I should have forgot my
habitual calmness of exterior & hugged you heartily, and I know well, should
have felt the greatest confusion & blushed the moment after, as if in overstep-
ping the line of custom I had committed a heinous fault. Just so it was in the
absence of your bodily self. I wrote & then spoilt all by re-consideration. I was
not afraid you would think them forced, for they were honest heart utterings
that to you would have carried their own guarantee of honesty. But I thought
they were not worth the postage, that they were more valuable to me than
you, & that it needed no such professions (genuine, though they might be)
to convince you of what you already know. I care not that you <u>at least</u> should
find me contradicting myself in each succeeding letter. I will tell you truth &
every one[4] shall be sincere in itself. With others I dare not yet do so, but you
are <u>my</u> surveyor & will plot the section of my life on a natural scale & its angu-
larities will then vanish—And now for my bit of Biography since I last wrote.
You expressed some anxiety to know the result of the ordeal I thought then
before me Yet you must have known that it would be just nothing at all—that
it too would slip through my fingers when I grasped hardest—indeed your
advice—to work—was just calculated to bring about that same, and this has
been the case. —I was in a poor way then Tyndall, perfectly miserable, looking back on it at least; though then, I grieved that I could not feel miserable
enough, your letter seemed to hint at a bitterness that I really did not taste;
you seemed to credit me with a harder trial than existed; and thus I tortured
myself with the additional bugbear that my effort & trial was not sufficiently
real: You told me not to sit looking into myself, but to <u>act</u>. I thought it cowardly to do so. True I found relief by so doing daily, but when night came,
instead of satisfaction I had a sense of duty, shirked, & set to once more at my
crimino-judicial task. Well what came of it? I wish I could say; but the fact
is I cannot recall anything, but gross blackness, everything seemed hollow
and unreal, life but a useless misery & struggle, between conscience and the
senses. I thought I ate too much, then smoked too much, & so on with most
indulgences, & I could see little in life but struggles to keep such in abeyance,
the world looked drear & lonely enough.

'So lonely 'twas, that God himself
Scarce seemed there to be'.[5]

About this time my Aunt[6] the last relative on my Mothers side, a good kind soul as ever breathed, sent for me to Bristol. She could not pass the Christmas happily unless as usual, we met round the same fire. Her letter to me was simple enough, I have received many such before & never thought more of them; but its honest loving words struck a chill through me now: 'I long' she said 'to see your dear faces once more about me. We shall have such a Merry Christmas. Wont I make you a nice plum pudding & give you a hearty welcome, & what's more <u>you shall have a pipe after it</u>. Do come' The contrast 'twixt this & me was painful at first, it seemed as if I alone were miserable, so had thought <u>shed its sickly cast over me</u> and all I once had loved.

I have just been downstairs to Mr Wright who is drunk, and very disorderly. Poor Mrs Wright was standing between him & another person (whom he professed he had some intentions of killing on the spot) her hair & cap all torn and greatly afraid of disturbing her revered Lodgers.[7] What a miserable existence she has & what a pitiable specimen of humanity for a husband. I have sat with him for an hour & by the greatest difficulty dismissed his enemy & turned his mind on artists & painting I hope he'll be quiet for the night; at any rate he has turned the train of my thoughts so I must continue my narrative tomorrow.

Well at first I refused to go. 'What use' I thought 'any more attempting to realize that phantasm, a Merry Christmas: have I not always failed to do so, could I leave this devil 'self' at home, I might! I changed my mind however and went, not with the intention of escaping myself; but of breathing a short time the atmosphere of contentment and love, in fact as an experiment, to see how they could <u>live</u>. How vain the effort, you may guess; what means had I for judging this case. From my standpunkt they might look fair enough; but from their own?— It was the verdict of intellect apart from the moral emotions. It is almost useless to continue further to recall such unhealthiness; I probably seemed to enjoy myself as much as usual but to myself I seemed villainous. When all sat together on that Christmas Eve, when every lip seemed to express unrestrained the hearts emotion, all restraint withdrawn, & mutual confidence established, I watched them narrowly, and seemed to myself a cool spectator, an excluded member. Now they would recall the memory of those who once had been amongst us, and seemed to heave as it were <u>one</u> long heartrending sigh and their countenances bore witness of its sincerity. But I— Good God I could not join them here. Their grief I could not get near enough to me, & I was in agony that I could not sigh—so in their prospects for the future—all were as one to me: <u>indifferent</u>. The thought that I ought to enjoy myself was continually before me, but the more I grasped the more it slipped through my fingers. In fact I had lost <u>myself</u> & seemed to suspect someone had stolen me. I am better now but how I became so, I am at a loss to know, perhaps it might be accounted for on Pharmaceutical principles. This only I am conscious of, that your letter at that time only served to make me worse; for I could not help contrasting you with myself, and foolishly thinking that I was too weak in my powers of resolution. I thought of M'Arthur[8] & other instances nearer home, where this very feeling & internal antagonism had led to consequences most frightful, & instead of manfully facing these goblins, & as Carlyle says sending them to their own Hell by action;[9] I trembled like a coward at an idea that once flashed on me that they might conquer me too.

Well—we grew exceedingly busy at the office, a geological paper too, that must come off had yet to be prepared, so that luckily I was <u>forced</u> to work; one night or rather morning at 3:30, when I had been engaged in my essay,[10] it flashed on me all on the moment that I was somewhat changed to what I was a few weeks back. I opened your letter again Tyndall—read it & as I said could have kissed you. My essay at length was finished & delivered, and here again I had to discipline myself to avoid a relapse. I had thought to accomplish something but as usual it has escaped me & the thing when done is next to worthless. Well! what matter? It has been employment and personally has done much good, though intrinsically valueless, & I will be satisfied & replace it as quickly as possibly with other employment. 'To fill the hour—that is happiness: to fill the hour & leave not a crevice for repentance or approval,'[11] or as some old classical writer says:

> He alone is happy, who can say
> When night comes: I've lived well to day.[12]

The same old struggles between Reason & Will still go on, yet the regret that such antagonism must <u>be</u> under gone does not so often recur. I wish I could say the task grows lighter, but I cannot: yet I will have hope & courage. Tyndall, I have a mind to burn this letter too; for I cannot faithfully tell you either what I am or have been. I don't know myself. I'm in a maze, a labyrinth & confused jumble of principles & actions. Surfaces & Realities—Beliefs & Scepticisms, which all belong to me, but are not arranged. Before it was all dark, now I see but dimly, though I do see somewhat; one thing I see, that there is a great tendency in me to exert my intellect to the neglect of my heart: (no, that is not what I mean) I spend too much time in the ideal, & trying to perfect it, to the neglect of the actual. Can you understand me for I can't? As to religion, I won't say a single word: I should frighten you, and myself too, were I to put my awful scepticism into black and white. The fact is I avoid the subject at present, though I find it difficult to do at all times—Better for Orthodoxy & my own comfort, if it had never rechallenged me. I'm not in a proper temper to meet it now, but its turn will come.

Phillips and I keep up our intimacy,[13] indeed it has strengthened much since I last wrote. He is a good fellow, but I like the <u>man</u> much more than his literary attainments: with the latter I don't much sympathize, but I feel he understands me & although I do not return him brilliant thoughts or actions I am conscious of possessing somehow his regard. You will like him I am sure, and I shall like you to meet him. It is different with Hutchinson. I don't know what to say of him. I admire his attainments but cannot get nearer him: there is always something of coldness, that I cannot banish in our interviews; with far less intellectual effort, I feel more exaltation & warmth of emotion, with Phillips than with him. He is sarcastic & too critical, all his reading seems to have ended in critical opinions of the <u>talents</u> of his authors, as if never undertaken as a guide to his own actions. I may do him injustice, I hope so truly; but he never seems sincere enough, all is outside with him, & cold. This is the only way I can account for his religious opinions: he views it objectively & never in reality seems to bring it home to himself. The great questions of human life & destiny, he seems to consider as questions in which he has no voice: but has merely to judge between the claims of this or that religionist. Pity one's first estimate should fall thus. I had expected different & with reluctance even yet give way to my conclusion. It is his worst trait, however; I still like him, but not so much. Have you seen the Truth-seeker for December. What do you think of the 'Mask of Life'.[14] I don't understand it fully.

The other & former part of my letter has lay'n (?) by me some time in order to make some final arrangements before sending this—I should certainly have been cross with you had you not told me the import of that letter to Carter.[15] Why for goodness' sake Tyndall are you so ceremonious? I tell you candidly I have not delivered that letter, nor ever will; it is quietly stowed away in my pocket & waits only your orders to thrust it unopened into the fire. I do not think Carter has the money to spare just now, nor if he had, should I like you to be under any obligations to him, when others that know, esteem & understand you better have it to spare. Probably before this reaches you, you will have received £20[16] through Mr Schwann (a Merchant of Huddersfield[17] whom Phillips in dedicating a book of his to him addresses 'Thou great Gods nobleman of Huddersfield') He has kindly undertaken to convey it by a Bankers order through his Agent M[r] C. E. Alder of Frankfort. Write me immediately if you receive it all safe, it will serve as a receipt & <u>security</u>. I hope it will serve your purpose at present, so be 'asy';[18] & if you don't wish to offend you'll be kind enough to drop all ceremony & write when it is finished, for there's more at your service when you want it. Meantime you must work & repay by instalments of M.S.S. Your last contains some genuine sterling *<1 or 2 words missing>* my boy. I have sent it to the Peoples Journal,[19] & though I have not *<yet received>* a reply have little doubt about its insertion; it is perhaps <u>scarcely popular</u> enough for their readers taste; but it is the most suitable publication I know & from the general tenor of their late numbers, I predict it will be acceptable They are sure to ask me, how <u>many</u> of the sort you will furnish—Suppose I promise 6—Will it suit you?—and how much would you like for them? or will you leave it to me to bargain, without again communicating with you as to their offer? Write a book by all means, if you have the time; But you should have told me your subject & further particulars, before I could estimate the number of subscribers. What is it?—I'll be bound as security against all loss, I'll go about canvassing for subscribers, write glorious reviews in the Peoples Journal, Truth Seeker &c or any other mortal thing that shall be required—Don't fear about its success—set to it & leave the rest to Providence & Tom Hirst. Whose plumes are you & Professor Knoblauch (by the bye he's my optical friend 'nicht wahr'[20]) about to pluck—Liebig's?[21] I shall wish to see your transactions,[22] write them to send me a copy or tell me where to get them. Where is Frankland & How is he—& friend Knoll[23] & <u>Gustave</u> (Is he a reality or an <u>ideality</u>?) What are your future intentions; do you still intend to visit Berlin & France? I should like you to do so—Ask your Landlady to reserve your garret for me. I shall want it soon: in 6 months I shall be at liberty, & feel more inclined than ever to take your place, perhaps this time 12 months or earlier.

We do not find our present Laboratory convenient, & are just looking out for a detached one. I often translate D[r] *[Mill]*[24] & have got a good part completed. The Halifax Mech: Inst[n] & Mutual Imp. Soc:[25] have at last amalgamated and are getting on <u>pretty</u> well. In your remarks in the Preston Chronicle[26] you just anticipated a remarkably similar intention I had in view with respect to the Halifax Inst: In these democratical institutions I find office bearing, speechifying &c warping many weak but well designing students from the path of their own improvement. They lack individual exertion & seem to think that breathing a literary atmosphere merely is required. The mere necessary <u>means</u>, Government &c is usurping the legitimate object, <u>self</u>-improvement: not Halifax alone but many other similar institutions are on the decline & depend upon it this is the cause: from Mutual Improvement they denigrate into 'Sprouting' Societies, then come Banterings & petty personalities & finally from sheer disease they die a natural death. My class has fallen off gradually until lately I had only one, <u>the</u> one. I still go regularly & hope to make it better by and bye. Jemmy is at a Cavalry Ball this Evening, figuring away in a waltz probably at this moment. I have read Emerson more than ever lately or rather better than ever not more. His essay 'Experience'[27] I am especially fond of, it has been of great value to me lately. If I had not written so much I had intended saying a word about him & also another authoress an anonymous one though privately known as a young lady (28) living at Haworth near Keighley (you know the spot). She has written a novel called 'Jane Eyre'[28] (a rival of 'Yeast' even)[29] that would do you good to read. By the law, I did not think it possible that we could possess such a precious lump of feminine flesh, & such a heart in it, too! I have fallen positively in love with her. Phillips & I are going to prostrate ourselves at her feet some day soon. Ada[30] has forsaken me, I have not seen her for some time, she must have gone to school. I wrote to Ginty but he knew nothing about Jack:[31] I wish much to hear about him.

John Tyndall | care of Prof. Bunsen | Marburg- | Hesse Cassel, Germany
P^d in England- | Jan. 29th 1850.[32]

RI MS JT/1/H/142

---

1. *20[–28] January 1850*: this letter was completed on 28 January (see Hirst, 'Journals', 28 January 1850).

2. *long dear letter*: probably letter 0390, in which Tyndall discussed his own life experience and religious pilgrimage.

3. *'Life is a string of Moods'*: R. W. Emerson, 'Experience', *Essays: Second Series* (London: John Chapman, 1844), p. 34. Hirst misquotes; the passage reads 'Life is a train of moods like a string of beads'.

4. *every one*: that is, every letter.

5. *'So lonely … to be.'*: S. Coleridge, 'The Rime of the Ancient Mariner' (1798), lines 600–601.

6. *Aunt*: Hirst was probably referring to his mother's sister Battersby. Hirst's mother died in early September 1849, and in his journal he recounted that his aunt's face 'brought back to me so vividly her whom I had so lately left recovering' (Hirst, 'Journals', 12 September 1849).

7. *her revered lodgers*: Hirst had lodged with Mrs Wright since his mother's death (see letter 0382).

8. *M'Arthur*: referred to in letter 0406 as having killed himself by excess drinking. The implication here is that the root cause of his disastrous end was some inner distress, perhaps religious doubt.

9. *Carlyle says . . . action*: specific reference not identified. Ridding oneself of demons and fears by action or work is a constant theme in Carlyle.

10. *my essay*: Hirst presented a paper to the Franklin Society in early 1850 on the geology of Halifax (Hirst, 'Journals', 26 September 1849).

11. *'To fill the hour . . . approval'*: Emerson, 'Experience', *Essays: Second Series* (cited n. 3), p. 40.

12. *'He alone is happy . . . to-day'*: from Michel de Montaigne's essay, 'Of Prognostications', *Essays*, Vol. I Essay XI. Montaigne's essays were first published in three volumes in 1580. Hirst quoted from a 17th century translation by Charles Cotton. Various editions were available to him: for example, *The Essays of Michael de Montaigne* (3 vols, 1811) and William Hazlitt's *Complete Works of Michael de Montaigne* (1842) which incorporated the Cotton version of the *Essays*.

13. *Phillips . . . our intimacy*: contact between Hirst and Phillips was instigated by Carlyle (see letter 0378).

14. *the Truth-seeker for December . . . 'Mask of Life'*: *The Truth Seeker in Philosophy, Literature, and Religion* was a transcendentalist periodical edited by F. R. Lees and G. S. Phillips. 'The Masque of Life' was an article by Phillips, writing under his regular pseudonym of January Searle.

15. *letter to Carter*: the letter Tyndall had asked Hirst to post (see letter 0392, esp. n. 11).

16. *you will have received £20*: Hirst began arranging the loan on 22 January. It was finalised on 27 January (see Hirst, 'Journal', for those dates). Tyndall received the money on 6 February (see letter 0395).

17. *Mr Schwann . . . of Huddersfield*: Frederic Schwann (1799–1882) was a wealthy textile merchant committed to social improvement, and first president of the Huddersfield Mechanics' Institute. Huddersfield was a textile town 7 miles south-east of Halifax.

18. *'asy'*: a deliberate use of dialect for 'easy'.

19. *People's Journal*: see letter 0392, n. 3.

20. *he's my optical friend 'nicht wahr'*: 'is he not' (German). Hirst was asking if Knoblauch was the life-model for the character, 'my optical friend', in Tyndall's sketch which he had submitted for publication. Hirst was correct (see letter 0395).

21. *Whose plumes . . . Liebigs?*: in letter 0392 Tyndall had indicated that his investigation would refute a celebrated theory; in letter 0395 he informed Hirst that it was a theory propounded by Plücker.

22. *your transactions*: see letter 0392, n. 15.

23. *Knoll*: probably F. G. Noll.

24. *D*^r *[Mill]*: or possibly 'Dr Hill'. The book was a volume on chemistry (Hirst, 'Journals', 24 January 1850), but is not further identified.

25. *Halifax Mech: Instn & Mutual Imp. Soc*: the Halifax Mechanics' Institution (1825) and the Halifax Mutual Improvement Society (1847) shared rooms on Horton Street from 1849. According to LT, Hirst had written an anonymous letter to the *Halifax Guardian* in late 1849, where he argued for their amalgamation. 'This suggestion bore fruit before long, with the result that the two societies became one early the following year' (LT, 'Biography', Vol. 1., p. 352).

26. *Preston Chronicle*: Wat Ripton, 'Thoughts of Preston, Entertained in Germany', *Preston Chronicle* (29 December 1849), p .3.

27. *'Experience'*: in *Essays: Second Series* (cited n. 3), pp. 30–56.

28. *young lady . . . 'Jane Eyre'*: Charlotte Brontë wrote *Jane Eyre: An Autobiography* in 1847 under the pseudonym Currer Bell. Her hometown of Howarth was just 10 miles north of Halifax.

29. *a rival of 'Yeast' even*: *Yeast: A Problem* (1848), by Victorian social and religious controversialist Charles Kingsley. Both texts explore religious institutions and hypocrisy.

30. *Ada*: see letter 0392, n. 12.

31. *Jack*: John Tidmarsh.

32. *John Tyndall . . . 1850*: address and writing on envelope.

From James Craven   [20–28] January 1850[1]   0394

Horton Street, Halifax | January 1850

Dear Tyndall,

Changes come round and it is now my turn to give you a few lines though the curious stage of existence in which I am now placed renders me not exactly in such a fitting train of thought as I could wish to reply to your letters.[2] However for a start or beginning I will (having already made you in part my confessor) relate the ideas which have for a very great number of years possessed me concerning my individual self. Though not generally much given to self esteem, for my failing lies more to the organ of approbation, yet the following savours more of the former. Well then—somehow a feeling of superiority over all that I come across possesses me—I feel as if I have powers (don't laugh) which did I only take the determination and bring the whole of my mind to bear on must be successful. I feel in fact the lawyer like scheming, contriving and cunning is settled within me, and nothing would gratify me more than being immediately connected with some case of great intricacy requiring powers of discrimination etc. I do not know how it is but the slow, and in a great measure the almost useless occupation of a Surveyor is such as never to have excited this feeling within me—in fact I cannot say that I take to it with that interest I should, and the prospect opening forth to me is not very inviting seeing the number who are in it and the general unprofitable ness attending it. These are ideas which have long filled me, and late as it is I almost meditate a change. Yes, late as it is—the feeling thus engendered has received particular impulse from my brother's having just passed his examination,[3] and as a seemingly good opportunity arises were I to join him (seeing that he is likely to have a first rate introduction in London) I find I could be of assistance here. And I am now considering whether it is too late or not to change. These turnings to and fro—these worldly schemes will have no interest to you except as an insight into character; however, as I stated above, having made you my confessor I must let you know my inmost ideas, fully confident in your directing me when you deem it necessary and fully aware of the secrecy which you will maintain in all the little matters I so freely, so trustingly confide in you. Of my power to keep myself for the necessary term of years and the expenses which would necessarily be incurred I can manage well, so that my plotting brain only considers the chances of luck or no luck in the undertaking. I mentioned this to my father partially the other day, when he seemed to think it too late; however I told him frankly that as to making a living out of Surveyorship was humbug. However he considered what with what I have already, together with my expectations, I need not fear. Indeed I do not at all, on this score, but I feel that I am throwing myself away in a profession in which the greatest scamps and vagabonds often as a last resource take to, and who often by gullibility gull the people—a profession wanting in respectability, wanting in honour. I am severe you see but it is what I have arrived at through a long series of reasonings, and while I write, I almost pass the word that after the expiration of my clerkship I go to some far distant town (my pride could not bear the reproaches which might be raised here) and there quietly by saving and perseverance obtain that which I at present think will suit the bent of my disposition.

After all I may be running after a shadow—perhaps so; however is it worth the experiment? Tom is reading his Algebra, I almost covet him his easy, unscheming and uncaring life. What a contrast, he and me—I diving into all the ins and outs of futurities—weighing chances for and against—while he without a look or glance at future makes use of time present. Well, well, as usual I comfort myself with the old hackneyed consolation 'it's natural, and what's nature is nature.'

But what is this—I have been bothering you—the philosopher—the uncaring and heedless you? you who don't care one bean stalk about the whole matter? Aye—Well, only this, let's have your opinion and if it sanctions the course intimated above—by George, if my opinions on the matter do not undergo a supernatural change I will do it. And now to leave this speculative and weighing subject, I refer to your letter to answer it more in form.

So you were pleased with my last letter.[4] I question what verdict this will receive. Methinks if allowed to enter your hands just now, the material views which are entered would give you a sickening idea of the thoughts of James Craven as a whole. But amidst all these forebodings and looking forwards I do think I frequently catch myself in the act of thinking on more immaterial questions. And now I will with your leave go on in a more systematic and regular course than what I have done in this letter, for I question whether it is altogether understandable to you.

Well here commences—You say that Buffon and Sir Isaac Newton say that genius and patience are the same[5]—I don't agree with it. A man may have as much patience in the world, but can never never instil into me that consequently he is a man of genius, and therefore I hold that genius and patience are not the same. But I do not deny that genius often smiles most when under the restriction of patience. I have noticed latterly in my general conversation that a rule seems established to obtain from the individual you are conversing with to get all the knowledge he possesses and at the same time to return as little as possible. I trust this is not our case. However nothing seems to present itself to me just now and so with very best wishes, only take care not to stay too long in Germany[6] for my dear fellow I want to see you, and

Believe me, | Your little fellow what confesses | James Craven. | John Tyndall Esq.

P.S. I have not heard as yet respecting the composition of English money;[7] however am seeing about it.

RI MS JT/1/TYP/11/3539–3541
LT Transcript Only

---

1. *[20–28] January 1850*: probably written at the same time (and sent with) letter 0393, which was written over several days. See letter 0393, n. 1. Letters from Hirst and Craven were often posted together.

2. *reply to your letters*: letters missing. The last extant letter from Tyndall to Craven is 0387 (21 October 1849) but Craven had replied to this in letter 0389 (29 November).

3. *my brother's . . . examination*: John Craven, further biographical details not identified (see letters 0374 and 0389).

4. *last letter*: perhaps letter 0389.

5. *genius and patience are the same*: 'Genius is nothing else than a great aptitude for patience' is attributed to George-Louis de Buffon, as narrated by Herault de Sechelles during an alleged conversation in 1785, though it does not appear in Buffon's published work. The short form, 'genius is patience', is attributed to Buffon by Samuel Smiles (in *Self-help*, chap. 4). Similar statements were widely attributed to Newton in the nineteenth century.

6. *take care not to stay too long in Germany*: Tyndall had been in Germany since October 1848.

7. *respecting the composition of English money*: possibly an allusion to the loan Hirst and Craven organized for Tyndall. Hirst borrowed the amount from Craven, and arranged for it to be paid to Tyndall. See letters 0392 and 0393.

To Thomas Archer Hirst   [7–c. 26 February 1850][1]   0395

My dear Tom,

I think the employment of Intellect is fair and honorable and not to be always <u>consciously</u> in love, or a lover, is no fault of yours or any body elses. If this feeling is within you it will flow out spontaneous where occasion offers, but dont hunt after it. There is an old proverb in Ireland which to this moment I have thought utterly without meaning 'a watched pot never boils' It springs from the nature of things in Ireland, as the vacuous native sits with anxious eye watching the potato pot, he no doubt imagines the boiling point one of tardy attainment; but if he quit his watching and leave the matter to itself it will boil in due time, and that all the quicker apparently for him. If within you, as I have said, it will flow out of you at proper times, and in choosing these times it is scarcely ever polite enough to consult your will in the matter, if not within you (which I utterly disbelieve) then good; be true to your own nature; if you are the Devil's child as somebody says[2] why by all means be true to the devil!—but preaching on these matters is of very little use, you can by no means shirk your present schooling, and a day will come when you will be better able to appreciate its value than you are at present.

A lover! A friend! Sweet words. Truly has somebody said that our unconscious acts are greatest. As a matter of curiosity I should like to know the verdict of 12 of the wisest men in Halifax upon the act you have just committed.[3] 'The children of this world are wiser in their generation than the children of light'[4] Go tell John Abbott[5] what you have done, he will call you a fool for your pains. Here I have been receiving a packet of motley currencies Louis, Thalers, florins and Kreuzers,[6] sufficient to break an asses back, sent to me by order of Thomas Hirst Esq. of Halifax on Mr Schwann[7] of Huddersfield. I never was more puzzled in my life than to guess the contents of that packet before I opened it. I had not the remotest dream of its contents, and imagined that some benevolent individual had sent me a lot of crystals knowing that I was experimenting[8] with the same. Well Tom I was never so deeply in debt before, and there is no man under heaven's umbrella to whom I would prefer being in debt. I wrote to Carter as I imagined that an act of the kind on his part would be mere reciprocity. I did not ask him to lend me 20 pounds but merely for the sake of enlightening myself as to what was best to be done, I wished to know in case of need what I had to fall back upon. The probability is that I should never have borrowed a penny from him, and my asking him was merely calculated to make my footing more sure. Burn the letter[9] by all means; you have rendered the question unnecessary. As I told you however I am in no immediate want of money. I have a little beside me here and have left 30 pounds in the hands of Mr Edmondson so that strictly speaking, I am not yet a pauper. I cannot say that I love you a bit the better on account of this remittance, and were the case otherwise, I should begin to doubt the purity of the connexion between us; but your act affords me very pleasing material for a Journal entry an entry that may warm my old heart when my 'pow' is frosty and if a young Tyndall should be in the way it may do him good. 'There lad' I can exclaim 'you see when your father was poor he had a friend that trusted him'.

Whence comes that deep interest which one feels on meeting unexpectedly a picture of some well known spot? There is the dwelling house, and the apple tree at the corner, and there is tiger the mastiff with his tail cocked to the left, and little Bill or Tom or Jimmy playing a penny trumpet,[10] at such times we detect the common nature of artist and lookeron and the discovery causes us pleasure, nor could the most familiar landscape that human pencil could draw be better recognised and understood than your picture of yourself, drawn by yourself; and I would wager £10 against a Hessian halter[11] that Emerson would understand it too, that he has gone over the same ground, and that he would agree with me when I say that if a man would know anything he must not expect to shirk their zwischenraum,[12] this strife gulph in his existence.

You have heard of Henry Kirk White, Byron wrote as follows about him.[13]

> 'Unhappy White while life was in its spring
> And thy young muse just shook her joyous wing
> The spoiler came, and all thy promise fair
> Has sought the grave to sleep for ever there.
> O! what a noble heart was here undone
> When Science Self destroyed her favourite son!
> Yes she too much indulged the fond pursuit
> She sowed the seeds but death has reaped the fruit.
>
> T'was thine own genius gave the final blow
> And helped to plant the wound which laid thee low
> So the struck Eagle stretched upon the plain
> No more through rolling clouds to soar again
> Viewed his own feather as the fatal dart
> And winged the shaft which quivered in his heart.
> Keen were his pangs but keener far to feel
> He nursed the pinion which impelled the steel
> While the same plumage that has warmed his nest
> Drank the last life drop from his bleeding breast.'

This is only a note of recommendation, the youth who drew that encomium from Byron <u>muss etwas taugen</u>.[14] The following is a scrap extracted from a larger scrap entitled 'my own character' written by White himself[15]

> Well, first I premise its my honest conviction
> That my heart is a chaos of all contradiction
> Religious, deistic, now loyal now warm
> Then a dagger drawn democrat hot for reform

This moment a fop, <u>that</u> sententious as Titus
Democritus now and anon Heraclitus
Now laughing and pleased like a child with a rattle
Then vexed to the soul with impertinent tattle
Now moody and sad, now unthinking and gay
To all points of the compass I veer in a day
I'm proud and disdainful to fortunes gay child
But to poverty's offspring submissive and mild
As rude as a boor and as rough in dispute
Then as for politeness, oh dear, I'm a brute!
I shew no respect where I never can feel it
And as for contempt take no pains to conceal it
And so in the suit of these laudable ends
I've a great many foes and but very few friends

There is a vein of pleasantry here but it is solid earnest at bottom—there never was a chosen soul who escaped this ordeal. 'Whom the lord loveth he chasteneth and scourgeth every son whom he receiveth'.[16]

I shall be very glad indeed to know Phillips; wherever I meet an article from him I am always sure to read it, and the outline I have gathered of the man from his writings is just such a one as your character of him would fill up, I have read the masque of life[17]—it evinces a good deal of poetic power.

You drew my attention once to an article[18] from the pen of Hutchinson, and at the time I formed my opinion of the writer. It was very cleverly written, much tact, clear sight, but all in one direction—the thing had no breath, and its cleverness barely rescued it from the tinge of bigotry—but how many honest simple souls are thus tinged, and estimating him as one of such I was glad to find that he had fallen in your way—his attainments may be very high Tom, but this critical way never led a man to <u>the highest</u>. Shakespeare has had a host of critics both in France and England, many of them able men, but no one among them ever derived half the benefit from Shakespeare that Goethe has derived, who sat down before the mighty man as docile as a child willing to accept all he had to give, and not caring to cavil about trifles.

The theory which our investigations contradict has emanated from Professor Plucker of Bonn,[19] you are aware that when light falls <u>perpendicularly</u> upon a glass surface it goes straight through but when it falls obliquely it is bent on entering the glass and bent again on leaving it, pursuing in the last case a direction parallel to its first course.

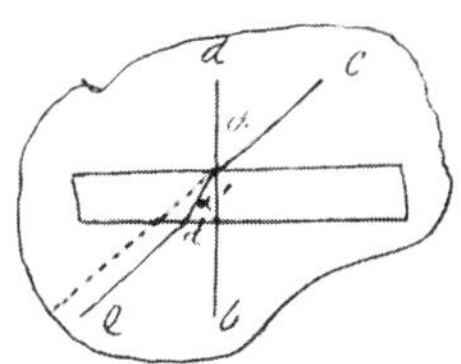

a b represents the perp. ray; *c d e* the oblique or refracted ray; now no matter what the angle is which the oblique ray makes with the perpendicular <u>without</u> the glass, the sine of this angle divided by the line of the angle which the oblique ray makes with the perpendicular <u>within</u> the glass, is a constant quantity; that is

$$\frac{\sin \alpha}{\sin \alpha'} = n$$

in the case of glass *n* is equal $\frac{3}{2}$ or $1\frac{1}{2}$ and this is called the index of refraction for glass. for other substances it is different; diamond has a very high refractive power; in this case *n* or the index is $\frac{5}{2}$. But there are many crystalline bodies in which the law of refraction is a little more complicated, for example let us take the case of Iceland spar which has the form of a cube pressed into an oblique position, that is to say, the cube is bounded by 6 squares but the spar is bounded by 6 rhombuses, looking at any object thro' this crystal (which as I have said is only an example of a numerous class) you see two images of that object—The ray on entering the crystal is decomposed into two, one of which follows the common law of refraction, that is, it has a constant n, the other on the contrary has a variable index the value of which depends upon the direction which this ray takes through the crystal; the former is called the 'ordinary ray,' the latter 'the extraordinary ray'—now there is one direction in which both fall together; and looking thro' in this direction you have only one image; this direction has therefore received the name of the 'optical axis'—in Iceland spar (you can readily procure a bit) it is the line running diagonally through the crystal from one obtuse angle to the other. These crystals where the ordinary ray is bent more than the extraordinary have received the name of negative, those where the extraordinary ray is most bent are called positive. Prof. Plucker 1st established the law;[20] that between the poles of a strong electric magnet the optical axis was always repelled; latterly however he has published in a letter[21] to Faraday a modification of the law, the result of his three years experiments namely that the optical axis of positive crystals are attracted, those of negative repelled; and it is the law thus modified which our enquiries have overturned—I sent a short abstract of our investigation with Frankland to England for insertion in the Philosophical Magazine,[22] but have as yet had no reply from him   Frankland is now Prof. of practical chemistry in the college of civil engineers at Putney near London, he left Germany about 3 weeks ago. The above explanation will render the abstract when you receive it intelligible. The term 'diamagnetic' refers to substances which are <u>repelled</u> by the poles of a magnet; 'magnetic' of course to those <u>attracted</u>. By the way Tom, there is a translation of a German work in Physics which I would recommend you to purchase; it is pretty dear, here in Germany it costs a pound, I believe the translation is cheaper, it is called Treatise of Physics[23] by Dr. John Müller founded on a treatise by Prof. Pouillet of Paris—a general acquaintance with this book which is very clearly written will be a great advantage to you should you come to Germany you will thus be able to derive twenty times more benefit from lectures on the same subject.

I had made up my mind to go to England at the end of next month, but with the intention of coming back again provided I could make any arrangement to enable me to spend the summer here, what you have done modifies this resolution. I shall probably remain here. To Berlin I intended to go for a month or so, but as to France that is improbable. Noll is in Berlin—he will return to Marburg in a month or two. I hope you will find him here when you come, he is a worthy fellow, no pretension but very clever.

By George I was on the point of forgetting all about the 'articles'[24]—if it be received I must say it is more than I expect as it is wanting in that very material point[25] to which you allude. Should it be received however I leave the thing altogether in your hands, whatever contract you make I will abide by. I will not think of a book just yet; it would take too much time. Gustave is the incarnation of a few hints derived from the persons and circumstances around me. Pof. Knoblauch <u>is</u> your optical friend.[26]

On this 6th day of Feb. received the sum of 20 pounds from Thos. Hirst thro the hands of Madame Sophie Adler of Frankfort.

John Tyndall

There thou great Ursa Major is receipt & <u>security</u>

Before I received your letter[27] I had another sketch written for the Chronicle[28] on the same subject; your remarks may cause an alteration or two or rather an addition or two.

Do present my vows <u>also</u> at the shrine of 'Jane Eyre'—and be sure to tell her that I am the nicest young man of the three! I expect this as an act of honour on your part and as an act of justice to myself.

A garret is not respectable—I have my eye on a room a stage lower down.

Will you ask Ginty whether he has received a letter from me addressed to Capt. Wynne[29] and whether he has forwarded it to the Gentleman?

And now my brother for the present good bye. I have not thanked thee much, nor do I intend to do so. I absolve myself from this by the conviction that were our necessity changed I would do precisely the same for you.

Lebewohl! Lebewohl![30] | <u>Tyndall</u>

Thomas Hirst | Horton Street | Halifax | Yorkshire[31]

RI MS JT/1/T/526

1. [7–c. 26 February]: date is uncertain. The earliest possible date is 7 February, the day Tyndall received Hirst's letter (0393) to which he replies here (see allusion to Carter's letter, for example). Postmarks indicate that this letter, posted in Marburg, reached Southampton on 1 March and Halifax on 3 March. Hirst wrote 'March 3—1850' on the envelope, though it took a further three days to reach him in Lincolnshire, where he was surveying (see 'Journals', 6 March). As it took around 5 days for a letter to travel from Marburg to Halifax, the latest date this letter could have been posted was c.26 February.

2. *Devil's child as somebody says*: perhaps someone harassing Hirst over his beliefs, possibly John Abbott below.

3. *The act you have just committed*: the loan of £20 to Tyndall.

4. '*The children . . . of light*': Luke 16:8, 'And the lord commended the unjust steward, because he had done wisely; for the children of this world are in their generation wiser than the children of light'.

5. *John Abbott*: probably John Abbott (1796–1870), who came from a family engaged in the woollen industry. In 1849 he had been appointed a Borough Justice in Halifax. He was active in numerous local organisations and, on his death, left charitable bequests of over £60,000. He was also mentioned in letter 0387.

6. *Louis, Thalers, florins and Kreuzers*: Hirst borrowed £20 from Craven and arranged for it to be paid to Tyndall. See letter 0394, n. 7.

7. *Mr. Schwann*: see letter 0393, n. 17.

8. *crystals knowing that I was experimenting*: Tyndall became fascinated by Julius Plücker's research on diamagnetism. Tyndall, H. Knoblauch, and M. Faraday exchanged crystals throughout this period (see letters 0396 and 0416).

9. *Burn the letter*: the letter to Carter concerning a loan, which Tyndall had enclosed in letter 0392 to Hirst.

10. *a penny trumpet*: a Victorian-era child's toy, similar to contemporary vuvuzelas.

11. *Hessian halter*: a well-crafted horse halter.

12. *zwischenraum*: an open or empty space between two things (German).

13. *Byron wrote as follows about him*: Lord Byron composed this eulogy to Henry Kirk White after the famed English poet's death in October 1806. See G. Byron, 'Eulogy on Henry Kirke White', *English Bards and Scotch Reviewers* (London: James Cawthorn British Library, 1809), lines 831–58.

14. <u>*muss etwas taugen*</u>: has to be good for something (German).

15. *written by White himself*: See H. White, 'My Own Character', *The Poetical Works and Remains of Henry Kirke White* (New York: D. Appleton & Company, 1869), p. 21, lines 11–28.

16. '*For whom the Lord loveth . . . he receiveth*': Hebrews 12:6, 'For whom the Lord loveth he chasteneth, and scourgeth every son whom he receiveth'.

17. *I have read the masque of life*: see letter 0393, n. 14.

18. *an article from the pen of Hutchinson*: not identified.

19. *The theory . . . of Bonn*: Hirst had asked Tyndall whose theory his investigation would contradict in letter 0393. He incorrectly guessed Liebig.

20. *Prof. Plucker 1ˢᵗ established the law*: Plücker found that the 'optic axes of crystals are repelled by the poles of a magnet, that the force is independent of the magnetic or diamagnetic condition of the crystal, and that it diminishes less than the magnetic or diamagnetic forces as the distance from the poles increases' (Jackson, 'John Tyndall and Diamagnetism', p. 439). Plücker's work appeared as J. Plücker, 'Über die Abstossung der optischen Axen der Krystalle durch die Pole der Magnete', *Poggend. Annal.* 72, no. 10 (1847), pp. 315–43; and J. Plücker, 'Über das Verhältnis zwischen Magnetismus und Diamagnetismus', *Poggend. Annal.* 72, no. 10 (1847), pp. 343–50.

21. *he has published in a letter*: see letter from J. Plücker to M. Faraday, 28 September 1848 (*Faraday Correspondence* 3: 2108). Plücker wrote another letter to Faraday on 20 May 1849, which Faraday subsequently had published in the *Phil. Mag.* See J. Plücker, 'On the magnetic relations of the positive and negative optic axes of crystals', *Phil. Mag.* 34, no. 231 (June 1849), pp. 450–52.

22. *short abstract . . . for insertion in the Philosophical Magazine*: published as 'On the deportment of crystalline bodies between the poles of a magnet', *Phil. Mag.* 36, no. 242 (March 1850), pp. 178–83. Tyndall implied that he had taken the initiative for this 'short abstract'. The English and German versions of the Knoblauch-Tyndall publications were not translations of one another; rather each wrote up the results in their own language (compare letter 0417, n. 17).

23. *Treatise of Physics*: J. Müller and C. Pouillet, *Textbook of Physics and Meteorology (Lehrbuch der Physik und Meteorologie)* (Brunswick, 1847).

24. *the 'articles'*: written by Tyndall concerning German University life. See letter 0392, n. 2.

25. *that very material point*: Hirst assured Tyndall (in letter 0393) of future publication success, confidently stating, 'set to it and leave the rest to Providence and Tom Hirst'.

26. *Gustave . . . optical friend*: Tyndall was describing the sources for his characters in his sketch which Hirst had sent to the *People's Journal*.

27. *your letter*: letter 0393.

28. *sketch written for the Chronicle*: Tyndall probably is alluding to an article published as Wat Ripton, 'Man and Magnetism', *Preston Chronicle*, 9 March 1850, p. 6. In it, Tyndall discussed self-improvement and made recommendations regarding the Preston Mechanics' Institute. He had written on the same subject in December (see letter 0393, n. 26) and Hirst had given his thoughts in letter 0393.

29. *letter . . . to Capt. Wynne*: letter missing, but Tyndall received a letter from Wynne in early April, which may have been part of the train of correspondence (see letter 0397, n. 1).

30. *Lebewohl! Lebewohl*: Farewell! Farewell! (German). 31. *Thomas . . . Yorkshire*: on envelope.

From Hermann Knoblauch   [18–28 March 1850][1]   0396

Dear Mr Tyndall

I have hastened to get the crystals for our experiments and I send you a number of them.

1. <u>Dioptas</u> (from the 'Kirgisensteppe') <u>rhomboëdric</u> (like spar)
2. <u>Diopsid</u>
      a) from the Mussa. Alp (Piemont)*
      b) from Reichenstein.
      cleavage parallel to the surfaces of the orig. prism.
     *(fine green crystals, together with granates)
3. <u>Eisenfreier Turmalin</u>. (from Elba) <u>rhomboëdr</u>.
     [3] cleavages (exactly the same with those in the common turma-
    line) converg. to the axis of the column.
4. <u>Topas</u> (from Saxony)
      1 peace without iron (colourless)
      1 " " " with " " " (yellow)
    cleavage parallel to the 'gerade Endfläche')
5. <u>Grauspiessglanz-Erz</u>
      1 cleavage parallel to the 'gerade Endfläche'
      an other " " " " *[Längesfläche]*.
    to be compared with antimony in respect of the induction
    phenomena.
6. <u>Sapphir</u>
      cleavage parallel to the 'gerade Endfläche'.
7. <u>Apatit</u>
      cleavage " " " " " " .
8. <u>Wolfram</u>
      cleavage parallel to the surfaces of the original prism.
9. <u>Grünbleierz</u>
      (2 peaces—) cleavage parallel to the 'gerade Endfläche'.
    <u>same</u> form <u>with Apatit</u> | Grünbleierz (Arsenikblei) contains 'Kalkerde'
    instead of lead.[2]
10. <u>Petalit</u>
      cleavage parallel to the surfaces of the orig. prism.
    11. <u>Hornblende</u>   <u>with iron</u>
      a) 2 little peaces
      b) a large peace
    12. <u>Tremolit</u>   <u>without iron</u>
      <u>same</u> crystalline <u>form</u> : (prismatic) to be compared.

13. <u>Augit</u> (from Bohemia).　　　augit <u>with iron</u>.
　　<u>same form</u> as diopsid S. <u>n<u>o</u> 2</u>.　　　diopsid <u>without iron</u>.
14. <u>Spinell</u>
15. <u>Automolith</u>.
　　<u>same</u> octaïdric form, the one containing iron, the other not so.
16. <u>Smaragd</u>　cleavage parallel to the 'gerade Endfläche'.

<u>Ewart</u> ist leider nicht hier, so dass ich wegen der optischen Eigenschaften keine Auskunft erhalten kann. Ebenso ist <u>du Bois</u> verreist. Er hatte unsere Abhandlung noch nicht der Physikalischen Gesellschaft mitgetheilt. Ich that dies daher selbst in der Sitzung am vergangenen Freitag. Das Heft der Philosoph. Magaz., worin Ihr Aufsatz, ist bereits hier verbreitet. Poggendorff's Februarheft, worin die deutsche Mittheilung, ist im Druck vollendet und erscheint in diesen Tagen.

Poggendorff ist hier der Einzige, welcher die Faraday-Plücker'schen Untersuchungen näher kennt. Zu meiner Verwunderung sagte er mir, dass er eine <u>Kupferscheibe</u>, beim <u>Öffnen</u> der Kette immer <u>angezogen</u> gefunden hätte. (A <u>copper disc</u> <u>attracted</u> (by induction), when <u>opening</u> the circuit. As I remember, we found it <u>repelled</u> *[2 words illegible]*. Be so kind as to repeat the experiment before <u>one pole</u> with copper as well as with antimony and give me notice of the result). Plücker hat die Erscheinung beim Öffnen der Kette gar nicht bemerkt, Poggendorff konnte mir von der Gesammtheit der Wirkungen keine genügende Erklärung geben. Es wird daher gut sein, wenn wir unsern Plan ausführen, die Inductionserscheinungen im Zusammenhange gründlich zu untersuchen.

Einen <u>Reostaten</u> und eine <u>Tangentenboussole</u> habe ich bestellt, den ersteren in der Form, in welcher Prof. Poggendorff ihn bereits für wissenschaftliche Untersuchungen angewandt hat, die letztere, wie sie Prof. Dove anwendet. (Die Dimensionen der Tangentenboussole sind von denen der Bunsen'schen etwas verschieden, aber ich habe keinen Grund gehabt, sie abzuändern, da sie sich als zweckmässig erwiesen haben). Auf Ihren Wunsch theile ich Ihnen mit, dass der Reostat 19 Thaler, die Tangentenboussole 12 Thal. kostet. Ich denke auch noch einige andere Apparate mitzubringen.

Der Vollständigkeit wegen bemerke ich, dass <u>Pinit</u> nicht zu haben war. <u>Lepidolith</u> kommt nicht krystallisiert vor. —<u>Rubellit</u> ist eine Art Turmalin. <u>Zircon</u> ist unbrauchbar für unsern Zweck. <u>Corund</u> haben Sie im Sapphir. — <u>Antimon</u> und <u>Arsenik</u>-Krystalle giebt es nicht in geeigneter Grösse. Ebenso *[wenig]* <u>Kupferkrystalle</u>.

Plücker soll 2 Arten von Topas haben, wovon er den einen für *[negativ]*, den andern für positiv hält. In <u>Venedig</u> beschäftigt man sich jetzt ebenfalls mit magnetisch-diamagnetischen Untersuchungen, wie ich aus einer Abhandlung ersehe, die mir so eben von Zantedeschi als Geschenk übersandt wird.

Es würde mir sehr erfreulich sein, wenn Sie die Güte hätten, <u>mich zu benachrichtigen, ob die Mineralien richtig angekommen sind, und welche Resultate Sie *[damit]* erhalten.</u>

Ich bin gern bereit, Ihnen sonst noch zu besorgen oder Ihnen Auskunft zu ertheilen, so weit dies irgend möglich ist. Ich bitte mir nur <u>Ihre Wünsche baldmöglichst</u> mitzutheilen.

Meine Wirthsleute (Gutmann) bitte ich zu grüssen und Ihnen zu sagen, dass Herr Krämer, an den Sie mir einen Brief mitgegeben hatten, vor einigen Wochen gestorben ist, weshalb ich den Brief nur seiner Frau übergeben konnte.

Mit den besten Grüssen an Sie und meine sonstigen Freunde in Marburg,
ganz der Ihrige
Herm. Knoblauch.

I join 2 *[1 word illegible]* crystals:
1. <u>Heulandit</u> (from Fassathal) (red, containing iron)
2. "   "   (from Island) (white, without iron).
    cleavage parallel to the 'schiefe Endfläche'.

It appears to me of a peculiar interest to compare with each other
1) antimony and Grauspiessglanzerz.
2) arsenik-blei and apatit.
3) hornblende   " tremolit.
4) augit   " diopsid
5) spinell   " automolit
6) 2 sorts of Heulandit.

they have *[resp.]* the <u>same crystalline form</u> (same cleavages) but <u>different chemical constitution. Don't confound the diff. crystals and keep them until I return</u>; because I wish to see <u>the experiments.</u>

<u>The lectures at the Berlin university</u> are not closed before the <u>end of this week</u>. —My father, who regrets that you have not accompanied me, sends you his best compliments. After the close of the lectures you would have no profit here (as I believe), especially I am afraid you would have no opportunity of working in a laboratory.

Dear Mr Tyndall

I have hastened to get the crystals for our experiments and I send you a number of them.

1. <u>Dioptase</u> (from the 'Kirghiz Steppe') <u>rhombohedric</u> (like spar)
2. <u>Diopside</u>
    a) from the Mussa. Alp (Piedmont)*
        b) from Reichenstein.
        cleavages parallel to the surfaces of the orig. prism.
    *(fine green crystals, together with garnets)
3. <u>Iron-free Tourmaline</u>. (from Elba) <u>rhombohedr</u>.
    [3] cleavages (exactly the same as those in the common tourmaline) converg. to the axis of the column.
4. <u>Topaz</u> (from Saxony)
        1 peace without iron (colourless)
        1 " " " with " " " (yellow)
        cleavage parallel to the 'vertical face')
5. <u>Stibnite ore</u>
        1 cleavage parallel to the 'vertical face'
        another " " " " 'longitudinal face'.
    to be compared with antimony in respect of the induction phenomena.[3]
6. <u>Sapphire</u>
        cleavage parallel to the 'vertical face'.
7. <u>Apatite</u>
        cleavage " " " " " " .
8. <u>Wolfram</u>
        cleavage parallel to the surfaces of the original prism.
9. <u>Pyromorphite</u>
        (2 peaces—) cleavage parallel to the 'vertical face'.
    <u>same</u> form <u>as apatite</u> | <u>pyromorphite</u> <u>(arsenic of lead)</u> contains 'calcium oxide' instead of lead.[4]
10. <u>Petalite</u>
        cleavage parallel to the surfaces of the orig. prism.
    11. <u>Hornblende</u>   <u>with iron</u>
        a) 2 little peaces
        b) a large peace
    12. <u>Tremolite</u>   <u>without iron</u>
        <u>same</u> crystalline <u>form</u> : (prismatic) to be compared.[5]
    13. <u>Augite</u> (from Bohemia).   augite <u>with iron</u>.
        <u>same form</u> as diopside S. <u>no. 2</u>.   diopside <u>without iron</u>.
    14. <u>Spinel</u>
    15. <u>Gahnite</u>.

same octahedric form, the one containing iron, the other not so.

16. <u>Emerald</u>	cleavage parallel to the 'vertical face'.

<u>Ewart</u>[6] unfortunately is not here, with the result that I cannot obtain any information about the optical properties. <u>Du Bois</u> is away too. He had still not communicated our paper to the Physikalische Gesellschaft. I therefore did this myself at the session last Friday.[7] The issue of the Philosoph. Magaz. in which your essay appeared has already been circulated here.[8] Poggendorff's February issue in which the communication in German[9] appeared, has been printed and will appear in the next few days.

Poggendorff is the only person here who knows the investigations of Faraday and Plücker in more detail. To my surprise he told me that he had always found 'a <u>copper disc</u> <u>attracted</u> (by induction) when <u>opening</u> the circuit. As I remember, we found it <u>repelled</u> [2 words illegible]. Be so kind as to repeat the experiment before <u>one pole</u> with copper as well as with antimony and give me notice of the result'). Plücker had not noticed the phenomenon at all when opening the circuit. Poggendorff could not give me a satisfactory explanation of the totality of the effects. It will therefore be good if we carry out our plan to thoroughly investigate the induction phenomena connected with it.

I have ordered a <u>rheostat</u> and a <u>tangent galvanometer</u>, the former in the form in which Prof. Poggendorff has already used it for scientific investigations, the latter as Prof. Dove is using it. (The dimensions of the tangent galvanometer are somewhat different from those of Bunsen's one, but I have had no reason to modify them, as they have proven to be suitable). As you wished, I am informing you that the rheostat cost 19 Thaler and the tangent galvanometer 12 Thal. I am considering bringing some other apparatus with me as well.

For the sake of completeness I note that <u>pinite</u> was not to be had. <u>Lepidolite</u> does not exist in crystallised form. —<u>Rubellite</u> is a kind of tourmaline. <u>Zircon</u> is unusable for our purpose. You have <u>corundum</u> in sapphire. —There are no <u>antimony</u> and <u>arsenic</u> crystals in suitable size. There were just as [few] <u>copper crystals</u>.

Plücker is supposed to have 2 sorts of topaz, of which he considers one to be [negative] and the other positive. People are engaged with magnetic and diamagnetic investigations in <u>Venice</u> now too, as I see from a paper which has just been sent to me as a gift by Zantedeschi.[10]

It would be very gratifying for me if you were so kind as to <u>inform me whether the minerals have arrived safely, and what results you then obtained [with them]</u>.

I am happy to supply you with anything else or to give you information, as far as this is at all possible. I ask that only you communicate <u>your wishes</u> to me <u>as soon as possible</u>.

I ask you to send my greetings to my hosts (Gutmann), and to tell you that Herr Krämer, whom you had given me a letter to take to, died some weeks ago, which is why I could give the letter only to his wife.

With best greetings to you and my other friends in Marburg.

entirely Yours

Herm. Knoblauch.

I join 2 *[1 word illegible]* crystals:

1. <u>Heulandite</u> (from the Fassa Valley[11]) (red, containing iron)
2.   " " "   (from Iceland) (white, without iron).
    cleavage parallel to the 'sloping face'.

It appears to me of a peculiar interest to compare with each other

1) antimony and stibnite ore.
2) arsenic of lead and apatite.
3) hornblende   " tremolite.
4) augite   " diopside
5) spinel   " gahnite
6) 2 sorts of heulandite.

they have *[respectively]* the <u>same crystalline form</u> (same cleavages) but <u>different chemical constitution. Don't confound the diff. crystals and keep them until I return</u>; because I wish to see <u>the experiments.</u>

<u>The lectures</u> at the <u>Berlin university</u> are not closed before the <u>end of this week</u>. —My father, who regrets that you have not accompanied me, sends you his best compliments. After the close of the lectures you would have no profit here (as I believe), especially I am afraid you would have no opportunity of working in a laboratory.

RI MS JT/1/K/27

---

1. *[18–28 March 1850]*: Tyndall's journal entry for 29 March 1850 (JT/2/13b/483), the first entry for nine days, recorded: 'Received a number of crystals, and a long letter from Professor Knoblauch'. Tyndall therefore received the letter between 21 and 29 March. Allowing one to two days postage time, it was posted between 18 and 28 March.
2. *same form . . . lead*: added in left margin.
3. *to be . . . phenomena*: added in left margin.
4. *same form . . . lead*: added in left margin.
5. *same crystalline . . . compared*: added in left margin.
6. *Ewart*: perhaps the Ewart who became laboratory assistant at Marburg (letter 0488).

7. *did this myself . . . last Friday*: Friday, 15 March (see *Die Fortschritte der Physik in den Jahren 1850 und 1851. Dargestellt von der physikalischen Gesellschaft zu Berlin* (1855), p. vii).

8. *your essay . . . circulated here*: the English language account of their joint research, just published in *Phil. Mag.* (cited letter 0395, n. 22).

9. *Poggendorff's February issue . . . the communication in German*: the German account of the same joint investigation (cited letter 0392, n. 13).

10. *Zantedeschi*: Francesco Zantedeschi (1797–1873), an Italian physicist who held a professorship at the University of Padua between 1849 and 1853.

11. *Fassa Valley*: a valley in the Tirol known to be a source of heulandite.

To George Wynne   [c. 7 April 1850][1]   0397

Perhaps no portion of your letter excited my interest more than the kind and friendly wish expressed towards the end of it. I notice this the more particularly as I would rather that you had never taken a step in my behalf than that you should do it under the conviction of my being that which I am not. This would savour of obtaining money under false pretences, perhaps be the less respectable of the two. With regard to my religion the tale is briefly told, were I asked for an epitome thereof I would lay my hand upon the writings of John in the New Testament,[2] and the religion there taught is not the religion of the Understanding but a deeper one. Since I came to Germany I have not got a single new notion on this subject, having occupied myself almost wholly with science, intending at a later period of life to test as far as my abilities permitted the fallacy or the truth of German philosophy. There are however two writers in the English language which have exercised and still continue to exercise a great influence upon me, and these are Carlyle and Emerson; I mean Carlyle the historian, the author of Past & Present, Heroes & Heroworship, Letters and Speeches of Oliver Cromwell[3] &c. Emerson is an American, doubtless you are acquainted with the writings of both, and if so will have observed that though they never attack the Church of England[4] nor any other church, their writing in many places breathe a spirit to which the church of England would scarcely subscribe. Heroes & Heroworship is a foolish kind of name for a very serious work which made a great impression on me, an impression which has not yet worn away. You have thus a complete view of the character of him whom you would befriend—if you blame me, then the plea of Martin Luther must be mine—'I cannot otherwise—may God assist me!'[5]

RI MS JT/2/5/386–7
JT Transcript

1. *[c. 7 April 1850]*: the source for this letter is Tyndall's copy in his journal (entry for the
    week of 1–7 April). He recorded receiving a letter (missing) from Wynne: 'At the end
    of M[r] Wynnes letter a hope was expressed that I still adhered to the pure doctrines of the
    church of England, wrote as follows in reply.' The journal version is a clean version, not a
    rough draft.
2. *writings of John in the New Testament*: Church tradition holds that the apostle John is the
    author of the Gospel of John, the three Epistles of John, and the Book of Revelation.
3. *Past & Present, . . . Oliver Cromwell*: T. Carlyle, *Past and Present* (London: Chapman
    and Hall, 1843); *On Heroes, Hero-Worship, & the Heroic in History* (London: Chapman
    and Hall, 1840); and *Oliver Cromwell's Letters and Speeches: With Elucidations by Thomas
    Carlyle*, 4 vols (London: Chapman and Hall, 1845).
4. *Church of England*: Carlyle did not openly attack the Church but he had little respect for
    it: 'is it not perishing rapidly enough, much too rapidly perhaps, without help of mine?'
    (Carlyle to J. G. Marshall, 18 November 1841, *The Carlyle Letters* 13: 297–99).
5. *'I cannot otherwise—may God assist me!'*: according to tradition, Luther said these words
    (in German) at his trial at the Diet of Worms in 1521.

From Thomas Archer Hirst   16[–18][1] April 1850   0398

Halifax | April 16th, 1850

My dear Tyndall,

Your last letter[2] reached Lincolnshire where I was engaged in some engineering business being 'lent' by M[r] Carter;[3] since then I have been knocked
about too much to attempt a reply; now I have got into still water once more
and gladly commence my task. But I recollect now I received another scrap[4]
since then forwarded from Preston, but have been unable to comply with its
wishes on account of the Preston Chronicles never having reached me altho I
had written expressly for them. I can only account for the circumstance in one
or the other of the following ways. They are full of other matters, or the tenor
of the article[5] does not suit them, if the former is true it will appear by and
bye, if the latter is the cause I want you to request the Editor to forward the
MS. to me. I have great curiosity to see it & have a suitable means for its publication enclose such letters[6] (should you think proper to write it) to me, so
that in the event of its appearance before arrival I may withhold it. The Editor
of the Peoples Journal[7] wrote me a very kind answer back saying that had he
not unluckily just engaged a continuous article on a similar subject he would
have been very glad to accept it for he thought very highly of it. I shewed it to
Phillips who also liked it & with his help I could have easily got it published
had I not had another scheme in view. A new paper called the 'Leader' has just
appeared a very able one indeed & conducted by a party with whom both of

us have much sympathy I mean that peculiar class to whom Carlyle has imparted some of his energy. He will write occasionally for it himself. Lewes, Thornton Hunt, Linton, Ballantyne, & Phillips are other contributors[8] that I know of for the rest I will send you a number for your own perusal. It is already supported by the best heads in England perhaps and as a means of effecting much good gives great promise. Now we can get access to this and I want you to extend your sketches[9] to 6 not striving so much for <u>mere</u> popularity as would have been requisite had the Peoples Journal been the medium of publication. The rest I shall leave to yourself only you must get to work quickly for should circumstances hinder its publication even here I am determined they shall appear. I have got amongst a good deal of literature & literary men here. Phillips, who has an extensive & <u>well selected</u> (intellectually not fashionably) acquaintance & who as I said, has taken a great liking to me, takes every opportunity to introduce me. I know his intention is good, but to tell the truth I don't much like it. I have got to know some able & right good fellows certainly, but I find its effect on me is not one of the best. In the first place I am naturally of a retiring disposition, and one of few words & being thrust forward as it were, is uncomfortable from the ever present notion that something is expected from me, and were there anything in reality to satisfy that expectation it is just the way to keep it back. In default thereof I feel dissatisfied with myself & find previous determinations to steady persevering effort at times weakened. All this shews me Tyndall that I am not ready and worthy to play such parts it arises in fact from a small conceited spirit of ambition which must be purged out of me, either by such meetings or without. But here I am again lamenting and exposing my own shortcomings as usual & filling up a melancholy letter when I have plenty of other things to say. It must be some secret influence in that long solemn face of yours that always seems to say 'Tom, confess or I'll tell you all about it myself', and there is an influence just as strong in me too that always & yet for the most part unconsciously prompts me to do so if I tell my conscience a secret it can't keep it but must share it with its brother John not that we want brother John to do for us what is our work, but because the desire is irresistible & we cannot help it. To change the theme therefore. Among the best of these acquaintances is a young fellow called Smith, formerly of Manchester but who has come to reside in Halifax. he is as clever & generous little fellow as one often meets, with some failings of course like all the rest of us but be-hanged to them at present. He has written a work a life history of Mirabeau,[10] of which Carlyle said 'it is a highly creditable book, good in itself but giving promise of far better' which is as short & significant a criticism as I can give you. Carlyle <u>awakened</u> him too, & this work was the result. Some of the reviews of course severely castigated it[11] & could see nothing or rather <u>would</u> see nothing but the Carlyle-isms with which its diction abounded, but to a deeper looker it displays a depth & ability sufficient to render Carlyle's predictions a moral certainty. Carlyle then advised him in a beautiful letter which I have seen to turn his attention to an English subject & try to extract something of value from his own life & experiences in Manchester. Carlyle of course saw he was able to do this & the result has proved he was not entirely wrong nor his advice without effect. He is publishing at present a work entitled 'Social Evils'[12] (I think) of which I have seen the MS & I consider it one of the most interesting & cheering signs of the day. It is just this—<u>The word that most wanted saying, well said</u>. It is the result of a varied experience in the Middle Classes, a keen eye for its shams & failings & a sincere honest love for a reality & truth.[13] With a broad liberal spirit he has manfully grappled with his subject & with an earnestness that sincerity alone can impart he has spoken what will meet with sympathy & energy from every thinking young man. You will see I like the book Tyndall amazingly. I cannot give you a just idea of it but you will soon see it for yourself, I hope. He is a cotton-spinner by trade & has just set up business here he is worth some little money, at any rate does not follow literature for a living but because it is his taste & forte. Well he's going to commence a new journal to be edited at least by him & Phillips together & to come out probably in September next; it is to be about the same size & of the same stamp as the Truth-Seeker (without its eternal Teetotalism, Critical Notes & Social Science of course)[14] in fact it will be more like the Leader in quality & from their extensive acquaintance they hope to gather together an able & staunch number of contributors. It will do much good for there is a want of a suitable medium for the thoughts of this class that are every day increasing. It was first mooted in my little room & after considering the matter for some time we met again & decided upon the project £75 was raised there and then by this little knot of fellows that we have found in Halifax, as a guarantee fund that might be calculated on should the first year prove a loss (which we calculated would in the worst circumstances be the greatest amount), should it prove profitable the surplus is to pay the editors or any contributors, so as to ensure talent. I mention all this because on your return & when we change places I should like you to keep it in view indeed I have promised them your help & I have no fear you will give it.- Good Gracious I have such a lot of things to say & such a number of questions to ask I don't know where to begin or what to say next. I wish we could fly & hold an interview for an hour on the top of the Drakenfels[15] or some central meeting place. I don't know what to make of you in one respect you must either be a man most lamentably without a fixed object in the world or else you're very close fisted about it. I want to know when you are coming home and what you are going to do <u>first</u> when you get here. Are you going to seek another situation as Tutor or Surveyor? you ought not to stop at either long. Have you got your degree?[16] You have never told us yet. Phillips has some sort of a scheme about establishing a Peoples College, but I won't attempt to describe it for I cannot do it justice. it is a favourite notion of his for giving first rate instruction to classes who could not afford to go to Cambridge etc. indeed establishing an University something after the German fashion[17] (with regard to expense especially) he will talk to you himself about it when he sees you for he is curious to know how you will regard it & how far you would fall in with it. But however that will not find you employment on your return and I guess you will not want to be resting long. What do you think to coming to Carter again a bit. I shall have done there in August. He has not said a word yet about my further stay nor has he made any arrangements towards filling our places (for we both leave near together) Now I don't care if I start immediately (that is in October next) to Germany indeed I should like, but if he at all seemed to wish or to rely on my stopping a little longer I should not mind for an additional 6 months, though I feel to be wasting the best part of my life almost at present. Again had it been possible I should have liked much to have commenced my studies with you before you left Marburg. Could that be managed? Then again failing that I should want a lot of instructions about costs &c from you, so as to see my course straight before me before embarking. Consider all this & give me your advice as also full information about yourself. If you would care to engage with Carter in the place of us two (I don't know what Jemmy will do, get a situation in London I fancy) I would sound Carter on the subject. Another idea strikes me, you said something about coming over to England and returning if you could do that I would wait a little longer & go back with you. But however I depend on you to be a way out of all this, recollect my actions must be such as will serve my own interests best, that is to say, there is no fear of going against the wishes of any of my relations and I am glad to be able to say that, for all to whom I have yet made known my intentions altho they could not enter sufficiently into the merits to <u>advise</u> the step, yet they express no objection, indeed have expressed their faith in my own <u>prudence</u>.

And now about the remittance you received; I must set you right in a small error you are unconsciously labouring under, & tell you that you it is not to me you are entirely indebted, though all, or such of your gratitude as is worthy my reception, I am inclined to receive on the same ground as you would absolve yourself, namely that there was the will if not the way. It happened thus. Your letter[18] to Carter, & your aprising me of its contents, rather made me uneasy; because I thought you were in need of the installment more, than you liked to confess, & also because I both feared Carter could not, & was persuaded he ought not, to do what I could do better; but as I had told you, I had not the money yet by me. I knew Jemmy to have some at his own disposal, with which he did not know well what to do. I asked him for a loan

& to guard from misinterpretation, told him what I intended to do with it; & that I should be able to pay him shortly, he jumped at the opportunity, lent me it, and I forwarded it, but the little rogue insisted upon considering it his loan, because he would have it he was quite as able, & more so than me. It was sent however in my name, and if there is the slightest need, or he will allow *[me]*, I shall take it in my own hands for it will be no inconvenience. I should not have submitted to this small deception my good lad, but for fear you should make any foolish 'boggles'[19] about the matter, I determined it should get into your hands, & then you might know all about it. And now thou mayest add a codicil to thy journal entry, & *[canst]* shew a future Tyndall, that instead of one thou hast <u>two</u> friends, who can trust thee not only with £20, but with far more & that, of another & superior quality. The little fellow has done it from the purest & most disinterested motives, & was moved to the act by a honest unassuming wish to be of use to one of his best friends—

Carlyle is in the field again! and is astonishing the world with some 'latter-day pamphlets',[20] on topics of the day; 3 have already appeared, and another is published in London today. 'The Present Time No 1' of which I sent a short article to the Truth Seeker,[21] which you have perhaps ere this noticed No. 2, 'Model Prisons' an enquiry into the Anti capital punishment mania of the day, and some right-solid remarks thereon; and No 3 'Downing Street', a thorough exposure of our so-called Government their miserable inefficiency, & what is new for Carlyle; a <u>decided</u> measure for their improvement; but in the next which is entitled 'New Downing Street', I expect a further elaboration of these, something tangible & fit for <u>immediate</u> application; and coming from such a quarter likely to create a great sensation. But I will not attempt further description here, as if all is well, I may perpetrate a few more remarks on the same for the Truth Seeker, which you can then see; I can tell you however that they are wonderful performances & such as you will read with no small interest when you see them. I have read a good deal of his works lately, from his first efforts, those splendid critical & biographical articles that appeared in the magazines of 1828 & 30[22]—those clear, calm & deep portraits, where Jean Paul, Goethe, Burns, Voltaire, Werner, Heyne & others are taken and literally turned <u>inside out</u>, imaged there before you, perfectly plastic, aye <u>breathing</u> with vitality:—Such Biography I never saw before or since. Pity there are so few to be had. —Next I take his Hero Worship[23] for a philosophy of Human Nature, and after that, that strange autobiography the 'Sartor Resartus'[24] the record of his life struggles, his glorious battles & victory, with the world & himself; I mark too, that at this point he turns from his own high & self-created sphere, & *[veteran]*-like places himself at the head of his fellow-comrades, to lead them too, to warfare & to victory. His 'Chartism',[25] his 'Past & Present',[26] & last of all his 'Latter-day Pamphlets', have this for their object. God bless him then I say, for the true & valiant Hero of the day! the finest <u>MAN</u> perhaps of this era. Thousands are indebted to him for an 'inheritance incorruptible & that fadeth not away',[27] and when his work shall be done, & he shall have rested from his labours (which may God prolong) he will carry with him our blessing, & along with Luther, Shakespeare, Jean Paul & Goethe, shall be reckoned as an inspired brother; sent from God, amid the downfall of Religious & the reign of blind Scepticism, to build up once more a true temple to his praise. Long live Carlyle!—

Amongst other literary news, or what will be such to you perhaps, is a new volume of Lectures from Emerson, called 'Representative Men', being the 7 lectures that he delivered when in England, and one of which I recollect you heard in Halifax ('Napoleon')[28] I venture to say if Cicero had only read these 7 lectures, before he expressed his opinion on the use of books, he would have defined them as something more than 'entertainment for leisure hours, charm for solitude, or alleviation for sorrow'[29]: at any rate if he had stuck to, and acted upon, his narrow hypothesis, he would have considered the work before us as so much unmeaning jargon, & would have advocated its immediate extinction, & perhaps a straight jacket for the Author. In fact, Emerson's is no rudimentary treatise; no superstructure raised gradually from a well-known axiom as foundation; he presupposes a reader of something of his own grasp of thought & he deals with <u>results</u> ready made, omitting the <u>processes</u> as superfluous. In a sentence may often be found the net result of volumes, & the student must with much difficulty & brow-sweating fill up these vacuities: must scramble down the one side & climb up the other of these chasms, unless luckily he be nimble enough to jump them. There is the sum-total here of 30 years of not common place but quite original life & experience, a saturated solution of it dissolved in concentrated <u>Thought</u> acid & then crystallized. It must be no tyro in intellectual chemistry that will analyze this compound; the <u>first</u> step towards this, its <u>resolution</u>, is a task that will occupy us yet for some time to come.—

But to cease from my chemical illustrations, which nevertheless will bear even more prolongation yet, this book is published for the small charge of ˢ1(!) by Bohn, who by some means or other issued his edition at the same time that the ˢ5 one by John Chapman[30] appeared, & by this means gave the book an immense circulation, it is the cheapest shilling's worth I know certainly.

Talking about Chemistry, we do very little in that line as yet; Surveying, and Lincolnshire trips don't allow it; we have taken a small room, however for a laboratory, & hope during these coming Summer mornings to renew our acquaintance with it. My studies indeed just now, are necessarily very desultory & consequently not effective. I feel convinced that to prosper, and to make progress in any given study, numerous others extraneous to it must be laid on the shelf, however tempting their interest may render them. Indeed genius as I take it, consists for the most part, in overcoming this temptation to examine the surfaces of many things, & with perseverance, concentrating these our otherwise scattered forces on a single aim; or better thus—all our aims are alike, one lode-star, (call it self-expansion if you like) is our beacon; the unsuccessful traveller runs here & there, now on the one side, now the other, crying for a Road, unconscious, poor fellow, that the <u>road</u> is ever by him, that if he will but strike in anywhere, <u>here</u> close to him, & only follow his nose, with patience, & unwaveringly, he shall prosper.- 'Well, but', says a Road-seeker, 'there is some one thing that every one can do best, and how important that we should discover it'. 'Quite a delusion this, Sir', is the answer, from the man with the nose; 'believe me, the indecision arises altogether from a want of energy to start, & perseverance to pursue. Strive for this, attain this, and your road is clear; you throw in your reagent <u>will</u>; the law of affinity does the rest'. From this you may gather where 'long Tom'[31] (as Phillips calls me) is, at this time; always struggling with some wave you see, & now & then getting a duck too. Amidst it all however, one pursuit remains firm, and is the fruit of more self content than all the rest; I mean my weekly class at the Mechanics;[32] I have got a score fellows round me every Thursday Evening that it does one's heart good to see & be amongst; fellows of all ages, from 16 to 35; hard-working, sharp fellows too, many of them; two or three from vill[ages] 2 miles distant, & one, my oldest pupil (i.e. been with me longest) already stands on the same platform, & as things are, will pass me soon. I have found out this fact there; that actual instruction (i.e. direct gift) is just the least part of a Teacher's use; provocation & encouragement are the principle; once kindle a <u>love</u> for their task, & the materials are ready enough. Again, this love for the Study is contagious for the most part, & the success for the Teacher will depend equally as much on its personal possession by him; as on his talent. And lastly, if he <u>do</u> possess it, he need not seek to inculcate it by words; it will flow out unknown to him in every action. —I have finished my news, and to finish my paper, I will give you a Sonnet I attempted when in Lincolnshire; I have not room to explain the circumstances that originated it, I can only say that I lived a fortnight at a beautiful Farm-House, past-whose garden gate the River ran, & inside whose walls was genuine hospitality, it was written in the Album of the fair daughter of mine Host. Now listen, and laugh at it if you like—

> Gentle River; peaceful, calm and Holy,
> Thou pure interpreter of this Glory,
> Who made alike thyself, all men, and me,
> Companions bound unto shoreless Sea;
> There Life upon thy banks, but like thy course,

So void of Strife, so full of silent Force,
Could thus reflect, along its fruitless march,
The lovely Sun-set, or the great high arch
Of Heavens Immensity: azure deep
Where twinkling stars their holy vigils keep:
Then would I live and die contented here,
Here learn from thee my onward path to steer,
And fearless, glide 'twixt Times green banks with thee,
Unto the unknown, vast Eternity.

—Tom—

RI MS JT/1/H/144
RI MS/JT/1/HTYP/68–73a

1. *16[–18]*: Hirst began the letter on 16th, finished it on the 18th, and posted it to Tyndall on the 19th (Hirst, 'Journals', 16–19 April 1850).
2. *Your last letter*: probably letter 0395, but this was not the most recent letter (see n. 4).
3. *being 'lent' by Mr Carter*: Hirst was 'lent' by his employer, Richard Carter, to assist with a drainage survey at Fenton in Lincolnshire from 26 February until 23 March 1850 (Hirst, 'Journals', 26 February to 23 March 1850).
4. *received another scrap*: Hirst remembered a short, more recent letter, which is missing. Tyndall had apparently asked Hirst to send him something, perhaps his own published article, from the *Preston Chronicle*.
5. *the article*: it was not published (see letter 0392, n. 2).
6. *enclose such letters*: that is, letters to editors. Not only would it prevent confusion, as Hirst explained, but it would save expensive postage.
7. *The Editor of the Peoples Journal*: John Saunders was chief editor for the *People's Journal* from its founding in 1846 (see letter 0392, n. 3).
8. *new paper called the 'Leader'… other contributors*: Hunt was Thornton Leigh Hunt (1810–73), an editor, who along with Lewes established the weekly *Leader* in London in 1850, selling at the price of *6d.*, with financial backing from noted liberal clergyman Edmund Larken. It ran until 1860. Linton was William James Linton (1812–97), wood-engraver, polemicist and poet; Ballantyne was Thomas Ballantyne (1806–71), radical newspaper editor. See Biographical Register for Lewes and Phillips.
9. *extend your sketches*. Tyndall had sent Hirst one such sketch of German University life, requesting that he try to get it published (letter 0395).
10. *a life history of Mirabeau*: J. Stores Smith, *Mirabeau: A Life-History* (London: Smith, Elder, and Co., 1848).
11. *some … castigated it*: for example: 'We notice this book only to stamp it with our severest reprobation. It is an echo of all the most offensive follies of Carlyle; and, like all copies, infinitely more offensive than its original. The Theory of hero-worship, that is, of the

adoration of what may be termed brute force separate from goodness, whether intellectual or physical, is here pushed to its vilest excess' ('V.-Mirabeau. A Life History', *English Review* 10, no. 19 [1848], p. 204).

12. *'Social Evils'*: J. Stores Smith, *Social Aspects* (London: John Chapman, 1850).

13. *shams & failings . . . reality & truth*: all these were characteristic moral metaphors from Carlyle.

14. *the Truth-Seeker (without . . . Social Science of course)*: Frederick Lees, Phillips' co-editor, was a campaigner for many unpopular causes, including teetotalism and Owenite socialism.

15. *Drakenfels*: Drachenfels, a high hill and hiking area, with a ruined castle and spectacular views, located near the Rhine, south east of Bonn.

16. *Have you got your degree?*: The certificate announcing Tyndall's doctoral degree is dated 5 March 1850, which is some Months after he completed his degree requirements. Perhaps he did not tell Hirst because the certificate was a formality. We thank Elizabeth Neswald and Martina Düerkop for finding and translating the certificate.

17. *German fashion*: People's Colleges were established in Germany to provide adult education for all social classes (see H. J. Arnold, 'The People's Colleges of Germany', *American Scholar* 3, no. 1 [1934], pp. 105–8).

18. *your letter . . . its contents*: in letter 0392 Tyndall told Hirst that he had written to Carter, asking to borrow money.

19. *'boggles'*: objections (*OED*).

20. *'latter-day pamphlets'*: a series of pamphlets under this title, published through 1850, were collected in Carlyle's book, *Latter-Day Pamphlets*. The four already published were: 'The Present Time' (1 February 1850), 'Model Prisons' (1 March 1850), 'Downing Street' (1 April 1850), and 'New Downing Street' (15 April 1850). Four more followed between May and August 1850.

21. *a short article to the Truth Seeker*: cited letter 0400, n. 7.

22. *those splendid . . . magazines of 1828 & 30*: the essays on German literature which Carlyle published before his first major work, *Sartor Resartus*, begun in 1831 (see n. 24).

23. *Hero Worship*: cited letter 0397, n. 3.

24. *'Sartor Resartus'*: T. Carlyle, 'Sartor Resartus: The Life and Opinions of Herr Teufelsdrockh', first published in *Fraser's Magazine* (1833–34) and later collected in a volume (London: Saunders & Otley, 1838).

25. *'Chartism'*: T. Carlyle, *Chartism* (London: James Fraser, 1840).

26. *'Past and Present'*: cited letter 0397, n. 3.

27. *'inheritance incorruptible & that fadeth not away'*: an abbreviated quote from 1 Peter 1:4, 'To an inheritance incorruptible, and undefiled, and that fadeth not away, reserved in heaven for you'.

28. *you heard . . . (Napoleon)*: Tyndall and Hirst both attended Emerson's lecture on Napoleon in Halifax on 5 January 1848; Tyndall bought some of his books and Hirst attended another lecture in February (Journal, 5 and 7–10 January 1848, JT/2/13b/288–9; Hirst, 'Journals', 5 January and 7 February 1848). On the publication of the lectures see n. 30.

29. *'entertainment . . . sorrow'*: quotation not identified.

30. *'1 . . . by John Chapman*: Emerson's lectures, delivered in Boston in 1845–46 and then in England in 1847–48, were published as *Representative Men: Seven Lectures*. In addition to these two editions, there was another 1850 London edition by Routledge, at 1s. (The superscript 's' placed prior to the number is an unusual way to denote 'shilling'.) We reference the more accessible Chapman edition although Tyndall and Hirst probably read one of the cheaper editions.

31. *long Tom*: an allusion to Hirst's height, 6 ft. 2 ½ in. (Victor L. Hilts, *A Guide to Francis Galton's* English Men of Science, *Transactions of the American Philosophical Society*, n.s. 65 (1975), p. 80).

32. *my weekly class at the Mechanics*: the Halifax Mechanics' Institution.

From James Craven [c. 16–19 April 1850][1]   0399

Dear Tyndall,

You would perceive on perusing my last letter[2] that it was written when I had been dissatisfied at something or other, and I must confess that circumstances had for some time been occurring which led one to look nearer than hitherto into the future which is opening before me. My brother's passing with his entrance on the great world being a general topic naturally led your humble servant into prospective thoughts, likewise, and I then felt that a nice career was opening to him which I could not flatter myself would present itself to me, and being not exactly in a self-reliant or sanguine state of mind, with a slight touch of dissatisfaction and desire for change, I certainly did somewhat warmly give vent to what had been in my mind with a few editions to yourself and to your expressed astonishment. Your letter[3] however came in proper time and on asking myself whether solicitorship would answer all the cravings I had, or was more likely to do so than surveyorship, I could not cordially or with certainty answer Yes, for I did not believe it to be a panacea to all my wishes, and since then I have settled into a species of torpor, having no ideas whatever on the subject further except that whatever may arise I must be governed solely and entirely by circumstances. The ideas however which do still float across me intimate that bread and water cannot be obtained unless some poor devils get out of it, and only they who are nicely ensconced in a cosy practice can make anything handsome from it. However, time works wonders they say, so let us see what she will do with me. I'm all ready for the rub and once let me have some settled place, I care not what, I will endeavour to do my duty with regularity and merit.

Here am I—great fool that I am—penetrating the misty future and wishful to know what I am destined for! Surely I think to myself what a strange mortal I am to do so, while all around heedless of it think only of time present—some in improvement, while others seek enjoyment and excitement. Yet I do not envy them in their nightly carousals, having but just returned from London and joined in their 'merry stirs', their sprees and pleasures which to me have become now insipid and more than ever disagreeable to my taste. I feel I could never, were I ever so wishful, make myself a 'jolly fellow' with them. Such is my decision after joining in what I have often wished to see. However, no more of it unless to please other parties. My dear fellow, you will by this think me a curious, strange, simple fellow; however we all have hobbies, but what mine is I don't know, unless music—though the somewhat harsh and grating sound which occasionally saluted your ears at the Northgate office[4] will lead you to suspect that I am wrong in my supposition. However you cannot think how much the sentence in your letter saying that at times you were dull and heavy, tired and almost inclined to give up the task on which you were engaged and then substitute another, which in time alike palled on you, gave me. It seems that <u>some</u> at all events have a common feeling along with myself in this particular, and you cannot think what pleasure it gave me to know that <u>you</u>, <u>the great indomitable you</u> sometimes experienced it.

To tell the truth, Tyndall, I never—let me take the easiest undertaking in hand with any responsibility connected with it—feel easy, and I can hardly express the astonishment I feel when one is surrounded by parties on whom devolves much and yet can lay aside so easily the remembrance of any particular responsibility attaching to them. This in particular is the greatest antagonist to me, and often (though you may laugh heartily) do I think myself a curious fellow who with application might, should some strange emergency require it, become tolerably clever and at least see through a millstone;[5] which at the same time I feel myself now most certainly misunderstood. Of what though am I rambling about? I am trying to tell you what I think, or rather <u>dream</u>, for certainly for some time back I have been doing little else than dreaming—it seems as if I was still subject to these occasional humours at times, seemingly able to undertake any quantity of labour underhand while after a time a depression comes in the which all the faculties seem in the most negative state of electricity requiring a corresponding amount of interest to bring them round again. At present I must confess I do nothing but eat, drink and smoke, while music (either piano or flute) fills up the remaining leisure from the office hours. Even reading has gone to the devil just now, and especially that of a sober and instructive character. Yet with all this laying aside I feel most miserable, out of humour with myself, and when I come to ask myself the cause it is only to increase the dissatisfaction having arrived at the conclusion that I know not why I am placed here. I know not what to believe in—neither do I know anything. Surely this is miserable—at least I feel it so—to be surrounded as one is by shams and hypocrisy (which become the more apparent in everything the more they are examined) and yet to be unable to substitute something real and good seems to be more like the machinations of a devil rather than a God—and one seems almost inclined to follow the customs of the day, regardless of truth or untruth in religion and at the same time in other respects follow the dictates and act consistently with the dictates of my own conscience. This appears to be the way in which the affair is managed and with so many precedents I cannot surely mistake besides it being carried out by the world at large we shall only be either <u>all admitted</u> in heaven or hell, where certainly we shall not be [fast][6] for company.

You see I treat the affair jocularly, but how strange it is that such must be the case, at least on its first opening, when parties must think of what they are acknowledging and declaring when they have arrived at 'years of discretion'.[7] I do not think though that they are all hypocrites, but rather mad than otherwise! seeing that by the frequent and the eternal recital of the same doctrines they have gradually been led to make themselves believe that they must be true and that they believe in them. Ask them however on the subject and they stare at you aghast, and declare you an infidel or other contemptuous name for doubting it, and finally open their eyes showing much 'the whites' and turn their back on you! 'Tis but too true, I believe there is more talk, more wrangling and discussion among the various sects than real pure Religion in the present day. Yet how much more beautiful and real was the religion of the Ancients[8]—despised and discarded as it is at the present day—what a reverence, awe and respect do they pay the Divine Creator, and yet what care do they take to answer and feel in accordance with the first questions of its reasonableness. Surely the mysteries and the humbug which at present is called on persons to believe would appear in their eyes, could they but examine it, but a poor squalid system not worthy of succeeding theirs. Tom read me a portion of a work which Philips has been reviewing[9]—it is one of the oldest books yet discovered and it is thought to have been written 4000 years before Christ, and consequently is older than the Bible! In it is nearly the <u>same</u> Philosophy and written in a most figurative and enigmatical style—shewing that the Bible must really have been only the record of the People a few centuries later whose opinions had not much changed. Here is a little more food for consideration as well as a new nut for our controversial and spiteful Christians (!) to crack.

I do not wish you to think from this that I want exactly as you have intimated in a previous letter,[10] 'A copper-headed Christ'—no, a little mystery, a little room for speculation is all very well, but too much of these and you reduce it to nothing but a vain shadowy speculation having neither length breadth nor thickness. Methinks however that though certain parties are very wishful that you should believe all this they themselves use it but as a means of gaining golden treasures by the way. And yet to give them credit for their sincerity in the cause of God and present them with memorials and the like in proof of the gullibility of John Bull,[11] especially in religious zeal!

Talking of zeal and fault-finding you will say I think that I can do a good deal under this head, so to change the subject. Carlyle has been writing some 'Latter-Day Pamphlets', which excite great attention coming from him. I have read all of them up to the present time, but his meaning generally is so choked up and so incumbered with towering gigantic and almost meaningless words that it requires an old head to understand him. In his last however, tho' in a similar strain (yet it is not quite so bad) he fires away at 'Downing Street'[12] and draws some laughable though pitiable descriptions. He is very sarcastic when I can understand him, which in the crowd of 'Universal Shams', flunkeys,[13] with a variety of adjectives before and after of a very questionable nature, is not very easy. Tom however as usual lauds him and of course understands him. As for me I will say 'It is high, I cannot attain unto it!'[14]

Parties are all busy here and preparations are being actively commenced relative to the Grand Exhibition about to take place in London in 1851,[15] Prince Albert being the schemer and deviser of that undertaking.

Now my dear fellow when are you coming back? Don't be long—let that German fellow[16] speculate alone; and now good-bye until I next hear from you, and

Believe me, | Yours very sincerely, | James Craven.

John Tyndall, | Marburg.

RI MS JT/1/TYP/11/3543–3546
LT Transcript Only

---

1. *[c. 16–19 April 1850]*: Tyndall recorded receiving this letter and letter 0398 from Hirst in the same journal entry (Journal, 28 April 1850). They were probably posted at the same time to save postage, hence this was posted on 19 April (see letter 0398, n. 1).
2. *my last letter*: letter 0394.
3. *Your letter*: letter missing.
4. *The Northgate Office:* the office in Halifax where they had worked together as surveyors.
5. *through a millstone*: used ironically to mean 'extraordinarily acute' (*OED*).
6. *fast*: LT transcription; possibly the word should be 'lost'.
7. *'years of discretion'*: old enough 'to be capable of exercising sound judgement, and therefore to be fully responsible for his or her words and actions' (*OED*). Theologically, which may be the meaning here, it has specific reference to moral responsibility.
8. *the religion of the Ancients*: reference to the work that Phillips was undertaking concerning the Hindu religion. See n. 9.

9. *a work which Philips has been reviewing*: Phillips wrote a three-part article (January Searle, 'Theosophy of the Hindoos', *Truth-Seeker*, n.s. 2 (1850), pp. 167–79, 250–81, 282–84), which was an exposition of the Bhagvat Geeta as translated by Charles Wilkins. The article was reprinted in *Essays, Poems, Allegories, and Fables: with an Elucidation and Analysis of the 'Bhagvat Geeta'* (London: John Chapman, 1851), pp. 87–133 (Hirst, 'Journals', 11 April 1850).

10. *a previous letter*: letter missing.

11. *John Bull*: a personification of England, or sometimes the United Kingdom.

12. *'Downing Street'*: probably the third of the *Latter-Day Pamphlets* (see letter 0398, n. 20).

13. *'Universal Shams' flunkeys*: Carlyle condemned a multitude of shams and flunky-like behaviours in Latter-Day Pamphlets.

14. *'It is high … it'*: Psalm 139:6: 'Such knowledge is too wonderful for me; it is high, I cannot attain unto it'.

15. *the Grand exhibition … in London in 1851*: 'The Great Exhibition of the Works of Industry of all Nations', as it was formally called, was the first fully international industrial exhibition ever held. It took place in Hyde Park, London, from 1 May to 11 October 1851.

16. *that German fellow*: Knoblauch, with whom Tyndall was performing investigations in diamagnetism.

To Thomas Archer Hirst   [18 April–12 May 1850][1]   0400

My Dear Tom

Were I your sweetheart I should begin to call you faithless, and if I refrain from doing so it is a proof of my faith, a faith which refuses to measure you by the accomplishments of your steelnib. are you dead my boy or dumb? I have incurred the displeasure of a kind friend of mine here (Prof. Waitz[2]) purely on account of my unorthodoxy as regards the paying of visits—what would he think of you—I doubt not his philosophic brain would discover some special purgatory where you and similar transgressors must one day luxuriate on fire and brimstone.

So Emerson has given out his <u>Representative Men</u>.[3] I have read snatches of these lectures here and there: they appear to be excellent. but Emerson that <u>is</u> is somewhat different from Emerson that <u>was</u>—his writings have latterly approached what the Germans call the <u>objective</u>. Shakspeare is the best specimen of this style, and Goethe next; in these writings you cannot discover the predilection or bias of the writers—their individuality is totally merged in their office of reporters to nature. they are mirrors reflecting back what they see without increase or diminution, and permitting their own feelings (the subjective) to impart no shade or coloring to the picture. True such men are not without their deep convictions as may be gathered from an occasional burst now and then, as in Emerson when he says of Napoleon 'It was the nature of things, the eternal law of man and the world that ruined him'[4] this objective way of thinking appears to me to be higher and happier than the subjective, tho' in comparison with the latter it may appear cold and unenthusiastic. It is more like a natural force accomplishing the sublimest ends without apparent effort. I have watched between times the transition from the subjective to the objective in my own individual case. The last 6 months have taught me a good deal in this respect—for some time I struggled against it, loving my former self as fondly as a tory loves an old institution; I often exclaimed in spirit 'how can I give thee up Ephraim!'[5] but in truth Ephraim is not given up at all—he was the substratum of something else which in its turn will doubtless become a substratum also.

What has Carlyle been doing lately—he has written a pamphlet on the times?[6] I have read a half and half, apologetic article on the subject in the Truth Seeker,[7] and turned with great pleasure from the writers remarks to his extracts. They appear to me to need no apology—They appear to me to be the full free utterance of a brave soul, By Heaven I should die of inanity in this world of fustian[8] were I not aroused by such voices now and then!—But you have read that article from Caliban;[9] Caliban one would gather is the writer of both articles, but if so, Caliban <u>here</u> is not Caliban <u>there</u>—he wrote the review after dinner[10]—I have not read a more refreshing scrap for some time than that 1st letter of his[11]—I shook my ribs with laughter to see myself so exquisitely pictured, and my eye glistened as the boy's intuition predicted—it is ludicrous in the extreme, but I felt an irresistible desire to throttle Caliban when done and maul him soundly out of pure love. <u>Es Kann vielleicht /mit/ meiner Irlandischen Natur zusammenhangen.</u>[12]

Early in July I expect to be in England. I have just finished a solid piece of work Tom which will appear in the Philosophic Magazine for June.[13] A short memoir has already appeared in the March number.[14] I have lately sent, at his request, a statement of qualifications, accompanied by testimonials to the Government Inspector of Railways;[15] and hope by his agency to creep into some berth where I may have an opportunity of following up my studies. my last investigation has been of infinite value to me, nature has stood here and there at the gaps and corners of the enquiry pointing in this direction and in that, and seeming to smile down upon me the promise 'I will reward you my son if you persevere' to speak plainly I see a rich unexplored field before me and my object now is to fix myself in a position to cultivate it. Mr Edmondson has written to me to know when I will return,[16] and whether I am engaged. If I can avoid it, however, I will not go to Queenwood.[17]

RI MS JT/1/T/893

1. *[18 April–12 May 1850]*: this letter was enclosed with letter 0401, and written at least a week earlier. Hirst wrote 'May—1850' on the outside, indicating when he received both. We are not sure about the date of writing, but believe it to be late April. Tyndall's complaints about Hirst's not using his steelnib (writing utensil) suggest that he had not recently received any letter, placing this letter before 23 April when 0398 should have reached Marburg. Allusions to Carlyle and Emerson read like replies to 0398, putting this after 23 April, though the discussion could reflect Tyndall's and Hirst's ongoing common interest in the authors.
2. *Prof. Waitz*: Theodor Waitz (1821–64), professor of philosophy at Marburg, usually described as a psychologist and anthropologist. He became unhappy with Tyndall for his failure to attend his psychology lectures (Journal, 15 March 1850 (JT/2/13b/480)).
3. *Representative Men*: cited letter 0398, n. 30.
4. *'It was . . . ruined him.'*: from 'Napoleon; or; The Man of the World' in Emerson, *Representative Men* (ibid), p. 252.
5. *'how can I give thee up Ephraim!'*: Hosea 11:8, 'How shall I give thee up, Ephraim?'
6. *Pamphlet on the times*: Tyndall probably referred to the first of Carlyle's *Pamphlets* (see letter 0398, n. 20).
7. *a half and half . . . Truth Seeker*: an article on Carlyle written by Hirst under the pseudonym 'N.E.P.' in *Truth Seeker*, 2 (1850), pp. 101–5 (see also letter 0398, n. 21). Here Tyndall believed this article, along with another, to have been written by 'Caliban' and only later realized that it was written by Hirst. In letter 0404, Hirst confirmed his authorship.
8. *fustian*: inflated, turgid, or inappropriately lofty language (*OED*).
9. *that article by Caliban*: see n. 7. Tyndall recorded reading the article in his journal entry for 18 April 1850: 'Received some Truth Seekers. In one is an article from Caliban on Carlyle. The fellow riddled me out bodily before my own eyes' (JT/2/13b/486).
10. *he wrote the review after dinner*: see n. 7. Tyndall was indicating the change in tone between the two articles.
11. *1st. letter of his*: Caliban's article.
12. <u>*Es Kann vielleicht . . . zusammenhangen:*</u> 'It can perhaps have something do with my Irish nature' (German).
13. *a solid piece . . . for June*: cited letter 0403, n. 2.
14. *A short memoir . . . march number*: cited letter 0395, n. 22.
15. *a statement of qualifications, . . . Inspector of Railways:* the Government Inspector of Railways was Captain George Wynne, Tyndall's mentor and patron, whom he had asked for employment advice and assistance.
16. *when I will return*: Edmondson hoped that Tyndall would return to Queenwood, where he had taught in the year 1847–48 before going to Germany
17. *Queenwood*: letter ends abruptly without closing. Tyndall broke off and continued in letter 0401, as that letter explains.

To Thomas Archer Hirst   [8/9–16 May 1850][1]   0401

My worthy Tom,

Since the enclosed 4 pages[2] were penned to thee, I have seen the necessity of acting upon your suggestion. I have written an article[3] which I send you, for [which] and for others similar I am ready to accept money, though God knows I thought little of the cash while writing it. Six Tom would take up too much time from me, that is 6 with the possibility that all six might fail: get me an order for six of the same calibre as the enclosed and you shall see how quickly they shall appear I will make one more spasmodic effort to remain here a little longer, if it fail I shall dry my eyes like David over his dead child[4] and reconcile myself like a philosopher to the necessities of my [case.] I send thru your hands the continuation of my magnetic investigations,[5] you can read them if you like or let [it] alone if you like but in any case forward it without delay to the gentleman to whom it is addressed. Stick two postage stamps on it for me, and fasten it with a seal and drop in the post office.

The above was written a week ago. Unexpected obstacles [have] turned up to prevent my sending the memoir on magnetism. It must remain by me for 10 days longer. I will how[ever send it] through your hands. In the mean time you can try the Leader for [me. I shall] probably send another sketch along with the memoir. Write [to me as] soon as you can Tom. My present arrangement is as follows[. I leave] Marburg at the end of June. Spend a short time in Berlin and return to England in July. I dont intend to accept any situation until I try whether it is not possible to establish some kind of literary connexion sufficient to allow me to return for a [term to] Germany.[6] Give me the reply of the Leader as soon as possible. Like a flying straw it will shew me how the wind blows.

Your affectionate | Tyndall

A word[7] more Tom—Jimmy may want that money sooner than I can well pay it[8]—I can soon discharge the debt be assured when I once make up my mind to settle down—When do you think he expects payment?

[You are[9] a good boy to send me the Guardian.[10] I read the 'new outline' of Mr Smith's] speech[11] with much pleasure, [And with more pleasure still stumbled over] that hardy 'no' with which he stopped the mouth of that noisy fellow who wished to draw him into a discussion on American morality.

RI MS JT/1/T/527
RI MS/JT/1/HTYP/74

1. *[7/9–16 May 1850]*: this letter was finished on or just before 16 May (Marburg postmark, 16 May 1850, and Halifax postmark, 21 May 1850). It was begun 'a week' previously (second paragraph), after Tyndall received letter 0398 from Hirst, to which it alludes. It was posted with letter 0400 (see n. 2). The order of letters 0401 and 0402 is not absolutely certain. Both were responses to letter 0398 from Hirst. We consider this to be Tyndall's first response, because he deals with urgent matters (such as the *Leader* article), and 0402 to be the later, more leisurely response.

2. *the enclosed 4 pages*: letter 0400 (4 mss pages) was written before Tyndall received letter 0398. He continues here, still somewhat abruptly, but responded to Hirst's suggestion about the *Leader* and his information about the money loaned by Craven and Hirst (both in letter 0398).

3. *I have written an article*: probably the philosophical article about 'Propensities' referred to in letter 0406 (see n. 7), rather than a letter about student life in Germany such as he and Hirst discussed in letters 0392, 0393 and 0398.

4. *I shall dry my eyes … dead child*: see 2 Samuel 12: 18–23 on David ceasing to mourn for his child by Bathsheba.

5. *my magnetic investigation*: as the following paragraph makes clear, Tyndall did not enclose the memoir on magnetism. It was sent with letter 0403.

6. *literary connexion … return to Germany*: as many other letters in January–June 1850 make clear, Tyndall was attempting to write for money for the *Leader* and other magazines so that he could continue his research in Germany.

7. *A word*: this first postscript is squeezed into the bottom of the page, to the left of the signature.

8. *Jimmy … pay it*: see letters 0393 and 0398 for details of the loan.

9. *You are*: this second postscript is written vertically in the left margin.

10. *the Guardian*: the *Preston Guardian*, a noted liberal newspaper established in 1844 by Joseph Livesay.

11. *'new outline' of Mr Smith's speech*: specific speech not identified, but several speeches on the Ten Hours Bill by George Smith of Manchester were reported in the *Preston Guardian* in April–May.

To Thomas Archer Hirst   [mid–late May 1850][1]   0402

Marburg

My dear Tom

I'm almost afraid to tell you how precious your letter[2] was to me lest you should strive to make the next equally so and thus spoil it by conscious endeavor. It spun out of you with far greater freedom than I can hope to attain in the production of this. It is strange that the intellect by long looking at one object should lose the power of seeing others clearly. In the domain of natural science, which to me has ever been a secondary domain I was never more at home than at present, but as a compensation my insight on subjects which I deem of still higher importance is comparatively dim. This I know is only temporary, the day I know will come when I shall be able to make use of natural science as the handmaid and expositor of this higher insight; but until that day comes you must bear with my feeble outflow as a writer. I will now take a dash at your letter and run through it as quickly as its temptations will allow me; when a little boy on my way to school I have often lurked among the birds nests and primroses, and here in a moral sense I may possibly do the same.

I'm glad you did not send that piece further than the People's Journal,[3] indeed the editor has passed an opinion on it to which I could not subscribe. That there was some truth in it I admit: but it was not well knitted together. It lacked that spontaneity which is the elemental grace of all composition; it was harsh and disjointed and so I am very glad that it has not appeared. With regard to the Leader your plan is excellent, but my time in Germany is now so short which you have already learned,[4] and I have so many other matters on hand, that it is not probable that I shall be able to turn my thoughts to this subject. I'm rejoiced to see such a paper, I breathe more fully in the hope that I shall one day be able to lay myself at full length in its columns, unswathed by the cramping trammels of conventionality. My dear Tom, these small annoyances are at bottom healthful exercises. I am troubled with the same disorder myself. I should shrink at being pushed prominently forward by a friend, I should much rather that this were done by a foe for then the spirit of defiance would come to my aid and relieve me of all embarrassment. At no time is a man so free to utter a fair thought as when he knows it is not expected from him. This consciousness is the greatest meddler and busybody that I know, the greatest drawback to the full overflow of the soul. Dont shrink from these little encounters my boy, use them rather, it is possible to make them profitable. Do not at all be afraid of being misinterpreted, no man was ever judged by his words. Were you to sit silent as a Sphinx in Egypt for a thousand years there is an instinct in human nature which would penetrate you. People are not so thickheaded as not to be able to discern when silence flows from backwardness and hesitation and when from stupidity, a thousand winged messengers flutter about your personality and declare the fact without asking your leave. What man, you dont know how wide that little organized germ now beginning to unfold in Halifax[5] may spread itself. Too much intercourse would undoubtedly be prejudicial, you will yourself know what are the wholesome proportions.

It is really refreshing to see such a cotton spinner as Smith. He has however taken up a perilous task, To write books and to govern a cotton factory in the manner it ought to be governed at one and the same time is no easy matter. I have no doubt I shall read his book[6] with deep interest. Such as he prove the wisdom of Carlyle's plan to choose men from the twenty seven millions.

There is a noble implication in this proposal, and that is, a doubt in the agencies of education, and a reliance upon the productions of nature. Here is a man who waters no narrow hotbed or parterre but scatters the refreshing element over the nation at large and whereever a healthy plant exists he gives it a chance. He recognizes the odour of the wild wood blossoms, and the strength of the oak's untutored arms: all ability at bottom is a natural gift to which education gives architectural proportions. 'But behanged to them at present!'[7] Bravo! my boy, there is more philosophy in the exclamation than you are yet aware of. Phillips likes you, does he? well for that very reason I like him; he is a decent fellow—greet him for me with brotherly kindness. Somehow or other I feel a community with all your designs there in Halifax. You did perfectly right in promising all the assistance in my power to give to your projected periodical;[8] that £75 is a substantial looking fact, quite a refreshing fact indeed. I should be glad to subscribe also but in the language of Peter and John I am forced to exclaim 'silver and gold have I none but such as I have give I thee.'[9] With regard to your kind proposal as regards Mr Carter I must say Tom that the thought of spending 8 hours a day measuring and plotting is not very inviting: besides I dont think Carter could afford one any thing worth accepting. I will throw myself into some private school or other till I have time to look around me, dont be afraid my boy, I shall come out [with] some thing. Your coming to Germany has often occurred to me, it will be impossible to remain until you come, but I will so arrange matters in England that I may accompany you and see you comfortably installed: make your mind easy on this subject, it will be nicely managed. I believe I have never given you any advice one way or the other upon this matter, having as great a faith in your ability to decide as other people; but this I do know, that long before you came to visit me it was a settled thing in my mind that you should come; I had counted the cost and found that it might be met, on this point I can only say that your unexpected good fortune has deprived me of the deepest gratification that my life could taste; and I now say without fear of consequences, careless of your upbraiding should these consequences disappoint you, that it would be deliberate moral suicide on your part to neglect the cultivation of those powers which God has given you; to my view you would have a far better chance of trotting safely over Mahomets cobweb[10] into paradise, had you gone like Judas and hung yourself upon a rotten string.[11]

Phillips's thought is 'holy and wholesome'[12] It appears that such must be our means of advance in England, as government is too feeble and too cowardly to take the initiative. Our poverty in our educational respect is an object of astonishment among Germans. I cannot express strongly enough my sense of the value of such an institution as that spoken of by Phillips;[13] but the means, the means—it will cost money! 'Lamentably without an object

Tom' most lamentably 'consider the lilies of the field, they toil not neither do they spin'.[14] It would be difficult for me perhaps to set down a tangible object, appreciable to a prudent man, and of sufficient importance to justify the loss of time and money which my sojourn in Germany has caused. Miss Carter asked me once 'and pray Mr Tyndall what is all this study for, what object have you in view?' I was gravelled[15] instantly, I knew I could not satisfy her and that any attempt to do so would have rendered me ridiculous, I therefore contented myself with saying that I had no object or something of that kind, do you want a special answer look round you and see, is there not opportunity enough for a strong man to exert himself: give me the strength only and if I don't find opportunity why then I shake salt on my tail and transmute me into a fossil cocksparrow! My object here then is to gather a little force, well knowing that the world affords opportunity enough for its special applications. Had I in my wanderings met a little girl sufficiently near my ideal I might on her account have been a little more definite and prudential, but I must defer the exercise of worldly wisdom till such comes in my way.

I received the Truth Seeker, but dont know what to think of Dr Lees. Sometimes a free and beautiful utterance breaks out upon you as if truth had met a smooth spot in the Doctor's intellect and was reflected faithfully; but the next moment you find yourself in a cross light. His subject seems to be illuminated by glares, like a landscape by sheet lightning—It flashes and he sees a part, flashes again and he sees another, but there is a gap of unimaginable darkness between, which the Dr. fills as he can. Over teetotalism he screams like a frightened plover that would guard its young, his energy at times approaches ferocity. A strong argument of his, if I recollect aright, is, that alcohol is an artificial product, exactly in the same sense however is the Doctor's shirt an artificial product, fermentation is brought about by the arrangement of chemical forces, the shirt is due to mechanical—in the latter case, inertia, friction, adhesion, &c are cunningly taken advantage of, but the Dr nevertheless is silent on the subject of cotton spinning. My poor Nep! I have given you a hard knock all unconsciously,[16] I hope it will be of service to you and make you lie closer to your feelings in future. I must say that I had a dim suspicion that the review was yours, 'can it be the N E P. of Tidmarsh?' I asked as I saw the initials. I will just take two antithetical specimens from the thing, one of which I like, the other I dont like. I find experimental wisdom here 'How often does it happen that in the forthcoming crop we expect the speedy growth, the rankness &c of our weedy species'[17] I take this as expressing the same idea as Emerson when he says 'every body knows what a strong moral girdle is required to supply the place of a superstition'[18] again 'Think of the terrible significance of that word and if custom or indifference have not turned your heart to stone it will bleed indeed'.[19] There is a kind of wire-drawn agony in this sentence which does not please me. Such a sentence ought to be the fresh flower of a feeling, deeper than that which it would insist upon—it strikes me however that the heart of the writer was by no means suffering from the effects of ph*<1-2 words missing>* when he penned it.

You are now able to estimate the value of that mighty thing called 'common sense' and the sagacity of the plan proposed by Volney[20] to decide on the merits of all religions namely the gathering together of an indiscriminate horde of 'peoples, nations, and languages' and laying the matter before them, 'common sense' as its preachers use it appears to me to be uncommon nonsense, for to pronounce upon the subjects which they refer to this faculty presupposes, as you say, culture. There are instincts almost infallible, common to the ignorant and to the uncultivated, and as long as a man abides by these he is safe, but the decisions of common sense are those of deliberate judgment and entirely out of the range of these instincts if rude and undeveloped. There are many subjects of human experiences upon which at the present moment I would not pronounce an opinion, nay I have one or two articles of faith which I could if called upon by no means sufficiently defend, but my inability I know arises from the withdrawal of my intellect from the contemplation of such subjects and not from any lack of intrinsic evidence, A wise remark of Dr. Lees here occurs to me, speaking of Macaulay he says 'His psychological experience <u>must</u> be the limit of his belief in human nature.'[21]

Thank Jimmy for me. I am proud of the confidence of both of you, though I dare say there are people in the world who would think that I ought to be ashamed of myself.

Your little sonnet pleases me exceedingly, the ruling thought is a very fine one.

> 'Were life upon they banks but like thy course
> So void of strife, so full of silent force.'[22]

That is an ideal worth striving after, and involving many conditions physical and moral for its attainment. My own experience induces me to share in the word physical, for example, I have had no exercise since I left Switzerland, and have been working daily in an unwholesome atmosphere, namely that of Nitrous acid this escapes from the galvanic battery which I use in this magnetic investigation.[23] One day I had been experimenting with 20 cells and the fumes were very strong, at night I could scarcely breath or speak. I walked up and down for three hours in the fresh air to recover my respiration, and next morning I spat blood for the first time in my life, the effects of the acid had settled in the lungs and small of my back—the spine had been affected. I usually experimented with 6 or 8 cells the vapour from which is small and

I am so used to it I dont perceive it; but I can see the effects of all this—I am in a worse condition to realize the above ideal than I should have been had I attended to physical conditions and drank the fresh air oftener. But for all this there are sanitary measures in store, and I will wrestle you for a pound on my return in England!

There is scarcely a remark in your letter which pleases be better than that which alludes to your class at the Mechanics Institute.[24] Why you had only one pupil when you were here, 'verily I say unto you if ye have faith as a grain of mustard seed you shall say to yonder mountain "remove" and it shall obey you'.[25] Why by the Lord there is no stupidity that can withstand the force of a man if he be brave and patient. Your remark on this subject argues a deep knowledge of it; by holding fast to the view you have expressed I was enabled at Queenwood to make my occupation a delight when teachers of 10 times my experience were complaining. Some people set about education in a ram-down-cartridge kind of way: stuffing their unfortunate pupils with crudities without ever once enquiring after their powers of digestion, or struggling to promote a healthy appetite—not a doubt of it Tom, not a doubt of it—I have said the same twenty times in our teachers meetings 'It will flow out unknown to him in every action.'[26] To every one of your remarks regarding 'selfexpansion' I most cordially subscribe. It is only another way of stating the aphorism of Solomon 'What thou doest do it with thy might'[27] and this is impossible without a certain exclusiveness; dig one ridge deep and well before you commence another. There is some difficulty with words to draw a distinction between this mode of cultivation and <u>onesidedness</u> but they are heaven-wide asunder.

I conclude my dear Tom by praying the gods to prosper you and the little band to which you belong. That is what I call cooperation, the cooperation which leaves me free and yet at the same time binds me to you all irresibly, a cooperation in whose services I can apply my whole force not because it is the demand of certain stipulated conditions, but because it is the dictate of my own will and pleasure; could we but spread the leaven how miserable would our conventionalities appear; then would society first feel the sublime truth of our great teacher 'to faith all things are possible'.[28]

Tyndall

RI MS JT/1/HTYP/75–78
LT Transcript Only

---

1. *[mid–late May 1850]*: written after letter 0401 (see that letter (n. 1) and n. 4 below).

2. *your letter*: letter 0398.

3. *that piece further than the People's Journal*: see letter 0392, n. 2 and letter 0398.

4. *you have already learned*: Tyndall assumed that Hirst already knew when he would be leaving Germany; that is, that Hirst had received letters 0400 and 0401.

5. *that little organized germ . . . in Halifax*: probably Hirst's new circle of G. S. Phillips and his friends. In letter 0398 Hirst expressed hesitations about participating in the group.

6. *his book*: cited letter 0398, n. 12.

7. *'But behanged to them at present!'*: quotation not identified.

8. *projected periodical*: see letter 0398 for the plans to establish a journal to rival the *Leader* and *Truth-Seeker*.

9. *'silver and gold . . . I give I thee'*: Peter gave something better than money: Acts 3:6, 'then Peter said, silver and gold have I none; but such as I have give I thee: in the name of Jesus Christ of Nazareth rise up and walk'.

10. *Mahomets cobweb*: Islamic oral traditional maintains that during the *Hijra*, Muhammad's journey from Mecca to Medina, he and his companion Abu Bakr were relentlessly pursued by Quraysh soldiers and took refuge in a cave. Legend holds that Allah commanded a spider to weave a web across the cave's opening, which caused the Quraysh to pass by, thinking that the cave was uninhabited.

11. *Judas and hung yourself upon a rotten string*: accounts of Judas Iscariot's death vary. Matthew 27:3–10 holds that he committed suicide by hanging, while Acts 1:18 describes how he fell headfirst into a field and 'burst asunder'. Augustine harmonized these differing accounts to suggest that Judas hanged himself, but the rope rotted, causing his body to fall and burst open. Tyndall could be referring to Augustine's interpretation.

12. *'holy and wholesome'*: perhaps an allusion to the liturgy for All Saints Day, which quotes 2 Maccabees 12:46: 'It is therefore a holy and wholesome thought to pray for the dead'.

13. *an institution . . . by Phillips*: see letter 0398 concerning Phillips' desire to establish a People's College after the German model.

14. *'consider . . . do they spin'*: Matthew 6:28, 'Consider the lilies of the field, how they grow; they toil not, neither do they spin'.

15. *gravelled*: perplexed or puzzled (*OED*); or perhaps the word is 'gavelled', i.e. gavelled into silence.

16. *My poor Nep!*: under the pseudonym, N.E.P., Hirst had published a review of Carlyle's first *Latter Day Pamphlet* (cited letter 0398, n. 20). Tyndall criticised the review in letter 0400, wrongly thinking that it had been written by Caliban.

17. *'How often . . . weedy species'*: quotation not identified.

18. *'every body knows . . . superstition'*: R. Emerson, 'History', *Essays* (London: James Fraser, 1841), p. 29.

19. *'Think . . . bleed indeed'*: quotation not identified.

20. *plan proposed by Volney*: Constantin François Volney (1757–1820), French philosopher, historian, and politician, who argued that history would end with the final union of all religions by recognition of common, underlying truths (*Les Ruines, ou méditations sur les révolutions des empires*, 1791).

21. *'His psychological . . . human nature'*: quotation not identified.

22. *'Were life . . . silent force'*: see end of letter 0398 for Hirst's 'little sonnet', of which these are lines 5–6.

23. *galvanic battery . . . magnetic investigation*: Tyndall's work with galvanic batteries stemmed from his research with Knoblauch on diamagnetism (see letter 0395, n. 22).

24. *your class at the Mechanics Institute*: in letter 0393 Hirst had lamented that he had only one student. In his last letter (0398) he reported that attendance had increased to 20.

25. *'verily I say . . . obey you'*: Matthew 17:20, 'And the Lord said, "If ye have faith as a grain of a mustard seed, ye shall say to this mountain, Remove hence to yonder place, and it shall remove"'. This is the closest of many similar biblical sources.

26. *It will flow . . . every action*: source not identified. It could be a paraphrase of a sentiment from Carlyle or Emerson.

27. *'What thou doest do it with thy might'*: Ecclesiastes 9:10, 'Whatsoever thy hand findeth to do, do it with thy might'.

28. *'to faith all things are possible'*: Mark 9:23, 'Jesus said unto him, "If thou canst believe, all things *are* possible to him that believeth"'.

To Thomas Archer Hirst   [31 May or 1 June 1850][1]   0403

My Dear Tom,

I send you as I promised the Memoir on magnetism,[2] you will not understand it all, but certainly so much as to convince you that we have made out a strong case. I have had you in my eye at the commencement and therefore have thrown in a few definitions to render the matter intelligible which otherwise I probably should not have done. I have derived a great deal of profit from this investigation, not on account of the results won, though these are of highest interest in a scientific point of view, but from the conviction which I have gained of the power of resolute patience to break down difficulty. The investigation has further shewn me on what an insecure foundation the most brilliant discoveries of physical science rest, and I would wager a thousand to one that many of them will be overturned. I could not have labored in a better field. The subject lies at the bottom of all others, this cutting and cleaving has almost rendered the very constitution of crystals transparent and I seem almost to see into their very molecular arrangement—this as regards light magnetism and electricity is the fundamental point. Then there are one or two undecided questions at the present moment before my mind which have baffled the attempts of men of profound sagacity and I am convinced they are capable of the simplest solution. The subject has by no means absorbed me. I have watched with the deepest interest the workings of my own mind throughout and my deepest satisfaction is that it at the end is stronger, abler to face another difficulty tomorrow. Had I not promised I would not have sent you the Memoir as thereby time is lost where none is to spare. Read it speedily and forward it to M[r] Taylor the moment you have got through it. <u>Pay the postage of it</u>—I shall see you soon, my son—and tell you my speculations as regards the future. I have now to make a few apparatus here. Spend some days in Berlin, and then Westward ho! full of trust in God and empty pockets.

Love to Jimmy—love to Philips—love to Smith—and take the last pull of the becher[3] yourself.

Your affectionate | Tyndall

If you write here within a fortnight I shall receive your letter on my return from Berlin.[4]

Thomas Hirst | [M[r]] Carter's office | Horton Street | Halifax | England[5]

If TH be from home, to be opened and the enclosed forwarded to M[r] Taylor without delay[6]

via Frankreich[7]

Paid at Marburg June 1st 1850[8]

RI MS JT/1/T/528

1. *[31 May or 1 June 1850]*: '1|6|1850' is the Marburg postmark.
2. *Memoir on magnetism*: at last, Tyndall sent the memoir promised in letters 0400 and 0401. It was published as J. Tyndall and H. Knoblauch, 'On the Magneto-optic Properties of Crystals and the relation of Magnetism and Diamagnetism to Molecular Arrangement', *Phil. Mag.* 37, no. 247 (July 1850), pp. 1–33.
3. *becher*: mug or tumbler (German).
4. *return from Berlin*: there is no evidence that Tyndall made this intended visit to Berlin.
5. *Thomas Hirst . . . England*: address from envelope.
6. *If TH be . . . delay*: at the top left corner of the envelope (in two lines).
7. *via Frankreich*: at top right corner of envelope. The letter was sent via France, possibly to save postage, and has a French postmark, '4 JUIN 50'.
8. *Paid . . . June 1st 1850*: at the bottom left corner of the envelope.

From Thomas Archer Hirst   3 June 1850   0404

Halifax. June 3rd, 1850 | By the bye, just put dates
to your | letters,[1] will you, my lad? —I like it.

My dear Friend.

I have delayed writing, thinking to send you the Leaders Answer, but I have unfortunately been delayed in sending it off, & have not heard from them yet:[2] but I shall write again to you very soon. Your intentions are much clearer to me now, & I have thought a good deal about your coming home and returning; I did not like it at first, but on reconsideration, I think it is perhaps best for I should most decidedly like to see you here in England, before I leave. But mind I don't want to lose much time. I have mentioned my going to Carter, he shewed very little interest in the matter, that is, he <u>made</u> no remark, Jemmy too I think will like to stop longer with him, there is not sufficient work for two <u>paid</u> clerks, & therefore I shall walk off immediately that is early in September. So I'll tell you what it is, if you want to spend more than a month or six weeks here, before you return, you had better get home as soon as you can (I use the word <u>home</u> for Harrison Road Halifax[3] remember). You talk about establishing some literary connexion,[4] it is a good notion. I have enquired & we may possibly manage it. A fine (somewhat extraordinary!) Lincolnshire clergyman,[5] a true man, from what I hear; (he has been a kind friend to Phillips) is at the head of the Leader, chief Proprietor, &c! now Phillips is going to spend a week with his wife down there, (at the house of a fine old quaker you may have heard of at Queenwood a Tho⁵ Shaser; he was instrumental I believe in getting Yates there, which latter person is an acquaintance and schoolfellow of Phillips's {I could run off into parenthetical observations for a week but I must not}) he has made me promise to join him for a day or two there, he wants me to know these two fellows, especially Larken, we will then /sound/ him on the matter of a German correspondent, they are making every effort to make their paper attractive, & it is just possible they might agree to it. But any way it does not matter much; I am rich enough (oh! I have persuaded Jemmy and have paid him the £20[6]), & thine is mine & mine thine,[7] besides I consider thee a safe investment, I was forced to lend to some one who did not want it the other day and I should be very cross indeed Tyndall if I were doing so when it would do you service. Speak then, theres a good fellow, & <u>treat</u> me as a brother. The circumstances are just these and I tell you them, to let you have a proper view of the thing; we (self & brothers) have been selling off some cottage etc. property we have in Heckmondwike, a thing that has been contemplated for some years; the purchase money is not all yet paid up, only a deposit of ten per cent, the remainder in August next; this puts a sum of money into our hands, not needed for income, & at the same time of too small amount to re-invest, until the whole purchase money shall have been paid up. The £20 you have is part of that floating & at present useless capital. There is yet another £20. Will it do thee good?

I had a hearty laugh at your <u>first</u> mention of my first review,[8] & I liked your last remarks[9] better still; I don't think you are the fellow to withhold such even if you knew me to be its author, at least I hope you wouldn't, for the criticism will do me more service than the exercise. But thou art a very wizard

I think, it seems as if thou could read my nature & feelings better than I could at the time; that is I was not <u>conscious</u> then, of wandering from my feelings, I am so now. The apologetic strain of the article was intentional, it was not written for Carlyles disciples & admirers, but for that class which I have just left in a measure, those who could or would not see through the mere garb into the heart & truth of the matter. The appearance of Calibans letters[10] in the same number was unexpected to me of course, but I decided at once to seeing that he had set himself a pretty similar task, & was far better able to do it, to leave it in his hands. This was better on many accounts, for after completing my first, I looked far from cheerily at the prospect of having to continue them. I haven't enough steam in me yet for criticism, that I saw clearly. Caliban is a very genial fellow indeed, his name is Sterling, a surgeon, of Merthy Tydvil,[11] Wales, and a friend of Phillips (of course). There is a natural & unaffected Hero-worship[12] about him, that I like very well. His letter for the present month[13] is also interesting, some of it excellent, some middling, some I don't like. That allusion to George Sand[14] is a specimen of the last. I should hesitate a good deal in finding fault with a man for a too ardent admiration of Carlyle, indeed I don't know where I should draw such a line. Nevertheless, I would not be warped out of my own way of thinking on a matter and moreover by not so doing, I believe I should be shewing best my appreciation of Carlyle. From what I have seen of George Sands writings I think Carlyles epithet & insinuations are gross libels,[15] moreover I have grounds for believing that Carlyle is as unable to judge as Caliban himself; & still he confesses that Carlyle 'has clinched his thinking.' He unconsciously misinterprets a provincialism I used with respect to that quotation about flogging and shooting for idle scoundrelism; I am not only of opinion that it was perfect sincerity on the part of Carlyle, but moreover, that there is a stern divine justice at the bottom of it. My expression was a provincial one which is ironically used. I have read that 'Theosophy of the Hindoos' in Manuscript it is wonderfully interesting, the present one is only an introductory article,[16] there will be more soon. I had Phillips to see me yesterday, we had a long and delightful day among our valleys, he made me a present of a little book he has just got out on the 'Life & Characteristics of Elliott'.[17] Did you know much of the old Corn law Rhymer?[18] He was a brave & affectionate old fellow, & Phillips has here culled some of his best poems. His chief characteristic is his touching pathos; which runs through all his poems, he loved Nature as a mother, but the tyranny of man & the miseries of his fellow labourers, stung him deeply, & threw a gloomy shade on his other & sunnier side. Did you ever see the following 'Plaint'[19] written by him in extreme physical agony some time before his death?

1

Dark, deep, and cold, the current flows,
Unto the sea where no wind blows,
Seeking the land which no one knows.

2

O'er its sad gloom still comes and goes,
The mingled wail of friends and foes,
Borne to the land which no one knows.

3

Why shrieks for help you wretch who goes,
With millions from a world of woes,
Unto the land which no one knows.

4

Tho' myriads go with him who goes,
Alone he goes, where no wind blows,
Unto the land which no one knows.

5

For all must go where no wind blows
And none can go for him who goes;
None, none return, whence no one knows.

6

Yet why should he who shrieking goes
With millions, from a world of woes,
Reunion seek with it or those?

7

Alone with God, where no wind blows,
And Death, His shadow, doomed he goes,
That God is there, the shadow shews.

8

Oh! shoreless deep! Where no wind blows!
And thou, oh Land, which no one knows!
That God is all, His shadow shews.[20]

Hear another verse or two from a very striking poem called 'Leaves and Men'[21]

How like am I to thee, Old Leaf!
We'll drop together down;
How like art thou to me, Old Leaf!
We'll drop together down.

I'm grey, and thou art brown, Old Leaf!
We'll drop together down, Old Leaf!
We'll drop together down.

Drop, drop into the grave, Old Leaf,.
Drop, drop into the grave;
The acorns grown, thy acorns sown,—
Drop, drop into the grave.
December's tempests rave, Old Leaf,
Above thy forest grave, Old Leaf;
Drop, drop into the grave!

The last he published was another of the same sad melancholy cast, it appeared only a short time ago in the Truth Seeker & was called 'Let me rest'. where he says most touchingly yet sweetly --

'But lay me low
Where the hedge-side roses blow;
Where the little daisies grow;
Where the winds a-maying go &c

There beneath the breezy west
Tired and thankful, let me rest
Like a child, that sleepeth best
On its gentle Mother's breast.—[22]

And poor Ebenezer soon got his wish, & now rests, that 'all-sought, all-known Unknown'[23] as he called it, troubles him no longer, he has solved his own problem.

You once said something about me writing a book on 'Nep in search of a Religion'[24] and strange enough, one of a similar stamp has just appeared by F.E. Newman (late a graduate of Oxford & brother to the famous Puseyite) called 'Phases of Faith or the History of my own Creed'.[25] He is the author of another splendid book, called 'The Soul its sorrows and aspirations,'[26] a review of which if I recollect, was in one of those Leaders I sent you.[27] He is a fine fellow & coming from Oxford a kind of prodigy. Carlyle has had a good influence undoubtedly on him, with some such expansive views, a different experience & more systematic, scientific mind, he has stated his opinions more consecutively & perhaps popularly; they will do much good. The best part of them is their unpretending, simple style, their sincerity & faithfulness: In his 'Phases of Faith' I can recognise many correspondences to my own. I wrote to Carlyle some time ago (when I sent him Zig-Zag[28]) I had long wished to write to the man. I felt it almost a necessity, and what I said, I believe was perfectly natural and spontaneous. I didn't of course either thank him, or laud him much, I knew better; whatever I might feel; I merely told him who & what I was, what I was doing, & what I intended to do, & so forth, mentioned having written an article for The Truth Seeker,[29] & my object in doing so. He did not answer me personally (I found out after I had forgotten to give him my address) but in one of his later pamphlets The 'Stump Orator',[30] one of the most deeply significant, and valuable perhaps of the series, he contrived to answer me. It is of course made universal in its application, but from a few circumstances I am morally certain that it was written at the suggestion of my letter. First the few paragraphs in question, are in a measure isolated from the context, & added to the tail end of the Pamphlet. Second, there is a remarkable correspondence in subject, and even in the very expressions which I used, it is too long to fully explain, you must see it when you come. You know his doctrine of silent work, & his intense hatred of <u>talk</u>, whether with the pen or tongue. He evidently thinks I may err in this respect, and it is a word of caution, given in a manly & even affectionate manner. 'Bravo young friend', he concludes, 'dear to me and <u>known</u> too in a sense (I had expressed an intention & wish to be on a nearer acquaintance with him) though never to be seen by me, know that you are in the happy position, which I am not, of <u>doing</u> something, instead of eloquently talking about what has been, is doing, & what ought to be. Englands sole hope is in such as you. Macte i fausto pede,[31] & may future generations acquainted once more with divine truths & silences look back on <u>us</u> with incredulous astonishment'.[32] (I quote from memory only[33])

    T.A.H.

    On re-reading[34] the other part of my letter, Tyndall, I am afraid you will not understand, or will hesitate again at that cash portion, by thinking that I shall want repayment in August. It is no such thing, for a year or two I am going to keep back about a hundred for contingencies on first going to Germany it will make little difference to the plans for investment, & will be safer for me as additional expenses & part of my premium to Carter which was just due have swallowed up any thing that in other circumstances I might have saved from my income—What will be the best way of forwarding it should you desire it? You will know best.

    Yours affectionately, | <u>T.A. Hirst</u>

    Private Student | | John Tyndall | ^{care of} Prof. Bunsen | Marburg | Hesse Cassel | Germany

    paid at Halifax | June 4th 1850[35]

RI MS JT/1/H/146

1. *just put dates on your letters*: the volume editors whole-heartedly and exhaustedly concur with this request!
2. *Leaders Answer . . . not heard from them yet*: Tyndall had sent his manuscript with letter 0401.
3. *home for Harrison Road Halifax:* Hirst lodged with Mrs Wright on Harrison Street (see letter 0393, n. 7). Tyndall had previously lodged with Mrs Wright (see letter 0375) so this had been his home address as well.
4. *literary connexion*: many letters (e.g., 0408) from Tyndall concern his interest in publishing in literary journals; he also used this phrase in letter 0401 (at n. 6).
5. *Lincolnshire clergyman*: Edmund Larken, the financial backer of the *Leader*.
6. *have paid him the £20*: see letters 0398 and 0401.
7. *thine is mine & mine thine*: a proverbial saying, which may derive from Cicero or from John 17:10.
8. *your first mention of my first review*: see letter 0400, third paragraph.
9. *last remarks*: see letter 0402, fifth paragraph.
10. *Calibans letters*: there was only one letter in the issue (cited letter 0400, n. 7).
11. *Merthy Tydvil*: Merthyr Tydfil, a town in central Wales about 25 miles north of Cardiff.
12. *Hero-worship*: an allusion to Carlyle's praise of heroes and advocacy for hero-worship.
13. *His letter for the present month*: see letter 0407, n. 15.
14. *George Sand*: pseud. of French novelist, Aurore Dupin (1804–76), notorious for her romantic affairs.
15. *Carlyles epithet & insinuations are gross libels*: Carlyle's criticisms of Sand had been interpreted in the light of his misogyny.
16. *'Theosophy of the Hindoos'*: cited letter 0399, n. 9.
17. *'Life & Characteristics of Elliott'*: *The Life, Character, and Genius, of Ebenezer Elliott, the Corn Law Rhymer* (London: Charles Gilpin, 1850), written by Phillips under his pseudonym, January Searle.
18. *Corn law Rhymer*: Ebenezer Elliott (1781–1849), political radical and worker poet, who fought to repeal the Corn Laws.
19. *'Plaint'*: Hirst probably transcribed this poem from the book gifted to him by Phillips (*Life of Elliot,* pp. 21–22).
20. *Dark, deep, and cold, the current flows, . . . That God is all, His shadow shews*: see n. 17.
21. *'Leaves and Men'*: another Ebenezer Elliott poem (*Life of Elliot,* pp. 64–65).
22. *'Let me rest'*: another Ebenezer Elliott poem (*Life of Elliot,* pp. 66–67). In Phillips' book he directly mentioned its previous publication in the *Truth Seeker*.
23. *'all-sought, all-known Unknown'*: line 9 from Ebenezer Elliott, 'Oh, Tell Us!' (*Life of Elliot,* pp. 65).
24. *'Nep in search of a Religion'*: see letter 0392 for Tyndall's suggestion. Hirst here adapts Tyndall by using his pseudonym, N.E.P.
25. *'Phases of Faith . . . my own Creed'*: F. W. Newman, *Phases of Faith; or, Passages from the*

History of My Creed (London: John Chapman, 1850). Hirst abbreviated both this title and the following one.

26. *'The Soul... aspirations'*: F. W. Newman, *The Soul: Her Sorrows and Her Aspirations: An Essay Towards the Natural History of the Soul* (London: John Chapman, 1849).

27. *one of those Leaders I sent you*: *Leader* 1, no. 2 (6 April 1850), p. 38.

28. *Zig-Zag*: an article by Tyndall, published in the *Preston Chronicle* (20 April 1850, p. 3), under the pseudonym Wat Ripton.

29. *article for The Truth Seeker*: cited letter 0400, n. 7.

30. *'Stump Orator'*: No V. dated 1 May 1850 of the *Latter-Day Pamphlets*, pp. 146–81.

31. *Macte i fausto pede*: Carlyle wrote 'Macte; i fausto pede'. Any translation is heavily context-dependent. Macte, can be translated, 'Great fellow!', 'Bravo', 'Good luck', 'Go man!' depending on context. 'I pede fausto', literally, 'go with a fortunate foot' (Horace, *Epistles*, II.2.37), was a phrase Carlyle often used in letters to friends.

32. *'Bravo young friend' . . . incredulous astonishment!'*: from the last page (181) of the 'Stump-Orator' (cited n. 30).

33. *I quote from memory only*: Hirst is mostly accurate in quoting Carlyle, although he confused the order of some phrases and omitted others.

34. *On re-reading*: this postscript follows the last page of Craven's letter (0405).

35. *John Tyndall... 1850*: address and note (in bottom left corner) are both from the envelope.

From James Craven 4 June 1850 0405

Horton <u>Street, June 4</u>th 1850

Dear Tyndall.

Tom has just allowed me the use of this half page[1] tho' really the slip is so very small that to a long winded monster like myself I do not know how or where to begin however as a commencement let me thank you for the <u>long</u> letter[2] you favoured me with—I am glad that your opinion of me has not sunk very materially indeed I may say that however dissatisfied I was at one time it has decreased to a large extent by the encouraging letter you wrote me.

Tyndall I can well understand the result which you wish to bring about which my own inner man tells me is the more noble & proper career for me,[3] I feel this still small voice[4] continually hinting that there is a screw loose & that I am not pursuing the track which is most in accordance with the <u>majority</u> of my faculties & if I come to ask myself, as I often do what subject or thing to engage myself in I feel at a loss remembering well the multifarious schemes & things I had commenced aforetime with but one result of mortification & dissapointment. The strength of mind & will to serve & <u>continue</u> in the course allotted & marked out for myself is in spite of long opposition seen to grow more and more feeble until now after the repeated experiments made I shrink from any similar undertaking. 'Contentment is great gain'[5] says the wise man & a happy mortal I add is he that is naturally of a free & easy disposition—with no aspirations of beyond the enjoyment of time present, & who can cosily sit under his own Porch & within the shade of his own Fig tree and the enjoyment of social 'comforts' <&> blessings, and enjoy the favor & affection of all around him—These are persons I am sure that <I> thus live well without a distressing thought to mar or hurt the quiet of their lives & though you may look on them with silent disdain be assured that there are few very few that can afford to compare the real happiness they have with them.—

Its as well the paper is done for I see your'e laughing fit to split those doctor's sides of yours so that afterall Tom's curse proves a blessing.

<u>John Tyndall</u>   <u>James Craven</u>

RI MS JT/1/H/146

---

1.   *this half page*: written on a half page, which Hirst had left blank on letter 0404.

2.   *the <u>long</u> letter:* letter missing.

3.   *proper career for me*: Craven had discussed his uncertainty over his career-path in his previous two letters (0394 and 0399).

4.   *still small voice*: a biblical allusion (1 Kings 19:12).

5.   *'Contentment is great gain'*: 1 Timothy 6:6: 'But Godliness with contentment is great gain'.

From Thomas Archer Hirst   18 June 1850[1]   0406

Halifax | June 18th, 1850

My dear Friend,

I wrote to you some time ago and sent the letter[2] by a new method[3] to lessen the postage, and consequently am not quite certain of its safe arrival. To ensure this I send this as formerly, and hope it will come to hand before you start for England. I expect you to land in Hull and come straight to Halifax;[4] if you do not I shall be forced to find you out for I am anxious to see that old face, and have another grip of that fist of yours, so I tell you I shall be dreadfully jealous and I would advise you not to provoke it. I sent that MS.[5] of yours on the day it arrived after reading it with great interest though as you anticipated I did not enter into its merits; but it gave me a good idea of the nature of your labours and what was very curious I attended a lecture of the West Riding Polytechnical Society[6] that very day, and unexpectedly saw some of the same experiments in Diamagnetism illustrated. Your paper on Propensities and their Equivalents[7] was an excellent one about which we will have a jaw together when I get you into my snuggery here at Halifax. I showed it to Phillips who was equally pleased with it and offered to send it to Larken, the head proprietor of the Leader. He did so, but I have heard nothing since. I shall be down in Lincolnshire next week and shall see him about it most likely, together with other things. Its subject is similar to one I was engaged on, and as usual there are many coincidences between us. I am about to give them a farewell paper at the Franklin Society[8] (I should like you to be here at the time) and intend giving them the results of a few experiences gathered from them. I have called it 'Societies considered in their relation and adaptation to Intellectual Culture and Individual Improvement'. Chemistry is proceeding prosperously just now, but we shall incur your rebuke about our hours of study, it will jar against your habits of early rising, for we work there generally from 8 P.M. to 1 A.M. Intellect in this case however without being at all bribed says that is our most convenient time. I am not in a good writing humour to day and shall not say much more. I am in fact rather low spirited at a small religious dispute or rather difference I met with yesterday. I have an Aunt[9] here in the neighbourhood, my father's sister who is very fond of me, very religious & sincerely so I'm certain, but also rather shrewd and preju-diced, it seems she has been frightened, poor soul, at my conduct and writings lately. How she has got hold of the latter I do not know. My bad conduct consists in Sabbath breaking and neglecting M[r] Pridie's Gospel.[10] I have great affection for her too & always found something beautiful about her religion, but I always carefully avoided satisfying her love of argument and disputation. Yesterday however she took me to task most seriously, seemed determined to have it all out of me, the consequence of it was, that she got such a flood of heterodoxy or what she calls infidelity on her head, that completely staggered her. I did so with great reluctance but she was wilful & would have it. I have been sorry though since for I believe it will cause her much anxiety, and cer-tainly it opens an old wound in me, that I flattered myself was almost healed, namely, whether my belief is or is not under the control of my will? When she wrung my hand at parting and with tears in her eyes beseeched me for the Love of God, for everything I held dearest, to consider my sin, to repent & return to Christ like the Prodigal Son to his fathers house: I felt myself to be a monster and a sinner before her and almost frightened her with the earnest-ness of my answer 'Would to God, and for your sake, that I could Aunt'. For some hours after I thought I could & would try to be orthodox, and paced about my chamber until 2 O Clock this morning quite restless & quite mis-erable, as if I had committed a heinous crime. Why can I not wear the gar-ment that brings her such consolation? Reason soon returned and shewed me the impossibility of this, nay the religion which made her believe me to be threatened with eternal damnation & to tremble for my sake seemed far from enviable. Yes, she positively told me that she dare not encourage the idea of my salvation should I be cut off without repentance. One thing she told me surprised me much, viz. 'That my words were the echo of what she had heard my father say a few years before his death' Now my father died when I was 12 years old, and his opinions on these matters I have long wished to know, for it has often crossed me what he would have thought of me had he been living at this day. My curiosity was excited all the more by certain curious documents (MSS)[11] that fell into my hands at Mothers death,[12] which were tossed about uncared for, but from which I could gather that he too was a thinker. I could meet with no one that could enlighten me on the point until yesterday & I must say the information was encouraging that I was not a wanderer altogether, but treading in my Fathers footsteps was <u>some</u> consolation. And to*[-day]* I have found myself speculating upon the changed circumstances had he been living yet, and besides the tie of kindred we had been bound also by sympathy in our beliefs and aspirations. But, unluckily for my Aunt's peace, my Father died in sin, (like poor M'Arthur[13] he killed himself with drinking) His life I feel certain now is worth investigation and I have an anxious desire to know it better. Meanwhile may the son while following his fathers aspirations, conquer his frailties <and> avoid his errors.

John Tyndall will say Amen.

John Tyndall | ^{care of} Professor Bunsen—| Marburg | Hesse Cassel Paid at Halifax | June 18th, 1850[14]

RI MS JT/1/H/147

---

1. *18 June 1850*: Hirst posted this letter to Marburg, although Tyndall had already left. He may not have received it until he returned to Marburg in October, as there is no allusion to its content in later letters.

2. *the letter*: letter 0404.

3. *the new method*: not determined.

4. *expect . . . straight to Halifax*: Tyndall had other priorities. He landed in London and spent a week making literary and scientific connections there before going north.

5. *MS*: the English translation of Tyndall and Knoblauch's German article on diamagnetism (cited letter 0403, n. 2), sent with letter 0403 to Hirst.

6. *West Riding Polytechnical Society*: established primarily as a place for coal owners to discuss geological questions, but over time, it offered lectures on diverse scientific subjects. From 1849 to 1859, the Society held meetings in Leeds, Sheffield, Halifax, Doncaster, and Barnsley (see *Proceedings of the Yorkshire Geological and Polytechnic Society of the West Riding*, 2 (1849–59), p. 370).

7. *Propensities and their Equivalents*: the article was accepted by the *Leader* and published anonymously. See 'Propensities and their Equivalents', *Leader* 1, no. 14 (29 June 1850), pp. 332–33. This may have been the article sent with letter 0401 (at n. 3).

8. *farewell paper at the Franklin Society . . . at the time*: Tyndall attended when Hirst read his paper on 'Societies: considered in their relation to Intellectual Culture' (Journal, 26 July 1850).

9. *Aunt*: Hirst's aunt Allatt (Hirst, 'Journals', 17 June 1850).

10. *Mr Pridie's Gospel*: probably a reference to Rev. James Pridie, an Independent Minister in Halifax, and the father of Hirst's good friend Roby Pridie (see Hirst, 'Journals', 30 June 1850).

11. *curious documents (MSS)*: documents about, or even written by, his father, that revealed he was 'a thinker'.

12. *Mothers death*: Hirst's mother died on 2 September 1849, while Hirst was with Tyndall in Marburg.

13. *M'Arthur*: see letter 0393.

14. *John Tyndall . . . 1850*: on envelope.

To Thomas Archer Hirst [21 June 1850] 0407

Friday—I don't know the date![1]

My dear Tom

Here I am at length in the heavenly region of Hammonds Hotel St Martins Court, Saint Martins Land Covent Garden London, that is to say in the third story up near the stars. I left Marburg on Wednesday week, and through that propensity to lying common to the whole race of German carriers[2] I was held 3 days upon the road. It was guaranteed to me that my luggage should be in Frankfort on Thursday evening, it was not delivered until Friday at noon and the steam vessel was then away. Thus half a week was lost. I should have been satisfied could I have spent the time in Bonn and thus been able to visit Drachenfells[3] and that neighbourhood but I found this attendant with so much risk to the safety of the luggage that I was compelled to give it up. In Frankfort I spent half my spare time, in Rotterdam and Leyden the other half I visited the picture gallery the Ariadne,[4] and the theatre at Frankfort. The latter was a milk and water affair.[5] The second is a delightful piece of sculpture. The lady is represented sitting on a tiger's back, her attitude is difficult for the artist but he has reduced it to one of perfect ease and flexibility. The light admitted into the small chamber which contains the statue passes through a net screen, and thus a rosy freshness is thrown upon the white marble. It is a beautiful achievement, and you must see it when you go to Frankfort. The painting which pleased me best was that of Huss defending himself before the Cardinals.[6] There he stands in the midst of foes which a few days afterwards burnt the life out of him. His intellect seems keenly awake. There is neither hope nor fear in his countenance, the latter is thin and pale, and you can see the veins traversing his emaciated hands. a face here and there among the audience has a gleam of benignity if not of approval in it, but the majority wear the expression of deadly hatred.

We had a beautiful passage from Rotterdam. The sun and blue heaven by day, and the moon and stars by night. We reached the custom house on Wednesday evening—it is now Friday morning. I went to Putney yesterday to see Frankland, and he came yesterday evening and remained with me till 10 o'clock. I have a little business here in London which will probably detain me two or three days, and afterwards I steer northward. Should you wish to say any thing to me you will have time if you reply promptly. Did the magnetic investigation reach you?[7] and did you send it punctually away?

Yesterday in moping through the Strand, religiously resolved not to spend a penny on any of the temptations there exhibited, I passed Chapman's window,[8] there dangling from a string hung No 6 Carlyle's Latter day pamphlets,[9] price one shilling, by a kind of instinct my finger ends found their way into the sanctum of my breeches pocket and in half a minute I was a shilling poorer. I moped through the crowd gulping down a mouthful at intervals and finally reached Trafalgar Square. then seating myself upon the basin's edge, right under the shadow of the Nelson Column, and chewed through the pamphlet from beginning to end. The matter afforded me an hour's delectable employment, and in some places, I laughed like a horse, for instance where he asks 'Peter do you know why the age of miracles is past? Because you are become an enchanted human ass (I grieve to say it)' and again. 'You will carry it, you, by your rating and your eloquencey and your babbling and the adamantine basis of the Universe shall bend to your third reading, and the paltry bit of engrossed sheepskin and dog-latin. What will become of you?'[10] On the whole there is nothing new in this pamphlet, its merit is that, the breath of life is in every line of it.

You made a remark in your last about Geo. Sand. I know nothing about the woman.[11] I have read a good deal of reviews and opinions concerning her, in the people's journal[12] and elsewhere and mostly laudatory; but I would not give three straws for all the liquorish syrup,[13] expended by those her admirers, and until I have an opportunity of judging her myself I must consider Carlyle's opinion of the matter, and not yours, as the most probable one. 'Gross libels' are they—'Caliban is just as unfit to judge as Carlyle' is he? hard words my dear disciple—but there is danger in throwing up your unfettered arms and asserting your independence of Carlyle that it is merely for the love of the thing and not as a matter of necessity that you do so. I have seen this spirit very well in Carlyle's disciples. January Searle tells us that Carlyle's books are not healthy books. Dr Lees attributes his earnestness to dyspepsia and the habit of smoking, and Thomas Hirst regrets his <u>temper</u>. Admirable all, very cool, very philosophic, most laudably independent! Until however I see better grounds for each and all these opinions I must be excused attaching very little value to them—they are so many swags and tassels with which the said disciples delight to adorn themselves. Carlyle doubtless smiles good humouredly on these turkey cock bristlings[14] well knowing that they have no deep cause. On the whole those utterances of Caliban[15] please me better than most that I have seen. They are evidently the transcript of his very soul and he appears in no way anxious to conceal the fact of Carlyle's chosen captive influence upon him; quite untroubled, he, as to what the world thinks about the surrender of his independence, the prostration of his own individuality before the man whom he recognises as his captain.

But I must stop. I shall soon see you. You ask me, will I accept another £20.[16] I answer 'not if I can avoid it', which latter is probable. I am glad you paid Jimmy, because Jimmy is not yet a fool; he will reason as a prudent man, and would perhaps be subject to a periodic qualm in his hours of reflection; again I repeat I will soon see you, but wont promise to stop long with you. I have another asylum in the north.[17]

J. Tyndall

RI MS JT/1/T/1015
RI MS JT/1/HTYP/88–89

---

1. *Friday . . . the date*: the relevant Friday was 21 June 1850. We have no independent evidence to support the LT annotation: 'begun Wed evening June 19'. Hirst had asked Tyndall to date his letters.

2. *German carriers*: an allusion to the road carriage of Tyndall's luggage.

3. *Drachenfells*: see letter 0398, n. 15.

4. *the Ariadne*: a sculpture of 'Ariadne on the Panther' by Johann Heinrich von Dannecker.

5. *milk and water affair*: something feeble, insipid, or mawkish (*OED*).

6. *of Huss defending himself before the Cardinals*: Karl Friedrich Lessing's 'Johann Hus auf dem Konstanzer Konzil'.

7. *the magnetic investigation reach you*: sent with letter 0403.

8. *Chapman's window*: the office of the publisher, John Chapman, at 142 Strand.

9. *No 6 Carlyle's Latter day pamphlets*: No VI, 'Parliaments' (1 June 1850), *Latter-Day Pamphlets*, pp. 182–215.

10. *'Peter do you know . . . become of you'*: passages from paragraphs 40 and 39, respectively, of pamphlet 6 (ibid.).

11. *your last about George Sand*: letter 0404. Tyndall had not yet received letter 0406 (18 June) which Hirst had sent to Marburg.

12. *the people's journal*: see letter 0392, n. 3.

13. *not give three straws .. liquorish syrup*: 'not give three straws'—a common phrase for 'care nothing for'; 'liquorish syrup'—a sweet sticky syrup.

14. *Turkey cock bristlings*: A 'turkey-cock' is a pompous or self-important person. In this case, the 'bristlings' of a 'turkey cock' refer to pompous and anger-infused rhetoric (OED).

15. *those utterances of Caliban*: Caliban had written three articles defending Carlyle's *Latter Day Pamphlets*, all under the title 'Letters on Thomas Carlyle', in the *Truth-Seeker*, n.s. 2 (1850), pp. 94–100, 148–63 and 245–49 (letter 0400 comments on the first article). The letters cannot be matched to dates as extant copies of the *Truth-Seeker* are not dated.

16. *You ask me, will I accept another £20*: letter 0404.

17. *another asylum in the north*: Hirst wanted Tyndall to spend time in Halifax, but Tyndall preferred to stay with Josiah Singleton, at his school in Over Darwen.

To [Charles Mackay]¹   [21 June 1850]²   0408

Hammond's Hotel, St. Martin's Court | Friday

My Dear Sir,

I have just returned to England after an absence of nearly two years in Germany and one of my first acts is to solicit your advice on a little matter which personally interests me. During my stay in Germany I devoted myself chiefly to scientific pursuits and in the winter of 1849 passed an examination and took out the degree of Doctor of Philosophy from that time to the present I have been engaged in magnetic investigations, and have been fortunate enough to obtain results which I believe scientific men will deem important. These results have already partially been made public both in England and Germany.³ I am anxious to follow the matter up,⁴ as I feel assured that the issue will prove more interesting than anything that has appeared in physical science for many years. This wish on my part has already induced me to decline offers which have been lately made to me,⁵ nor should I like to accept an engagement until I first ascertain whether it is not possible to follow up the line of thought already indicated. You are one of the counsellors which I have resolved to consult; and the necessary information would be a reply to the following question;—Do you know of any periodical, scientific or literary where I might have an opportunity of employing my pen and brains a little? I should be very happy to devote half my time to such a purpose, provided the other half could be left open to the completion of my enquiries.⁶ I am perfectly well aware that I bother and bore you nor shall I be the least surprised should you decline advising me. I have thought the thing worth a trial and have therefore done it.

Believe me Dear Sir | Very faithfully yours | <u>John Tyndall</u>

I leave town on Saturday morning and should the spirit prompt you to reply to me, the following address will always find me.

JT. | Care of Thomas Hirst | Harrison Road | Halifax | <u>Yorkshire</u>

If you like I will send you a specimen article.

Imperial College London, Papers of S. P. Thompson, 1.384

---

1. *[Charles Mackay]*: probable recipient. Although housed in the papers of S. P. Thompson, the letter is not to him as he was not born until 1851. From internal evidence, letter 0409 from Charles Mackay seems to be a response to this letter; moreover Tyndall had been in contact with Charles Mackay at the time of his departure for Marburg (see letter 0362). Possibly, but less likely, this letter might have been prompted by letter 0409 and addressed to William Little (letter 0409, n. 2), or written to some other person in publishing (for example, see n. 5 below). It may be a draft (see n. 6).

2. *[21 June 1850]*: probable date. Tyndall arrived from Germany on 19 June and stayed at the Hammond's Hotel, St. Martins Court, before leaving for Halifax on 29 June. Friday the 28th does not give time for Tyndall to receive the hoped-for reply in London. This supports MacKay being addressee (n. 1).

3. *made public in England and Germany*: Tyndall alludes to the paper published in *Poggend. Annal.* (cited 0392, n. 13) and in *Phil. Mag.* (cited letter 0395, n. 22). A second article was published in the *Phil. Mag.* in July (cited letter 0403 n. 2). See letter 0398, n. 16, on the award of Tyndall's doctoral degree.

4. *follow up the matter*: Tyndall wanted to return to Germany to continue his experimental work with Knoblauch.

5. *offers . . . made to me*: Tyndall had received a query from Edmondson about Tyndall's availability to return to Queenwood, which Tyndall answered in a manner 'that will probably throw cold water on the subject' (Journal, 2 June 1850, JT/2/13b/491). Tyndall may also have been exaggerating. In his Journal for 23 July 1850, he wrote: 'It has been a fortnight since my last entry. Wrote to Queenwood—Nichts; wrote to Halifax, to C[arter] Nichts; wrote to Oak Park, have heard Nichts; Wrote to Longman, enclosing 13 pages, Nichts' (JT/2/13b/502).

6. *enquiries*: At this point Tyndall has slightly and ineffectually crossed out the sentence, 'I am totally ignorant of the method of effecting such', suggesting that this letter might have been a draft.

From Charles Mackay   25 June [1850][1]   0409

22 Brecknock Crescent | Camden Road Villas | June 25th

My dear Sir,

I should have been glad to see you before your departure from town. The

competition in London for all kinds of literary employment is very great—
and it is only by being on the spot—and by being very active, very patient, and
very persevering that anything can be done. —I know one gentleman (who
however has not the direct means of being of service) to whom I would have
been glad to introduce to you—Mr Little of the Illustrated London News,[2]
who has lately been engaged, I believe in similar investigations to your own. I
will show him your letter on the first opportunity. Let me know your doings
from time to time—and if you are in London be sure to call. I leave town in
a few days for Birmingham and Liverpool—but letters addressed to me at
home will be duly forwarded.

    With best wishes for your success in all your undertakings
    Believe me | Yours very sincerely | Cha. Mackay.
    John Tyndall Esq.

RI MS JT/1/TYP/2/781
LT Transcript Only

---

1. *25 June [1850]*: year based on Tyndall's search for literary employment in mid-1850, and
   the match of dates with his stop in London, 19–29 June 1850 (see letter 0408, n. 2).
2. *Mr Little of the Illustrated London News*: William Little (1816–84), an accomplished
   chemist, who printed and published the *Illustrated London News* from 1843–58.

To Heinrich Debus   [3 July 1850][1]   0410

Spring Bank, over Darwen, Lancashire[2]

Mein Lieber Heinrich,

    Ich schreibe ohne Worterbuch, und wenn ich Fehler mache dann müßen
Sie ihre Güte hervorrufen und meine Unwissenheit entschuldigen. Ich war
8 Tage unterwegs weil der verfluchte /Fuhrman/ zu spät bis Frankfort kam
jedoch bin ich am Ende sicher angekommen und jetzt finde mich zum ersten
Mal seit meiner Ankunft ein Bischen ruhig. Ich blieb in London 10 Tage und
habe Frankland mehrere mal gesehen. wir werden zusammen nach Edinburg
reisen. die British association wird sich da versam<m>eln und wir werden
vielleicht etwas vortragen Gleich nacher wird er nach Deutschland reisen um
Fraulein F zu besuchen, und es ist möglich daß ich mit reisen werde. er hat
mir Gesagt, daß er an Sie geschrieben hat.

    Unsere Abhandlung ist schön öffentlich. Die Leute schienen sie für
wichtig zu halten. ich habe Faraday besucht und fand mich ganz herzlich
empfangen. Wir sprachen fast eine halbe Stunde zusammen, aber seine Mei-
nung uber die Abhandlung gab er mir nicht, er Sagte nur daß er die Sache für
wenigstens ein monat uberlegen muß.

Seit meiner Ankunft ist das Wetter sehr unfreundlich gewesen, bisweilen stark <u>nebelisch</u>. ist es nicht der Fall auch in Deutschland gew<esen>? Wenn nicht dann hat der Herr Major Recht gehabt, und wenn ich ihn wieder sehe, werde ich alle seine Einwurfe gegen unseres unglüc[k]liches Land ganz ruhig dulden.

Ich kann nicht sagen lieber Heinrich ob ich Sie wieder sehen soll oder nicht. Jedenfalls geht Hirst nach Marburg vielleicht ich auch aber bis jetzt ist es unbestimmt. Ich habe zwar mich noch nicht uber die Sache erkundigth wenn ich wieder schreibe werde ich etwas mehr darüber sagen können.

Gestern abend um 11 uhr haben wir unsern größten Man verloren; gestern Abend um <11> uhr starb Sir Robert Peel. Ich fühle es tief und schmerzlich. wir konnten ihn nicht sparen. Manche Augen waren auf ihn gerichtet aber er ist weg und wir wißen nicht wo seinen gleichen gefunden werden kann. Die Welt wird sich doch umdrehen und die liebe Natur wird hoffentlich einen Stellvertreter sehen.

Ich bin so ruhig hier Heinrich weit von allen [Städtten], ganz auf dem Lande, nichts zu hören mit ausnahme der [Windseufzer] und des Brausens des Wassers. ich werde hier ungefähr 3 wochen bleiben, So wenn Sie an mich schreiben und wie oben addressiren dann wird ihr Brief mich finden, aber Sie müßen keineswegs denken daß ich einen Brief erfordere, das muß gelegentlich kommen wenn der Geist Sie bewegt aber sonst gar nicht. Ich kann Sie wenigstens versichern daß wenn Sie schreiben so werden ihre Wörter immer einen Klang in meinem Brust finden. ich werde Sie nicht [bald] vergessen. wenn alle Kleinigkeiten die man von man <un>terscheiden weg geschafft sind dann stehen Sie immer vor meinen Augen [wie] eine gerader aufrichtiger Kerl der ich sehr gern habe. Es sind drei haupsächtlich die so stehen; die sind Friedrich, Karl, und Heinrich und zu den drei empfehle ich mich mein Freund Hirst wenn ich selbst nicht da bin ihn vorz<ustellen>

Grüßen Sie alle Freunde [herz] von mir, alle meine Tisch genossen [und Vetter]; den Professor habe ich schon durch Pro Knoblauch gegrüßt aber grüßen Sie ihn auch; ich will nicht viel sagen aber könnte ich [Gelegenheit] haben durch Thatsachen zu beweisen wie ich gegen ihn fühle! und nun was habe ich mehr zu sagen? nichts - Sie wißen Heinrich daß ich nicht wißenschaftlich schreiben kann weil bis jetzt meine Gedanken in einem wirbel geblieben sind, darum klagen Sie nicht und nun mein lieber fahre wohl

Your sincere friend | John Tyndall

Spring Bank, over Darwen, Lancashire

My dear Heinrich,

I am writing without a dictionary, and if I make any mistakes, then you will have to summon your kindness and excuse my ignorance. I was away for 8 days because the damn [carrier] came too late to Frankfort.[3] However, I arrived safely in the end and now find myself a bit calm for the first time since my arrival. I stayed in London for 10 days and saw Frankland several times. We shall be travelling to Edinburgh together. The British Association is going to meet there and we shall perhaps give a lecture on something.[4] Immediately afterwards he is going to travel to Germany to visit Fräulein F.[5] and it is possible that I shall travel with him. He told me that he has written to you.

Our paper[6] is already public, and people seemed to think it is important. I visited Faraday, and found myself quite warmly received. We spoke together for almost half an hour, but he did not give me his opinion about the paper. He said only that he will have to consider the thing for at least a month.

Since my arrival, the weather has been very inclement, from time to time quite <u>foggy</u>. Has it not been the case in Germany too? If not, then the Herr Major was right, and when I see him again, I shall quite calmly suffer all his objections about our unfortunate country.

I cannot say, dear Heinrich, if I should see you again or not. At any rate, Hirst is going to Marburg,[7] and perhaps I shall too, but up till now that is uncertain. Admittedly I have not enquired about the matter[8] yet, but when I write again, I shall be able to say somewhat more about it.

Yesterday evening at 11 o'clock we lost our greatest man; yesterday evening at 11 o'clock Sir Robert Peel died. I feel it deeply and painfully. We could not spare him. A good many eyes were fixed on him, but he is gone, and we do not know where his kind can be found. The world will turn, however, and dear Nature will hopefully find a substitute for him.

I am so calm here, Heinrich, far from all the [towns], completely out in the countryside, with nothing to hear except the [sighs of the wind] and the rushing of the water. I am staying here for approximately 3 weeks, so if you write to me and address any correspondence as above, then your letter will find me, but you must not in any way think that I am demanding a letter, that need only come sometime or other when the spirit moves you, but otherwise not at all. I can at least assure you that if you write, your words will always find an echo in my heart. I shall not forget you [quickly]. Once all those little things which distinguish people from one another have been removed, you always stand before my eyes [as] an upright and sincere fellow whom I like very much. There are mainly three who stand out as such; they are Friedrich, Karl and Heinrich,[9] and to these three, I shall commend my friend Hirst if I am not there myself to introduce him.

Give all my friends, all my table companions [and brothers], my [kind] regards. I have already given my regards to the Professor[10] via Prof. Knoblauch, but do give him my regards as well. I do not want to say much, but could I only have the [opportunity] to prove with facts how I feel towards him! Well now, what else do I have to say? Nothing. You know, Heinrich, that I am not able to do any scientific writing because up till now my thoughts have been in turmoil, so do not complain. And now, my dear man, farewell,

Your sincere friend | John Tyndall

RI MS JT/1/T/357

1. *[3 July 1850]*: Tyndall's reference to the death of Robert Peel the previous evening dates this letter.
2. *Spring Bank Over Darwen, Lancashire*: location of Josiah Singleton's school for boys. Tyndall helped with the teaching when he stayed there. See his account in letter 0424.
3. *damn driver came too late to Frankfort*: see letter 0407 (at n. 2).
4. *together*: in the event, Tyndall neither travelled nor presented with Frankland.
5. *Fraulein F.*: Sophie Fick, Frankland's fiancée.
6. *Our paper*: just published in the July issue of *Phil. Mag.* (cited letter 0403, n. 2).
7. *Hirst will go to Marburg . . . too*: as it turned out, Tyndall accompanied Hirst when he entered the University of Marburg in October 1850.
8. *the matter*: probably an allusion to financial arrangements that would permit his return to Marburg.
9. *Friedrich, Karl, and Heinrich*: Friedrich Stegmann (possibly), Karl Schmitt (probably), and Heinrich Debus himself.
10. *the Professor*: perhaps Professor Stegmann who supervised Tyndall, but more likely Professor Bunsen, who had given him laboratory space and for whom Debus managed the laboratory.

From James Craven   5 July 1850   0411

Halifax, July 5, 1850

Dear Tyndall,

I would have forwarded you a copy of 'Social Aspects' had not its author[1] presented you with a copy, as it is a book which somewhat suits and harmonizes with my own ideas and thoughts. As I cannot, however, I have substituted a work which I send you by this post,[2] which so far as myself is concerned, I am totally ignorant of; but since it is by a favourite writer for whom I believe you have some regard—a writer who has somehow slowly gained a noble little band of followers, composed too I believe of some of our best men—men who have thought much and hope much—I have taken the liberty to send it you and trust it may prove a source of pleasure to you.

Your inaugural lectures[3] were duly handed to me by faithful Tom; though most uninteresting and forbidding to some people, I shall always look on them, though ignorant of their <u>real</u> worth, with a feeling akin to veneration—to think that such cold subjects could have monopolized your sole attention for two years is strange. However accept my thanks for the gift. Yet it is by no means strange that after being pinned up so long you should find yourself inclined to study in a more genial atmosphere; however beware—folks say there's great things come out of little ones now and then—with which I will leave you to cogitate on and so again

Remain | Yours very faithfully, | James Craven.
John Tyndall, | Lancashire.

RI MS JT/1/TYP/11/3542
LT Transcript Only

---

1.   *'Social Aspects' had not its author*: John Stores Smith (letter 0398, n. 12).
2.   *substituted a work . . . post*: Craven enclosed a copy of Carlye's *Sartor Resartus* (Journal, 4 July 1850, JT/2/13b/502).
3.   *inaugural lectures*: possibly referring to Tyndall's publication on diamagnetism (cited letter 0403, n. 2).

From Thomas Archer Hirst   9 July 1850   0412

Northowram,[1] July 9th, 1850

My dear Tyndall—

I am out surveying here & have been overtaken by violent rain, and taken refuge in a little coal-weighing-machine house, where luckily to dissipate ennui the fellow happens to possess a sheet of writing paper: so that you must not think I am writing so much for an express object, as just to please myself. I got your letter[2] & delivered its enclosed one to Carter; likewise told Jemmy what you thought were his deserts. Poor fellow he is poorly in bed to day, with his Rheumatism. I had my weekly meeting last Saturday and as usual passed a very pleasant evening. They are a lot of good fellows and interesting too. Their idiosyncrasies are so varied and well worth studying I have put a short sketch of them all into my Diary for last Sunday week[3] which I will shew you some time. Carlyle's last pamphlet[4] formed the principal point of conversation as it generally does once a month. It contains some of his most powerful & valuable sentences though as a whole, and as an analysis of the phenomena of Hudson's influence, it is scarcely what one might have anticipated. I would not part with it however at any price. Jemmy was very sly in his present to you,[5] though when he mentioned having sent for it I must say I suspected it. Well it is a valuable present and one you will prize I know, you possess few books more prizable, and no library would be complete without it. We will have many a dip together into that in 'Vaterland'.[6] How are you getting on with your educational work?[7] There was a somewhat doubtful review of Smiths book[8] in the Leader last Saturday;[9] it had too much of a patronizing, self important air about it for me. He could not but acknowledge its great merits, but he seemed to do it grudgingly & what is more he lost sight a good deal of its real intention, which is an earnest protest against the rottenness and hypocrisy visible in our Society, viewed thus it is a worthy book, its protests are everywhere just, & immensely earnest, which as I take it is the most valuable style of writing for getting them remedied. Instead of taking such expressions to heart & using them for their real purpose, there is too much cant about originality stirring, in a new book we expect something <u>new</u>, as if the old were yet but into action. Carlyle suffers from this sort of Criticism & it is now wielded against Smith. Viewed as an analysis of our Social Aspects there is undoubtedly a crudity apparent, but as I said, it is not in this aspect that we should view it. I have arranged for you to be at Halifax <u>at least</u> on the Thursday & Friday the 25th & 26th of July the former is my night at the Improvement Society the latter my paper at the Franklin Society.[10] The Saturday you must stop also, & on Sunday pay Phillips your promised visit. This is what you have to do <u>without fail</u>; you can walk about your business as soon as you like after, you restless being—If you come through Manchester again remember me to M[r] & M[rs] Ginty, tell the former he never returned my Preston Chronicle with Zig-zag in it,[11] bring it with you if you can. Ginty has talked about coming to Halifax soon—I wonder if he meant it. Oh, I had nearly forgot, you must go & see little John[12] too, he invited Jemmy & I to go over & we promised, perhaps we can manage it altogether. Now I have done M[r] Carter is just saying it is clearing up so Goodbye.

Yours affectionately | T.A. Hirst

RI MS JT/1/H/148

---

1. *Northowram*: a village in Calderdale, about 1.5 miles east of Halifax.

2. *I got your letter . . . Carter*: letter missing.

3. *short sketch in my Diary for last Sunday week*: Hirst, 'Journals', 30 June 1850. Hirst not only described each member of the group present, including Phillips, Hutchinson, Pridie, Craven, Smith, and on that occasion Tyndall, but also included a sketch of the room, where each person sat, and the placement of all the furniture.

4. *Carlyle's last pamphlet*: 'VII. Hudson's Statue' was published on 1 July 1850. Carlyle lamented Britain's spiritual condition: spending thousands of pounds on a statue dedicated to 'railway king' George Hudson (1800–71) was to make Hudson, who had been widely accused of fraud, a hero or a demi-god (*Latter-Day Pamphlets*, pp. 317–64).

5. *present to you*: letter 0411, n. 2.
6. *'Vaterland'*: 'Fatherland' (German), an allusion to Hirst's hope that Tyndall would accompany him to Germany.
7. *your educational work*: the 'education tract', which Tyndall refers to in his next letter (0413).
8. *Smiths book*: cited letter 0398, n. 12.
9. *doubtful review . . . last Saturday*: 'Smith's Social Aspects', *Leader* 1, no. 15 (6 July 1850), pp. 353–54. The review was anonymous, but probably by George Henry Lewes.
10. *paper at the Franklin Society*: see letter 0406, n. 8.
11. *Preston Chronicle with Zig-Zag in it*: cited letter 0404, n. 28.
12. *little John*: not identified.

To Thomas Archer Hirst   [10–12 July 1850][1]   0413

Dear Tom.

Your note from the weighing machine[2] reached me this morning. I give you credit for always writing to please yourself as I do at present blessed be the Gods. There you spun like a magnetic bodkin[3] on a diamond pivot and pointed at length with your pole hitherward. Don't you see you dog, though you take pains to deny it that this inconscious avowal is infinitely flattering to me. I wish I had the physical education of Jemmy in my hands, I would extirpate that Rheumatism, root and branch and send him in 6 months home to his father as strong and handsome as I am myself. I'm glad to find your weekly meeting in a healthy condition; there are pretty experiences to be gathered from such contact; your analysis of character proves its reaction upon yourself. This I shall read with much interest. I have read Sartor Resartus[4] but wont say a word about it till I have more time. I must read Social Aspects[5] when I go to Halifax, and I can then compare my own convictions with those of the Leader. Originality is all humbug!—I like it myself but I know <u>when</u> I like it—when I'm idle and want excitement, and at such a time a glass of Stumpcross ale and a cigar would supply its place.

> '______ Why should I roam
> Who cannot circumnavigate the sea
> Of thoughts and things at home'[6]

You see I have been reading Emerson. What a superlative jingle of nonsense those poems are to some people—to those for instance who have made Pope and Scott their ideal. It requires a wonderful abandonment to get into them. His voice reaches me sometimes dull and feeble as if from the bottom of a mine—I listen and the tones heighten till at length they become strong as the rumble of an earthquake. Emerson has had his own trials, if not how could he write thus.

But man crouches
and blushes
Absconds and conceals
He creepeth and peepeth
He palters and stealst
Infirm melancholy
Jealous glancing around
An oaf an accomplice
He poisons the ground.[7]

and there the fine antithesis
'Erect as a sunbeam
Upspringeth the palm
The elephant browses
Undaunted and calm'[8]

There is a deep apprehensiveness in <u>Each and all</u>. I suppose everybody must have felt the truth of it. I felt it yesterday 'see' said a friend 'that beautiful tuft of brown grass'. Had the surrounding hazels and ferns been away. were the tuft for instance set erect upon his carpet it would not have been beautiful

The essence of the problem I conceive lies in the line, 'These temples grew as grows the grass'[9] one is the primal utterance of the universal soul and the other a secondary utterance through man as instrument. Emerson is a pantheist in the highest sense and so is Carlyle. I dropped an hour ago upon a very significant passage in the Sartor. 'Is there no god then, but at best an absentee god sitting idle ever since the first Sabbath at the outside of his Universe and <u>seeing</u> it go?'[10] At the <u>'outside'</u> of his universe. I imagine Carlyle's entire creed is folded in this Sentence. And here the difference between his faith, and that of Paley's is very distinct. According to the latter god bears the same relation to the Universe that a clockmaker does to the clock.[11] He is an omnipotent mechanic detached from his work. With Carlyle the universe is the blood and bones of Jehovah[12]—he climbs in the sap of trees and falls in Cataracts. The theist of Paley's class is I believe intrinsically the same is the atheist. It is impossible to decide which has the best grounds for his belief.

The same power of observation is evinced in Rhea[13]

'Thou shalt seem in each reply
A vixen to his altered eye.'[14]
    that's a bare fact.

I like such passages as the following in the Wo[r]ld Soul.[15] The same is true as regards <u>men</u> also. I have found it.

'For Gods delight in Gods
And thrust the weak aside
To him who scorns the charities
Their arms fly open wide'.[16]

'Gay'[17] appears to be a favourite character of Emerson's 'The winds are always favourable to the good seaman'.[18] He loves to exhibit this mastery over circumstance, and detect <u>law</u> where others would cry out <u>chance.</u> I remember once Prof Waitz[19] said to me that Goethe was one of those <u>lucky</u> fellows with whom everything prospered. I question whether Emerson would subscribe to the term '<u>luck</u>' 'The Rhodosa.'[20] This is an exquisite piece. As monkeys jibber among leaves so do the mass of men among facts. the origin of the foliage is unknown to the one. the origin of the fact unknown to the other. Emerson sits placidly where the trunk and the earth meet and sees unity of bough and twigs. Living in the primeval element he can trace its outgoings and thus reconciles puzzles. As before 'These temples grew as grows the grass'[21] so here

> 'I never thought to ask; I never knew
> But in my simple ignorance suppose
> The selfsame power that brought me there brought you.'[22]
> exquisite!

In §3 of Woodnotes[23] Mr Smith has changed 'mouse' into 'moose'. I think Emerson meant <u>mouse</u>. There is a great beauty and force, and truth, and trust, in that part which ends in

> 'For ever nature faithful is
> To such as trust her faithfulness'[24]

The second part of Woodnotes[25] is magnificent I cant select it is all so good. the leading ideas appear to be

> 'The primal mind
> That flows in streams and breathes in wind'.[26]

and the metamorphosis of the same

> 'I, that today am a pine
> Yesterday was a bundle of grass'[27]
> 'Halteth never in one shape
> But for ever doth escape'.[28]

I can never believe Emerson unhappy The fluidity of his philosophy guards him against disappointment. Nature cant trick him. Indeed he appears to have resigned himself heart and soul to her, and finds pleasure in her methods.

Thats a splendid passage marked by Smith in 'Monadnoc'.[29] Emerson's hope and prediction is here

'I await the bard and sage
Who in large thoughts like fair pearl-seed
Shall string Monadnoc like a bead'.[30]

But the finest poetry in the book I think is to be found at page 87.[31] He tosses the earth as an Indian flings his boomerang. Perhaps I deem it so because a similar thought made a profound impression on me in my passage from Rotterdam.[32] It is partly expressed in a scrap I sent to the Chronicle last week[33] and which if it [suits] them will appear on Saturday.

That's a fine piece marked by Smith in the Ode to Channing34

Things are in the saddle
And ride Mankind[35]

You remember that fine old song:-

If she be not fair for me
What care I how fair she be.[36]

The same sentiment is in 'Give all to love,'[37] most beautifully expressed, read the three last verses. when I'm an old man I'll tell you an anecdote which in some measure illustrates Emerson's advice. but I know I bore you. Ill copy this and have done.

Leave all for love
Yet hear me yet
One word more my heart beloved
One pulse more of firm endeavour
Keep thee today
Tomorrow for ever
Free as an Arab
Of thy beloved

Cling with life to the maid
But when the surprise
Vague shadow of surmise
Flits across her bosom young
Of a joy apart from thine
Free she be fancy free
Do not thou detain a hem
From her summer diadem

Though thou loved her as thyself
As a self of purer clay

Tho' her parting dims the day
Stealing grace from all alive
Heartily know
When half god's go
The gods arrive.[38]

This is brave and it is wise. nay politic—if ever fickleness is to be broken down and woman thoroughly subdued it must be by this unfearing readiness to yield her if she will. it enchants her, and magnetizes her, and holds her faster, and keeps her truer than 10 million reproaches coupled with the wringing of hands.

I shall be there God willing on the 25th.[39]

~~Don't think too much on the first part of this letter. I am half inclined not to send it, however I shall do,~~

~~as ever~~ |[40] Tyndall

I bought Carlyle's pamphlet[41] in Manchester and read it with a certain grimness of soul which was very refreshing to me. The great rolling snorting leviathan how he smashed up that Hudson. 'And then swung as a tragic pendulum admonitary to earth in the name of heaven &c &c.[42]

I must manage to see John for I have a great regard for him. Was Phillips with you on Saturday night. I find my education tract slow work. It requires thought to make any thing of it. I could write a deal very soon, but a deal of stuff.

May the gods bless you my son | Goodbye

In 'Merlin'[43] there is a fine limit drawn round the understanding.

Nor profane affect to hit
Or compass that by meddling wit
Which only the propitious mind
Publishes where its inclined[44]
& c.   _________

The poet cant sing off-hand like a blackbird, perhaps I do the animal wrong. It also may have its hours of inspiration— The nightingale loves the moon

In Bacchus[45] the high Pantheism again crops out.

That I intoxicated
And by the draft assimilated
May float at pleasure thru all natures
The bird language rightly Spell
And that which the roses say so well[46]

and again.
'The poor grass shall plot and plan what it will do when it is man'[47]

All this puts me in mind of Plato. It is unspeakable and yet Knowable— Who does not feel the heavy truth of the image—'Reason in nature's lotus drenched!'[48] There is a kind of Swampishness of the soul which this perfectly expresses. Somehow or other that beautiful piece 'the House'[49] does not please me as it used to do. There is confusion in the figure.

She lays her beams in music
In music every one—[50]

Perhaps its my defect. I used to love it—

Here in Saadi[51] is a passage similar to one already noticed

'Flee from the gods which from thee flee
Seek nothing; Fortune seeketh thee.'[52]

This is the moral gravitation between man's true necessity and the objects of them.

Emerson in one of his essays I think mention an occult relation subsisting between the man and the vegetable.[53] The apprehension of this fact lies at the bottom of most of his poems. The suggestion haunts him and he yearns for its fulfilment. The transmutation appears sometimes to be effected and both natures run into one.

'We coldly ask their pottage not their love'[54] I used to meet a jolly fat old fellow sometimes on the roads about Marburg. He always saluted me with the assurance 'now the coffee will smack <u>vortrefflich</u>[55] when you get home. Nun mein Freund Sie werden [sich] heute gut finden'.[56] He walked to get an appetite and nothing else—he had his reward. Nature is a kindly mother.

There is a sweet consolation in that verse in 'Musketaquid'[57]—

The polite found me impolite; the great
Would mortify me but in vain
I am a willow of the wilderness`
Loving the wind that bent me.[58]

all this verse to the end is beautiful—
Canst thou shine now then darkle
<u>And</u> <u>being</u> <u>latent</u> <u>feel</u> <u>thyself</u> <u>no</u> <u>less</u>
admirable!—[59]

The affirmative to this question would presuppose the entire eradication of egotism.

Thats beautiful in the Threnody[60]

'Hast thou forgot me in a new delight?'[61]
and here.
O child of paradise
Boy who made dear his father's home
In whose deep eyes
Men read the wellfare of the times to come
I am too much bereft
The world dishonoured thou hast left
O trusted broken prophecy!
O richest fortune sourly crost
Born for the future, to the future lost![62]

<u>And</u> <u>his</u> recovery is very fine

> 'And [thought^st] thou that such a guest
> Would in thy hall take up his rest
> Would rushing life forget its laws
> Fate's glowing revolution pause
> High omens ask diviner guess
> Not to be conned to tediousness'[63]

His hatred of immobility breaks out even here

> 'Not of adamant or gold
> Built he heaven Stark and Cold
> No, but a nest of bending reeds
> Flowering grass and scented weeds'[64]

The world and all that therein is *[is]*, with Emerson, a differential, a fluxion what the Germans call a Werdende—a becom<u>ing</u>. It is as it were an effort perpetually manifesting itself and not to be cabinned in an institution—no not in heaven of 'adamant or gold'

I heard a lady socialist lecturer[65] once assert 'For my part I shouldn't like to be caged up in what you call your heaven, singing an eternally Holy! Holy!'[66] <u>nor should I exactly</u> | J.T.

Mr Hirst | Harrison Road | Halifax | Yorkshire[67]

RI MS JT/1/T/529

---

1. *[10–12 July 1850]*: perhaps started on 10 July if Hirst's letter (n. 2) took only one day to reach Tyndall. There are a number of postmarks on this letter: the one from Darwen, where Tyndall was located, seems to be the 12th. Hirst has added, 'July 13—1850 | Tyndall Manchester', presumably the day he received it, implying that it was posted on 12 July.

2. *Your note . . . weighing machine*: letter 0412.

3. *bodkin*: a needle-like instrument with a blunt knobbed point and a large eye, used for threading cord or ribbon through hems or loops.

4. *Sartor Resartus*: cited letter 0398, n. 24.

5. *Social Aspects*: cited letter 0398, n. 12.

6. '____ *Why should I roam . . .*': Emerson, 'The Day's Ration', *Poems*, p. 177, lines 29–31. The blank on line 29 would have read: 'My apprehension?'

7. *But man crouches and blushes . . .*: Emerson, 'The Sphynx', *Poems*, p. 3, lines 49-56.

8. '*Erect as a sunbeam . . .*': Emerson, *Poems*, p. 2, lines 17–20.

9. '*These temples grew as grows the grass*': Emerson, 'The Problem', ibid., p. 9, line 45.

10. '*Is there no god then, . . .*': *Sartor Resartus* (cited letter 0398, n. 24), p.167.

11. *clockmaker does to the clock*: This was a reference to Paley's watchmaker analogy from his *Natural Theology*.

12. *blood and bones of Jehovah*: Tyndall was expanding on his vision of Carlyle as a pantheist.

For an explanation of Tyndall's interpretation of Carlyle, see R. Barton, 'John Tyndall, Panthiest: A Rereading of the Belfast Address', *Osiris*, 2nd Series, 3 (1987), p. 111–34.

13. *Rhea*: Emerson, 'To Rhea', *Poems*, pp. 13–16.

14. *'Thou shalt seem in each reply . . .'*: *Poems*, p. 14, lines 19–20.

15. *World Soul*: Emerson, 'The World-Soul', *Poems*, pp. 21–25.

16. *'For Gods delight in Gods . . .'*: *Poems*, p. 25, lines 93–96.

17. *'Gay'*: Martin Gay, a Harvard classmate of Emerson who was the subject of several poems, sometimes described as erotic.

18. *'the winds are . . .'*: a misquoted passage from Emerson, *Nature* (Boston: James Munroe and Company, 1836), p. 25. The quote from Emerson reads, '"The winds and waves', said Gibbon, 'are always on the side of the ablest navigators'". Emerson quoted from Edward Gibbon, *The History and Decline and Fall of the Roman Empire* (London: Strahan & Cadell, 1776–1789).

19. *Waitz*: see letter 0400, n. 2.

20. *'The Rhodosa'*: Emerson, 'The Rhodora', *Poems*, p. 44.

21. *'These temples grew as grows the grass'*: see n. 9.

22. *'I never thought to ask . . .'*: Emerson, 'The Rhodora', p. 44, lines 15–17.

23. *3 of Woodnotes:* 'Wood Notes I', Emerson, *Poems*, p. 53, line 79.

24. *'For ever nature faithful is . . .'*: *Poems*, p. 56, lines 149–50. Tyndall misquoted the passage that reads, 'For nature ever faithful is'.

25. *Woodnotes*: Emerson, 'Wood Notes II', *Poems*, pp. 57–72.

26. *'The primal mind . . .'*: ibid., p. 68, lines 293–94.

27. *'I, that to day am a pine . . .'*: ibid., p. 70, lines 337–38.

28. *'Halteth never in one shape . . .'*: ibid., p. 70, lines 333–34.

29. *'Monadnoc'*: Emerson, 'Monadnoc', *Poems*, pp. 73–90.

30. *'I await the bard and sage . . .'*: ibid., p. 85, lines 305–7.

31. *Page 87*: ibid., p. 87.

32. *similar thought . . . my passage from Rotterdam*: see Journal, 18 June 1850 (JT/2/13b/417) for similar musing during his trip from Rotterdam. -

33. *I sent to the Chronicle*: published as Wat Ripton, 'Day-Book Splinters', *Preston Chronicle* (13 July 1850), p. 3. In it, Tyndall described his recent journey from Frankfurt to London.

34. *Ode to Channing*: Emerson, 'Ode', *Poems*, pp. 92–96.

35. *Things are in the saddle . . .*: ibid., p. 94, lines 50–51.

36. *If she be not fair to me . . .*: from G. Wither, 'Shall I wasting in despair', *Fidelia* (London: Nicholas Okes, 1619), lines 7–8.

37. *'Give all to love,'*: Emerson, 'Give All to Love', *Poems*, pp. 111–13.

38. *Leave all for love . . .*: ibid., pp. 112–13, lines 26-49.

39. *I shall be there . . . 25th*: a reply to Hirst's insistent invitation (letter 0412).

40. ~~*Don't think . . . as ever*~~ †: Tyndall crossed this out with only a single line. When he did not want a correspondent to read something he crossed it out thoroughly, with a looping scrawl. Hirst responded in letter 0414.

41. *Carlyle's pamphlet*: 'Hudson's Statue', discussed in letter 0412 (cited n. 4).

42. *'And then swung as a tragic pendulum. . .*: ibid., p. 232.

43. *'Merlin'*: Emerson, 'Merlin I', *Poems*, pp. 143–46.

44. *Nor profane effect . . . inclined*: ibid., p. 146, lines 66–69.

45. *Bacchus*: Emerson, 'Bacchus', *Poems*, pp. 149–52.

46. *That I intoxicated*: ibid., p. 150, lines 21–25.

47. *'The poor grass shall plot and plan . . .'*: ibid., p. 151, lines 41–42.

48. *'Reason in nature's lotus* drenched': ibid., p. 152, line 56.

49. *'the House'*: Emerson, 'The House', *Poems*, pp. 155–56.

50. *She lays her beams in music*: ibid., p. 156, lines 17–18.

51. *Saadi*: Emerson, 'Saadi', *Poems*, pp. 156–63.

52. *'Flee from the gods which from thee flee'*: ibid., p. 162, lines 143–44.

53. *essays I think mention an occult relation . . .*: Emerson, *Nature* (see n. 18), p. 13.

54. *'We coldly ask their pottage not their love'*: Emerson, 'Blight', *Poems*, pp. 178–80, line 39. The full passage reads, 'We devastate them unreligiously, And coldly ask their pottage, not their love'.

55. *vortrefflich:* excellent or superb (German).

56. *Nun mein Freund Sie werden [sich] heute gut finden*: 'Well my friend, you will find yourself well today' (German).

57. *'Musketaquid'*: Emerson, 'Musketaquid', *Poems*, pp. 181–84.

58. *The polite found me impolite. . .*: ibid., p. 183-4, lines 67–70.

59. *Canst thou shine now then darkle*: ibid. p.184, lines 79–80. 'Admirable' is Tyndall's addition.

60. *Threnody*: Emerson, 'Threnody', *Poems,* p. 188–99.

61. *'Hast though forgot me in a new delight?'*: ibid., p. 189, line 35.

62. *O child of paradise*: ibid., p. 194-5, lines 166–75.

63. *'And [thought*ˢᵗ*] thou that such a guest. . .'*: ibid., p. 197, lines 224–29.

64. *'Not of adamant or gold'. . .*: ibid., p. 199, lines 272–75.

65. *lady socialist lecturer*: Tyndall heard the notorious Emma Martin twice in 1844 (Journal, 1 and 3 September 1844, JT/2/13a/60–1).

66. *'singing an eternally Holy! Holy!'*: Martin alluded to biblical accounts of heaven; compare *Isaiah* 6:3 and *Revelation* 4:8.

67. *Mʳ Hirst . . . Yorkshire*: address from envelope.

# From Thomas Archer Hirst	[18]¹ July 1850	0414

Halifax | July 12th 1850

My dear Tyndall—

I have just returned from my lads at the Improvement Society,² and as I generally do, I can yet feel its healthy, refreshing influence; since then I have had a stroll on the moor³ with my eldest lad, & been listening to his rude & stubborn speculations on things in general. He is an eccentric lad, with some eccentric & crude notions, at the bottom of all which however is a sturdy individuality and a brave self-reliance. The rascal has some poetry in him too, and some faint notion of what we should call religion but he has not any suspicion of its being such. He carries a copy of Burns[4] in his pocket generally and has been known to get up at 2 AM and disappear until noon on to Beacon Hill[5] & so forth, at which spots he has been found asleep mesmerized by the scenery perhaps though what is the best & shews a good deal of this to be genuine, he seldom talks about it, for he is too blunt to put up with affectation, & no doubt bears it a great hatred. When I meet such fellows I am always impatient to see them develop themselves at a quicker rate than they seem to be doing, and have accordingly to keep crying patience to myself, for I well know Nature does not like hurrying: still when I see capacities so valuable for making a fine man, I grudge old Nature the time she needs to manufacture them. I liked your long spasmodic letter[6] very well, why did you consider to burn it you rascal, did you fear my criticism or think that anything you had to say would not be welcomed? But in truth some such sly notion affects us all many a time, I have found it I know. A certain fear of compromising or detracting from an opinion you consider that others possess towards you: a certain hesitation at saying or writing what you may consider commonplace, to one who you know could appreciate something better. But as we do not expect others always to be talking brilliantly neither should we always attempt, nay what is more we should never attempt it, for it is by the <u>commonplaces</u> and not by the <u>brilliances</u> that we shall all be judged. Just as it is not by a few notions (good or bad) but by the generality that we must be estimated and estimate. You have read Emerson well, I have not, & what you say shews this. There is only one remark (& that you scratched out intending me not to see); which I could have wished unwritten because it is useless. 'Don't think too much of the first part of this letter'. Why John? Do you anticipate harm thereby? and if so to whom? Yourself or me?—Now I <u>shall</u> think just what I like about them, and to punish you I shall think much of them, so much namely, that they are <u>plain</u> John Tyndall's ideas on the matter; that the <u>superfine</u> John may have some more elaborate ones in his cranium does not matter to me until they get out. I shall please myself whether I give him credit for them or not. You say[7] truly 'Emerson cannot be an unhappy man. Nature can't play <u>him</u> any tricks' yet for all that I can't envy him yet I see something beautiful in such a state but it is a cold beauty; too impassible for my taste of 20 years growth. There is something grand about such a <u>balanced</u> man but I sigh sometimes when I think it must be purchased at the price of these impulsive ecstacies. Nevertheless I feel myself journeying

<Handwritten letter ends; hereafter LT Transcript Only>
in that direction, and at every step I sigh for the beautiful unrealities I

must bid adieu to. I see Emerson, Carlyle and yourself (perhaps more besides if I looked carefully) all on the road before me there at different distances; and I find I must follow also. Why is it, Tyndall? Is it because a love of the Infinite is gradually absorbing and seemingly drowning the finite: a love for the Ideal Whole to which the Actual Part must succumb?

> Leave all to love
> Yet hear me yet
> . . .
> . . .
> Keep thee to-day
> To-morrow for ever
> Free as an Arab
> Of thy Beloved.[8]

Is this possible to Man? You say[9] 'it is politic', and that may be true when at the same time the freedom may be <u>assumed</u> only, not real. But the freedom meant seems to me as if purchased at the sacrifice of Love. True it is a wiser love, but I could not yet part with the embodiment in Human Shape without sadness. A thought strikes me Emerson could not either; but despite the sadness he <u>would</u> do it. So would I. - - The sadness would add to its value. I should like to hear your personal illustration of this, so when you get old <u>enough</u> don't forget to tell me. The sorrow of Emerson and his conquering it, or the result of it, is evidenced in the Threnody.[10] I have been reading to-day a sad but beautiful poem of Longfellow's, 'Evangeline'.[11] The metre, 'Hexameter', he has moulded to his purpose beautifully; despite its unwieldiness in English I confess to like it, or rather his use of it. There is a certain musical plaintiveness throughout, like AEolian harp melody made articulate. I cannot pick out any that shall represent the poem, for it is the <u>whole</u> that you admire, but the concluding lines I give you on the small slip enclosed. Evangeline the heroine has been banished with her father, betrothed, and his father, from their native land in Acadia (it is founded on fact). Her father dies before their embarkation, her betrothed is put into a different ship and they are landed at different parts of the American coast; when there, homeless and friendless, she sets out on her weary pilgrimage to find him, often escapes him narrowly, and thus through her sad pilgrimage you are carried until she gets reconciled to her failures, turns a Sister of Mercy,[12] and loses her own afflictions by trying to soothe those of others. At last in an Infirmary she sees her betrothed, now old and dying in a fever; he dies without speaking to her on her bosom. (Turn over)[13]

Come early on the 25th. | T.A. Hirst.

All was ended now, the hope, and the fear, and the sorrow,
All the aching of heart, the restless, unsatisfied longing,
All the dull, deep pain, and constant anguish of patience!
And, as she pressed once more the lifeless head to her bosom,
Meekly she bowed her own, and murmured, 'Father I thank thee'.[14]

Mr J. Tyndall | Spring Bank, Over Darwen, Lancs.[15]

RI MS JT/1/H/149
RI MS JT/1/HTYP/91–92

1. *[18]*: the date given by Hirst (July 12th) is inconsistent with the date for letter 0413, to which this is a reply. Moreover, it is clear from his journal (entries of the 11th and 18th) that the walk on the moor took place on the 18th. The letter was possibly finished early on the morning of the 19th.
2. *Improvement Society*: Halifax Mutual Improvement Society, where Hirst taught a class on Thursday evenings (see letter 0393 and 0398).
3. *moor*: Skircoat Moor (see letter 0617).
4. *copy of Burns*: Scottish poet, Robert Burns.
5. *Beacon Hill*: a hill that overlooks Halifax from the East and the site of the town's beacon (see letter 0617).
6. *your long spasmodic letter*: letter 0413.
7. *you say:* letter 0413. Hirst does not quote precisely.
8. *Leave all to love . . . Beloved*: Hirst repeats (inaccurately) the quote of letter 0413 (at n. 38).
9. *you say . . . politic*: letter 0413.
10. *Threnody:* 'Threnody', in *Poems*, 236–49.
11. *'Evangeline'*: an epic poem published in 1847.
12. *Sister of Mercy*: a female member of a religious order; also a sisterhood founded in Dublin in 1827 (*OED*).
13. *(Turn over)*: this note makes no sense in its position on the LT transcript but we assume that, on the missing manuscript, it indicated that the reader had to turn the page for the lines of poetry.
14. *All was ended now. . . I thank thee.*: 'Evangeline', lines 1376–80.
15. *Mr. . . . Lancs*: presumably from the envelope.

From John Phillips   18 July 1850   0415

18 July 1850 | 4 Castle St | Edinburgh

Sir

Your notice of a paper on the Magneto-Optical properties of Crystals[1] has been handed to me by Prof. Kelland.[2] As the paper is printed you wish of course to give an Abstract of it perhaps to illustrate by diagrams or experiments. To <u>read</u> a <u>printed</u> Memoir would be somewhat out of course. I advise you to make an abstract of the main points[3] & to prepare some illustrations if needed. The subject is curious & there will be persons here worthy of some pains being taken to make the paper satisfactory. It is <u>better</u> to send it before the Meeting,[4] but not absolutely demanded[5]

Yours very truly | John Phillips | Assist[t] Gen[l] Secretary

RI MS JT/1/P/82

1. *paper on . . . Crystals:* his recent paper, co-authored with Knoblauch, published in the *Phil. Mag.* (cited letter 0403, n. 2).
2. *Prof. Kelland:* Philip Kelland (1808–79), Professor of Mathematics at the University of Edinburgh, was one of the local secretaries for the 1850 BAAS meeting.
3. *I advise . . . the main points:* Tyndall followed this advice (Journal, 23 July, JT/2/13b/503).
4. *Meeting:* The BAAS meeting held, 1–7 August, in Edinburgh.
5. *It is <u>better</u> . . . demanded:* Tyndall did send a copy of his paper to Phillips before the meeting (Journal, 7 August, JT/2/13b/503).

From Michael Faraday   19 July 1850   0416

Upper Norwood,[1] 19 July, 1850

Dear Sir,

I am very much obliged to you for the specimens of calcareous spar,[2] and I shall very likely turn to the matter again some day soon. At present I am working upon another part of the great subject, and withal am out here for a little change of air and rest.

I am, my dear Sir, | Very truly yours, | M. Faraday.

J. Tyndall Esq. | &c. &c. &c.

JT/1/TYP/12/4125

1. *Upper Norwood:* a salubrious area of southeast London.
2. *calcareous spar:* calcite; a cleavable and lustrous mineral partly composed of calcium carbonate.

From Karl Hermann Knoblauch   21 July 1850   0417

Marburg, 21 July 50

Verehrter Herr Tyndall,

Den innigsten Dank sage ich Ihnen für alle Nachrichten welche Sir mir in Ihrem Briefe v. 3.ⁿ [o.] (den ich vor 8 Tagen empfing) zukommen lassen. Wir haben vom <u>Stauffenberg</u> an dem Tage, an welchem ich Sie begleitet hatte, eine fabelhafte Rückfahrt gehabt. Die Postillone waren nämlich in dem Grade betrunken, dass sie nicht reiten konnten, Neumann den einen Wagen fahren musste, und die beiden andern völlig unbenutzbar wurden. Sie können nun denken, wie sich auf dem einzigen Rettungsboot die ganze Schiffbruch leidende Mannschaft zu vereinigen suchte, so dass auf dem einen für 15 Personen berechneten Leiter-Wagen schliesslich 33—ich kann nicht sagen: Platz, aber doch eine Art von Unterstützung ihres Schwerpuncts—gefunden hatten. Obgleich <u>nicht</u> überzeugt, dass in einen Raum, in dem sich schon 18 Ueberzählige befinden, nicht auch noch beliebig mehr aufgenommen werden könnten, musste sich ein Theil der Gesellschaft entschliessen, bis zur Neh-Brücke zu <u>Fuss</u> zu <u>gehen</u>, wo sich endlich Gelegenheit zu fahren darbot. Die Uebrigen aber gelangten nachdem sie mehrmals gegen Stein-Haufen und in den Chaussee-Graben gefahren, einen entgegenkommenden Omnibus (bei der Unmöglichkeit auszuweichen) bedeutend beschädigt hatten, in der beständigen Gefahr umzuwerfen, bei stockfinsterer Nacht um 2^h nach Marburg.

Ich habe mich sehr gefreut zu hören, dass Sie Faraday's Bekanntschaft gemacht haben, und es ist mir sehr lieb, ein Stück seines <u>schweren Glases</u> zum Behuf der electrisch optischen Erscheinungen zu erhalten. Von meinem Vetter aus Bonn erfahre ich, dass Plücker unerschütterlich an seiner Theorie festhält und die Absicht hat, uns auf eine <u>mathematische Abhandlung</u> zu verweisen, welche nächstens in <u>Crelle's Journal</u> erscheinen wird, und worin er die Kräfte so definiren will, dass sie zur Erklärung aller Erscheinungen ausreichen. Attendons! Mir kommt das Festhalten an der optischen Axe bei diesen Wirkungen so vor, als ob man mit den Newton'schen 'fits of light' gegen die Undulationstheorie ankämpfen wollte.—

Plücker hat (wie mir mein Vetter <u>vertraulich</u> mittheilt) u. U. 2 Experimente angestellt, welche—seiner Ansicht nach—uns widerlegen: 1) habe ein—mit der <u>Axe vertical</u> gehängter Kalkspath nicht 3, sondern <u>4</u> Gleichgewichts-Lagen. 2) Von 2 <u>magnetischen</u> Kalkspathen stellte sich der eine mit der optischen Axe <u>axial</u>, der andre <u>äquatorial</u>; von 2 <u>diamagnetischen</u> Kalkspath<en> der eine <u>äquatorial</u>, der andre <u>axial</u>.—

Ich habe starken Verdacht dass es sich bei dem 1ⁿ Exp. um eine <u>lokale Wirkung</u>, bei dem 2ⁿ um einen Einfluss der <u>vielleicht nicht ganz reinen Oberfläche</u> handelt. Aber gesetzt auch, es wäre dies bei dem 2ⁿ Exp. nicht der Fall, so ist es sehr leicht möglich dass bei der Beurtheilung z.B. des Magnetismus aus der <u>Anziehung</u> von <u>einem Pol</u> bei gemischten Substanzen (wie der magnet. Kalkspath ist) die <u>magnetische</u> Resultante überwiegen kann u. doch, beim Aufhängen zwischen den flachen Polen, die <u>diamagnetische</u> (Entsprechend dem sogenannten: eisenfreien Turmalin etc.) In allen diesen Fällen kann nur die chemische Analyse Aufschluss geben, indem sie die Verhältnisse angiebt, in denen die magnetischen u. diamagnetischen Bestandtheile mit einander gemischt sind.

Mitscherlich hat mir seine Abhandlung aus den Academie-Berichten geschickt. Sie enthält aber nichts wesentlich Verschiedenes von dem, was wir bereits aus Poggend. Annalen kennen. Gauss soll (wie ich von Bunsen höre) Winkel an Krystallen bis auf Bruchtheile von Secunden gemessen haben. Aber er so wenig wie Gerling (Gauss's Schüler) konnten mir etwas über die Methode (welche nicht veröffentlicht worden ist) mittheilen.) Es wäre vortrefflich, wenn Sie auf Ihrer Rück-Reise über Göttingen gehen und sich dort über die Sache orientiren könnten. Ich wäre gern bereit die Mittel zur Anschaffung der Instrumente herzugeben, falls Sie die Ueberzeugung gewinnen, dass wir damit sicher die uns zunächst liegenden Ausdehnungsversuche (namentlich mit Schwerspath, Turmalin, Bergkrystall etc) ausführen könnten.—

Ich habe meine Dreh-bank aus Berlin kommen lassen, welche beim Bearbeiten der Krystalle eine sehr wesentliche Erleichterung darbietet. Mein Bediente<r> (den ich seit July angenommen habe) weiss bereits damit umzugehen und verfertigte mir neulich auf derselben einen Cylinder von Schwerspath innerhalb 10 Minuten. Mein Vater hat mir eine Wheatstonesche Polar-Uhr zum Geschenk gemacht. Sie kennen gewiss dieses sinnreiche Instrument: eine Sonnenuhr ohne Sonnenschein, eine Sonnenuhr auch bei Nacht.

Von Plücker ist im letzten Heft der Annales de Chimie et Physique eine lange Abhandlung erschienen; aber sie enthält für uns nichts Neues.

Die Formel, von welcher Dove seine Methode abgeleitet hat, das posit od. negat. Verhalt. der Krystalle zu erkennen ist die von Airy (Pogg. Ann. XXIII p. 228—Jahr 1831. —Transact. of the Cambridge Philos. Society):

$$I = \frac{c^2}{2}\left\{1 + \cos 2\beta \cdot \cos 2\varphi \cdot \cos 2(\beta + \alpha + \varphi) + \cos\frac{2\pi}{\lambda}\theta \cos 2\beta \sin 2\varphi \sin 2(\beta + \alpha + \varphi) - \sin\frac{2\pi}{\lambda}\theta \sin 2\beta \sin 2\varphi\right\}$$

worin: I die Intensität des Lichts;

$$\theta = T\left(\frac{a^2 - b^2}{?h}\right)\vartheta^2$$

ϑ, Einfallswinkel des Lichts auf die Krystallplatte,

$\frac{1}{a}$, [Brechungsverhältniss] des ungewöhnl. Strahls

$\frac{1}{b}$, " " " " " " " gewöhnlich. " " "

T, Dicke der Platte.

$c_\sim^2$, Intensität der einfallenden Strahlen.

$\alpha_\sim$, Winkel der Reflexions-Ebene des analysirenden Spiegels
    mit der   "   "   "   "   " polarisirenden   "   " .

β, das Azimuth des, zwischen dem polarisirend. Spiegel u. der Krystallplatte eingeschalteten, Fresnel'schen Parallelepiped. (β, v. der *[1ⁿ]* Polarisat.
Ebene an *[gezählt]*)

$\varphi_\sim$, Winkel, den der *[Hauptschnitt]* des Krystalls, (gelegt durch die Richtung der einfallenden Strahlen) mit der Polarisations-Ebene bildet.

$\lambda_\sim$, Wellenlänge in der Luft für eine bestimmte Farbe.
    Setzt man in der allgemeinen Formel $\beta=45°$, so wird:

$$I_\sim = \frac{c^2}{2}\left\{1-\sin\frac{2\pi}{\lambda}\theta\sin 2\varphi\right\}$$

$$\sin\beta=135° \text{ wird } I = \frac{c^2}{2}\left\{1+\cos\frac{2\pi}{\lambda}\theta\sin 2\varphi\right\}$$

Diese Werthe von *I* stellen die Intensität des Lichtes dar wenn dasselbe
(z.B. durch ein Glimmerblatt) circular polarisirt ist, durch eine Krystallplatte
hindurchgeht und alsdann linear analysirt wird

Für das Experiment stellen sich im circular polarisirten Lichte, statt der
einfachen Ringe, welche—auf gleiche Weise bei negativen u. posit. Krystallen
im linear polarisir. Lichte sich darstellen. - discontinuirliche Kreise: wie:

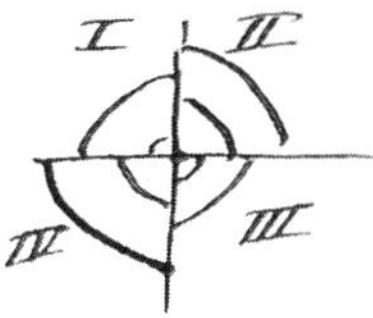

oder

dar; je nachdem der Krystall negativ od. posit. ist.
    Stellt z.B. ein zu prüfender Krystall (einaxig.), vertical *[geschnitten]*,
gegen die Axe, diejenige dieser Figuren dar, welche auch Kalkspath zeigt, so
ist er negativ; stellt er die des Bergkrystall dar, positiv.

Bei optisch 2 axigen Krystallen lassen sich die Ringtypen negativer u. positiver Krystalle im linear polarisirten Lichte ebensowenig wie die der einaxigen von einander unterscheiden. —Im circular polarisirten Lichte erhält man Figuren wie:

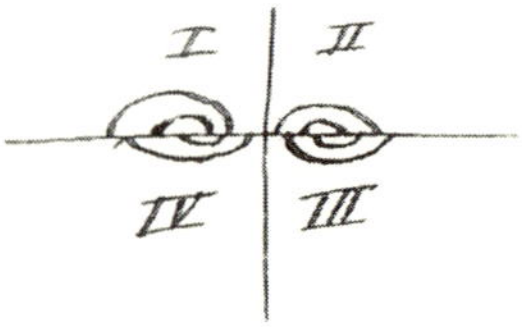

oder:

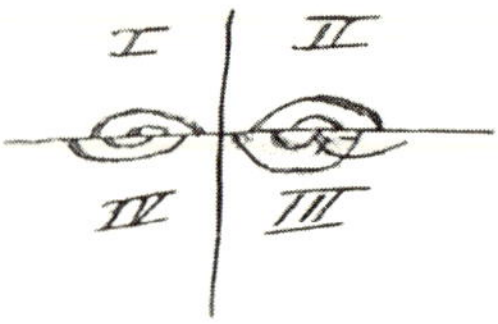

(worin immer I und III sich II <und> IV entsprechen) je nachdem d. Krystall negativ od. positiv ist. Stellt z.B. ein zu prüfender Krystall (2 axig) vertical geschnitten gegen die Mittellinie, diejenige dieser Figuren dar, welche auch Glimmer oder Arragonit zeigt, so ist er negativ, stellt er die Figur des Topases dar, positiv.

Ich habe hier die Hauptsachen mitgetheilt, das Nähere findet sich im Original: Pogg. Ann. XL., p.457 für einaxige; p.482 für 2 axige Kryst. Der Beschreibung des Experiments bei den letzteren ist keine Formel hinzugefügt worden. Die Airy'schen sind, meines Wissens, überhaupt nicht weiter als auf einaxige Krystalle ausgedehnt worden.

In diesen Tagen habe ich eine neue Beobachtung hinsichtlich der strahlenden Wärme gemacht, nämlich dass dieselbe—wie ich vermuthete—den Turmalin längs der Axe in geringerem Verhältniss als nach andern zu durchdringen vermag.

Die unter unser beider Namen erscheinende Abhandlung über die letzte Versuchs-Reihe habe ich leider noch immer nicht fertig machen können. Ich bin jetzt gar zu sehr beschäftigt; indess ist meine Zeit von der nächsten Woche an freier und ich kann alsdann ungestörter daran arbeiten. —Am Mittwoch hielt ich meinen Antritts-Vortrag in der Naturforschenden Gesellschaft.

Fräulein <u>Fulda</u> hat sich mit meinem Collegen Prof. <u>Vorländer</u> verlobt. —Sie selbst sind Allen hier in bestem Andenken u. Alle werden sich sehr freuen, Sie hoffentlich bald wieder zu sehen.

Ganz besonders wünscht dies von Herzen

Ihr treuer Freund | Herm. Knoblauch.

Ich war ganz überrascht zu hören, dass meine Arbeiten über Wärme ins Englische übersetzt *[worden]* sind. Ich kann nicht zweifelhaft sein, dass dies auf Ihre Veranlassung geschehen ist und sage Ihnen meinen besten Dank dafür. Gestern machten wir eine Parthie nach d. Frauenberg.   HK.

Marburg, 21 July 50

Dear Herr Tyndall,

I give you my deepest thanks for all the news which you send me in your letter of the 3rd *[inst.]*[1] (which I received 8 days ago). We had a splendid trip back from <u>Stauffenberg</u>[2] on the day I accompanied you. The coachmen were drunk to the degree that they were unable to ride; Neumann had to drive one carriage, and the other two became completely unusable. You can just imagine how the whole ship-wrecked crew sought to assemble in the single lifeboat, with the result that on a cart meant to take 15 people, 33 eventually found—I cannot say: a place—but rather a kind of support for their centre of gravity. Although <u>not</u> convinced that any number more could still not be accommodated in a space in which there were already 18 people in excess of the limit, part of the company went and decided to g<u>o</u> by <u>foot</u> as far as Nehbrücke,[3] where the opportunity of driving finally presented itself. The rest, however, after driving several times into piles of stones and into roadside ditches, seriously damaging an oncoming omnibus[4] (there being no possibility of getting out of the way), and being in constant danger of overturning, eventually reached Marburg in pitch-dark night at 2 o'clock.

I was very pleased to hear that you have made Faraday's acquaintance, and I am delighted to receive a piece of his <u>heavy glass</u>[5] with a view to the electrical optical phenomena. I hear from my cousin in Bonn that Plücker is holding unshakably to his theory,[6] and intends to refer us to a <u>mathematical paper</u> which will appear shortly in <u>Crelle's Journal</u>,[7] and in which he will define the forces in such a way that they will suffice to explain all the phenomena. Attendons! Holding fast to the optical axis with these effects seems to me as if one wanted to do battle with Newton's "fits of light" against wave theory.[8]—

Plücker has (as my cousin <u>confidentially</u> informs me) possibly done 2 experiments[9] which—in his view—contradict us: 1) a calcite crystal with the axis suspended <u>vertically</u> does not have 3, but <u>4</u> equilibrium positions. 2) of 2 <u>magnetic</u> calcite crystals, one sets <u>axially</u> with the optical axis, the other <u>equatorially</u>; and of 2 <u>diamagnetic</u> calcite crystals one <u>equatorially</u>, the other <u>axially</u>.—

I have a strong suspicion that with the 1st exp. it is a question of a <u>local effect</u>, and with the 2nd of an influence of a <u>perhaps not completely clean surface</u>. But assuming also that was not the case with the 2nd exp., then it is very easily possible that with the finding, e.g., of magnetism from the <u>attraction</u> of <u>one pole</u> in mixed substances (as magn. calcite crystal is), the <u>magnetic</u> resultant can predominate, and yet, when suspended between flat poles, the <u>diamagnetic</u> (corresponding to the so-called: iron-free tourmaline etc.) does. In all these cases, only <u>chemical analysis</u> can give information, in that it gives the <u>proportions</u> in which the <u>magnetic</u> and <u>diamagnetic</u> constituents are mixed with one another.

Mitscherlich has sent me his paper from the Academie-Berichte.[10] However, it does not contain anything substantially different from what we already know from <u>Poggendorff's</u> Annalen. <u>Gauss</u> is supposed (as I hear from Bunsen) to have measured the angles on crystals to fractions of seconds. But neither he nor Gerling (Gauss's pupil) could tell me anything about the <u>method</u> (which has not been published). It would be excellent if you could go via Göttingen on your trip back and while there put yourself in the picture about the matter. I would be quite happy to put up the money to purchase the instruments, <u>in the event you become convinced</u> that we could <u>definitely</u> carry out with them the expansion experiments we need to do first (particularly with <u>baryte, tourmaline, quartz</u> etc).—

I have had my lathe sent from Berlin, which makes things very considerably easier when machine-working crystals. My servant (whom I have employed since July)[11] already knows how to use it and recently made me a cylinder of baryte on it within 10 minutes. My father has given me a <u>Wheatstone polar-clock</u>[12] as a present. You will certainly know this ingenious instrument: a sundial without sunshine, a sundial even at night.

A long paper by Plücker has appeared in the <u>latest</u> issue of Annales de Chimie et Physique;[13] but it contains nothing new for us.

The <u>formula</u>, from which Dove has derived his method of identifying the <u>posit.</u> or <u>negat. behaviour of crystals</u> is that of <u>Airy</u> (Pogg. Ann. XXIII p. 228—year 1831[14]—<u>Transact.</u> of the Cambridge Philos. Society)[15]:
in which: I is the intensity of light;

$$I = \frac{c^2}{2}\left\{1 + \cos 2\beta \cdot \cos 2\varphi \cdot \cos 2(\beta + \alpha + \varphi) + \cos\frac{2\pi}{\lambda}\theta \cos 2\beta \sin 2\varphi \sin 2(\beta + \alpha + \varphi) - \sin\frac{2\pi}{\lambda}\theta \sin 2\beta \sin 2\varphi\right\}$$

$$\theta = T\left(\frac{a^2 - b^2}{2b}\right)\vartheta^2$$

ϑ, angle of incidence of light onto the crystal plate,

$\frac{1}{a}$, [refractive ratio] of the extraord. ray

$\frac{1}{b}$, " "   "   ordinary   "

T, thickness of the plate.

c^2, intensity of the incident rays.

α, angle of the reflection plane of the analysing mirror
  with " " " " " polarising " .

β, the azimuth of the Fresnel parallelepiped inserted between the polaris. mirror and the crystal plate. (β being counted from the [1ˢᵗ] polarisat. plane onwards)

φ, angle which the [plane] of the crystal (<u>laid through</u> the direction of the incident rays) forms with the polarisation plane.

λ, wavelength in air for a specific colour.
  If one inserts $\beta=45°$ in the general formula, one obtains:

$$I = \frac{c^2}{2}\left\{1 - \sin\frac{2\pi}{\lambda}\theta\sin2\varphi\right\}$$

$$\sin\beta = 135° \text{ obtains } I = \frac{c^2}{2}\left\{1 + \cos\frac{2\pi}{\lambda}\theta\sin2\varphi\right\}$$

These values for *I* represent the intensity of light when it is circularly polarised (e.g., by using a sheet of mica), goes through a crystal plate, and is then analysed linearly.

For the <u>experiment</u>, discontinuous circles like

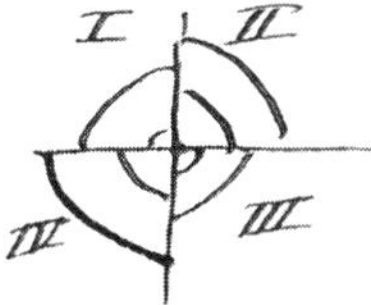

or

appear in <u>circularly polarised</u> light, depending on whether the crystal is negative or posit., <u>instead of simple rings</u>, which appear in the <u>same</u> way with <u>negative</u> and <u>posit.</u> crystals in <u>linearly</u> polarized light.

If, for example, a crystal to be tested (<u>uniaxial</u>), [cut] vertically against the axis, displays those of these figures which <u>calcite</u> also shows, then it is <u>negative</u>; if it displays those of <u>quartz</u>, <u>positive</u>.

With optically <u>2-axial</u> crystals, the <u>types of rings</u> of <u>negative</u> and <u>positive</u> crystals can be distinguished from one another in lineally polarised light just as little as those of uniaxial ones can. —In <u>circularly polarised</u> light one obtains figures like:

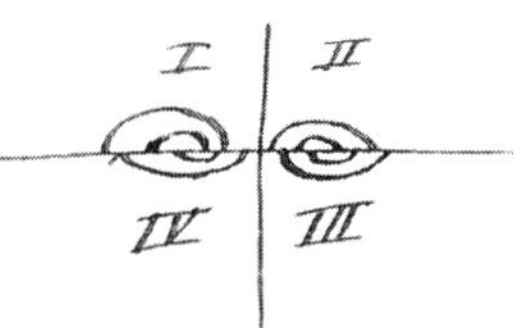

or:

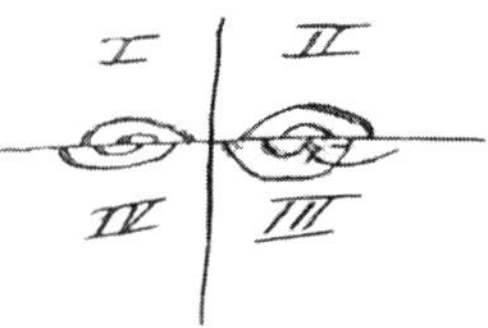

(in which I and III always correspond to II <and> IV), depending on whether the crystal is negative or positive. If, for instance, a crystal to be tested (<u>2-axial</u>), cut vertically against the middle line, displays those of these figures which mica or <u>aragonite</u> also show, then it is <u>negative</u>, if it displays the figure of <u>topaz</u>, <u>positive</u>.

I have communicated the main points here; further details can be found in the original: Pogg. Ann. XL., p. 457 for <u>uniaxial</u> cryst<als>; p. 482 for <u>2-axial</u> ones.[16] No formula has been appended to the description of the experiment with the latter. Airy's ones have, to my knowledge, not been taken any further at all than uniaxial crystals.

The other day I made a new observation with regard to <u>radiant heat</u>, namely that it—as I suspected—is able to penetrate <u>tourmaline</u> along the axis at a <u>lower</u> rate than in other directions.

I have unfortunately still not been able to finish the paper about the last series of experiments which is appearing under both our names[17]. I am much too busy now; however, my time will be freer from next week onwards and I can then work on it with less interruption. —On Wednesday I gave my inaugural lecture in the Naturforschende Gesellschaft.—

Fräulein <u>Fulda</u>[18] has got engaged to my colleague Prof. <u>Vorländer</u>.[19]—Everyone here has the fondest memories of you and everyone will be happy to see you again, hopefully soon.

With the most heartfelt wishes for this
Your loyal friend | Herm. Knoblauch.

I was quite surprised to hear that my works on heat have *[been]* translated into English.[20] I can be in no doubt that this has happened at your instigation and give you my best thanks for this. Yesterday we went on a trip to the Frauenberg.[21] HK.

RI MS JT/1/K/14

1. *letter of the 3rd [inst.]*: letter missing.
2. <u>*Stauffenberg*</u>: misspelling of Staufenberg, a town in the district of Gießen, Hesse, Germany along the Lahn River.
3. *Nehbrücke*: situated about 6 miles south of Marburg along the Lahn River.
4. *omnibus*: a large public vehicle carrying passengers by road, running on a fixed route and typically requiring the payment of a fare (*OED*).
5. *heavy glass*: this refers to the glass Faraday used to discover the Faraday effect. It contained traces of lead also known as boro-silicate of lead.
6. *his theory*: see letter 0395, especially n. 20.
7. <u>*mathematical paper*</u> . . . *Crelle's Journal*: this paper never appeared. Plücker may have been referring, not to his own paper, but to W. Thomson and G. Green, 'An Essay on the Application of mathematical Analysis to the theories of Electricity and magnetism', *Journal für die reine und angewandte Mathematik* 39 (1850), pp. 73–89.
8. *Newton's "fits of light" against wave theory*: Newton believed in the particle nature of light. Yet, his 'theory of fits' seemed compatible with (indeed, supportive of) a wave theory as well.
9. *Plücker has (as my cousin <u>confidently</u> informs me) possibly done 2 experiments*: not identified.
10. *Mischerlich . . . Academie-Berichte*: Eilhardt Mischerlich's work, but the paper is not identified.
11. *My servant (whom I have employed since July)*: not identified.
12. *Wheatstone polar-clock*: Wheatstone presented his instrument to the BAAS meeting in 1848 in a paper entitled, 'On a means of determining the apparent Solar Time by the Diurnal Changes of the Plane of Polarization at the North Pole of the Sky'. It was published in *Brit. Assoc. Rep.*, 1848 (pt. 2), pp. 10–12.
13. *Long paper by Plücker. . . Annales de Chimie et Physique*: J. Plücker, 'Sur le Magnétisme et le Diamagnétisme', *Annales de Chimie et de Physique, Ser 3, T29* (1850), pp. 129–60.
14. *Pogg. Ann. XXIII p. 228—year 1831*: G. B. Airy, 'Ueber die Natur des Lichts in den beiden

durch die Doppel brechung des Bergkrystalls hervorgebrachten Strahlen', *Poggend. Annal.* 23, no. 10 (1831), pp. 204–81.

15. *Transact. . . . Philos. Society*: G. B. Airy, 'On the Nature of the Light in the two Rays produced by the Double Refraction of Quartz,' *Cambridge Phil. Soc. Trans.*, 4 (1831), pp. 79–123, 199–208.

16. *Pogg. . . . crystals:* H. W. Dove, 'Ueber den Unterschied positive und negative einaxiger Krystalle bei circularer und bei elliptischer Polarization,' and 'Erscheinungen an zweiaxigen Krystallen in circular polarisirtem Licht,' *Poggend. Annal.* 40, no. 3 (1837), pp. 457–62, 482–84.

17. *finish the paper . . . under both our names:* the paper, as written up by Tyndall, had appeared in the July number of the *Phil. Mag.* (cited letter 0403, n. 2); Knoblauch was referring to the German version, which was not published until December (see letter 0458). Knoblauch's wording indicates that they each wrote an account of their joint results, one in English and one in German, but that neither version was a direct translation of the other.

18. *Fräulein Fulda:* a close friend of Sophie Fick, whom Frankland married in 1851 (see Russell, *Edward Frankland*, p. 84).

19. *Prof. Vorländer:* Franz Vorländer (1806–67) was professor of philosophy at the University of Marburg.

20. *my works . . . English:* this might refer to the translation of two papers by Knoblauch entitled 'Investigations on Radiant Heat,' first published in *Poggend. Annal.* in 1847, that were eventually published as part of volume 5 of the *Scientific Memoirs* in 1852 (pp. 188–237 and 383–434). Both papers were translated by J. W. Griffith.

21. *Frauenberg:* a high hill 4 to 5 miles south of Marburg. Its panoramic views and 13th-century castle made it a popular scenic site.

To Thomas Archer Hirst 4 August 1850 0418

Manchester Sunday Aug. 4th 1850

My dear Tom

Well the ordeal is over and here I am again. I started to Edinburgh on Wednesday morning last and reached there the same night, taking up my quarters in a temperance hotel, being converted thereto by one of my fellow travellers. On Thursday morning I enrolled myself as 'an associate' by paying a pound and finding that Saturday was intended for excursions I knew that if I could not get my paper read on Friday I should have to wait till the next week. I saw the Secretaries of the Mathematical and Physical Sections[1] and expressed my wish to have the opportunity of reading the paper next day. One of them, the professor of natural philosophy in the University of Glasgow[2] spoke of the great interest the subject excited, but could not bring himself to accept my explanation in toto. For the most part he agreed with me fully but on certain points of importance he felt compelled to dissent;—He strongly wished us to have a conversation together before the paper was read, but this except for a few minutes next morning was not attained. These few minutes however were invaluable to me, as I had time to turn his objections over in my mind instead of having to combat them impromptu. Well on Friday morning the name of your friend[3] appeared upon the public list in the respectable society of Sir David Brewster and others. I sat among the audience, a paper was read by some gentleman upon meteorology and then the President announced that the next paper was by M[r] Tyndall on the Magneto Optic Properties of Crystals.[4] I arose fumbled towards the rostrum where I soon found myself with lord [Rochly][5] to my immediate left, next him Sir David Brewster, next him professor Forbes the president of the Section and other members of the committee, on my right were two professors of my own age perhaps,[6] the one he whom I have already mentioned the other a Cambridge fellow. Well I felt either the God or the devil, or perhaps the abstract strength belonging to each very full in me at the moment, I was never more at home in my life and from beginning to end felt no trace of alarm or hesitation. I had looked at the case before-hand and determined not to be cowed. The paper was listened to with great attention and just before me were a bunch of ladies with mild brown eyes and every time I raised mine I found theirs fixed on me as if I had been reading the story of Jack and the beanstalk or something else equally interesting—It must have been the half dare-devil way in which I spoke and not the subject itself that interested them. Well at the conclusion the president[7] arose, passed a high eulogium on the paper, said however that he had been in Bonn himself very lately, had seen Plücker and seen his experiments which appeared in the highest degree conclusive and satisfactory, said moreover that he knew how impossible it was to form a correct conclusion as to the theory from seeing merely a few experiments. I arose and said that I had experimented for 10 hours a day during the space of three weeks under the conviction that Prof. Plücker's theory was fully established. The president then invited any gentleman present to make any remarks which he thought called for upon the paper just read, and if any part appeared objectionable to him to state his objections. The Glasgow professor to my right arose, and commenced by introducing Poissons theory of magnetism, and said that he had not the shadow of doubt that the paper would corroborate the theory fully and that the theory would support the views advanced in the paper. He was profuse in his praises of the beauty and ingenuity of the experiments; but one thing appeared to him to be not at all established and that was that the action of the dough was due to the proximity of the particles and not to the optical axis, 'for' said he 'the powder of which the dough is composed is a crystalline powder and the pressure exerted upon this is exactly what might be expected to cause the little component crystals to take up a position in which their optic axes would come into play. Indeed he exclaimed rather triumphantly this very specimen of bismuth dough is the strongest proof of what I say. Look at it you see the particles are all minute plates or scales and these being pressed flat cause the action.

He then went on to contend that Faradays view of a directive force was the correct one—it may said he differ from the views of Mr Tyndall in some minor particulars but they are substantially the same.

I have proved he continued and I intend to lay the matter before the Section at a future day, that crystals have three lines of equilibrium, at right angles to each other and no matter how the crystal is hung one of these lines will stand from pole. I have worked out the formula for the magnetic and diamagnetic forces under the circumstances. He concluded by repeating the eulogium which had at commencing passed upon the paper.

Of course I cannot repeat the 10th part of what he said nor the arguments he adduced in support if his views.

I arose again but the president anticipated me and asked whether any other gentlemen had a remark to make. one present proposed a question or two to my antagonist and there being no other questioner present I was called upon to defend myself.

At this time people kept thronging into the room. the president had to rise several times and request them to ascend to the upper seats and not to block up the passages Well I got up again and said that with regard to Mr Faraday (my antagonist appeared anxious to stand up for him) I had only to express my unqualified admiration of his experiments. he had described not only every action but every shade of action with perfect exactitude, I had had the pleasure of seeing Mr Faraday a few weeks before and then told him that we had been compelled to dissent from his views, 'no matter' he replied 'you don't differ from me as a partisan but because your convictions lead you to differ from me'—'thus encouraged by Mr Faraday himself,' I proceeded 'I feel rather inclined to stick to my old notion in this matter notwithstanding what has been urged by Professor Thomson—Prof Thomson has remarked on the necessity of care and caution in coming at the bare facts in a case like the present; I fully agree with him, and I could tire the audience with a recital of the experiments necessary to obtain these facts. We have examined nearly 100 natural crystals and in some cases from 10 to 20 different specimens of each. while a vast number of artificial crystals have also passed through our hands. So much for care. Prof. Thomson has stated a hypothesis of his own that magnetic action in crystals can be reduced to three lines of equilibrium and that there are only three. unfortunately for him the hypothesis is against facts. Take a disk from Calcareous Spar cut perpendicular to the Optic axes, such a disk suspended horizontally has three lines of equilibrium thus suspended; the optic axis is another line of equilibrium, here we have four instead of three as affirmed by Prof. Thomson.' 'Have you made that experiment Mr Tyndall?' asked he 'yes' said I 'repeatedly' [']has your disk been perfectly circular?' 'perfectly circular' I replied 'and your crystal pure?' said he 'that also' I responded 'well I can only say' said he 'that it is most extraordinary and entirely contradicts my notions on the matter' 'Prof. Thomson' I continued has raised an objection against my experiments with the magnetic and diamagnetic dough supposing that the arrangement of the minute crystals and not their proximity caused the action—'Now' said I 'it is a well known fact that if a precipitate of carbonate of lime be washed out with boiling water['] (here Thomson interrupted me—'dont you think Mr Tyndall that if the precipitate was examined by a microscope it would be found to be composed of small crystals?') 'I'm just coming to your question' I replied—The washing with boiling water induces sometimes a crystalline structure, but when these crystals are examined under the microscope as Prof. Thomson suggests they are found to be of the form of arragonite[8] not of calcareous spar (which I ought to have remarked was the case under consideration) now it is utterly incredible that a jumble of small arragonite crystals could exhibit exactly the deportment of calcareous spar; but more than that' I continued 'the precipitate has been washed with cold water and has no crystalline structure whatever, this mass' said I 'holding up a model made of your carbonate between my finger and thumb—this model is a perfectly amorphous mass, there are no crystals there['] said I 'squeezing the thing to a fine white powder and scattering it over the table. and even in the case of this bismuth dough if you look at it transversely you will see as many plates as you see upon its surface'. I broke the disk and handed it to Prof. Thomson—he shook his head as if but half convinced—Sir David Brewster arose and said it appears to me that the supposition of Professor Thomson is exceedingly improbable, but it can be decided at once—does the mass cleave?'—'no Sir David' I responded, 'not at all, it is a purely amorphous mass['']. The old fellow evidently thought as I thought. Thomson got up again and said that the bismuth was not at all convincing but that the carbonate of lime according to the facts stated met his objection fully. He said to me at the end that I should find his opinion and the theory of Poisson in entire harmony with my views. This is only a bare outline of the discussion Tom, I wish you had been there. I was never more at home in my life. I enjoyed full freedom of thought and utterance and the whole matter was to me the keenest amusement—The vote of thanks was right warmly responded to, and an old gentleman a member of the committee said to me when done—go in to the committee room Mr Tyndall and make me an abstract of your paper. I want to send it off to the Athenaeum[9]—In passing thru' the street towards home, another gentleman met me. 'Oh Mr Tyndall' he exclaimed—I looked at him I had never seen him before—'I have just heard your paper—I have taken notes but dont think I can represent the case fairly, would you oblige me with an abstract putting your argument in what appears to you to be the strongest light. I want it for the Literary Gazette[10]—But the old door keeper which I encountered on leaving the Section was most characteristic of all—'Well Sir' said he as I passed him 'you have got your matter done' yes I said. 'Well Well' he said chuckling 'it was most interestin, I have na heerd any thing like it—Ah but you made out a strong case—Why Professor Thomson could make nothing of you'

Thus the affair passed Tom. I had a clear stage and an impartial hearing and had you been there I think you would have admitted that I did as much as could be expected from a <u>poor</u> devil like me.

To stop longer in Edinburg would have been against the interest of my exchequer. tho' to remain would have been a pleasure and a benefit to me. yesterday excursions were made. and certainly the country around Edinburgh affords glorious opportunity for such. The evening meetings were also most interesting—but the stern fact was before me and I reconciled myself to it half lovingly. took my portmanteau cheerily in my hand and bade Edinburg good bye yesterday morning. this is the reason I have not written to you from Edinburg, because I have arrived here as soon as a letter could have done -

With best love my dear fellow— | Your affectionate Tyndall

Kiss Jimmy for me! if he lets you—

RI MS JT/1/T/530
RI MS JT/1/HTYP/98–100

---

1. *Secretaries of the Mathematical and Physical Section*: W. J. Macquorn Rankine, C. P. Smyth, John Stevelly, and G. G. Stokes.

2. *Professor . . . Glasgow*: William Thomson

3. *The name of your friend*: Tyndall was referring to himself.

4. *Mr. Tyndall on the Magneto Optic Properties of Crystals*: this was an adaptation of Tyndall's recently published paper (cited letter 0403, n. 2.); see Phillips advice in letter 0415.

5. *lord Rochly*: John Wrottesley (1798–1867), second baron, a leading astronomer who had set up a private observatory. He was President of the Royal Society from 1854 to 1858

6. *professors of my own age*: Thomson and Stokes.

7. *the president*: David Brewster.

8. *arragonite*: 'A carbonate of lime, crystallizing in orthorhombic prisms and many derived forms, whence several varieties are distinguished' (*OED*).

9. *Athenaeum*: summarized in 'Twentieth Meeting of the British Association for the Advancement of Science', *Athenaeum*, 1189 (10 August 1850), p. 842.

10. *the Literary Gazette*: reported under J. Tyndall, 'Magneto-optic properties of Crystals, and the relation of Magnetism and Diamagnetism to Molecular Arrangement', *Literary Gazette,* 1751 (10 August 1850), pp. 555–57.

From Heinrich Debus 4 August 1850 0419

Hirst werde ich genau so aufnehmen wie ich John Tyndall aufnehmen würde wenn er wieder nach Marburg käme. Ich werde Hirst als den Freund meines Freundes J. Tyndall behandlen. Bunsen, Knoblauch, Noll grüssen Sie herzlich. Wann ich wieder an Sie schreibe lieber John dann hoffentlich mehr und dies bald.

Gott sei mit Ihnen | Ihr | treuer Freund | H. Debus
Marburg IV. VIII. 50.

Hirst I shall receive just as I would receive John Tyndall, if he were to come to Marburg again. I shall treat Hirst as the friend of my friend J. Tyndall. Bunsen, Knoblauch, Noll send you their warm regards. When I write to you again, dear John, then hopefully there will be more, and this soon.

God be with you | Your | loyal friend | H. Debus
Marburg IV. VIII. [Aug] 50.

RI MS JT/1/HTYP/103
Transcript Only[1]

---

1. *Transcript Only*: this extract from a Debus letter was copied out by Tyndall in letter 0424 to Hirst (and in that letter Tyndall gave his translation). The source here is the further transcription by LT's assistant.

To Elizabeth Dawson Steuart 5 August 1850 0420

Spring Bank, Over Darwen, | Lancashire | 5th Aug. 1850

Dear Madam,

Though I date this from Spring Bank which is my standing address for the present I write it from Manchester.[1] Had I followed my impulse I should have written at once to thank you for your last truly kind letter,[2] but I have been in the habit of curbing these impulses a little, not by any means quenching them, but keeping them like a little fire within me for the warming of my feelings when these grow cold.

Well I have been to Edinburgh and shall now give you a short abstract of my three days life there.[3] I started on Wednesday morning last and reached the City the same night, put up at a temperance hotel, being converted to this act by one of my fellow travellers. Next morning on payment of a pound I was enrolled an associate of the British Association; as Saturday was devoted to excursions I saw that if I could not get my paper read on Friday that I must remain until the next week. I saw the Secretaries of the Mathematical and Physical Section and expressed to them my desire upon this point. Next morning accordingly I found my name in the published list in the respectable society of Sir David Brewster and others. I was among the audience on Friday morning, one gentleman read a paper on Meteorology and the President standing up announced 'the next paper is by Mr Tyndall on the Magneto-Optic Properties of Crystals'. Well I fumbled forward to the rostrum and soon found myself there with Lord Wrottesley at my immediate left, next to him Sir David Brewster and further on the Chairman Professor Forbes with other members of the Committee—at my right were two professors about my own age, the one Prof of Natural Philosophy in the University of Glasgow the other a Cambridge gentleman. Well I never felt more at home in my life, from beginning to end I did not experience a trace of alarm or hesitation there was something genial in the atmosphere and calculated to draw out the best power of a man. The paper occupied about 3 quarters of an hour and at the end the Chairman arose, and said that he had been at Bonn lately and had seen the experiments alluded to in my memoir himself. They seemed to be quite conclusive and satisfactory as regards the proof of the theory which Prof. Plücker their originator sought to establish, at the same time he said he knew how impossible it was to form a correct notion of the theory from seeing a few experiments merely. I arose and said that I had experimented for 3 weeks 10 hours each day before I discovered that the theory was defective. The Chairman proceeded, 'Here['] he said 'is an entirely new branch of scientific enquiry the researches of Plücker and Faraday have been very elaborate but Mr Tyndall's memoir goes directly to invalidate the views of both, if any gentleman present has any remark to make or any objection to urge against the reasoning we shall be happy to hear him'.

The Glasgow professor arose and spoke at some length on the interesting nature of the inquiry and the beauty and ingenuity of the experiments described in the memoir. He introduced Poisson's theory of Magnetism which had been abandoned for want of experiments to confirm it 'Now' he said 'this is the very thing itself. I have not the shadow of a doubt but the theory will be established by this investigation and that the investigation in its turn will be supported by the theory.' But he could not agree in toto with my explanations and proceeded to combat the views advanced at some length. It would be totally void of interest to you to go over the ground after him, suffice it to say it came to a regular hand to hand contest between us, but I believe it was the opinion of every one present that he had not made a single point good against me. Once Sir David Brewster arose and though he did not pronounce a decided opinion it was manifest from what he said that he thought as I thought. The discussion endured for about 3 quarters of an hour and I firmly believe would have taken up the entire time of the Section had not the Chairman felt compelled in order to make way for other claims to shorten it. It was not a task to me, nothing like a painful conflict, for I had thoroughly mastered my subject beforehand, and found little difficulty in settling objections—it was in point of fact the keenest piece of amusement I ever enjoyed.

The vote of thanks was right warmly responded to, and an old gentleman, a member of the Committee said to me when done 'go in to the Committee-room Mr Tyndall and write out a good abstract of your paper for me—I want to send it off to the Athenaeum'. In passing through the street afterwards a gentleman a perfect stranger accosted me—'I have just heard your paper' said he 'and have taken notes of it but I don't know whether I can do it justice. Would you oblige me with an abstract from it for the Literary Gazette?' But the old door keeper of the Section amused me most - he had peeped in during the controversy and heard it all—'Well Sir' said he as I passed him 'you have got your business settled' 'Yes' I replied 'Weel weel' said he 'it war most interestin I have heard nothin like it, really Sir Professor Thomson could make nothing of it He war completely under'.

Saturday was to be devoted to excursions, but Saturday saw me on my way southward, where I shall now have time to gather up my thoughts and to contemplate the possibilities that await me.

You ask me what my plans are, well that is easily answered I returned from Germany with the intention of borrowing a little money so as to enable me to return and pursue my studies for six months longer: Circumstances have turned up to render this difficult of accomplishment—I now see no probability of being able to return. Were I singlehanded I might manage it but I must not forget those depending on me in Leighlin[4]—I must look out for some employment and hand over my task to those in luckier circumstances than myself.

With best wishes and remembrances to Captain Steuart | believe me dear Madam | most faithfully yours | John Tyndall.

RI MS JT/1/TYP/10/3327–3328
LT Transcript Only

---

1. *from Manchester*: Tyndall was staying with Ginty. He travelled to Spring Bank the following day, 6 August (Journal, 7 August, JT/2/13b/506).

2.   *Your last truly kind letter*: letter missing.
3.   *abstract . . . life there*: this letter is largely a repetition of letter 0418, which should be con-
     sulted for annotations.
4.   *Leighlin*: Leighlin Bridge, Ireland, Tyndall's hometown.

From Thomas Archer Hirst   6 August 1850   0421

Halifax | 6 Augst 1850

My dear Tyndall,

Thank you for your full and faithful account of your successes at Edinburgh[1] I read them with as much interest as if they had been my own and in
imagination slapped you on the back and cried 'Bravo, my boy'. The only thing
I found fault with was your returning so soon It would have been useful for
you to have seen more of those men and unless you returned to Manchester
to <u>earn</u> some bread I don't know why it might not have been as easy for you
to have bought it at Edinburgh as at Manchester You're a queer, independent,
and perhaps also a partly insane man, and although I want to be cross with
you I am forced to admire partially some of y<our> independencies & insan-
ities which makes it all the more uneasy to me. In one of your letters from
Germany[2] which I have read I should think a score times and which I cherish
most you mentioned an <u>ideal</u> Friendship for us both to aim at, and attempt
to realize: it was a ideal for me that and many times its soft soothing light has
brightened these dark nooks and crannies—after our meeting I rejoiced that
it was progressing to a reality, but when I left you to walk home from Pye Nest
Lodge[3] last Monday week I was sad, sad because partly you had left a blank
behind you and I had nothing else to do but think of it, sad also because I
thought proper to notice that this my beautiful ideal of Friendship was far
from a reality yet and I cursed (foolish fellow) the world & its prudencies
because they were a dead weight that held our performance fast Earth-bound
and would not let it join its high soaring mate Fantasy. Fantasy said 'ye two
shall be as one, there shall be no mine and thine, but the cup shall pass freely
between you', & the mocking World laughed 'Ha! Ha! what a pair of fools—
look how one bucks the other the one a knave the other a fool They throw
dust about their eyes, call it Love & fancy themselves half-Gods' So that poor
performance stood irresolute, confused & helpless—I was just opposite Bernard Hartley's[4] then and my pace had slackened to about half that of a snail I
felt something wet on my cheek I believe & then as if the Devil had kicked me
up I sprang & walked away briskly crying 'Fool, will all this come for whining
after, you have as much as you can bear, and are rewarded according to your
worth; go along stupid Carter wants thee' And I did go and forgot you; now

and then over my desk, I thought I saw you trudging on & up Blackstonedge,[5] but I lost you among my Building Lots at 50 to an inch,[6] at night I tried to lose sight of you in the German Dictionary, but I did not succeed as well & at Evening I went to bed rather restless & thought as I set my alarum how it would growl out at 8 oclock and make me go to my Building lots, I fell asleep however and in the Morning when it sang out it connected itself with some confused dream or other, and I thought it didn't growl as I expected, but called me up more cheerfully & bade me jump up & spoke something about better days if I did, and how all these difficulties would settle themselves this way and no other. We are fearfully & wonderfully made[7] Tyndall aren't we? Who dare say that Alarum did not speak to me? It is not a bit more wonderful than that I can speak. John[8] is in the neighbourhood trotting about, we are as cheerful & friendly as ever—He read your letter handed it over to me, laughed & like Tom Perkinton[9] said you were mad, at which I laughed too and said I thought you must be, and I do veritably think so yet, but then I also think that it is an enviable sort of madness and he's a fool who once knowing it would wish to be sane—There are no letters here for you—Jimmy is in Glasgow, Little John is leaving Leeds it appears, he said he wrote to you at Edinburgh. If you can send me that Carlyle's pamphlet[10] of mine they are done now I will get them bound—send me Zig-zag[11] also by post, I want it particularly, don't neglect. —Let me know what you are doing with yourself repeatedly, and where to write to you & don't forget the Commentary on 'The Sphinx'[12] | T.A. Hirst

Mr Ginty— | Engineer's Office | Lancas^{re} & Yorks^{re} R:way | Manchester For J. Tyndall, to | be forwarded (if | absent) to him.[13]

RI MS JT/1/H/150

---

1. *your full… Edinburgh*: letter 0418. Note that Hirst omitted punctuation at the edge of his paper and, in this position, as in many others, there is therefore no full stop.

2. *one of your letters from Germany*: possibly letter 0390, where Tyndall wrote 'Let us endeavour to be friends Tom in as high a sense as any pair of fellows as ever lived'.

3. *Pye Nest Lodge*: the point to which Hirst walked with Tyndall (Hirst, 'Journals', 29 July 1850).

4. *Bernard Hartley's*: possible reference to a resident of Allengate near Halifax.

5. *Blackstonedge*: misspelling of 'Blackstone Edge,' the route which Tyndall walked to Manchester.

6. *Building Lots at 50 to an inch:* referring to the scale of the maps produced by Hirst's surveying work.

7. *fearfully… made*: quotation from *Psalms* 139:14.

8. *John*: Hutchinson.

9. *Perkinton*: probably the same 'Perkinton' who surveyed with Tyndall in Halifax; see letter
    0340, n. 8, and, for example, Tyndall's journal entry for 29 June 1846 (JT/2/13a/131).
10. *Carlyle's pamphlet*: one of Carlyle's 'Latter Day Pamphlets'. It is unclear to which pamphlet
    Hirst refers.
11. *Zig-Zag*: letter 0412 explains this request.
12. *Commentary on 'The Sphinx'*: see letter 0426.
13. *Mr Ginty . . . to him*: address from envelope; and forwarding note in bottom left corner.

To William Thomson   7 August 1850   0422

Spring Bank Over Darwen Lancashire | 7th Aug. 1850

Would professor Thomson have the kindness to write down in the margin
the title of any book where I might find a statement of the magnetic theory
of Poisson?[1] If after having done this, the professor would be good enough to
return me this leaf[2] I should feel very much obliged indeed.
     J Tyndall.

Cambridge University, Kelvin Correspondence, Add.7342/T623

1. *in the margin . . . theory of Poisson*: at the BAAS meeting in Edinburgh, Thomson had sug-
    gested that Tyndall's investigation was consistent with Poisson's theory (see letter 0418).
2. *return me this leaf*: Tyndall left the right hand half of his sheet free for the references,
    but Thomson responded with a long letter (0425), in which he included the Poisson
    references.

From George Wynne   8 August 1850   0423

Rostrevor | 8 Aug^r, 1850

Dear Tyndall,
     I received your first letter[1] yesterday telling me of your plans and your wish
to be employed for the present on the Ordnance Survey, and your second
today giving an account of the reception of your paper by the British Association,[2] I am sure you could not have gone before any tribunal where your paper
would be more fairly tried on its merits than there, and from your account it
appears to have met with a very favourable and encouraging reception.
     With regard to your first letter I will with much pleasure if you continue
to wish it write to Captain Yolland[3] but from my acquaintance with him I
think a personal application would be better, I would advise your writing to
him[4] simply to say that having returned from Germany you would be glad

to avail yourself of his former offer of so & so if he has a place for you, and I would avoid appearing too anxious for employment. I have reason to believe that you could not choose a more favourable time for making the application as the Ord[n] has just had thrown upon it the towns Survey under the Board of Health which later I believe is to be conducted by Capt[n] James and if your application is unsuccessful with Capt[n] Yolland I will write to him and to Capt Hale.

I have got a kind of leave for six weeks and hope to be able to remain at this pretty place[5] 'till about the 10th Sept[r] when my boys vacation will terminate. They and Mrs Wynne desire their kind remembrances.

If your exchequer runs low I shall be very happy in giving you a lift till times mend with you. | Believe me very truly yours | Geo. Wynne

RI MS JT/1/W/92

---

1. *first letter*: letter missing.
2. *your letter . . . the British Association*: the letter is missing, but it would parallel letters 0418 and 0420 of 4 and 5 August.
3. *Yolland*: William Yolland (1810–85) was a senior Officer in the Ordnance Surveys of England and Ireland.
4. *writing to him*: following Wynne's advice, Tyndall did write to Yolland, asking if his offer of employment made two years ago still stood (Journal, 19 August, JT/2/13b/506). Letter 0428 suggests that Yolland replied with an offer of employment, although by that time Tyndall had decided against accepting it.
5. *at this pretty place*: in Rostrevor, a village in County Down, Northern Ireland.

To Thomas Archer Hirst   [10 August 1850][1]   0424

Manchester Saturday

Dear Tom.

The 'incomprehensibility' is easily made clear. I wrote the note quickly and did not read it over; the word <u>Halifax</u> was set down for <u>Manchester</u>.[2] The substitution of the latter would make all clear. If its a proof of any thing it is that I was thinking too much of your locality and you ought to feel yourself flattered on that account. If I am insane there is certainly a certain method in my madness.[3] I remained in Edinburgh until Saturday morning—that day was devoted to excursions and in some measure to feasting, in neither of which I felt myself qualified to take a share—next day was Sunday. here were two days, requiring their adequate amount of board and lodging and nothing in return for it. True I might as well have bought my bread in Edinburgh as in Manchester but the great difference is that in Manchester there is a certain mendicity institution established on a friendly basis,[4] where I can obtain temporary shelter without paying for it. —In Edinburgh I paid 1/6 a night for my bed and 1/s a day to servants. In Manchester I get half a bed for nothing. M[rs] Ginty has put a table and one chair in a garret for me at the top of the house where I can retire when I will and speculate to my heart's content. I have not stopped in Manchester ever since, I have been at Spring Bank whither I return on monday—I venture to say if I were to set down on paper all my acts since I saw you that a prudent man; I mean a plain, practical, worldly wise man would not censure them, but the contrary. My simple rule has been to live within my means, and my only moments of insanity, (which thank the gods are now few) are those moments when I shrink from complying with the restrictions which limited means necessarily impose; I calculated probabilities long ago, and as is my want took the worst into account, and I should be a selfdeluding theorist did I now repine when the test-hour is come: But the test is not very heavy after all, indeed it would be a mockery to call it a test—My friend at Spring Bank[5] has an Academy. It is nicely situated in the Country, really a very beautiful place to live. Dont you see boy that I can make myself worth my board and lodging with him. I return there on Monday under those conditions and there I shall remain until something turns up for me. There my three diurnal meals can be purchased by 3 hours labour and I do feel a certain satisfaction in the attitude of independence which this enables me to assume. With regard to our one-ness if it be real it is a one-ness established by the fates to which we must resign—for my own part, in all the moods and mental phases through which I have passed since the writing of that letter to which you allude[6] I never for the fraction of a second felt a desire to revoke a single word then and there penned—The 'ideal' is one from which I have never shrunk—and it will, please the gods, be perfected without any <u>present</u> commingling of cups. There is a certain state of mind and feeling shadowed by the words

> Keep thyself now and free for ever
> Free as an Arab
> From thy beloved.[7]

A state which you seem to think must be assumed, a state which I know may be real, and a state which may exist between man and man as well between man and woman. Benefits throw a kind of chain round a man, and I have at the present moment, as far as you are concerned, just as much metal about my ancles as I can conveniently carry—any more might meddle with my freedom, and that I wont sell if I can help it. You may think all this inconsistent inasmuch as I have endeavoured to borrow money from others—I have, but were you to see my letters to those parties you would not say that there was any relinquishment of freedom on my part, and further these were parties well skilled in the calculus of prob-abilities, who knew something of insurance offices and the laws of mortality and of accidents; and if they made the loan would do it with their eyes wide open which I am inclined to think is more than you could do. You exclaim repiningly 'Fantasy saith so!' if she be not backed by <u>fact</u> Fantasy may go to the devils as soon as it suits her convenience—I don't thoroughly understand this part of your letter.[8] Which of us does the mocking world frighten by its 'Ha! Ha!' —As far as I am concerned the world has less to do in the affair than you imagine. —I enclose you a scrap from Debus,[9] received yesterday. The part referring to you may be thus translated, 'Hirst will I exactly so receive as I should receive John Tyndall should he again come to Marburg. I will treat Hirst as the friend of my friend John Tyndall'

I heard a lecture last night on the electric light in the mechanics institute[10]—it was crowded—My God what a maggot the fellow would appear beside our Bunsen! The man talked incessantly of the necessity of getting to the bottom of the subject but it is a region which he himself had never approached. I may possibly lecture on the subject myself in the course of the winter.

John Tyndall

Knoblauch wrote me a long letter,[11] kind hearted fellow, his last words are 'I wish from my heart you were once more here!' Plucker still holds on by his former views. he threatens to explode us by an appeal to mathematics. in Crelle's Journal we shall soon see a Memoir which is to overthrow us and place the matter on everlasting foundations. We shall see!

[. . .][12]

RI MS JT/1/T/531

---

1. *[10 August 1850]*: Hirst wrote at the top of the letter, 'Aug. 10—1850', a Saturday, the day he determined the letter was written. This is consistent with Tyndall receiving letter 0419 from Debus 'yesterday', the 9th, 5 days after Marburg date of 4th.

2. *Halifax . . . for Manchester*: this refers to a missing letter or 'note', written quickly. It does not match the most recent extant letter (0418). It may be in reference to the bag (see letter 0427) which Tyndall seems to have requested Hirst to send.

3. *method in my madness*: Tyndall defends himself from the accusation by Hirst in letter 0421.

4. *institution . . . friendly basis*: the Gintys.

5. *My friend at Spring Bank*: Josiah Singleton.

6. *that letter to which you allude*: in letter 0421 Hirst alluded to Tyndall's hope (in letter 0390) that they would become friends in a 'high' sense.

7.   *Keep thyself . . . beloved*: Emerson, 'Give All to Love', *Poems*, pp. 111–13, lines 31–34 (Hirst and Tyndall discussesd this poem in letters 0413 and 0414).

8.   your letter: letter 0421.

9.   *scrap from Debus*: letter 0419 (see n. 12 below).

10.   *lecture last night . . . institute*: William Edwards Staite's lecture in Manchester (Journal, 17 August, JT/2/13b/506).

11.   *long letter*: letter 0418 (the quoted extract is from the second page of a six-page letter).

12.   *[. . .]*: this second postscript is an extract (or 'scrap') of the Debus letter which appears as letter 0419.

From William Thomson   14 August 1850   0425

Row, Helensburgh | Dumbartonshire | Aug 14, 1850

Dear Sir

I regret not to have been able to answer your letter[1] of the 7th sooner. I received it last week in passing through Glasgow, and I should have written to you immediately if I could have put down the titles of some works, as a sufficient answer. In reality however I cannot give you any satisfactory reference for an account of Poisson's Theory of Magnetism except his own Mémoires which are published in the Mémoires de l'Institut,[2] if I remember right between the years 1810 and 1815. There are two, I believe published in the same volume, on the Theory of Magnetism, and one, published in a later volume (I believe consecutive to that which contains the first two) on Magnetism in Motion. In the last mentioned Mémoire, Poisson makes a brief statement to nearly the following effect. If, as we may conceive to be the case in crystalline substances, the magnetic elements are non-spherical and are symmetrically arranged, the induction of magnetism would follow different laws from those which have been obtained in the first two Mémoires, on the supposition that the elements are spherical or destitute of symmetrical arrangement. It would result that a spherical portion of such a substance would experience different inductive action according to the way in which it is turned when placed near a magnet 'but such circumstances not having been observed I confine my present researches to the cases in which the magnetic elements are spherical, or if not spherical, non-symmetrically arranged'.

The preceding statement and quotation are, as nearly as I can recollect, all that Poisson says regarding the theory of the phenomena which might possibly be presented by crystals, in his third memoir. I know that in one or both of the first memoirs the same idea is started but passed over in a similar cursory manner.

It may be added to Poisson's statement that, according to his theory of magnetic elements, the phenomena which he hints at as to be expected from non-spherical magnetic elements, symmetrically arranged, would also result from an arrangement of spherical elements such as you produce of the particles of powder by the compression of the dough containing them; that is, an arrangement in which a finite line drawn in a <u>certain direction</u> will cut across a greater number of the small spheres, than a line of equal length in any other direction. A ball of such substance, if free to turn round its centre of gravity would, if in the neighbourhood of a magnet, only be in stable equilibrium with the direction mentioned above (underlined in the preceding page) along the lines of force. If this direction were inclined to the lines of force at any oblique angle, there would be a directive 'couple' exerted upon the ball tending to bring this direction to be parallel to the lines of force, in virtue of the magnetism induced in it, by the magnet acting on the magnetism which it has induced.

Regarding the 'magnetic elements' the existence of which is the hypothesis adopted by Poisson, they will differ essentially from the particles of powder enclosed in dough which you use as an illustration, in this important respect. In each magnetic element there are two magnetic fluids (I make use of the language of Poisson's hypothesis of the falseness of w^h, however, I am convinced) capable of moving about freely within the element but incapable of leaving it and passing to another element or into a space of the body external to the magnetic elements. Each element contains an enormous quantity of the northern fluid, and an equal quantity of the southern fluid. Hence a single magnetic element, if it could be detached from the body of which it is a part, would be susceptible of magnetic induction, by the actual separation of the fluids within it which the presence of a magnet would produce. The law which the separation of the elements within it w^d follow is this—The resultant magnetic force at any point P within the element due to the separated fluids at the surface of the magnetic element,

must balance the resultant force at the same point due to the influencing magnet. —(so that a small compass needle placed within the element would experience no directive action) From this inductive magnetic action the detached magnetic element would exhibit a 'capacity for magnetic induction' greater than that which is possessed by any <u>real substance</u>; & the phenomena of magnetic induction observed in actual substances result from the separation in this way of the magnetic fluids within the infinitely small elements, but are not dependent on any separation of the fluids through finite distances in the substance. In considering analogous electrical problems, I convinced myself a long time ago that a row of such spherical magnetic elements (corresponding precisely to insulated, naturally non-electrified conducting spheres) if placed in a uniform field of magnetic force, would, by their mutual actions when subjected to magnetic influence, produce a greater magnetization when their line is parallel to the lines of force than when it is perpendicular to them, and that when it is oblique to the lines of force there would be a resultant couple acting on the group w^h would tend to turn their line towards the direction of the lines of force. If the diameters of the balls be very small compared with their distances asunder, this directive action will be excessively feeble. If instead of imaginary detached magnetic elements we conceive balls of soft iron to be substituted, similar phenomena will be presented, to nearly an equal degree, since the capacity for induction of a ball of soft iron is so intense that it does not fall far short of what it would be if the magnetic fluids were freely separable throughout its entire extent. But if instead of balls of soft iron, we have a row of balls of a very feeble ferromagnetic substance (balls containing very weak solutions of sulphate of iron for instance) their mutual action when subjected to the influence of a magnet would be so slight that any directive tendency that would result (although it would certainly be towards the lines of force, if the field be uniform) would be practically inappreciable.

I have also convinced myself that a row of balls of a diamagnetic substance, placed in a uniform field of magnetic force, would experience a directive tendency of such a kind that there would be a stable equilibrium with the line joining their centres parallel to the lines of force; and, generally, that a long bar of a diamagnetic substance, free to move round its centre of gravity, would take the same direction as a bar of ferromagnetic substance, that is, the length of the bar parallel to the lines of force, provided the field be absolutely uniform; but the capacity for inductive magnetization in all known substances which are diamagnetic being excessively feeble, this directive tendency would I believe be absolutely inappreciable in the most refined experiments that could possibly be made with the most powerful uniform field of force that could be obtained.

It has struck me that some of the very remarkable phenomena which you have discovered in your experiments upon the cakes of dough may have been due to a <u>quasi</u>-crystalline structure induced in the dough, or stiff from water, by compression, and become permanent when the substance has dried in that constrained state. This occurred to me in consequence of the curious optical experiments performed by Mr. Clerk Maxwell Junr having been noticed, and shown to some of the members in Section A. From these it appears that isinglass jelly[3] dried in a state of constraint (very moderate forces having been sufficient) presents a permanent crystalline structure as far as optical properties can indicate; why not also for induced magnetism? I believe every transparent solid that has been magnetically experimented upon which has the optical properties of a crystal has been found to possess the peculiar properties regarding induced magnetism. It is therefore to be expected that a small sphere of Clerk Maxwell's dried isinglass, when placed near a powerful magnet would experience a directive tendency. I have requested him to prepare some small spheres and discs with a view to having experiments of this kind made, but I fear my apparatus will not enable me to arrive at any satisfactory results by experimenting myself. I should be extremely glad if the suggestion I have made should induce you, or M[r] Knoblauch if you communicate it to him, to make some such experiments. There are several distinct arrangements I could indicate which I think might lead to a complete elucidation of the very remarkable experiments you have already made. If you wish it I shall be glad to communicate all I could suggest. I shall conclude this rambling letter by mentioning that Poisson's Theory of Induced Magnetism is briefly described in Lame's Cours de Physique,[4] and probably in many other similar treatises; & that the elementary mathematical treatment of the subject is given in Green's Essay on the Application of Mathematical Analysis to the Theories of Electricity and Magnetism,[5] Nottingham 1828 (now out of print unfortunately) and in Murphy's Treatise on Electricity, Cambridge 1833.[6]

Believe me, Dear Sir, | Your's very truly | William Thomson

P.S. I enclose a paper which I published some time ago in the Cambridge and Dublin Mathematical Journal,[7] and which contains some remarks closely connected with the subject of this letter. | John Tyndall Esq.

RI MS JT/1/T/9
RI MS JT/1/TYP/5/1523–1526

---

1. *your letter*: letter 0422.

2. *his own Mémoires . . . l'Institut*: probably S. D. Poisson, 'Mémoire sur la distribution de l'électrité à la surface des corps conducteurs', *Paris, Mém. de l'Inst.* (1811), pp. 1–92 and 'Second mémoire sur la distribution de l'électrité à la surface des corps conducteurs', *Paris, Mém. de l'Inst.* (1811), part 2, pp. 163–274 for the first two, and 'Mémoire sur les surfaces élastiques', *Paris, Mém. de l'Inst.* (1812), part 2, pp. 167–226 for the later volume.

3. *isinglass jelly*: a firm whitish semitransparent substance (being a comparatively pure form of gelatin) obtained from the air-bladders of some fresh-water fishes, esp. the sturgeon (*OED*).

4. *Lame's Cours de Physique*: G. Lamé, *Cours de Physique de l'École Polytechnique. Tome troisième, Electricité-Magnétisme-Courants électriques-Radiations* (Paris: Bachelior, Imprimeur-Libraire, 1837).

5. *Green's Essay . . . Magnetism*: G. Green, *An Essay on the Application of Mathematical Analysis to the Theories of Electricity and Magnetism* (Nottingham: Printed for the Author, by T. Wheelhouse, 1828) contains a generalization and extension of Poisson's work.

6. Murphy's Treatise . . . 1833: R. Murphy, *Elementary Principles of the Theories of Electricity, Heat, and Molecular Actions. Part I on Electricity* (Cambridge: Pitt Press, 1833).

7. *Cambridge and Dublin Mathematical Journal*: most likely 'On the forces experienced by small spheres under magnetic influence; and on some of the phenomena represented by diamagnetic substance', vol 2 (1847): 230–40.

To Thomas Archer Hirst   19 August 1850   0426

Aug^{st} 19th 1850

Dear Tom,

Thrice you have mentioned that paraphrase: I made that promise rashly unwarily[1] for the paraphrase of such a thing demands time and thought and a mood of mind favourable to abstraction. That I have understood sufficient of it to make it delightful to myself is quite a different thing from an endeavour to excite the same delight in another by an analysis of the poem![2] Besides, I read many things which I don't understand and which nevertheless benefit me, give me a certain airiness and courage which is profitable even in common affairs. But I will take a ramble over the poem as you seem to like that I should, but dont measure it by me, study it for yourself if you want to get any thing out of it.

The Sphynx I take to be nature, nature external and internal, the universe that envelops us. Space and time, the stars, the earth, animals trees and flowers, the sciences even, and Man placed in the centre of all, all dumbly questioning him. Whence are we? What mean we? <u>What is your business there?</u> This latter appears to be the main question.

> 'The fate of the man-child
> The meaning of man'[3]

The poet discovers here a discordance in the harmonies of nature. Page 2 and the 1st verse of page 3 are one side of an antithesis; the second verse of page 3 is the opposite side. There is a contentedness in the sprouting palm, in the browsing elephant, in the singing thrush, in the leaves that cover him in the waves, in the breezes, in the sea the earth, sound, silence.[4] All these are so to speak so many natural acts and own the charm which belongs to such. they are because they are. Their appearance is the announcement of their right to appear. They all have their roots in nature, are the mere spontaneous expression as it were of her will. Now I find it exceedingly difficult to make myself clearly understood here, for I can sometimes fancy a certain close alliance between as it were the soul of nature and such objects as those mentioned but I cannot hope to make myself understood if the same apprehension be not shared by him whom I address. I believe this powerlessness on my part to be partly a defect, or rather a want of clearer and more certain insight, and partly a deficiency of material to illustrate what little insight I have.

The leading idea of Emersons mind and indeed of almost all /philosophic of minds/ an idea which is embodied in the religion of the Brahmans thrusts itself forward here—the unity of the universe.

> By one music enchanted
> <u>one</u> Deity stirred[5]

each of the parts that he recounts fitting in with architectural symmetry and beauty, living stones in this great edifice of God.

But man is a block out of square, he does not fit into the edifice—he is an alien from his fathers house.

> He crouches and blushes
> Absconds and conceals
> He creepeth and peepeth
> he palters and steals[6]

The right apprehension of this verse will enable us to understand the foregoing. For Emerson cannot thus condemn without reference to a standard from which man has swerved; this standard he endeavours to illustrate in the foregoing verses—'The babe by its mother'[7] is perhaps the most intelligible of these illustrations—look at the content of a child's eyes, its fearlessness, its spontaniety—I take pleasure in noticing this every morning. Two little boys sleep in the same room with me: I watch them when they are stripped washing themselves and when they rub their little cheeks and look out from behinds the folds of their towels they look as independent and happy as if the whole universe was their mother's lap—There is something of the 'blowing clover and the falling rain'[8] about their movements—There is a phase in every mans progress which contrasts painfully with this spontaneity—Emerson has felt it as deep as any, otherwise he could not have hit the nail so directly on the head.

> Infirm melancholy
> Jealous glancing around
> An oaf, an accomplice
> He poisons the ground.[9]

Do you not notice Tom how often Emerson alludes to the sinking of the eye. I think his frequent use of this allusion has caused a thousand eyes to sink which would otherwise never have done so. He has set people thinking of it and to think of it is to be tormented by it for long or short. I cannot help attributing it to mere <u>consciousness</u> operating upon a wakeful and unimpassioned state of the mind. When a man is in love there is none of it, when a man is enraged there is none of it, in both these states consciousness though present is not able to make the eye tremble on its axis. But where a man unimpassioned speaks to his fellow and <u>thinks</u> that they are looking at each other he straightaway forgets his subject he stares willfully and gets into confusion. I have noticed this between the best of friends Tom and I notice it with concern—you know whom I mean I dare say—it is a very delicate point[10]—how long will two fellows play cat and paw on this subject before either dares to mention it—I may be rash in mentioning it now but the acknowledgement of the fact may assist in the eradication. I don't notice this uneasiness among fools, they look with most good natured stupidity into each others countenance, and never think of the qualms which agitate the prying intellectual man. I find the honestest men subject to it—It is in fact a form of bashfulness and by no means an indication of guilt. It is however a bashfulness which a man will think beneath him and hence the uneasiness it causes. But Im forgetting the Sphynx—there is nothing however like falling back habitually on experience.

Nature is concerned for the man, for he also as well as palm and elephant is her child—

> Who has drugged my boy's cup?
> Who has mixed my boy's bread?[11]

The poet solves the question

> The fiend that man harries
> Is love of the best
> Yawns the pit of the dragon
> Lit by rays from the blest[12]

You complained to me once of your sins and iniquities and were almost in despair about them. This is a case in point. What was this but the 'pit of the Dragon' illuminated as expressed above, and you will solved the question and cast out the fiend before you read Emersons receipt. Remember ever that your life must be a <u>flux</u>—not stationary—you would have it so at the time alluded to and hence your misery, but there is no standing still;

> 'Profounder profounder
> Man's spirit must dive

> To his aye rolling orbit
> No goal will arrive'[13]

A profound philosophy underlies these verses. Emerson is saturated with it—You remember he calls a man a mystic who adheres to one symbol and refuses to translate it into other symbols. I open his first lecture on the uses of great men[14] and find the same. 'Rotation is the remedy of nature',[15] 'the soul is impatient of masters and eager for change' again—'On and for ever onward'—again 'In the moment the genius ceases to help us as a cause he begins to help us an effect', again 'Our strength is transitional'[16]—'There is no thought in any man but it quickly tends to convert itself into a power'.[17] again that great utterance I must call it in Montaigne 'The philosophy we want is one of fluxions and mobility'.[18] Here we have the same expressed in verse.

> The heavens that draw him
> With sweetness untold
> Once found for new heavens
> He Spurneth the old[19]

> Eterne alternation
> Now follows, now flies[20]

It is very necessary to a thinking mans happiness to see this and to lay it practically to heart. If he dont do so he will ever have some cause for repining. Some dear idol will depart from him daily and he will be left to mourn over the loss of his household gods. Let him offer them freely, and compensation instantly occurs. Shall I complain because one wave passes me when a thousand others are ready to bear me on. The parable of the 10 talents[21] is another way of expressing the same thought. the 'wood notes'[22] are full of it. When we would make a thing plain to the understanding of others, or even to our own we are driven to clothe the thing in words or images. Now for certain subjects though felt by an indubitable instinct it is exceedingly difficult to find imagery—I call to your mind that thing noticed by Emerson in his second series of essays[23]—in Experience I think, which every fine spirit has endeavoured to <u>name</u> or to define by a symbol—For the same reason it is difficult give a clear idea of our notion of god or the universe—People may talk about God very glibly, but the longer I live the more eager I am to break through his varnish of words and grapple with the thing itself. Locke talks very majestically of the improper notions which illiterate people have of god, he has not however taken the trouble to give us his own. I have mentioned this to you before. His reasoning on the being of a god certainly leads to the idea that he is merely a man of behemoth proportions, and so of Paleys god.[24] He compares the universe to a watch—see you not design he exclaims—If I assent what is to follow? Why simply that the difference between God and man is all a matter of <u>bulk</u>. but I'm running away from my subject—you see what I mean dont you? Paley looks at a watch and says certainly it had an intelligent maker. He looks at the universe and says the same. The maker of the watch being however a being detached from the watch, by analogy the maker of the universe is what Carlyle would call <u>an out-side God</u>[25]—I cannot look at god in this light. I cannot extricate him from the universe* (This very idea is implied in the term omnipresence if people would only take time to think of it.)[26] no more than I can extricate <u>myself</u> from my material habiliments my eyes arms legs & brain. You see Fichte's idea of the ego and nonego[27] comes in here. The ego can only manifest itself thru the nonego, yet to call the nonego the ego would be a transcendental absurdity. I think the universe is best illustrated by a human body.

> All are but parts of one stupendous whole
> Whose body nature is and god the soul.[28]

The universe is a body with a life within it, and among it, and thro' it, permeating its every fibre. Man is one form of that life, vegetables are another—herein consists that occult relationship between the former & the latter mentioned by Emerson in one of his essays.[29] The human skin illustrates the 'eterne alternation', deep down in fat layers below the surface a microscopic nucleus is formed—a point—it gathers other points around it and we thus have an aggregation of nuclei, these aggregations collect together and form cells; while all this is doing other nuclei are forming underneath, the cells are pushed upwards, are squeezed together and appear in the form of minute scales on the human skin—in the hair these scales are dandruff. In this way the black mark on your bruised nail is pushed forward. last week it was at the root, next week it will be at the outward edge, thus your hair grows and your whiskers (when you get them!) will sprout. Look at the process here going on! the Universe is typified by this microcosm; not an instants rest—every thing in nature is in the act of <u>becoming</u> another thing—

> Eterne alteration
> Now follows, now flies.[30]

Man is an illustration of the universe. A tree is also an illustration—a blade of grass. the currents of the universe so to speak flow thro' each. The dew drop solves the Sphere,—both are built on the same principles Who solves the one solves the other

'Who solveth one of my meanings
Is master of all I am'[31]

The universe is life, rendered so to speak concrete; as the cathedral is the concrete thought of the architect—Indeed Emerson uses this very figure in another place,

'These temples grew as grows the grass'32

Man is an offshoot from this eternal stock. his spirit is the spirit of the universe. Who solves it solves him and <u>vice versa</u>.

Thou art the unanswered question
Couldst see thy proper eye.[33]

This unity of the questioner with the questioned Emerson notices else-where. In the Woodnotes for instance

'Thou askest in fountains and in fires
<u>He is the essence that enquires</u>[34]

The gladdening effect of this recognition of life in the universe, that all these outward wrappages are a web by which we see the invisible [as noticed] by Emerson in the last verse but one. From a dead dull mechanism, a cold block of stone the Sphynx rises merrily—

Uprose the merry Sphynx
And crouched no more in stone
She melted into purple cloud
She silvered in the moon[35]

Who dare say saith Tom Hirst that that clock did not speak to me! How long stumbling on the boundary line between the highest wisdom and the very lowest absurdity!—

I have thus [ran] over the Sphynx imperfectly I am aware. probably when you come to think of the matter you will make out a better solution yourself. A great deal more could be written upon it, in fact whole systems of philoso-phy are wrapped up in it—See you not a direct connexion between the verse
Profounder profounder &c
and Fichte's doctrine[36] of the end and vocation of the scholar, perfection is his end, <u>eternal perfecting</u> is his vocation—'Let us rejoice' says he in another place 'that our work is infinite'[37]—

Your last letter[38] has set me thinking, thinking, what I have thought I will tell you by and by. aye come and see us. we have no beds however. they are all used up and even if they were not would be too short for you. We can get you a shop however. possibly we may make a little excursion somewhere, but really you must work hard at German.

    Your Tyndall

    Mr Hirst | Harrison Road | Halifax | Yorkshire[39]

RI MS JT/1/T/532

1. *Thrice you have mentioned . . . unwarily*: Hirst had reminded Tyndall of his promise in letter 0421. There are at least two missing letters (see letter 0424 n. 2, and n. 38 below) in which Hirst may have repeated the request. Tyndall's offer to paraphrase the poem is not made in any extant letters; we assume it was made over the long weekend that Tyndall spent in Halifax at the end of July (c. 25–28 July; see letter 0412).

2. *the poem*: Emerson, 'The Sphynx', *Poems*, pp. 1–6. In his interpretation Tyndall quotes directly from about one third of the poem, usually accurately. (In later editions (2nd in 1850 and 5th in 1856) Emerson introduced changes. We presume Tyndall used one of the 1847 editions.)

3. *The fate of . . . man*: ibid., lines 9–10.

4. *the sprouting palm . . . silence*: in this sentence Tyndall paraphrases lines 18–19, 22–23, 27 and 33 of the poem.

5. *By one . . . stirred*: ibid., lines 35–36.

6. *He crouches . . . steals*: ibid., lines 49–52.

7. *The babe by its mother*: ibid., line 41.

8. *Blowing clover and the falling rain*: from an address to divinity students, published in Emerson, *Nature; Addresses, and Lectures* (Boston: James Munroe and Company, 1849), pp. 113–46, quote from p. 125.

9. *Infirm melancholy . . . ground*: 'The Sphynx', lines 53–56, missing a comma between 'infirm' and 'melancholy' in line 53.

10. *I have noticed . . . very delicate point*: this indirect allusion may refer to intimacy. Hirst did not understand the allusion, as his reply (letter 0427) makes clear.

11. *Who has drugged . . . bread*: 'The Sphynx', lines 61–62.

12. *The fiend . . . blest*: ibid., lines 73–76.

13. *Profounder profounder . . . arrive*: ibid., lines 81–84.

14. *his first lecture on the uses of great men*: Emerson, 'Uses of Great Men', *Representative Men*, pp. 1–26.

15. *'Rotation . . . nature*': Tyndall seems to have conflated two phrases 'Rotation is her remedy' (ibid., p. 13) and 'Rotation is the law of nature' (ibid., p. 15). The next two quotations appear in ibid., pp. 13 and 22 respectively.

16. *'Our strength is transitional'*: Emerson, 'Plato', *Representative Men*, p. 39.

17. *'There is . . . power'*: ibid., p. 49.

18. *Montaigne . . . mobility*: quoted by Emerson, 'Montaigne; or, The Skeptic', *Representative Men*, p. 118.

19. *The heavens . . . old*: 'The Sphynx', lines 85–88.
20. *Eterne alteration . . . flies*: ibid., lines 97–98.
21. *parable of the 10 talents*: Matthew 25: 14–30.
22. *wood notes*: Emerson, 'Woodnotes', *Poems*, pp. 50–72.
23. *Emerson in . . . essays*: Emerson, *Essays: Second Series* (cited letter 0393, n. 3). 'Experience' is pp. 30–56.
24. *Paleys God*: W. Paley explained the watchmaker analogy in *Natural Theology* (see letter 0413, n. 11).
25. *Carlyle would call an outside God*: not identified.
26. *(This very idea . . .)*: Tyndall inserted this sentence as a footnote at the bottom of his sheet.
27. *Fichte's idea of the ego and nonego*: J. G. Fichte, *The Vocation of the Scholar* (Jena and Leipzig: Christian Ernst Gabler, 1794), trans. W. Smith (London: John Chapman, 1847). The first lecture, 'The Absolute Vocation of Man', addresses ego and nonego.
28. *All are but . . . soul*: A. Pope, *An Essay on Man: Epistle I* (1733), section IX, lines 9–10.
29. *occult relationship . . . in one of his essays*: Emerson, *Nature; Addresses, and Lectures* (cited n. 7), p. 8.
30. *Eterne alteration . . . flies*: 'The Sphynx', lines 97–98.
31. *Who solveth . . . I am*: ibid., line 131-2. Tyndall substituted 'solveth' for 'telleth' in line 131.
32. *These temples . . . grass*: Emerson, 'The Problem', *Poems*, p. 11, line 45.
33. *Thou art the . . . eye*: 'The Sphynx', lines 113–14.
34. *Thou askest . . . enquires*: Emerson, 'Wood Notes II', *Poems*, pp. 376–77.
35. *Uprose the merry . . . moon*: 'The Sphynx', lines 121–24.
36. *Fichte's doctrine*: see n. 27.
37. *Let us rejoice . . . infinite*: J. G. Fichte, 'Lecture V: Examination of Rousseau's Doctrines Concerning the Influence of Art and Science on the Well-being of Man' (cited n. 27), p. 72.
38. *Your last letter*: letter missing, but the most important contents are recorded in Tyndall's journal: 'Tom complains that I don't accept his cash, he tells me that I have disappointed him' (19 August, JT/2/13b/506); probably the same letter as Hirst wrote to Tyndall on 13 August, according to his journal entry. Other unexplained allusions (the opening of this letter and of letter 0421) also imply a missing letter. Tyndall sent Hirst's letter to Captain Wynne who had also offered him a loan (letter 0423 and Journal, ibid.).
39. *Mʳ Hirst . . . Yorkshire*: address from envelope.

From Thomas Archer Hirst 21 August 1850 0427

Halifax | 21st Aug. 1850

My dear Tyndall,

I got your very interesting letter yesterday[1] on 'The Sphinx' and have read it a good many times over, and the poem itself since. You have certainly opened up a good deal of it to me, though I shall yet have to ask you a question or two when I have more time. There is only one part of your letter which I cannot understand: your commentary on 'Jealous glancing around'. You refer to ourselves in it as far as I can make out, here is something you have noticed between us that I have not—you say it is a delicate point, but no matter, I shall treat it as it ought to be I hope, and I feel certainly anxious that you should speak more plainly. You say 'I may have gone too far already'. But how, and in what? I cannot for the life of me see. I know no one coming to Manchester just now—can you do without your bag until I come? If not I will send you it immediately.[2] I have received an invitation to a Tea-party[3] given by the Mutual Improvement Society to me before I go. They have sent for my <u>name in full</u>, by which I make a rough guess that they have some present or other in view. It's very kind on their part certainly, but puts me in a rather uncomfortable state of mind.

You have forgot my Pamphlet again.[4] If you haven't it by you it's no matter at all, only just say so.

Yours affectionately | T.A. Hirst.

You say my letter set you thinking.[5] That is what you ought to do and be a decent reasonable fellow. Let me know your conclusions early.

Phillips was over last Sunday, and was very glad to see your success at Edinburgh. In the Morning Chronicle and the Literary Gazette[6] I am told there are long accounts. I shall see the Gazette on Saturday.

I had a very interesting letter from Thomas Spencer the Quaker[7] this morning, which I will shew you when I come.

Dr. Tyndall, | Spring Bank, Over Darwen, Lancs.[8]

R1 MS JT/1/HTYP/108
LT Transcript Only

---

1. *very interesting letter yesterday*: letter 0426.

2. *can you do . . . immediately*: Hirst had the bag delivered (see letters 0431 and 0432).

3. *Tea-party*: a Mr. Hill of the Mechanics' Institute and Mutual Improvement Society of Halifax sent Hirst a letter suggesting that the Society hold a tea party for him before he went away (Hirst, 'Journals', 19 August 1850). The party took place on 23 August.

4. *my Pamphlet*: see letter 0421, n. 10.

5. *my letter set you thinking*: the missing letter of 13 August, see letter 0426, n. 38. Tyndall was re-thinking Hirst's offer of a loan.

6. *Morning Chronicle and the Literary Gazette*: *Literary Gazette* article, cited letter 0418, n. 10. We have found no mention of Tyndall's presentation in the *Morning Chronicle*, although on 5 August it noticed the Friday sessions of the BAAS.

7. *Thomas Spencer, the Quaker*: not identified. -

8. *Dr. Tyndall … Lancs.*: presumably from the envelope.

To Thomas Archer Hirst   [c. 22][1] August 1850   0428

Dear Tom,

I enclose you a letter received this morning from M[r] Edmondson.[2] I wrote to Queenwood as a matter of prudence, which in some circumstances becomes a matter of duty and this is the reply. Your last letter[3] set me thinking as I said, for on weighing the matter I felt convinced that so far from our being together in Germany being of injury to either of us we should reap reciprocal benefit from each other. This feeling was so strong within me that had I permitted myself I should have regretted writing to Queenwood at all—I intend to meet M[r] Edmondson in London and should hardly hesitate even now if I could arrange the matter to his satisfaction to put off my engagement for 6 months—I say I <u>ought</u> to be a proud fellow and I <u>am</u>—in here are two fellows in my hour of need coming forward and offering me their purse. You are one, and the Inspector of Railways for this Kingdom[4] is another a true born aristocrat but one whose aristocracy finds the noblest *[issue]*, a man whose very appearance gives the lie to any thing mean or unworthy. A man who knows in his heart that I care nothing for his aristocracy, that I in a measure defy him and it but who is still my friend. I have sent him your letter—after correcting the blunders of your style a little. You must pay a little more attention to this. No man has a right to eat before he has proved that he can do without eating and no man has the right to blunder without first proving that he can do without blundering.

I have not the least expectation that I shall go to Germany with you so dont dwell on that—I will see you before you go, and write out a record of hints and admonitions—besides we must constantly write to each other. I will certainly visit you there!

An hour ago your last letter reached me.[5] Can you find the date of the Morning Chronicle[6] which contained that thing? I should like to get a few as I have failed to provide myself with Athenaeums—'More plainly' Tom[7]—I'm like Dogberry[8] and could wish myself written down <u>ass</u>! 'More plainly' my beloved boy it is here—you wrote a letter to me to Germany once mentioning the thing itself this made me curious and I fancied I saw the same at Halifax. I knew it could be readily trampled down by a little more intercourse I knew it had no deep cause and therefore did not trouble myself about it, but I have been fighting a shadow, and I declare to the immortal Gods that I believe the defect is mine and not yours. I'm glad I mentioned it though I am an ass; for it fosters a tendency that I ever feel and that is to weld myself more closely to you and crush all barriers to our free intercourse—this time will come: I speak with[9] the authority of a prophet—this time will come.

I shall be in Manchester on Saturday I expect and shall remain there over Sunday.

tie <u>all</u> those things together in <u>one bundle</u> and send them to Manchester—directed to Ginty, 9 Howard St, West Broughton, Manchester.

I have had an offer of 80 or 90 pounds a year in the Ordnance[10]—I believe if I gave myself the trouble I might make it more but this I wont do.

Send me Mr Edmondon's letter back by return—I want it particularly. You shall receive your latter day pamphlet soon.[11]

Mr Thos Hirst | Harrison Road | Halifax | Yorkshire[12]

RI MS JT/1/T/533

1. [c. 22]: this letter may have been begun earlier, as a reply to a missing letter of the 13th, but during its writing ('an hour ago') Tyndall received letter 0427, which would not have reached him until the 22nd. It is postmarked Darwen 'AU22'.

2. *letter … Edmondson*: letter missing; probably a reply to the letter Tyndall sent 19 August. He recorded a reply in his journal (20–24 August, JT/2/13b/507). Edmondson had responded favourably to Tyndall's inquiry, saying that 'A brain and 10 fingers such as thine is just what I want'.

3. *Your last letter*: missing letter of 13 August (see letter 0426, n. 38).

4. *Inspector of Railways for the Kingdom*: G. Wynne (see letter 0423).

5. *your last letter reached me*: letter 0427.

6. *Morning Chronicle*: see letter 0427, n. 6.

7. *'More plainly' Tom*: Tyndall quotes and replies to letter 0427.

8. *Dogberry*: a character in the Shakespeare play *Much Ado about Nothing* known for his inflated view of himself and his comically blundering ineptitude.

9. *I speak with*: from this point the letter is squeezed into margins, which suggests that Tyndall did not want to start a new sheet. The rest of this sentence is in the left margin of the final (third) sheet; the next paragraph in the right margin and the following along the top of the same sheet. The final two paragraphs are cross hatched on the second sheet. Tyndall did not sign the letter.

10. *an offer … in the Ordnance*: probably from Yolland, to whom, on Wynne's advice (letter 0423), Tyndall wrote on or before 19 August (Journal, 19 August, JT/2/13b/506).

11. *latter day pamphlet soon*: Hirst had requested the return of the pamphlet in two extant previous letters (0421 and 0427).

12. *Mr Thos … Yorkshire*: on envelope.

To Thomas Archer Hirst [30 August 1850] 0429

Spring Bank, Friday Morning[1]

Dear Tom,

Give me an answer to the following question—Supposing I go to Germany with you are you without any inconvenience to yourself able to place 40 pounds at my disposal between now and this day 6 months? I wish to know this before I meet Mr Edmondson. It may have an effect upon my negotiations with him, possibly induce me to defer them to another season. The advantage of entering upon an engagement at present may in the long run prove a disadvantage—just give me an answer to this by return—

Your affectionate friend | J Tyndall

We must have 10 days together as you say at all hazards.

I have received the Guardian[2] accompanied by your note[3]—The reward was well earned and may be legitimately enjoyed for the simple reason that it was not for it that you laboured but under a higher influence Your remarks are quite to the point, I cannot add to them—A man cannot be concealed Tom. Stupidity ever finds him out in the long run and /bruises/ him. A man is often unconsciously an organic nucleus which holds the babblers of society together—Where we should drift to, were it not for these fixed points nobody knows. It was a pleasant illustration of human nature—go on my boy—care nothing about it.

> For Gods delight in Gods
> And cast the weak aside
> To them who scorn their charities
> Their doors fly open wide—[4]

RI MS JT/1/T/1014

---

1. *Friday morning*: 30 August 1850. Hirst replied immediately in letter 0433, written on 1 September.

2. *the Guardian*: a report of Hirst's farewell party, organised by the Halifax Mechanics' Institution and Mutual Improvement Society (see letter 0427, n. 2), was published in the *Halifax Guardian* under the title 'Testimonial of Respect'. Hirst copied the report into his journal (24 August 1850).

3. *your note*: missing.

4. *For Gods delight in . . . open wide*: Emerson, 'The World-Soul', *Poems*, pp. 21–25, lines 93–96, misquoted and should read: 'For gods delight in gods, | And thrust the weak aside; | To him who scorns their charities, | Their arms fly open wide'.

To William Thomson   31 August 1850   0430

Spring Bank 31[t] Aug 1850.

Dear Professor,

I return you my sincere though tardy thanks for your long and valuable letter and the pamphlet which accompanied it.[1] I have waited this long hoping to be able to reply to you at some length. But a reply on such a subject if not elaborated by reflection is worse than none at all and this reflection it has not latterly been in my power to bestow upon it—I may return to Germany in a few weeks. If not I will certainly procure the proper apparatus for myself and shall feel truly happy to make whatever experiments you may require—with best wishes

believe me Dear Sir | very sincerely yours | John Tyndall
Prof. W. Thomson | Glasgow

Cambridge University, Kelvin Correspondence, Add.7342/T624

---

1. *letter and the pamphlet which accompanied it*: letter 0425, which included an article by Thomson (cited letter 0425, n. 7).

From Thomas Archer Hirst   31 August 1850   0431

Halifax | 31st August, 1850

My dear Tyndall,

How are you coming on there? As usual I suppose or you would not be silent. January[1] and I were very near coming to Manchester last Sunday, he was there the day before and had promised to take me over to see some literary friend or other. But the day was too wet or we should have met with you. Did you get your bag safe?[2] I have read the Literary Gazette[3] carefully, and your case is perfectly clear to me now—the subject is called intricate by yourself and by Thompson;[4] if so there is great credit due to you, for it is as plain and simple as a pike-staff,[5] as you have presented it. I have not been able to get hold of the Morning Chronicle[6] I spoke of. Baines[7] told me he had seen it there and to-night I will ask him the date. To-night is my last 'Saturday evening'[8] and the sorrow at their conclusion is alleviated most by the fact that when viewed from a point in time their interest and value will be increased, or rather made more visible to me. For these things when viewed near have never that steady clearness which they possess when viewed from their proper focus in Time.

I am working away at Ollendorf's Exercises.[9] I find it a thorough good system. I progress slowly but begin to feel the ground under me. I have been reading some poems of Carlyle's—he has only written about 7 or 8, which I almost wonder at, seeing the worth of those he has done; it must be that he finds verse-making is not his proper task—and why he thinks so is a significant question to me and one to which his poems have not yet yielded an answer. I am so pleased with them, and they are so few, that I have copied them in my Log Book,[10] and we will read them together. Another thing I have read that pleased me also is an old Critique by John Sterling on Carlyle.[11] It was written in the London and Westminster Review many years ago, immediately after the appearance of 'Sartor Resartus' in the Magazines.[12] Sterling is a man of deep clear insight and wide liberal sympathies; he appreciates Carlyle thoroughly and did not hesitate at that early period to characterize him as our modern Luther. He intermingles his own individuality strongly in his writings, and so clearly presents himself that he as it were holds up his candle and illuminates both Carlyle and himself. He is a sceptic much after Emerson's definition, a fellow that can always see there are two sides of a question, and is anxious to place himself across the fulcrum of the balance.[13] Carlyle has in a measure passed through this, and moreover has been so constituted by nature as to feel more keenly than most; he has on <u>cool</u> subjects just as much insight as Sterling—perhaps more, but it is impossible when his heart is heaving for him to weigh his assertions and define their limitations; if it is applicable to his case he uses it: careless whether one who feels less keenly and thus cannot see its deepest truth applies it to circumstances of less pressing moment, generalizes it, and finds it wanting. In this peculiarity of the two men, Sterling and Carlyle, lies all their differences. The critique is truly an instructive one, such as we seldom see in this day.

How are you getting on with Edmondson, and when do you expect going to meet him in London? Tell him however he may want your services someone else[14] wants them also, and perhaps more, and that you wish to keep yourself as free as possible should a more advantageous situation present itself in the course of 6 months. It would be prudent for you to make yourself as much known as possible through your magnetic experiments, that Edinburgh affair[15] will be of service in this respect, and could you extend it, it would be of more. Tell me what would be the best address to find me in Marburg. Continue for a week or ten days to direct my letters to Halifax. I shall be away a good deal but will leave all instructions for having them forwarded.

Yours as ever, | T.A. Hirst

Dr Tyndall, | Spring Bank, Over Darwen, Lancs.[16]

R1 MS JT/1/HTYP/110
LT Transcript Only

1. *January*: Hirst and Tyndall usually referred to G. S. Phillips by this pseudonym.
2. *Did you get your bag safe?*: see letters 0427 and 0431.
3. *Literary Gazette*: see letter 0418, n. 10.
4. *Thompson*: William Thomson. The error may be LT's, though Hirst often misspelled names.
5. *plain and simple as a pike-staff*: a proverbial phrase referring to something obvious (*OED*).
6. *Morning Chronicle*: see letter 0427, n. 6.
7. *Baines*: a member of Hirst's Saturday evening circle. Not otherwise identified.
8. *'Saturday evening'*: see letter 0412.
9. *Ollendorff's Exercises*: Heinrich Gottfried Ollendorff wrote several language lesson books. Hirst probably alludes to either *A New Method of Learning to Read, Write, and Speak a Language in Six Months, Adapted to the German*, 2 vols (London: Whittaker & Co., 1838 and 1841) or *Key to the Exercises in Mr. Ollendorff's Method of Learning German* (London: Whittaker & Co., 1840).
10. *copied them in my Log Book*: Hirst copied 7 poems by Carlyle into his journal between 28 August and 3 September.
11. *old Critique by John Sterling on Carlyle*: J. H. Sterling (pseudonym '£'), 'Carlyle's Works', *London and Westminster Review* 33, no. 1 (October 1839), pp. 1–68.
12. *'Sartor Resartus' in the Magazines*: T. Carlyle first published 'Sartor Resartus' as a serial in *Fraser's Magazine* between November 1833 and August 1834 (cited letter 0398, n. 24).
13. *a sceptic . . . of the balance*: Hirst alludes to Emerson's essay, 'Montaigne; or, the Sceptic', in *Representative Men*, pp. 109–38, especially p. 114.
14. *someone else*: that is, Hirst.
15. *the Edinburgh affair*: the recent BAAS meeting (see letter 0418).
16. *Dr Tyndall . . . Lancs.*: presumably on the envelope.

To Thomas Archer Hirst　　　　1 September 1850　　　0432

Sunday morning Sept 1ˢᵗ 1850
<u>Churchtime!</u>

My Dear Tom,

Your letter[1] reached me 10 minutes ago and I have read your remarks about Carlyle and about Stirling[2]—how I like to follow those gropings of that soul of yours!—There is a great difference between such men as Carlyle and such men as Stirling the one has faith and the other none—Stirling must <u>see</u> to believe, Carlyle is able to appreciate 'the evidence of things not seen'[3]—and however this faith may be decried if exhibited by our Contemporaries it must be confessed that it has formed the great substratum of achievements which even sceptics admire—Even in an intellectual view there must be something solid here. that principle is not to be slighted which enables a man to set hunger and cold at defiance and enable the martyr to rejoice amid the flames that are choking him—I confess to a certain obstinacy on this point—a willfulness, if you like to call it such, that will not listen to reason—I know all the intellectual men have to urge, and grant it what I conceive to be its proper value; but I know also that there are sources of power open to me which they have never fathomed—there is something in nature, name it I cant, but <u>know</u> it I do an habitual falling back upon which gives a man sinews of iron. This is a fact—a fact proved and reproved by a hundred experiences—Let the intellect make what it will of it there it lies as incontestable as the fact of my own existence—Shall I allow my logical faculty then to bamboozle me by a sneer—or if the said logical faculty weakly suffer itself to be bamboozled by the logic of another shall I cut myself adrift and follow it in its tossings? Not I faith. I confess to a certain stubbornness here and have after found its value—

There appears to be enough rude primal element of power rendered accessible to man by this thirst. But the intellect was not given for naught It must throw its torch light upon this power and guide it aright. It is the waggoner who by bit and rein guides the strong horse to his work—If the horse be wicked the waggoner must be brave—and if the power aforesaid be strong the guiding power must be proportionate—This constitutes what you yourself call a balanced mind. It is pitiful to hear the logical jibber of some people, but even more to witness the waste of force in other cases arising purely from want of balance between force and insight.

Carlyle is the begotten son of the Universe[4]—and he knows it—Stirling is an alien guest and he knows it—The one is Isaac, the other Ishmael[5]—'The son of the bondwoman shall not be heir with the son of the free woman'.[6]

Your remarks upon the paper in the Literary Gazette[7] have induced me to turn to it and to read it over. It strikes me that the reasoning is very fair, and if I had it to do over again I dont know where I should alter it. But you would not believe what [an] influence the simple reading of that paper has had upon my thoughts—turning from it the expression 'Carlyle is the begotten son of the universe' appears mere bombastic nonsense—why? Because I don't see the truth of the expression. and all that about sinews of iron has no meaning simply because I don't feel it—my intellect has been otherwise engaged it has been looking at other objects till the impression of the first mentioned has [become] dim—but it would be just as rational to accept this as an argument as to infer that there is no building called the exchange in Manchester because I dont just now see it—Is it not there to be seen any time I take the trouble of walking in to Manchester; and with the same perspicuity will my old experience arise if I only give myself the trouble of contemplating the matter—knowing this I do not call what I have written nonsense—I know that it had as good a right to be written as my scientific paper. The claims of both are recognized and thus a brotherly harmony is established between them—but I'm rambling Tom—I have written in a rigmarole style and it is a question whether you will extract anything from what I have written—

Your affectionate friend | John Tyndall

All the address necessary will be Thomas Hirst | Marburg | Hesse Cassel

Every little boy in Marburg will know you in a week! <u>der grosse Englander</u>![8] the expression will be the property of every household in 10 days—

I should have liked to see you and January—in fact I half expected you and told Ginty so and felt a kind of disappointment as the day past without any intelligence of you—the bag came safe[9]—

RI MS JT/1/T/534

---

1. *your letter*: letter 0431.
2. *Stirling*: the correct spelling is Sterling (see letter 0431, n. 11).
3. *'the evidence of things not seen'*: Hebrews 11:1.
4. *only begotten son of the Universe*: this metaphor makes a Christ-like claim for Carlyle. The phrase 'only begotten son [of the Father]' is used in the New Testament (see John 1:14 and 3:16) and Christian liturgy for Jesus. Tyndall criticizes the metaphor in the following paragraph. We cannot find any other source for the quote, thus he appears to withdraw his own grandiose claim.
5. *The one is Isaac, the other Ishmael*: Ishmael was Abraham's first son from Sarah's bond-woman Hagar, and Isaac his second, more-privileged son, because from Sarah, his wife. See Genesis 16:1–15, 21:1–20 and Galatians 4:21–31.
6. *'The son . . . free woman'*: Galatians 4:30.
7. *paper in the Literary Gazette*: cited 0418, n. 10.
8. *der grosse Englander!*: the tall Englishman (German).
9. *bag came safe*: see letters 0427 and 0431.

From Thomas Archer Hirst   1 September 1850   0433

Halifax | Sept. 1st, 1850

My dear Tyndall,

January Searle is just opposite me and come to spend his last Sunday with me. If he knew I was writing he would send his love to 'that independent old devil Tyndall'. He has been looking at your portrait and expressing his regard for you with all manner of questionable oaths. Your letter[1] came to me this morning and I am glad to answer it by return, for it is a cheering prospect. I can say without any exaggeration that in a fortnight's notice at any time I can lend you £40 as easily as I could lend you a penny. I shall take as much with me from here besides my income whether you accept it or not, so you may reckon me as safe as if it were your own purse. In the event of your accepting it you may leave Spring Bank and must tell me where will be the best place for us to spend a week together before we go. I cannot leave England very conveniently before the first week in October, and should like us to go together if possible—so write to me soon, old fellow.

Yours affectionately, | T.A.Hirst.

The last week of September I must spend in Bristol. The first week or 10 days of it I must spend here, and with my brother[2] at Badsworth. The interval must be spent with you. | T.A.H.

Dr Tyndall, | Spring Bank, Over Darwen, Lancs.

RI MS JT/2/14/174
LT Transcript Only

---

1. *your letter*: letter 0429.
2. *my brother*: William.

To George Edmondson   [2 September 1850]1   0434

I have come to the conclusion of making the proposal to defer further negociations to 6 months from the present time. Now if this arrangement suit you it will answer me admirably. There is a friend of mine going to study at Marburg and it will be a great pleasure to me if I can so arrange matters that we may spend the first 6 months of his time of study together.

If you have made any arrangements with reference to my coming to Queenwood I shall feel it my duty to see that you are not disappointed; but if it be all the same to you, or better still if it be a convenience to you, I should certainly prefer spending another 6 months in Germany.

Will you be good enough to grant me a reply to this by return.[2]

RI MS JT/1/T/954
JT Transcript Only[3]

---

1. *[2 September 1850]*: on 2 September 1850, Tyndall noted in his journal that he 'wrote to Mr. Edmondson proposing to defer negociations' (JT/2/13b/508).
2. *reply to this by return*: letter missing.
3. *Transcript Only*: the original letter is missing. We know it only through this extract, copied out by Tyndall in letter 0436 to Hirst.

To William Ginty   [2 September 1850][1]   0435

Dear Ginty,

The following is the manner in which I would propose treating the subject of electro-chemistry.

1. As subjects of this nature are generally mystical to a popular audience; lecturers I imagine introduce too many unexplained terms, and these when used serve only to arouse perplexity in the hearer's mind. This cannot be remedied by appending a dry definition of each term, as this would render the lecture insufferably tedious—besides the definitions would, by nine persons out of ten, be forgotten as soon as uttered.

This I think as regards the subject before us might be obviated by running shortly through the rise and progress of these terms. This process would form an interesting part of a lecture, indeed it should be introduced as such, and the terms explained incidentally as they occur. For example, to my mind the very best way of obtaining an intelligent notion of electric action is to begin by a short illustrated description of magnetic action—A few experiments with permanent and on[2] the action of the earth upon soft iron would make this sufficiently clear; the phenomena of attraction and repulsion would be exhibited, the terms poles and fluids would be made familiar and their adaptability to the case in hand exhibited in the manner which recommended the use of them to the mind of their inventor—this would undoubtedly be the most natural mode of representation.

2. From these easy beginnings we should go into electric action, making our experiments run as far as possible parallel with those already made in the case of magnetism, the point where both forces diverge and assume peculiar characteristics would in this way be strikingly exhibited.

3. Having obtained a general notion of electricity we would more strictly define its sources and here introduce the form it assumes in the Voltaic Circuit. We should begin by exhibiting an example of the simplest current possible, which any child could make, and from this ascend to the various contrivances which man have made use of and to obtain a copious and intense current.

4. These contrivances, however different in outward form depend all upon the same simple principles, and bear the common name of the galvanic battery. Grove's battery[3] would be explained, Daniells[4] also, and the coal battery of Bunsen, being that used by the lecturer, would occupy his especial attention.

5. The phenomena of the Stream would be dwelt upon - how to detect its presence and direction, the laws of its action upon a fully suspended magnet and the application of these laws to the determination of the power of the current.

6. The relation of electric action to chemical action, illustrated by the decomposition which invariably attends the passage of the current through a fluid. A series of very interesting experiments might be made here; various salts might be decomposed—the component acid being <u>visibly</u> exhibited at one pole and the base at the other; the decomposition of water and the recombination of the gases. Electrotype, electroplating, &c would follow here as illustrations of these laws of decomposition, medallions would be made and chains gilded before the audience. To a select audience it would be of deep interest to exhibit the connexion between these phenomena and the atomic theory of Dalton. I may also remark that the mathematical treatment of the laws which regulate voltaic action would be more interesting to a similar audience than the best experiments a lecturer could exhibit.

7. But I must not forget <u>le peuple</u>.[5] For these the heating and illuminating power of the current would be amply exhibited - metals would be smelted and the application of galvanism to blasting operations explained—the round down cliff[6] and other experiments would be illustrative. The electric light would be shown and many beautiful phenomena of interference—Newton's rings for instance—exhibited—If the Committee[7] thought fit, a comparison of lights might be made and the photometer of Bunsen, a simple and beautiful method for obtaining the relative intensities of any two lights might be applied and explained.

If a select number of gentlemen wished to look more deeply into the matter, then we might introduce the various apparatus used for determining its intensity and so forth—the Tangent [Galvanometer] of Weber[8] and the Rheostat of Wheatstone[9] might be practically applied. By means of a combination of these two instruments and a knowledge of the laws of voltaic action we are able by the decomposition of a drop of water to calculate the amount of the earth's magnetic force at any point upon her surface.[10]

RI MS JT/1/TYP/11/3702–3703
LT Transcript Only

---

1. *[2 September 1850]*: based on Tyndall's journal entry for 2 September 1850, where he wrote, 'Got a short lecture ready for the Athenaeum. Wrote to Ginty concerning it'.

2. *permanent and on*: this does not make sense. The LT transcript is incorrect; at the least the word 'magnet' should follow 'permanent'. LT perhaps omitted a line.

3. *Grove's battery*: named after William Grove (1811–96).

4. *Daniells*: named after British chemist John Frederic Daniell (1770–1845).

5. *le peuple*: the people (French).

6. *round down cliff*: on 26 January 1843, in a spectacular feat of engineering, 18,500 pounds of explosives were used to remove a large section of Round Down Cliff near Dover. (H. M. Noad, *Lectures of Electricity* (London: George Knight and Sons, 1844), p. 192.)

7.	*Committee*: n. 1 implies that this was the Athenaeum Committee.

8.	*Tangent [Galvanometer] by Weber*: an instrument used to measure minute electric currents. This refers to the galvanometer perfected by Wilhelm Weber in the early 1840s. See letter 0465 for Tyndall's use of such an instrument in his own experiments.

9.	*Rheostat of Wheatstone*: a rheostat is a variable resistor used to control the current in a circuit. This refers to the rheostat invented by Charles Wheatstone.

10.	*surface*: letter incomplete.

# To Thomas Archer Hirst	[2 September 1850][1]	0436

Monday

Boy!

I have written to Mr Edmondson and my letter contains the following words [...][2]

Boy I will pay you if the Gods grant me life and if I go with you—and boy if I don't go, the memory of this act of thine[3] shall cling to me until it is effaced by a nobler. Were I to tell thee my faith in thee it would make thee proud, and were I to mention the hopes I entertain concerning thee, it would make thee vain. Were I rogue I should not dare thus to write to thee for it might arouse suspicion—But by the Heavens I care nothing at bottom for thy 40 pounds—If I accept it I shall accept it as a free man. And yet the gods know that I should deem this fair act of thine cheaply purchased by 7 years probation upon bread and water.[4] I have met many who professed to love me in this world but two only can I now turn to with confidence. In my quiet moments I draw sweet solace from the thought of them—they are worth the universe to me

J.T.

poor January[5]—there's no love lost—I like the fellow heartily—I believe he is an Israelite in whom there is no guile.[6]

RI MS JT/1/T/954

1.	*[2 September 1850]*: based on Tyndall's journal entry, dating his letter to Edmondson to 2 September 1850 (JT/2/14/174). It could have been written a day or two later, but as it served also as a thank you letter for the loan, Tyndall probably did not delay.

2.	*[...]*: Tyndall copied out part of his letter (0434) to Edmondson at this point.

3.	*act of thine*: a reference to Hirst's willingness to loan £40 to Tyndall for the trip to Germany (letter 0433).

4.	*7 years... bread and water*: a type of extreme hard fare, as of a prisoner or penitent (*OED*).

This may refer to Jacob serving seven years to 'pay the price' of his bride, thus implying long service and harsh conditions.

5. *poor January*: the entire postscript is written in the left-hand margin of the first page.

6. *poor January . . . love lost . . . an Israelite . . . no guile*: three allusions which refer to Hirst's letter 0433. There are later letters in which Hirst and Tyndall express sympathy for January but we have not determined the specific allusion here. 'Israelite. . .' is an allusion to a statement of Jesus about Nathanael (John 1: 47).

To John George Roberts　　　5 September 1850　　0437

Now what can I say on this matter?[1] 'I have been intimately acquainted with Mr John Roberts for many years.[2] He has a gentlemanly deportment and address, superior natural powers which he has most wofully neglected to cultivate. I believe conscientiously that the restlessness and indolence which makes his present situation distasteful would accompany him to the continent, and it is my firm conviction that he would not remain in the situation at Geneva for 6 months'.

RI MS JT/2/5/43[43]

1. *say on this matter?*: Roberts, who was applying for a position in Geneva (see Tyndall journal entries for 3 and 5 September, JT/1/13b/508), had requested a letter of reference from Tyndall.

2. *acquainted with Mr John Roberts for many years*: Tyndall worked with Roberts on the Irish Ordnance Survey until he was discharged on 9 January 1841. Tyndall's opinion of him at this time was decidedly negative. 'When Roberts was in Leighlin I firmly believe he was an honest man, but now I regret to say he is devoid of every vestige either of principle or honour'. His pseudonym was Sam Weller, for a character in C. Dickens' *Pickwick Papers* (see letters 0038, 0338, and 0351).

3. *JT/2/5/434*: this fragment, from Tyndall's journal, is either a draft for or a transcript of the letter sent.

To Thomas Archer Hirst　　[6 September 1850]1　　0438

Friday

My Dear Tom.

I enclose you Mr Edmondson's reply.[2] The matter is settled and we shall work together in Germany. I have written to Knoblauch and to Debus announcing the fact. If I mistake not the Semester will commence on or before the 1st of October. It used to commence <u>later</u> but I think its changed and I dont know the precise nature of the change. I shall soon have all necessary information upon this head as I have requested Debus to write to me <u>immediately</u>.

Now your wish is gratified and mine also—I believe it will be to our mutual benefit to be together. It will not be necessary to spend the week or ten days together before hand, but if you like we will visit Queenwood together—What say you to this. —It will be nearly a fortnight before I can hear from Marburg. | J. <u>Tyndall</u>

~~*[Do you]* expect to have much baggage? if you want *[any information]* on this head ask it—~~[3]

Thomas Hirst | *[M*r *Thorps]* | Badsworth Pontefract[4]

RI MS JT/1/T/535

---

1. *[6 September 1850]*: a Friday. Hirst received this letter on Sunday 8 September (see letter 0441). It took two days to reach him because it was originally addressed to Halifax and was forwarded to Badsworth (see n. 3). Hirst noted on the outside of the envelope that he received it on 8 September.

2. *Mr. Edmondson's reply*: letter missing. Edmondson had agreed to Tyndall's proposal (letter 0434) to defer negotiations over his return to Queenwood for six months.

3. *[Do you] expect . . . ask it—*: postscript inserted, then crossed out in looping over-writing, in the left margin of this page. Hirst read it and replied, in spite of thorough crossing out, in letter 0441 (written before he received the same enquiry in letter 0440 from Tyndall).

4. *Thomas Hirst . . . Pontefract*: this address was written, in a different hand, after 'Harrison Road, Halifax' (in Tyndall's hand) was thoroughly crossed out. The postmarks are 8 September, at Wakefield and Pontefract.

From Thomas Archer Hirst   [6 or 7][1] September 1850   0439

Mr Thorp's | Badsworth | nr Pontefract | 6th September

My dear Tyndall—

I am here among the Farmers, at their Harvest Time, and a gladdening sight I find it. They all look so active and joyful that it infects me too & I seize a fork and work away with spirit. No doubt there is as much Poetry in Long Chimnies and Smoke as in these green Fields and Stoops of Corn, but it is something deeper than mere custom that hides the one & makes the other so apparent and pleasing. I wonder what it is. I have stipulated however to have my mornings to myself, on pain of damnation to all intruders, as I tell them I am composing sermons. I shall stop here until Wednesday next[2] then spend Thursday and Friday in Halifax packing up, and if all be well start for Lancashire on Saturday. We have all to get drunk on Friday night at Smiths[3] & then I have done with Halifax. I shall leave my luggage I think with Ginty until we return from Over Darwen. Which way must I come to find you? or is it necessary to come there at all, or will you be leaving before? Write to me here by return, as your answer may alter my present intentions

Yours as ever—| T.A. Hirst

Dr Tyndall | Spring Bank, Over Darwen, Lancs.[4]

RI MS JT/1/H/154

1. *[6 or 7]*: the letter is headed 6th, but according to his journal Hirst wrote to Tyndall and Jemmy on the 7th and, according to letter 0441 (at n. 3), he recollected that he wrote on a Saturday (that is the 7th). The postmark is Pontefract, 7 September.

2. *Wednesday next*: Hirst visited relatives in Badsworth (a village about 20 miles east of Halifax) and Wakefield between 6th and 12th September; he returned to Halifax on Thursday, 12 September (Journal, 6–12 September).

3. *Friday night at Smiths*: the party (according to Hirst's journal entry, 13 September) included his Carlylean, 'Saturday' friends and others. Smith was probably John Stores Smith.

4. *Dr Tyndall . . . Lancs.*: address from envelope.

To Thomas Archer Hirst [9 September 1850][1] 0440

Monday morning

My Dear Tom.

Here I am with nobody up by myself—I have paced the carpet for the last hour thinking of what I should do with myself till breakfast appears. The morning air is chilly and a fellows intellects are awake. without employment at such a time they are apt to become insurrectionary—This is the reason I write to you—you are a makeshift, a <u>tub</u> thrown to amuse the whale till the boat is out of danger. What shall I say to you? What spirit is to give me utterance? Well there's your luggage—will you have much to carry with you? We can make the journey much pleasanter and quicker if we can go thro' Belgium but then if the luggage be heavy the fare would be high. You see going <u>up</u> from Rotterdam we have the river <u>against</u> us which makes slow work of it—In the way of clothes I would recommend you to bring one thing and that is a very rough warm winter-coat—some rough pilot or Kersey[2] is the best thing you can have—let it reach down a little below your knees - you will find the want of such an article before the winter is over—You can also bring with you a few pairs of Comfortable woollen gloves. there is no advantage in buying these articles in Germany—but on the contrary a disadvantage.—

You talk of not being able to start till the first week in October[3]—but that was said under the impression that we were to stop a week or ten days together—now this time can be handed over to your brother in Bristol. I expect an answer to my letter in 10 days or less[4] and I should like us to be able to start as soon as possible afterwards. Even if we are there too early we can find profitable occupation for ourselves.

I hope you are enjoying yourself—Give up your speculation for the time being, and become a son of the earth once more, rattle about and amuse yourself—dine liberally[5] but dont injure your stomach—take a glass of wine if any body offers it—Strengthen the muscles of that frame of yours—for by gods help the Strength so acquired shall not lie latent yonder

J. Tyndall

RI MS JT/1/T/536

1.   *[9 September 1850]*: Hirst wrote this date (which was a Monday) at the top of the letter. It seems that here Hirst deduced the date of sending rather than recording the date received. This date is supported by Tyndall's allusion to letter 0439 of 6 or 7 September (n. 5 below).

2.   *pilot or Kersey*: both 'pilot' and 'Kersey' are types of woollen cloth (*OED*).

3.   *You talk of . . . first week in October*: see letter 0433.

4.   *my letter . . . or less*: probably the letter to Debus mentioned in letter 0438.

5.   *son of the earth . . . dine liberally*: Tyndall is responding to Hirst's letter (0439) about country life.

From Thomas Archer Hirst   9 September 1850   0441

Badsworth— | 9 Sepr 1850

My dear Tyndall—

I received your letter[1] and the good news[2] in it yesterday when going to the village church here. This Morning I have been walking down to the post office (or the apology for one) to see if you had written again yesterday but I did not expect it as yours was an answer to what I asked on Saturday.[3] On Saturday then instead of coming to Spring Bank I will go to Bristol and on my way from there to London will call at Queenwood and spend a day or two with you there. I understood you to say before that the Semester began at the

<u>latter end</u> of October.[4] I must wait until after the 1st of October in England and then I don't care how soon we go. On Thursday and Friday I shall be in Halifax let me have a letter from you then & say whether these arrangements will suit I am glad your returning to Germany suits Mr Edmondson also, for now we go without sacrificing any body's interest and I must confess I look upon my journey with much more satisfaction. Tell me on Thursday when you will start for Queenwood & then I can write you all further intentions from Bristol If you should write again before I answer your next a letter will find me directed to <u>Tom</u> Hirst. 9 Blenheim Square. Marlborough Hill. Bristol—I shall have 2 trunks—1 Box a Carpet bag and a Hat-box, which I think will not be too much. There will be some books which I had better perhaps get, but you will be able to tell me in London. You will also know about passports. I have your umbrella with the Eudiometer[5] (?) in it, will you take it back to Germany? or if not where will you have it left? I have borrowed Ollendorfs Exercises[6] in Halifax and am working at them here. At Bristol I shall be stopped by not having one. I hardly think it worth while to buy one now, as I can get hold of yours so if you go through London before I leave Bristol I wish you would send it me by post—Now I think I have finished all my practical matters & driven all others of a more kindly nature out of my head so no more till we meet. I am reading Tennysons 'In Memoriam'[7] in these green Fields.

Yours affectionately | T.A. Hirst.

Dr Tyndall, | Spring Bank, Over Darwen, | Lancs.[8]

RI MS JT/1/H/155
RI MS JT/1/HTYP/118

---

1. *received your letter*: letter 0438.

2. *the good news*: that Edmondson had agreed to defer negotiations over his return to Queenwood for 6 months, allowing him to travel to Germany with Hirst.

3. *what I asked on Saturday*: this implies either a visit or letters written on Saturday. Possible letters are letter 0431 (of Saturday, 31 August) in which Hirst asked Tyndall about his negotiations with Edmondson or, more likely because more recent, letter 0439 (possibly Saturday, 7 September) in which Hirst asked about meeting up with Tyndall. Tyndall answered both queries in letter 0438.

4. <u>*latter end*</u> *of October*: in letter 0439 Tyndall had said he thought the semester started on 1 October or even earlier.

5. *Eudiometer*: an instrument used to measure changes in a specific volume of a mixture of gas after a physical or chemical change. It can also be used to test the purity of the air or the quantity of oxygen it contains (*OED*). The question mark may indicate that Hirst was unsure of the spelling. Tyndall may have stored it in his umbrella for safe-keeping.

6. *Ollendorfs Exercises*: see letter 0431, n. 9.
7. *Tennysons 'In Memoriam'*: A. Tennyson, 'In Memoriam' was completed in 1849, a tribute to his friend Arthur Henry Hallam who died suddenly at the age of 22.
8. *Dr Tyndall . . . Lancs.*: only in LT transcript, presumably from the envelope.

To Thomas Archer Hirst   [10 or 11 September 1850][1]   0442

My Dear Tom—

Bring all that you have belonging to me to Manchester on Saturday and leave them with Ginty—Do you know where he lives? When you come down from the station turn to your right, pass under the railway bridge and walk straight forward till you come to Broughton toll bar—about 100 yards beyond the bar Howard Street branches off to the right—N° 9 is Gintys abode[2]—

If you could conveniently spend Sunday in Manchester I should rather like it and I would meet you there;[3] I can promise you a hearty welcome from Ginty but perhaps it would suit us both best to have a cozy chop together at Wovendens[4] or somewhere else; this point is easily settled if we agree to meet—Say the word then shall we meet or shall we not? if the affirmative then I promise to be in Manchester sometime on Saturday—you have nothing to do if you arrive before me but to sit quietly down at Gintys—they will give you Pickwick or Holts poems[5] to amuse you—Tell Mrs Ginty that I said she was to take good care of you and to treat you tenderly—she will do this— I'll vouch for her.

We can talk over many little matters when we meet which it would be a bore to write about. If you can conveniently manage it therefore I wish you would remain over Sunday—tell me your mind by return for if you cannot remain there I have no business in Manchester.

I dont think I shall spend any time at Queenwood—I should certainly like to take a run down from London stop there a night and come away next day—this time would be sufficient to see the place and to talk over business matters. We shall perhaps be able to manage it together.[6]

I am already busy with my investigation, that is busy in my brain, for every step ought to be verified by experiment—looking at the matter thoroughly I hope to be able to cram as much into the coming half year as any other chap of my age and opportunity in these Kingdoms—I have a fair field before me on the very outskirts of science looking from the known out upon the unknown and carving a track through the latter for myself. I shall endeavour to make my half year a profitable one and I believe I shall succeed. It is the strong conviction that abiding results will flow from the coming 6 months, backed by another conviction that our union for this time will be of mutual advantage to us in other respects that induces me accept a certain moral pressure which my present position imposes on me—My philosophy cannot entirely dissipate this, for the position is new to me—Reason however tell me that the pressure will be transitory and the advantages flowing from it permanent.

I dont know whether I mentioned to you before that I have been favoured with a letter 18 pages long from the Professor of Natural Philosophy in Glasgow,[7] my opponent in Edinburgh—He is working at the same subject, in fact every body will be having a trial at it as they see that a new field of speculation and experiment is opened. But it takes long preliminary discipline before a man can get thoroughly into such a subject and in this respect I am a certain distance ahead, which advantage by the favour of the immortals I intend to maintain—

I dont know exactly when the Semester begins—I shall know in a few days—One thing however I know that neither of us shall need profitable employment even should we arrive too soon. If business matters detain you until after the first of October in England I am content. I am anxious ere we leave to visit the Polytechnic Institution in London, and should like us to do it together.

Last saturday night at 9 oC I called at Ginty's house in Manchester, he was at the house a friend whither I went to seek him—I thought he and his friend were indulging in a social chat together, but on entering the room where they sat they formed a small fractional part of the company present—the table was loaded with brandy and whiskey bottles and the room crammed with tobacco smoke—I sat me down and remained there for nearly 3 hours. the[8]

Bid your friends farewell for me also. I feel a certain identity of fate and doom with them which induces me to make this request. —If Smith and January be finally sent to the devil where am I to go? certainly to share the same infernal hospitality—Well there's comfort in the prospective sympathy of such companions—tell January not to forget his Meerschaum![9]—

Commend me to the kind remembrance of Hutchinson, Roby and Booth—

truly thine | Tyndall—

RI MS JT/1/T/1013

---

1. *[10 or 11 September 1850]*: Hirst wrote 'Sep–11–1850' at the top of the letter. Whether this is date received or date of writing is uncertain; we consider the latter more likely.

2. *No 9 is Gintys abode*: Ginty's address was 9 Howard Street, West Broughton, Manchester.

3. *spend Sunday … meet you there*: Tyndall spent Sunday 15 September with Hirst and Craven (see Journal, 21 September (JT/2/13b/510) and Hirst, 'Journals', 14 and 15 September).

4. *Wovendens*: a lodging house in Manchester, England (see letter 0619).

5. *Pickwick or Holts* poems: C. Dickens, *The Posthumous Papers of the Pickwick Club* (London, 1836). Holt may refer to David Holt (1766–1846) and his *A Lay of Hero Worship and Other* Poems (London: William Pickering, 1850).

6. *I don't think . . . manage it together*: Tyndall and Hirst visited Queenwood together on 29 September; Hirst stayed only one night, Tyndall stayed two nights (see Journal, JT/2/13b/512, and Hirst, 'Journals', 29 and 30 September). This is an example of Tyndall's constantly changing plans to which Hirst had to adapt.

7. *a letter . . . in Glasgow*: see letter 0425.

8. *Last saturday . . . 3 hours. the*: there is a large X drawn through the middle of this incomplete paragraph and the LT transcription (RI MS JT/1/HTYP/120a) did not include it. Whether the X was made by LT or JT is unclear from the ink. Perhaps Tyndall had second thoughts about discussing what seems like excessive drinking by Ginty, as the last sentence is incomplete. But this form of deletion is not typical of Tyndall and it does not prevent the paragraph from being easily read. Certainly, Tyndall did not intend to prevent Hirst from reading it; if he did not want something to be read he scrawled out lines very heavily. Equally, Louisa may have decided that this paragraph was inappropriate for transcription.

9. *Meerschaum*: a tobacco pipe having a bowl made of meerschaum, a soft white, gray, or yellowish mineral resembling a hardened clay (*OED*).

From Thomas Archer Hirst   19 September 1850   0443

9 Blenheim Square | Marlborough Hill | Bristol | 19th Sept. 1850

Dear Tyndall,

I am labouring under a bad cold, and have not been out of doors for a day or two. I have felt it coming on for some time and am only thankful that it has overtaken me here where I can meet and conquer it conveniently. After a sleepless uncomfortable night, I have jumped out of the sheets wet with perspiration, dashed cold water over me, scrubbed myself well and am now all in a glow, but at the same time very weak and feverish; after which statement of my state and symptoms you must not be surprised if they make themselves apparent in my letter. The fact is after vainly attempting to read or sit still I have seized the pen to write to you, under the impression that if I could but faithfully record my symptoms, I should be doing a service not to Physiology or Medicine merely but to Metaphysics—Yes, just before I began and what made me do so, was a fleeting glimpse that I thought I had got into my own inner construction; instead of my <u>Ego</u> and <u>non Ego</u>[1] being one and indivisible, I thought I could feel the influence of both in antagonism. The latter felt as a dead weighty mass, no longer at the disposal of the former, but cramping and crushing it somehow. I assure you it was a strange feeling that: and I feel it now to be a disappointment to myself as well as a loss to the

science of Metaphysics that I can no longer recall it. And that is just the way
we are always kept in the dark, if Lunatics and others we call insane could
only be persuaded to record their sensations, so as we could understand them,
what an insight we might obtain! But this is impossible for the evidence that
their senses would give, or more properly the relation between their Ego &
non ego is <u>different</u> to our own. <u>Different</u> mind you not necessarily <u>diseased</u>.
Who knows but that we may be the insane. I thought so a few moments ago
when all was dark after what appeared an illumination. But I must conclude
or I shall cause you to believe that I am in reality delirious, indeed I feel it
necessary to be quiet & cease thinking if possible about such matters—On
Tuesday morning next I shall be ready to start from here,[2] but shall wait your
further orders.

Yours affectionately | <u>T.A. Hirst.</u>

RI MS JT/1/H/156

----

1. <u>*Ego and non Ego*</u>: possibly a reference to Fichte (see letter 0426, n. 27).
2. *On Tuesday . . . from here*: Hirst left Bristol for London on Friday 27 and met Tyndall there
    the following evening (Hirst, 'Journals', 27 and 28 September).

To Thomas Archer Hirst   [20 or 21 September 1850][1]   0444

My Dear Tom

Frankland has just returned from Germany and has brought me the
enclosed.[2] You will be able to make out some of Debus's but Noll's production will be a riddle to you I'm afraid. Debus says the Semester commences
about the 20th of October. Noll mentions that some cases of cholera have
occurred at Cassel, Debus who writes 8 or 10 days later from Cassel says
nothing about it—Marburg is free from it—I believe it has never yet been
visited by this pest. The country is in a state of war,[3] that is the ministry are
making military demonstrations to check the democratic tendencies of the
Hessian parliament, both Debus and Noll however concur in saying that it
does not interfere in any way with study, commerce, or the common habits
of the people. Bunsen has been invited to take the Chair of Chemistry in the
University of Halle[4] a much more considerable place than Marburg—he has
however declined and will remain where he is. I am glad of this resolution on
his part as we should both undoubtedly feel his loss. Poor Debus is very much
rejoiced to hear that we shall enter Marburg together he is a kind hearted
sterling fellow and with a head of the first calibre. One of the rooms that I had
looked at previous to leaving Marburg has been taken by a Dr. Hoffa, but the

hostess will induce him to withdraw so that we can have one story completely to ourselves. I will throw dice with you in the Ritter[5] about the rooms. Knoblauch is making a tour through Bavaria and Austria with his father so that I have not heard from him.

I sent you the Leader to day. Present my kind remembrances to your brother.[6]

Believe me dear Tom | Your affectionate | Tyndall

RI MS JT/1/HTYP/124
LT Transcript Only

1. *[20 or 21 September 1850]*: postmark 'Darwen 21 September' (from LT's note). It could have been written the previous day. Hirst, in Bristol, received the letter on 22 September (see n. 2), and only one day from the north to Bristol would be unusually fast. Given that Tyndall does not mention writing two letters on the same day, we think that 20 September is the likely date for this letter (compare letter 0445 which we believe was written 21 September).

2. *the enclosed*: Hirst noted in his journal (22 September): 'Received a letter from Tyndall having three other letters from Frankland, Debus, and Noll enclosed'.

3. *The country is in a state of war*: in autumn 1850, Austria and Prussia came into open conflict over Hesse-Cassel.

4. *Chair of Chemistry in the University of Halle*: in his journal Tyndall noted that Bunsen had been offered positions at Halle and Breslau (21 September, JT/2/13b/510).

5. *Ritter*: a hotel in Marburg where Tyndall ate inexpensive meals on Sundays, often in the company of professors at the University. He first met Heinrich Debus at such a Sunday dinner (see LT, 'Biography', Vol. 1, pp. 227 and 296).

6. *your brother*: John, who lived in Bristol.

To Thomas Archer Hirst   [21 September 1850][1]   0445

My dear Tom

Your letter[2] reached me this morning. You seemed to have a presentiment of your cold in Manchester. Well its a good job that you will get it over before we start. Although even did such a thing occur in Marburg I think I could nurse you out of it. I made the discovery myself 3 years ago after inhaling a quantity of sulphuric ether. As I sucked the gas into my lungs I was pondering on the connexion between mind and matter, when suddenly the full solution flashed on me. I started up clapped my hands and exclaimed 'I have it! I have it! I would not have missed it for a thousand pounds!' Like you unfortunately the thing escaped me. I will cram you with two large extracts from my journal Extract 1.[3] 'Beauty comes not at the noddings of the will it will not be coerced but flows freely into the open heart. I often think that it is possible to pierce this mystery, by the discipline of chaste living and mental exercise that an individual man may come to discover his exact relationship to the visible world, and a consoling knowledge this must be, for once it is fixed the guidance afterwards is sure. Human knowledge appears to me to reduce itself into the question what are laws and what are not, this once discovered, content is the consequence—I know I am to die but the fact is settled, I cannot escape it—it shall not frighten me—<u>I feel the resistance to the attainment of this insight sometimes as palpably as a mechanical push</u>'.

Extract No 2. entries made at Rotterdam[4] 'Were I doomed to a life of idleness I would shoot myself without scruple! The very occupation of writing this unorthodox sentence has been medicinal. I have been wading through the Illustrated News[5] till my intellect has become as weak as a half drowned rat. I would sooner starve than live by writing such stuff and still this is the pabulum of thousands. I am sour I confess it, but the paper has made me so. I sat down to it in a placid benevolent mood, but I have taken too large a dose and now recoil against it. There is Angus B. Reach[6] treating us to a dish of most delectable scum; were I at liberty I would kick the fellow into better writing. I am now pacified and look with greater calmness on the matter though still with feelings of implacable enmity; determined to accept half rations or no rations sooner than seek a living by writing such stuff'.

'Loosed from the quay at 5 o'clock, dropped down the Maas[7] __ at first slowly until the bar was crossed when the steam was allowed to exert its full power. Calm and beautiful, becoming still calmer as we proceeded. The sun sunk a huge red rayless orb beneath the horizon, the moon rose in the opposite heaven and brightened as the sun sank. The projection of a line from the earth to the moon traversed a field of silver glory, the light quivering and sparkling from the bosom of the trembling sea. All around this path of brightness was dim gray twilight silent as the grave. There was the earth like a gymnast twirling the moon around her like a dumb bell, while the great orb just descended twirled earth moon and planets around <u>him</u>. Surely some principle permeates this stuff. Who created it? What is it? The soul yearns over the mystery, retires baffled but will try again. Encompassed by such thoughts, revelation seems common place, for whoever listens with reverential ear, will not he also detect the spirit voices speaking in melody to his soul. Supernal whispers which fitly uttered would be as good and true as any revelation of them all'.

Now are you sick of the extracts—you see the 'mechanical push' is very like your 'dead heavy weight'. By the Lord you will soon know me as well as I do myself and I shall be tormented with a perpetual intruder into my most private thoughts.

I am preparing a Mathematical Paper for the Philosophical Magazine.[8] This and other matters will keep me here until Thursday perhaps. I will write to you again on Monday fixing the day on which I will meet you in London.[9] There is a nice short article on Hesse Cassel in to day's Illustrated News.[10] Read it if you can it will show you the state of matters there. I send you a letter received from Frankland[11] this morning. On the whole you have got a precious bag of trash.

Your affectionate | Tyndall

RI MS JT/1/HTYP/122–123
LT Transcript Only

1. *[21 September 1850]*: dated by references to the *Illustrated London News* (n. 9).
2. *your letter*: letter 0443.
3. *journal Extract 1*: entry for 15 June 1850 (JT/2/13b/498).
4. *entries made at Rotterdam*: entry for 18 June 1850 (JT/2/13b/499–500).
5. *Illustrated News*: the *Illustrated London News*, first published in 1842, was the world's first illustrated weekly newspaper.
6. *Angus B. Reach*: Angus Bethune Reach (1821–56) was the journalist in charge of the 'Town and Table Talk' gossip column in the *Illustrated London News*.
7. *Maas*: the Maas River, also known as the Meuse River, which flows from France through Belgium and the Netherlands into the North Sea.
8. *Mathematical Paper . . . Magazine*: not identified; could be a translation or a paper that was never finished.
9. *meet you in London*: they met in London on Saturday evening, 28 September (see letter 0443, n. 2).
10. *short article on Hesse Cassel*: *The Illustrated London News* (21 September 1850), p. 242.
11. *letter received from Frankland*: letter missing.

From Karl Hermann Knoblauch　　25 September 1850　　0446

Marburg, 25 Sept. 50

Mein verehrtester Freund,

Diesen Augenblick bin ich von einer Reise nach Marburg zurückgekehrt; und ich beeile mich, Ihnen die <u>grosse</u> <u>Freude</u> auszusprechen, welche mir Ihr Brief vom 6ten d. bei meiner Rückehr bereitet hat. —Bei der Eil, in welcher ich Marburg verliess, um—sogleich nach dem Schluss der Vorlesungen—meinen Vater nach Studtgardt, München und der Schweitz zu begleiten; hatte ich Ihren vorigen Brief noch unbeantwortet gelassen. Ich wollte dies jetzt thun und Ihnen sagen, wie unendlich ich es bedauerte, die Aussicht Sie wiederzusehen, in eine unbestimmte Ferne gerückt zu sehen, als Ihr lieber Brief, den ich heut hier vorfinde, mir die Gewissheit Ihrer baldigen Rückehr bringt:—Ich kann Ihnen nicht sagen, wie froh mich diese Nachricht macht!—

Fast die volle Zeit, welche mir im verflossenen Semester frei blieb, habe ich auf meine neue Untersuchung über den ungleichen Durchgang der Strahlenden Wärme durch Krystalle nach verschiedenen Richtungen verwendet, eine Arbeit welche auch jetzt noch bei weitem nicht abgeschlossen ist. Ich komme daher erst <u>von morgen an</u> dazu, die letzten Versuche über die Wirkung des Magnetismus auf die Krystalle <u>hintereinander</u> aufzuschreiben. Es wird dies jedoch von der Hand <u>meine ausschliessliche Arbeit</u> sein, so dass ich sie bis zu Ihrer Ankunft hier in Marburg beendet zu haben und Ihnen—vor der Absendung an Poggendorff, der bereits danach gefragt hat—vorzulegen denke. Sollten Sie einen Apparat zur Bestimmung des <u>pos.</u> u. <u>neg. Verhaltens der Krystalle</u> nach der Brewster'schen Methode (deren Praxis Sie jetzt gewiss näher kennen gelernt haben, so wie—nach Rücksprache mit ihm selbst—den Grund unserer Abweichung von ihm in einzelnen Fällen), oder ein <u>feines Goniometer scharf genug</u>, um damit die <u>Mitscherlich'schen Versuche</u> über die <u>ungleiche Ausdehnung der Krystalle</u> nach verschiedenen Richtungen <u>bei verschiedener Temperatur</u> anstellen und weiter ausdehnen zu können— oder überhaupt <u>irgend ein für uns brauchbares Instrument</u> (vielleicht zur Ermittlung des Gesetzes der <u>Abnahme</u> der <u>Diamagnetismus</u>) Krystalle od. dergl. <u>von England mitbringen</u> können, <u>so würde ich Sie recht sehr darum bitten</u> (namentlich damit wir mit Warten auf den hiesigen Mechanicus keine Zeit verlieren). Ich stelle Ihnen zu *[gedachtem]* Zweck 50 Thaler (die ich Ihnen sogleich bei Ihrer Ankunft hier wiedergeben könnte) zu völlig freier Disposition.—

Beim Durchlesen des Briefes erscheint mir die Summe von 50 <Th.> für <u>ausgezeichnete</u> <u>grössere</u> Apparate zu unbedeutend. Sollten Sie daher für <u>unsre Untersuchungen geeignete Instrumente</u> finden, oder bestellen wollen so überlasse ich Ihnen auch <u>bis zur Höhe von 150</u> <Th.> zu disponieren—

Sehr lieb wäre es mir, das <u>Faraday'sche schwere Glas</u> und ein <u>englisches Exemplar</u> meiner <u>Wärme</u>-Abhandlungen zu erhalten. —Kennen Sie vielleicht die kleinen Farbentafeln von <u>Wheatstone</u> (Carreaus auf lebhaftem Grunde und dergleichen) zur Darstellung der unter dem Namen: ,<u>fluttering hearts</u>' bekannten optischen Täuschung? Sollten sie in England leicht zu haben sein, so bäte ich, sie mir zu besorgen. Indess ist dies <u>unbedeutend</u>.

Mein Vetter aus Bonn hat mich vor 4 Wochen besucht. Plücker wird wahrscheinlich das Gesetz mit den <u>posit.</u> und <u>negat.</u> Krystallen fallen lassen, hält aber immer noch daran fest, dass <u>alle Erscheinungen</u> welche die Krystalle zwischen den Magnetpolen darbieten, aus dem <u>Gesichtspunct</u> der <u>optischen</u>

<u>Axe</u> abzuleiten seien. Ferner legt er <u>grossen</u> Werth (als <u>Gegen-Versuch</u> gegen unsre Resultate) auf die Beobachtung, dass ein <u>diamagnetischer</u> Kalkspath (d.h. ein von 1 Pol abgestossener) mit der opt. Axe <u>von Pol zu Pol</u>, ein <u>magnetischer</u> bisweilen mit der opt. Axe <u>äquatorial</u> stehen könne.

———————

Sie sehen leicht, dass dies unsrer Erklärung (sofern man es hier mit <u>gemischten</u> (diam. u. magn.) <u>Substanzen</u> in ungleichem Verhältniss der diamagn. u. magn. Bestandtheile zu thun hat) keine Schwierigkeit darbietet. Ich denke <u>Modelle</u> hierzu anzufertigen.—

Leben Sie recht wohl! Sobald Plücker's mathematische Abhandlung in Crelle's Journal erschienen ist, soll die physikalische Erwiderung auf unsre Abhandlgn. in Poggendorf nachfolgen.—

Auf ein <u>baldiges glückliches Wiedersehen</u>!

ganz der Ihrige | Hermann Knoblauch.

Entschuldigen Sie die Flüchtigkeit dieser Zeilen; aber ich wollte Sie nicht auf die Antwort warten lassen. Sie finden mich jedenfalls in Marburg!

To Doctor John Tyndall | Hammond's Hotel | Leicester Square | London

Marburg, 25 Sept. 50

My dearest friend,

I have just this moment returned to Marburg from a trip; and hasten to express the <u>great</u> pleasure which your letter[1] of the 6th gave me on my return. — With the haste in which I left Marburg—immediately after the end of lectures— in order to accompany my father to Stuttgart, Munich and Switzerland, I had left your previous letter[2] unanswered. I was wanting to do this now and to tell you how endlessly I regretted seeing the prospect of seeing you again postponed to an indefinite future, when your kind letter, which I find here before me today, brings me the certainty of your imminent return. —I cannot tell you how glad this news makes me!—

I have used almost the whole time which remained free to me in the past semester on my new investigation of the uneven transfer of radiant heat through crystals in different directions, a task which even now is still not finished by a long way. I am therefore not going to get around to writing up my last experiments on the effect of magnetism on crystals <u>one after the other</u> until from <u>tomorrow onwards</u>. This will, however, be <u>my exclusive work</u>, with the result that I intend to have it finished by your arrival here in Marburg, and to show it to you before sending it off to Poggendorf, who has already asked after it. Should you be able to <u>bring with you from England</u> an apparatus for determining the <u>pos.</u> and <u>neg.</u>

behaviour of crystals according to Brewster's method (the practical experience of which you have certainly got to know in greater detail now, as well as—after consulting with him himself—the reason for our differing from him in individual cases), or a fine goniometer[3] precise enough in order to be able to carry out and extend Mitscherlich's experiments on the uneven expansion of crystals in different directions at different temperatures—or indeed any instrument we could use at all (perhaps to investigate the law of the decrease of diamagnetism), crystals or suchlike, then I would very much ask you to do so (particularly so that we do not lose any time waiting for the instrument-maker here). I am putting 50 Thaler (which I could give back to you immediately upon your arrival here) at your completely free disposition for the *[intended]* purpose.—

On re-reading this letter the sum of 50 Th. appears to me to be too insignificant for excellent larger apparatus. Should you therefore find, or wish to order, instruments suitable for our investigations, then I will also leave it up to you to make arrangements up to a maximum of 150 Th.—

I should be very glad to receive Faraday's heavy glass[4] and an English copy of my papers on heat. —Do you happen to know Wheatstone's small colour tables[5] (squares on vivid background and the like) which show the optical illusion known by the name of "fluttering hearts"? Should they be easy to be had in England, then I would ask you to get them for me. However, this is not important.

My cousin from Bonn visited me 4 weeks ago. Plücker is probably going to abandon the law of posit. and negat. crystals, but is still holding fast to the idea that all the phenomena which the crystals exhibit between magnetic poles are to be derived from the perspective of the optical axis. Further, he lays great emphasis (as counter-experiment to our results) on the observation that a diamagnetic calcite crystal (i.e., one repelled from 1 pole) could stand from pole to pole with the opt. axis, and a magnetic one sometimes equatorially with the opt. axis.—

You will easily see that this presents no difficulties to our explanation (insofar as one is dealing with mixed (diam. and magn.) substances in unequal proportions of the diamagn. and magn. constituents). I am considering making models for this.—

Farewell! As soon as Plücker's mathematical paper has appeared in Crelle's Journal[6] the physics response to our paper in Poggendorf should follow.[7]—

To a happy meeting before too long!

entirely Yours | Hermann Knoblauch.

Please excuse the hastiness of these lines; but I did not want to keep you waiting for a reply. You will at any rate find me in Marburg!

To Doctor John Tyndall | Hammond's Hotel | Leicester Square | London[8]

RI MS JT/1/K/15

1.　*your letter:* letter missing.

2.　*I had left your previous letter*: letter missing.

3.　*goniometer*: an instrument for measuring angles, especially the angles between the faces of crystals. Tyndall 'saw Ross [Andrew Ross, optical instrument maker] regarding goniometer' on 5 October, just before leaving London (Journal, 5 October, JT/2/13b/513).

4.　*Faraday's heavy glass*: see letter 0417, n. 5.

5.　*Wheatstone's small colour tables*: these tables cause an optical illusion which is due to the attempt of the eye to focus for adjacent spaces of colors of unequal refrangibility, which could not, therefore, be in distinct focus at one time. This yields the appearance of fluttering or movement.

6.　*Plücker's mathematical paper has appeared in Crelle's Journal*: see letter 0417, n. 7.

7.　*our paper in Poggendorf should follow*: see letters 0417, n. 17 and 0458, n. 8.

8.　*To Doctor . . . London*: on the envelope.

From George Wynne　　　　26 September 1850　　　　0447

Melton | 26 Sept 1850

My dear Tyndall

If you knew how much I am tossed to & fro' you would not blame me for leaving your letters so long unanswered. I hope I have not delayed so long that this will find you gone to Germany.[1] I do not know what to say of your determination to pursue your magnetical studies there for another six months as I am not sufficiently informed on the matter to be able appreciate the value of the wealth you hope to attain, but I must nevertheless admire your determination to do or die and I cannot help believing that so much energy must at last meet its reward. But you must bear in mind that this is a strictly utilitarian age that we live in and tho the pursuit of the pure and abstract sciences will bring you fame they are not so certain of bringing the more substantive things of life. I therefore think when you say you are going abroad for six months hard study in the one subject you should steadily limit your time to that period and then determine with equal energy to obtain some employment. I do not think you have much to regret in not succeeding in the application you made for employment on the survey[2] as it would lead to nothing. As you must be in London before going to the Continent I will expect you to pay me a visit.[3] I must now tell you of the fate of your letter to my boy[4] he was greatly delighted at receiving it, he was dressing at the time, and put it aside until he had completed his toilet. I suppose he put it into some boy's nook or corner and was never able to able to find it & consequently never read of which he has been greatly disappointed. My train is just going to start so I must say good bye.

Most truly y[r] | Geo Wynne.

RI MS JT/1/W/93

---

1. *gone to Germany*: Tyndall did not leave England until 7 October.
2. *application you made for employment on the survey*: Tyndall had written to Wynne in August asking advice on employment in the Ordnance Survey (see letter 0423).
3. *to pay me a visit*: Tyndall visited Wynne at Harrow on 5 October (Journal, 5 October, JT/2/13b/513).
4. *your letter to my boy*: letter missing (perhaps still in the boy's nook).

To William Francis   17 October 1850   0448

Marburg 17 Oct. 1850

My Dear Sir.

Here with you will receive the translation,[1] I hope it will please you. I have not in every instance stuck verbatim to the text but the deviations are inconsiderable—nothing however has been written which I did not see the meaning of and hence it is probable that what <u>has</u> written is understandable.

In page 1 an allusion to water is omitted, it adds nothing to the sense.[2]

In the 1st line of page 3 the letters '$[M]^3$H' occur. H^4 is omitted in the diagram. It would have been better to use merely the letter $[M]^5$ in the text.

In the same page I is used, but it is J in the diagram; this of course is unimportant. There is a hastily written sentence in page 4 (4-5 printed memoir) It will I think be better to omit it altogether. I have crossed it out in the manuscript but have not totally defaced it.

A few expletives in page 21 (25 mem.) are also struck out—their insertion would do more harm than good.

I have left the term Wassertrommel gabläse[6] to yourself to translate. There is probably a technical name for it in England.

The note in page 27 (33 memoir) is obscure.[7] The funnels cannot both be above the surface. It is quite right if we suppose the <u>interior</u> funnel to reach above the water for then if it were raised the fluid would flow downwards between them and carry air along with it. But if <u>both</u> are above the surface how is the water to get between them? I have translated it however nearly as it stands.

I have looked over the last numbers of Poggendorff. There is an exceedingly interesting paper in N$^{o.}$ 7 'Ueber die Unhaltbarkeit der bisherige Theorie der Newton schen Farbenringe von E Wilde'.[8]

I learn from private sources that a memoir from Plücker on the old subject is at present in Poggendorff's hands and will soon appear.[9] Should you wish for a translation of either or both of these <one word illeg> stehe zu Dienst.[10]

With best wishes | believe me Dear Sir | most truly yours | <u>John Tyndall</u>
I expect to have something of my own[11] to send you soon.

Might I beg of you to have the enclosed[12] dropped into the nearest post Office? | J.T.

—I believe the editor's box is in Crane Court. this will be nearer than the post office.

StBPL T&F, Authors' letters

---

1. *the translation*: Tyndall translated G. Magnus, 'Über die Bewegung der Flüssigkeiten', *Abhandlungen der Königlichen Akademie der Wissenschaften zu Berlin. Aus them Jahre 1848* (*Transactions of the Royal Academy of Sciences of Berlin. From 1848*) (1850), pp. 135–64. The translation was published as G. Magnus, 'On the Motion of Fluids', *Phil. Mag.* 1, no. 1 (January 1851), pp. 1–23.

2. *adds nothing to the sense*: at the beginning of this paragraph, and of the following two paragraphs, someone (Francis presumably) has added 'good'.

3. *[M]*: it appears as the letter 'M' in the translation.

4. *H*: Tyndall has written this with parallel lines above and below the letter, though it appears simply as 'H' in the published translation.

5. *[M]*: Tyndall has written this with parallel lines above and below.

6. *Wassertrommel gabläse*: it was translated as 'water bellows'. See for example the title of the appendix on p. 163 of the original paper and p. 20 of the translation.

7. *The note . . . is obscure*: this is on p. 21 of the translation.

8. *'Ueber . . . von E Wilde'*: E. Wilde, *'Ueber die Unhaltbarkeit der bisherige Theorie der Newton'schen Farbenringe'*, *Poggend. Annal.* 80, no. 7 (1850), pp. 407–21. For the translation see letter 0452.

9. *I learn . . . soon appear*: as letter 0446 makes clear, Knoblauch was receiving word of Plücker's intentions from his cousin in Bonn. We assume that this was further information from the same source. The memoir appeared in the September issue of Poggendorff: J. Plücker and A. Beer, 'Ueber die magnetischen Axen der Krystalle und ihre Beziehung zur Krystallform und zu den optischen Axen', *Poggend. Annal.* 81, no. 9 (1850), pp. 115–62. See letters 0452 and 0457 on Tyndall's translations of this paper.

10. *stehe zu Dienst*: I am at your service (German). The illegible word is presumably 'Ich', though it does not look like it.

11. *something of my own*: Tyndall was probably alluding to his intended investigations on water-jets which Magnus's paper (n. 1) prompted (see letter 0456).

12. *the enclosed*: not identified, but the first of many letters which Tyndall asked Francis to post.

To Michael Faraday   24 October [1850]   0449

Marburg | Hesse Cassel | Oct. 24th

Dear Sir.

A short time after I had the pleasure of seeing you last June I took the liberty of sending you a few specimens of calcareous spar which from their appearance I judged to be magnetic.[1] I had at the time no means of proving whether they were so or not. Since my arrival in Marburg I have tested specimens of the same spar—the chemical analysis proves the absence of iron and the magnet shews them to be diamagnetic. The wish to furnish you with the means of observing the complementary action of the magnetic and diamagnetic specimens of the crystal being thus far disappointed I have great pleasure in now sending you a sample of the proper kind—a small rhomboid of magnetic spar. You will find that the optic axis of the rhomboid will set from pole to pole, in contra distinction to the diamagnetic crystals of the same form which as you are aware set their axes equatorial.

These effects appear to be capable of the fullest explanation by reference to a principle which you were the first to hint at, that is to say 'the action of contiguous particles'.[2] Wheat flour for instance is pretty strongly diamagnetic. If we take a little ball of dough made from the flour and squeeze it flat the plate thus formed will when suspended vertically in the magnetic field set its <u>shortest</u> dimension equatorial, thus behaving like a magnetic body. Now this is evidently due to the peculiar arrangement of the diamagnetic particles, and if we imagine a similarly formed magnetic mass suspended between the poles it is reasonable to suppose that the arrangement which in the former case caused the <u>repulsion</u> of the line of greatest compression would now cause its <u>attraction</u>. This conjecture is verified, for if instead of flour we use in the composition of our dough a precipitate of oxide of iron, the squeezed plate formed from the latter will set its <u>shortest</u> dimension <u>axial</u>. This action appears to be strictly analogous to that exhibited by the spar. In the diamagnetic specimen the shortest dimension stands equatorial; in the magnetic specimen the same dimension stands axial. The molecules of both crystals are similarly arranged, but in the one case we have to deal with magnetic molecules and the other with diamagnetic.

I remain dear Sir | Most truly and respectfully yours | John Tyndall.

Dr. Faraday. | etc. etc.[3]

1. *sending you a few specimens . . . magnetic*: see letter 0416, where Faraday thanks Tyndall for the samples he had sent.

2. *'the action of contiguous particles'*: for example, see *ERE*, vol 1., § 1615, p. 514.

3. *Dr. Faraday. | etc. etc.*: this letter was addressed and forwarded by Francis (see letter 0450). The Transcript adds a note that 'Don't crush!' was written on 'the cover', suggesting that the packing around the crystal was more than a simple envelope.

To William Francis   24 October 1850   0450

Marburg. 24th Oct. 1850

Dear Sir,

Might I beg of you to put the address of Mr Faraday on the enclosed[1] and have it dropped into the nearest post office for me?

I mentioned to you in my last that a memoir from Plücker was in the hands of Poggendorff. It is now published and occupies 47 pages of the Annalen.[2] Plücker has taken Dr Beer as colleague in this new investigation, but an introduction from his own pen precedes it in which he endeavours to dispose of our first memoir;[3] the last memoir[4] he says arrived too late to be taken notice of. You will probably have received Poggendorff by this time, so it is needless for me to enter more fully into the precise nature of the paper. It simply describes a repetition and expansion of Plücker's former experiments. A sufficient reply to those parts which refer to us shall be forthcoming in due time.[5] —I shall devote a spare hour now and then to the translation of the memoir[6] until I hear from you, if you dont want it the loss of time will be very trifling.

Very sincerely yours | John Tyndall.

via Frankreich!

The Editor of | The Philosophical Magazine | Red Lion Court | Fleet Street London

Francs 4–4[7]

StBPL T&F, Authors' letters

1. *the enclosed*: letter 0449.

2. *I mentioned . . . . Plücker . . . the Annalen*: Tyndall mentioned the article in letter 0448 (cited n. 9).

3. *first memoir*: cited letter 0392, n. 14 (German version) and letter 0395, n. 22 (English version).

4. *the last memoir*: their 'last memoir' (cited letter 0403, n. 2) was published in the *Phil. Mag.* in July, so Plücker may have seen it, but close to his publication time; the German version (cited letter 0458, n. 8) had not yet appeared.

5. *A sufficient reply. . .time*: Tyndall directly engaged Plücker's work in 'On Diamagnetism and Magnecrystallic Action' (cited letter 0498, n. 6), published almost a year later.

6. *the translation of the memoir*: this is discussed over a number of letters; Tyndall sent the translation two months later with letter 0458. The letter here seems to be written chiefly to initiate discussion over a translation of Plücker.

7. *via Frankreich . . . 4–4*: via France! (German), was written at the top of the envelope; 'francs 4–4' in the bottom left corner. There is a Marburg postmark so, although the cost is given in francs, it was not posted in France.

To William Ginty [10 November 1850][1] 0451

My dear Ginty,

I promised to write to you but what am I to say? I got here safe and have tackled into my work earnestly, the result of which up to the present time I have no reason to complain of. Things are looking serious here, war is lowering—a conflict between Prussia on the one hand and Austria and Bavaria on the other seems unavoidable.[2] Marburg is now in possession of the Prussians—they entered in the night before last and were welcomed by the chief magistrate. The Bavarians have entered the Southern skirt of the Province. All the principal towns are in the hands of the Prussians including Cassel the Capital. If a conflict occurs it will probably be first in this neighbourhood.

A little incident singular in my experience occurred when I met Hirst in London—'you would not guess what I have for you' he said. He opened his pocket book and placed two pounds before me—sent to me by the proprieter of the Leader newspaper for a short article which had appeared upwards of three months previous[3] and for which I never expected a farthing.

I have got employment in the literary line: sufficient I think to pay for board and lodging[4] here; I will assuredly take advantage of it and remain in Germany longer than I anticipated when I left you.

How are you getting on at the Athenaeum? the bazaar of course is over—Has it been successful? How did the Album take?[5] If it be not too dear I should like you to purchase me a copy. I trust you like your new situation—The hydrocarbon process I saw at Queenwood—it really is very beautiful. Tell me everything when you write—cram your letter with politics, literature and domestics.

This last word leads me to Peb.[6] I will write to herself by and bye, but she must remember that heaven has given the chief honour to the man and I therefore write to you first. Kiss her for me and say there is nothing in the world so well calculated to promote a little woman's happiness as activity provided always there be a little method in it. To hear from all that Peb wants to make herself perfect is a little improvement in this line, for the girl I believe

has a precious little heart. How is Bill!—with a kind thought of Miss Roberts[7]
I am | Tyndall.

RI MS JT/1/TYP/11/3639
LT Transcript Only

1. *[10 November 1850]*: dated by reference to the Prussians in Marburg. The night before last
    was Friday, 8 November, when Tyndall noted in his journal: 'This evening proclamation
    was made of the arrival of the Prussians' (JT/2/13b/516).

2. *war . . . seems unavoidable*: conflict between Prussia and Austria/Bavaria over competing
    claims to rule Hesse-Cassel broke out in early November 1850. Bavarians entered Hesse
    Cassel on 1 November, with open conflict erupting on 8 November. The issues were mul-
    tiple: according to Tyndall's German friends (letter 0444) 'the democratic tendencies of
    the Hessian parliament' had initiated the recent conflict.

3. *when I met Hirst . . . three months previous*: 'Propensities and their Equivalents' (cited letter
    0406, n. 7) was published in June. Hirst gave Tyndall the money on the evening of 28
    September. It was, Tyndall noted in his Journal, 'my first literary earnings' (2 October 1850,
    JT/2/13b/513, and Hirst, 'Journals', 28 September 1850). On the *Leader* see letter 0398, n. 8.

4. *got employment . . . for board and lodging*: translation work for Francis. Tyndall was paid £2
    per sheet of 16 pages.

5. *How are you . . . at the Athenaeum? . . . How did the Album take?*: Ginty, who was a leader
    in the Manchester Athenaeum, had helped organise a bazaar, held 22–26 October 1850,
    to raise funds for the institution. Many of the stalls were organized by women, including
    one by Margaret Ginty (see 'The Athenaeum', *Manchester Courier*, 19 October 1850, p.
    7). Tyndall contributed 'A Mountain Dialogue' (full text in Journal, 10 September 1850,
    JT/2/13b/508–9) to *The Manchester Athenaeum Album* (1850), published to coincide
    with the bazaar. More eminent contributors included Currer Bell (Charlotte Bronte) and
    Tennyson.

6. *Peb*: a short (familiar) form of the name Margaret.

7. *How is . . . Miss Roberts*: probably family members. Ginty's second child and first son was
    William; Miss Roberts is probably a sister of Ginty's wife.

To William Francis   11 November 1850   0452

Marburg 11th November 1850

Dear Sir

I send you a translation of Wilde's memoir.[1]

With regard to Plücker's paper[2] I certainly think an abstract would be sufficient and I will therefore send you one. His 'introduction' as you are aware
is a reply to our first memoir;[3] this and the single experiment with calcareous spar where our views are also combated I should like to send you entire

leaving it to you to abridge if you thought abridgement necessary. The other part shall be condensed as I scarcely think anybody would have patience to read a full translation of it.

You have not intimated which paper you would like first, and in the absence of your instructions upon this head I will attack <u>Clausius</u> next.[4] I have looked cursorily over his paper—the matter appears to be ably handled.

Should you wish to have Plücker first a single line will set me to work in the right direction.

With regard to the abstracts for the new series I can only say that it will give me much pleasure to place a portion of my time at your disposal.[5] The idea strikes me as being very valuable. By the means you propose the essence of German investigation may be appropriated, and such an abstract uniformly introduced as it puts your readers at once in possession of all that is going on, would doubtless be a great satisfaction to them. I know it would to me were I in England.

The translation of Magnus's paper has schooled me a little[6] and I trust Wilde will be found more flexibly rendered. He handles a subject which requires close accuracy of expression, and in one or two places he might have made himself plainer by putting a slight damper upon his enthusiasm—a little reflection however makes his meaning clear.

Since sending away the translation of Magnus's paper accident has drawn my attention more closely to it. I have repeated many of his experiments in a variety of ways and with different fluids. If you allow me so much space I will send you three or four pages on the subject[7] for it strikes me that Magnus has overlooked one of the most important circumstances of the case.

I return you my best thanks for the trouble you have so kindly taken with my letters.[8] Might I request a repetition of your favour as regards the enclosed[9]—would you be good enough to have it dropped in the nearest post office for me.

Believe me dear Sir | very faithfully yours | John Tyndall

Wm Francis Esqre.

Things are looking serious here at present. A war between Prussia and the members of the 'Bund' seems inevitable.[10] Marburg is now in possession of the former, the troops from Wetlzar[11] entered it the night before last—Prussia in fact has laid hold of all the important towns of northern Hessia, Cassel inclusive—| J.T.

StBPL T&F, Authors' letters

---

1. *Wilde's memoir*: cited letter 0448, n. 8. The translation was published as E. Wilde, 'On the Untenableness of the received Theory of Newton's Rings', *Phil. Mag.* 37, no. 252 (December 1850), pp. 451–62.

2. *Plücker's paper*: cited letter 0448, n. 9. See letter 0457, n. 2 for the translation.

3. *our first memoir*: cited letter 0392, n. 14 (German version) and letter 0395, n. 22 (English version).

4. *I will attack <u>Clausius</u> next*: the paper had been published in two parts: R. Clausius 'Ueber die bewegende Kraft der Wärme und die Gesetze, welche sich daraus für die Wärmelehre selbst ableiten lassen', *Poggend. Annal.* 79, no. 3, pp. 368–97 and 79, no. 4, pp. 500–24. The tone of this comment suggests that there is a missing letter (between 24 October and here) in which Tyndall had proposed translating Clausius. See letters 0454 and 0455 for the translation.

5. *With regards to the abstracts for the new series . . . at your disposal*: Tyndall is replying to a missing letter in which Francis proposed that 'Reports on the Physical Sciences' become a regular feature in *Phil. Mag.* Each issue would include 'an abstract of papers in German journals—here he invites my aid' (Journal, 5 November 1850, JT/2/13b/515–6). The first of the 'Reports' appeared in March 1851 (cited letter 0459, n. 1).

6. *translation . . . schooled me a little*: one of the indications that, in missing letters, Francis was commenting and advising on translation.

7. *I have repeated . . . on the subject*: when Tyndall sent the translation of Magnus to Francis he alluded to his intended paper (letter 0448); he sent it with letter 0456.

8. *my letters*: Tyndall often had Francis post letters for him, as for example, when he asked for a letter to be sent to Faraday in letter 0450.

9. *the enclosed*: probably letter 0451 to Ginty, and perhaps others.

10. *A war . . . seems inevitable*: the Bund, or 'Confederation', was a loose association of German states initially established by the Congress of Vienna in 1815. It was dissolved in 1848, revived in 1850, and lasted until the Austro-Prussian War of 1866. See also letter 0451, n. 2 on military actions.

11. *Wetzlar*: a Prussian-governed district about 20 miles southwest of Marburg.

From Michael Faraday   19 November 1850   453

Royal Institution | 19 Nov. 1850

Dear Sir,

I do not know whether this letter will find you at Marburg, but though at the risk of missing you I cannot refrain from thanking you for your kindness in sending me the rhomboid of calcareous spar.[1] I am not at present able to pursue that subject, for I am deeply engaged in terrestrial magnetism, but I hope some day to take up the point respecting the magnetic condition of associated particles. In the mean time I rejoice at every addition to the facts and to the reasoning connected with the subject. It is wonderful how much good results from different persons working at the same matter; each one gives views and ideas new to the rest. Where science is a republic, there it gains; and though I am no republican in other matters, I am in that.

With many thanks for your kindness, | I am, Sir, | Your very obliged servant, | M. Faraday.

John Tyndall Esq, | &c. &c. &c.

RI MS JT/2/12/4127
Transcript Only

---

1. *the calcareous spar*: sent with letter 0449.

To William Francis   25 November 1850   0454

Marburg 25th Nov. 1850

Dear Sir.

I forward you the first part of Clausius's memoir.[1]

Two things have conspired to delay it a little—first I was anxious to make myself acquainted with the literature of the subject, as this increases my confidence, and found some difficulty in obtaining the necessary books. Secondly the subject is a new one and to grapple with it an ejection of my old notions was to some extent necessary. To translate such a paper without understanding it would be a perilous adventure.

I don't know whether you will allow my <u>patching</u> to pass. I am rather merciless in striking out on a reperusal and generally find much that does not please me even after the fair copy has been made. To lessen the trouble of re-copying I have adopted the expedient of <u>piecing</u>—If you don't like it tell me and I will abandon the practice—indeed I should always feel thankful for any hints you might consider calculated to render the translations in any degree more suited to your wishes.

I am dear sir | most truly yours | John Tyndall
William Francis Esq[re]
The remainder of the memoir shall be forwarded in a few days[2] | J.T.

StBPL T&F, Authors' letters

---

1. *Clausius's memoir*: translation published as R. Clausius, 'On the Moving Force of Heat, and the Laws regarding the Nature of Heat itself which are deducible therefrom', *Phil. Mag.* 2, no. 8 (July 1851), pp. 1–21. See letter 0452, n. 4 for the original paper.
2. *forwarded in a few days*: see letter 0455.

To William Francis   [2–6 December 1850][1]   0455

Dear Sir.

I send you the 2nd part of Clausius.[2] and with it a 'notice' on an experiment of Thomsons which appears in Pogg N°. 9.[3]

I introduced the word '<u>not</u>' with a query in the translation of the 1st part.[4] in the errata which accompanies the 2nd part this error is also noticed—The ? and the ( ) may therefore be omitted.

At the bottom of page 505 he speaks of the '<u>Nebenannahme</u>'—I have now translated this '<u>incidental assumption</u>', would you be kind enough to introduce the same expression in the 1st part, as the mind will be led thereby to connect both at once.

In the first part I fear I have introduced '<u>steam</u>' inadvertently in some places where the <u>vapour of water</u> would be better as the former suggests the notion of the boiling point being already attained while no such limitation is made by Clausius. Could it not be altered?—

In the '<u>notice</u>' you will perhaps be kind enough to make the necessary references to the Phil. Mag.[5]

I shall now commence Plückers memoir[6] and hope to be able to send it to you soon.

yours most truly | John Tyndall

StBPL T&F, Authors' letters

---

1.   *[2–6 December 1850]*: dated by Tyndall's Journal entry for 2–6 December 1850 where he states 'all Clausius is away' (JT/2/13b/518).

2.   *2nd part of Clausius*: published as R. Clausius, 'On the Moving Force of Heat, and the Laws regarding the Nature of Heat itself which are deducible therefrom', *Phil. Mag.* 2, no. 9 (August 1851), pp. 102–19. See letter 0452, n. 4 for the original paper.

3.   *notice … Pogg. N°. 9*: R. Clausius, 'Notiz über den Einfluss des Druckes auf das Gefrieren der Flüssigkeiten', *Poggend. Annal.* 81, no. 9 (1850), pp. 168–72. This followed (and replied to) W. Thomson, 'Die Wirkung des Drucks, den Gefrierpunkt des Wassers zu erniedrigen, experimentell bewiesen', *Poggend. Annal.* 81, no. 9 (1850), pp. 163–68. The translation of Clausius's note was published in the *Phil. Mag.* as R. Clausius, 'On the Influence of Pressure upon the Freezing of Fluids', *Phil. Mag.* 2, no. 14 (1851), pp. 548–50.

4.   *translation of the 1st part*: cited letter 0454, n. 1.

5.   *references to the Phil. Mag.*: Thomson's note in Poggendorff (n. 3) had been translated from W. Thomson, 'The Effect of Pressure in Lowering the Freezing-Point of Water', *Phil. Mag.* 37, no. 248 (August 1850), pp. 123–27. Francis included the reference as suggested.

6.   *Plückers memoir*: cited letter 0448, n. 9.

To William Francis 9 December 1850 0456

Marburg 9th Dec 1850

Dear Sir

I send you the paper[1] which Magnus's memoir has prompted, and hope it will please you. It is a fruitful little subject though at first sight exceedingly barren looking. I cut away some *[score]* of experiments lest you might think the matter too much diluted.

When you have made up your mind as to the new series[2] you will perhaps let me know—Is Poggendorff the only Journal from which you make selections?

I will send you Plucker[3] in a few days—he is rather an obscure writer and therefore requires a little time.

very sincerely yours | John Tyndall

Wm Francis Esqre

would you oblige me by having one of the enclosed[4] dropped in the Leader letter box and the others[5] in the nearest post office

I have another favour to ask from you, and that is to have 20 or 30 copies of the enclosed[6] struck off for me. —and shd feel much indebted if you would send one to

Capt Wynne | Railway Commiss^rs Office | <u>Whitehall</u>

and another to

Geo Edmondson Esqre | Queenwood College | Stockbridge | Hants

J.T.

StBPL T&F, Authors' letters

1. *the paper*: J. Tyndall, 'Phœnomena of a Water-Jet', *Phil. Mag.* 1, no. 2 (February 1851), pp. 105–11.
2. *the new series*: see letter 0452, n. 5.
3. *will send you Plucker*: see letter 0448, n. 9 for the original paper and letter 0457, n. 2 for the translation.
4. *the enclosed*: not identified.
5. *the others*: Tyndall recorded sending letters to Wynne, Frankland, and Mrs Allen on 9 December (JT/2/13b/518), so this probably alludes to these.
6. *the enclosed*: the water-jet paper.

To William Francis   [mid–late December 1850][1]   0457

Dear Sir

I send you a translation of Plücker's paper.[2]

The experiments are fastened together in a short abstract, but I dare say you will consider it long enough as they dont appear to lead to any thing definite. He has given up his old notions but is evidently at a loss as to what is to fill their place.

The packet was sealed and ready for postage when the last two numbers of Poggendorff came into my hands. I have briefly run over their contents.

<u>No. 10.</u> There is a paper from Hankel 'Ueber die Electricität der Flamme' which is perhaps worth translating.[3]

A paper from Sondhauss 'Ueber die Schwingungs Gesetz [der] cubischen Pfeifen', a good paper apparently but very long—an abstract might be of interest.[4]

A description of the gyreidometer by Wilde[5]—it is short and might be given in full.

Doppler's paper[6] might also be given in full, it is interesting and has brevity to recommend it.

<u>N°</u> 11. We have a paper from Kolke 'Ueber eine neue methode Magnetismus zu bestimmen'[7]—and a very [sure] method it appears to be. An experiment or two might be of interest, just to shew the general increase of power from the centre of the pole to its edge.

I shall soon send you a paper on the subject myself and may have occasion to refer to Kolke. I shall certainly have to refer[8] to Dubbs investigation which appears in 8 and 9 Poggendorff.[9] Dubbs experiments have evidently been made with great care and many of them might perhaps be transferred with advantage to the pages of the Phil. Mag. For the sake of completeness it might be well to have at least those to which reference will be made. What do you think?

The paper from Kopp in N° 11. is clever as might be expected from the man.[10] There appears to be a law lurking somewhere and connecting the physical phenomena of bodies with their chemical constitution—I don't know whether you will consider the subject ripe enough for notice.

Wilhelmy[11] has a very interesting subject before him.

Riess[12] might be given in full. it is very short.

Foucaults memoir[13] has already appeared in the Comtes Rendus If you want it you will doubtless turn to this source.

Martins memoir[14] would interest many readers but it is not new.

While awaiting your instructions I shall venture on a translation of the papers of Riess and Wilde, and perhaps that of Doppler.[15] In doing this I think I cannot be far wrong.

very sincerely yours | John Tyndall
Wm Francis Esqre

StBPL T&F, Authors' letters

1. *[mid–late December 1850]*: letter is dated based on allusions to Plücker, and to Tyndall's Journal.

2. *translation of Plücker's paper*: J. Plücker and A. Beer, 'On the Magnetic Axes of Crystals, and their relation to Crystalline Form and to the Optic Axes', *Phil. Mag.* 1, no. 6 (June 1851), pp. 447–57. See letter 0448, n. 9 for the original paper.

3. *a paper . . . worth translating*: W. Hankel, 'Mittheilung einiger Versuche über die Elektricität der Flamme und die hiedurch erzeugten elektrischen Ströme', *Poggend. Annal.* 81, no. 10 (1850), pp. 213–35. The published translation (letter 0461, n. 3) was an abridged version of this article.

4. *A paper . . . be of interest*: G. Sondhauss, 'Ueber den Brummkreisel und das Schwingungsgesetz der kubischen Pfeifen', *Poggend. Annal.* 81, no. 10 (1850), pp. 235–57. The proposed abstract did not appear in the *Phil. Mag.*

5. *A description . . . by Wilde*: E. Wilde, 'Beschreibung des Gyreidometers, eines Instrumentes zur genauen Messung der Farbenringe', *Poggend. Annal.* 81, no. 10 (1850), pp. 264–67. See letter 0461, n. 8 for the translation.

6. *Doppler's paper*: C. Doppler, 'Einige weitere Mittheilungen und Bemerkungen, meine Theorie des farbigen Lichtes der Doppelsterne etc. Betreffend', *Poggend. Annal.* 81, no. 10 (1850), pp. 270–75. No translation was published in the *Phil. Mag.*

7. *a paper . . . zu bestimmen'*: H. Vom Kolke, 'Ueber eine neue Methode, die Intensität des Magnetismus zu bestimmen, nebst einigen mit Hülfe derselben gefundenen Resultaten', *Poggend. Annal.* 81, no. 11 (1850), pp. 321–47. No translation was published in the *Phil. Mag.*

8. *may have occasion . . . have to refer to Dub*: Tyndall referred to these papers by Dub in his memoir on magnetism (cited letter 0464, n. 2), but did not refer to Kolke.

9. *Dubb's investigation . . . in 8 and 9 Poggendorff*: J. Dub, 'Anziehende Wirkung der Elektromagnete', *Poggend. Annal.* 80, no. 8 (1850), pp. 494–520 and 81, no. 9 (1850), pp. 46–72. No translation was published in the *Phil. Mag.*

10. *The paper from Kopp in N° 11 . . . the man*: H. Kopp, 'Ueber Siedepunkts - Regelmässigkeiten und H. Schröders neueste Siedepunktstheorie', *Poggend. Annal.* 81, no. 11 (1850), pp. 374–402. No translation was published in the *Phil. Mag.* Hermann Kopp (1817–92) was a German chemist who completed his studies at Marburg and Heidelberg and moved to the University of Geissen in 1839, where he taught chemistry. In 1863 he was appointed professor of chemistry at the University of Heidelberg.

11. *Wilhelmy*: L. Wilhelmy, 'Ueber das Gesetz, nach welchem die Einwirkung der Säuren auf

den Rohrzucker stattfindet', *Poggend. Annal.* 81, no. 11 (1850), pp. 413–28. No translation was published in the *Phil. Mag.*

12. *Riess*: P. Riess, 'Ueber die Wirkung des einfachen Schliessungsdrahtes der Batterie auf sich selbst', *Poggend. Annal.* 81, no. 11 (1850), pp. 428–33. Tyndall discussed his translation in letter 0461. An abstract of the paper was eventually published in March 1852 (see letter 0514, n. 4 for details).

13. *Foucaults memoir*: L. Foucault, 'Allgemeine Methode zur Messung der Geschwindigkeit des Lichts in Luft und durchsichtigen Mitteln; relative Geschwindigkeiten des Lichts in Luft und Wasser; Project eines Versuchs über die Fortpflanzungs—Geschwindigkeiten der strahlenden Wärme', *Poggend. Annal.* 81, no. 11 (1850), pp. 434–42, was translated from the original French (see letter 0459, n. 3) and condensed for publication (see letter 0460, especially n. 2).

14. *Martins memoir*: C. Martins, 'Anweisung zur Beobachtung der Windhosen oder Tromben', *Poggend. Annal.* 81, no. 11 (1850), pp. 444–67. No translation was published in the *Phil. Mag.*

15. *the papers of Riess and Wilde, and perhaps that of Doppler*: see notes 12 (Riess), 5 (Wilde), and 6 (Doppler) above. Only for Wilde was a full translation published, although Tyndall wrote an abstract of Riess (n. 12).

# To William Francis					27 December 1850					0458

Marburg. 27th Dec. 1850

Dear Sir.

Your last kind letter[1] and my last packet[2] crossed each other on the way. You will of course take no further notice of the letter which accompanies it as I now understand what you require and shall do my best to fulfill your wishes—

It is exceedingly gratifying to me to know that the translations pleased you.[3] I certainly did my best to make them worthy of the magazine but my greenness as a translator caused me to doubt of the success of my endeavours I can only say that I shall endeavour to retain your first impression of my labours in this line.

With regard to the attaching of my name[4] I think I cannot do better than leave the matter entirely in your hands, merely remarking that whatever work I do for you I shall never be ashamed to own publicly. The idea of the reports of the progress of physics strikes me as an exceedingly good one and here I dare say the appearance of my name would be rather a benefit to me than otherwise[5]—The matter however rests with you; use my name in any way that you may think conducive to the interest of the magazine, and I have sufficient confidence in your good nature to believe that whatever you do in this respect will be for my interest also—

I thank you for the proof sheet.[6] I have read it but have remarked no errors. It is at the present moment in Knoblauch's hands otherwise I might have taken a second look over it—I think however it is quite as it ought to be. Knoblauch is kind enough to send a translation of it to Poggendorff.[7]

You must not be surprised if you see an article in the next number of Poggendorff bearing the names of Knoblauch & myself. The same, in a more expanded form, has already appeared in the magazine.[8]

There have been many investigations in Germany in <u>magnetism</u> lately. You are aware of course of the divergence which exists between Buff and [Lorimer] and Müller as regards the laws of Lenz and Jacobi.[9] the latter questioning these laws and the former supporting them. I have just commenced a review of the matter as it at present stands, one inducement being that in another month or so I shall send you an investigation of my own[10] on the subject and they way may be thus made clear beforehand. —The subject <u>per se</u> is highly interesting.

Unfortunately for Marburg it is too true that Bunsen goes to Breslau—he will leave next easter. This morning he left Marburg by rail for Breslau where he will remain for a week. I may also remark that this morning our beloved (?) 'Kurfürst'[11] returned to Cassel.

With best wishes believe me dear Sir | Most truly yours | John Tyndall
Wm Francis Esqre
might I beg of you to send a copy of the last paper[12] to
M^rs. Steuart | Steuart's Lodge | Leighlin Bridge | <u>Ireland</u>

StBPL T&F, Authors' letters

---

1. *last kind letter*: letter missing.

2. *my last packet*: see letter 0457, which was sent with a translation of Plücker's paper.

3. *the translations pleased you*: in letters 0452 and 0454 Tyndall had discussed the difficulties of translation and asked Francis if there was anything he should do differently.

4. *attaching of my name*: the translations appeared under the author's name, but also noted that they were translated by Tyndall.

5. *reports of the progress of physics . . . benefit to me than otherwise*: as Tyndall related in his journal for 31 December Francis 'proposes to publish a monthly Report of the progress of physics, bearing my name, and has handed the translation both from French and German over to me' (JT/2/13b/519). This was a development over the plans mentioned over previous months (see letter 0452 n. 5).

6. *proof sheet*: of Tyndall's paper on the water-jet (see letter 0456, n. 1); he mentioned

receiving the proofs in his journal entry of 31 December, which covered the previous ten days (JT/2/13b/519).

7. *translation of it to Poggendorff*: published as J. Tyndall, 'Ueber die Erscheinungen an einem Wasserstrahl', *Poggend. Annal.* 82, no. 2 (1851), pp. 294–303.

8. *The same . . . in the magazine*: H. Knoblauch and J. Tyndall, 'Ueber das Verhalten krystallisirter Körper zwischen den Polen eines Magneten', *Poggend. Annal.* 81, no. 12 (1850), pp. 481–99. The English version is cited in letter 0403, n. 2.   See letter 0417, n. 17 on their method of writing up their results.

9. *laws of Lenz and Jacobi*: see letter 0465, n. 4.

10. *I shall send you an investigation of my own*: Tyndall was referring to 'On the Laws of Magnetism' (cited 0464, n. 2).

11. *Kurfürst*: Elector, that is Frederick William (1802–75), the Elector of Hesse from 1847–66.

12. *the last paper*: his paper on the water-jet (see n. 6 and letter 0456).

1851

To William Francis   9 January 1851   0459

Dear Sir.

I forward you 'a report' on electromagnetism.[1]

I have looked thro all the nos. of the Phil Mag. for 1850 and find no notice of Foucaults experiments on the velocity of light.[2] This is a very interesting topic and directly I finish this note I shall plunge into it—it stands in the <u>Comptes Rendus.</u>[3]

I trust this will meet your approbation—you have given me a discretionary power and I must use it.

Before the expiration of another week I will send you some translations—

You should have received the accompanying report earlier had my investigations[4] not detained me. I thought to manage so that a portion of every day might be devoted to the translations and Reports, but I find this plan wont work well. I can get on better when I devote my entire time to one thing or the other. for this reason I worked at the investigation up to last Saturday. then quietly closed it and turned my attention wholly to you—It is almost certain that I shall be able to send a good paper on magnetism for the March N⁰[5] of the magazine. In a month I hope to have it in your hands—The results are very remarkable.

very sincerely yours | John Tyndall

9th January 1851

StBPL T&F, Authors' letters

<hr>

1. *'a report' on electromagnetism*: Tyndall, 'Reports on the Progress of the Physical Sciences. Recent Researches on Electro-magnetism', *Phil. Mag* 1, no. 3 (March 1851), pp. 194–205, discusses papers by Dub (*Poggend. Annal.*, Sept 1848), Müller (*Poggend. Annal.*, April 1850), and Buff and Zamminer (*Annalen der Chemie*, July 1850). This was the first in a series of 'Reports on the Progress of the Physical Sciences' that Tyndall wrote for Francis.

2. *Foucault's experiments*: these showed that light travels more slowly through water than

through air, and were important in establishing the wave theory of light over the corpuscular theory.

3. *the <u>Comptes Rendus</u>*: Tyndall discovered the article in German translation (letter 0457, n. 13) but translated it from the original French: Léon Foucault, 'Méthode générale pour mesurer la vitesse de la luminère dans l'air les milieu transparent. Vitesses relatives de la luminère dans l'air et dans l'eau', *Comptes Rendus* 30 (January–June 1850), pp. 551–60. Within the week Tyndall had decided an abridged version would be sufficient (letter 0460).

4. *my investigations*: Tyndall was investigating the attraction between an electromagnet and a mass of iron in relation to the distance between them and the strength of the electromagnet. He first mentioned his intention to undertake such an investigation in letter 0458 (n. 10).

5. *the March N°*: the article was not published until April (letter 0464, n. 2).

To William Francis   [16–19 January 1851][1]   0460

Dear Sir

I send you a second 'report'[2] and translation of two small papers from the <u>Comptes Rendus</u>.[3]

There is an unclear expression in each of the latter but it will scarcely damage the results given.

I at first intended to translate Focaults paper[4] entire but there is too much time spent in describing his apparatus, and in describing details he is sometimes doubtful. I therefore thought it better to write an introduction in which the principle might be developed and to append to this the <u>results</u> of Foucault and a short and well written paper by Fizeau and Breguet[5] on the same subject.

I will not detain these, as altho' I have promised you some more matter you will feel more at ease when you have it in your hands.[6]

I leave the title of the reports to yourself—perhaps nothing better could be introduced than the simple one which you have already proposed. To the general title of the report already in your hands might be attached 'electro-magnetism'—To that which I now send 'experimental proof of the theory of undulation' or some such thing[7]

I will send you some more matter in the course of the coming week—and shall be glad to receive notice of whatever French papers may strike you as being worth translating

most sincerely yours | John Tyndall

Wm Francis Esqre | &c &c—

StBPL T&F, Authors' letters

---

1. *[16–19 January 1851]*: this letter comes after, perhaps a week after, letter 0459 (9 January) and about 4 days before 0461 (23 January), that is, between 16 and 19 January. See n. 6 below for details.

2. *second report*: 'Reports on the Progress of the Physical Sciences. On the Velocity of Light.—Experimental Proof of the Undulation Theory', *Phil. Mag.* 1, no. 7 (supplement to vol. 1) (1851), pp. 544–48. In this report Tyndall discussed papers by Foucault, and Fizeau and Breguet that he had translated (nn. 4 and 5 below). See letter 0459, n. 1 for the first report.

3. *two small papers*: probably, Edouard Desains, 'On the Polarisation of Light reflected by Glass', *Phil. Mag.* 1, no. 4 (April 1851), pp. 335–36, translated from 'Mémoire sur la polarisation de la lumière reéfléchie par le verre', *Comptes Rendus* 31 (July-December 1850), pp. 676–77; and Mathurin-Joseph Fordos and A. Gelis, 'On the Sulphuret of Nitrogen', *Phil. Mag.* 1, no. 4 (April 1851), pp. 346–49, translated from 'Mémoire sur le sulfure d'azote', *Comptes Rendus* 31 (July–December 1850), pp. 702–5.

4. *Focault's paper*: see letter 0459, n. 3.

5. *paper by Fizeau and Breguet*: Hippolyte Fizeau and Louis Breguet, 'Notes ur la Vitesse comparative de la lumière dans l'air et dans l'eau', *Comptes Rendus* 30 (January–June 1850), pp. 771–74.

6. *hands*: at this point Tyndall deleted: 'before 4 days pass by I will post you an analysis of Poggendorffs October and November numbers'. Hence this letter is probably at least 4 days before 0461 (of 23 January), in which he sent the analysis.

7. *some such thing*: for the titles given by Francis see n. 2 above and letter 0459, n. 1.

To William Francis   [23 January 1851][1]   0461

Marburg, Thursday

Dear Sir

I have sifted N° 10 & 11 of Poggendorff[2] and now send you the result.

You should have had it two days earlier had not an attack of influenza suddenly dropped down upon me and disabled me for the aforesaid time from doing anything.

I have simply cut the kernel out of Hankel's investigation,[3] and excluded a good deal of superfluous talk.

I often imagine in translating that were the matter to be completely rewritten it might be made plainer, particularly those portions which deal with hypotheses—you must not blame me if some such portions appear darkly expressed, they are often dark in the original. I sometimes refer to the professors here but invariably find what is dark to me is equally dark to them.

However this refers to matters of secondary importance. The facts are at bottom the principal things and these I trust are so represented that he that runs may read.[4]

I sometimes use terms which I do not strictly remember to have seen in English for instance the term 'anchor' with regard to the mass of iron[5] attracted by an electromagnet—the terms 'crosspiece' and 'armature'[6] are used in English but neither seems appropriate. If the word 'anchor' had not been previously used its introduction will perhaps do no harm. On looking over the papers you will perhaps see other names of instruments which need a little anglicising.

The word 'neusilberdrahte' in Riess memoir[7] I leave to yourself to translate.

I have assumed that you have Pogg No 10 beside you and have therefore forbore sending a copy of the figure referred to in Wilde's paper.[8]

I trust this will reach you timely—there is one question I should like to ask before I close. would you not think it well to omit my name on the translation of Plücker's paper?[9] In a fortnight we shall commence an investigation[10] in this matter which will reply to Plücker and expand our former enquiry.[11]

Would you think a zusammenstellung[12] of the enquiries of Senarmont[13] as to the conduction of heat thro' bodies interesting. if you have not already translated his papers I certainly think it would. Now our next attempt will be to establish a unity between the magnetic action which we have observed and the results of Senarmont—and it will be well perhaps to have these results beforehand in the phil. mag.

Since writing the above I have read two papers from Provostaye and P Desains which appear in the Nov. number of the Annales de Chim. et de Physique—one on the Pouvoir rotatoire des dissolutions Sucrées and the other Sur la Reflexion de Chaleur.[14] Both these papers are highly interesting, the first especially is full of interest and I would therefore propose that it shall be translated in full.

Could it reach you in time for the February number I would translate it forthwith but you will doubtless have enough of matter. Tomorrow I shall commence the memoir on magnetism and hope to have it ready in a fortnight.[15] in the mean time you will perhaps be kind enough to impart any hints which you deem necessary for my guidance.

most truly yours | John Tyndall

Wm Francis Esqre

might[16] I beg of you to have the enclosed dropped in a post office for me.

StBPL T&F, Authors' letters

---

1. *[23 January 1851]*: possible Thursdays (between letters 0459 and 0462) were 16th and 23rd

(see letter 0460, n. 1 on the ordering of the letters). The memoir promised in a fortnight at the end of this letter was posted on 4 February (letter 0465) which suggests that Tyndall began this letter on 23 January. In letter 0465 Tyndall describes this letter as 'recent' which supports that 23 January date. Thus it appears that Tyndall was writing to Francis at approximately weekly intervals.

2. *Poggendorff*: the October and November issues of *Poggendorff's Annalen* 81, no. 10 (1850), pp. 177–320 and 81, no. 11 (1850), pp. 321–480.

3. *Hankel's investigation*: the abridged translation of Hankels' investigation (letter 0457, n. 3) was published as W. Hankel, 'An Account of some Experiments upon the Electricity of Flame, and the Electric Currents thereby originated', *Phil. Mag.* 2, no. 14 (1851), pp. 542–48.

4. *he that runs may read*: make clear, so that action is taken; Tyndall alluded to, but probably unintentionally reversed, the words of, Habakkuk 2:2 'And the Lord answered me, and said, Write the vision, and make it plain upon tables, that he may run that readeth it'.

5. *the term 'anchor'. . . the mass of iron*: there is no use of 'anchor' in the Hankel article (n. 3 above), but Tyndall used both 'anchor' and 'submagnet' in 'Reports on . . . Electro-magnetism' (letter 0459, n. 1). He explained in a footnote (p. 195) why he preferred 'submagnet' over the alternatives of 'keeper' and 'anchor', as used in German.

6. *armature*: a piece of soft iron placed in contact with the poles of a magnet, which preserves and increases the magnetic power (OED).

7. *Riess's memoir*: see letter 0457, n. 12 for the original. The translation was not published immediately, but formed part of one of Tyndall's later 'Reports on the Progress of the Physical Sciences' *Phil. Mag.* 3, no. 17 (March 1852), pp. 173–85 (see letter 0514, n. 4).

8. *Wilde's paper*: the translation (see letter 0457, n. 5 for original reference) was published as E. Wilde, 'Description of the Gyreidometer, an Instrument suited to the exact Measurement of Newton's Rings', *Phil. Mag.* 1, no. 7 (suppl., 1851), pp. 550–52.

9. *Plücker's paper*: cited letter 0448, n. 9.

10. *we shall commence*: Tyndall planned to start a joint investigation with H. Knoblauch in two weeks, when he expected to finish his memoir on magnetism (n. 15 below).

11. *our former enquiry*: see letters 0395, n. 22 and 0403, n. 2 for Knoblauch and Tyndall's earlier publications.

12. *zusammenstellung*: transl. 'compilation' or 'synopsis'.

13. *Senarmont*: Henri Hureau de Sénarmont (1808–62), French mineralogist and physicist.

14. *two papers . . . de Chaleur*: F. De la Provostaye and P[aul-Quentin] Desains, 'Sur le pouvoir rotatoire qu'exercent sur la chaleur l'essence de térébenthine et les dissolutions sucrées' and 'Mémoire sur la Réflexion de la Chaleur', *Annal. de Chimie* 30 ([Nov] 1850), pp. 267–76 and 276–86. See letter 0467 for the translations.

15. *the memoir . . . a fortnight*: Tyndall first referred to the proposed memoir on magnetism several weeks previously in letter 0458. It is a major topic in his letters to Francis over the following two months.

16. *might*: this postscript is written vertically, down the left-hand margin of the paper.

To William Francis   27 January 1851   0462

Marburg 27th Jan. 1851

Dear Sir

A friend of mine[1] is just writing to England and I seize the opportunity to enclose a line or two.[2] It is merely to tell you that I am just now engaged at the writing of the memoir on magnetism[3] which I mentioned in my last note[4] to you and I hope to have it posted within the coming week—It will therefore reach you in time for the March number of the magazine—If you can manage the thing conveniently you would oblige me by inserting it in the March number. It will occupy from 25 to 30 pages. probably 30. I thought a timely notice of this beforehand would not be disagreeable to you as it will enable you to take your measures accordingly.

I hope the matter I have sent you is suitable—after the completion of the memoir I shall gather a little fresh material together.

If it would be of any convenience to you D[r] Frankland of Putney college would correct the proofs of the memoir.

I have now only to express the hope that you will think it worth<y of> insertion, and to bid you g<oo>dbye for the present.

very sincerely yours | —John Tyndall

The Editor of | The philosophical Magazine | Red Lion Court | Fleet St. -- London.[5]

StBPL T&F, Authors' letters

1. *a friend of mine*: Hirst.
2. *enclose a line or two*: Tyndall saved postage by enclosing his letter with Hirst's. It was posted from Halifax (postmark) with a penny stamp.
3. *memoir on magnetism*: this memoir was important to Tyndall. He had already mentioned it twice to Francis (letters 0459 and 0461) and here he tried to ensure that Francis left space for it in the March issue of *Phil. Mag.* It appeared in April (cited 0464, n. 2.).
4. *my last note*: letter 0461.
5. *The Editor . . . London*: address from envelope.

From James Craven   28 January 1851   0463

Halifax, 16, Cheapside. | 28th January, 1851

Dear Tyndall

From the above heading you may give a shrewd guess at my present occupation and position. Yes, Tyndall, I am now what people call 'standing on my own bottom', which I as little contemplated when I served my time and when we last met that such a consummation was to take place so very early that I am placed in a position so novel and unexpected that I don't exactly know whether this can be a stern reality or merely a dream. Dreaming, however, or not dreaming, and I have almost a strong persuasion that it is the former, I perceive myself walking quietly down to the office with 'James Craven, Agent and Surveyor' painted on my glass doors immediately opposite the Post Office as regularly as I used to journey down into Horton Street.[1] Arriving in them I certainly fancy there is an air of greater comfort pervading them than those in Horton Street, especially in my little 'Private' room where I am at present located penning this:—Suffice it that I have comfortable offices and am ready to engage on any work. You will be glad to hear that I have already been employed and my first engagement was to take a set of levels which occupied the whole day and which I accomplished as well as anyone could, which I can assure you gave me great pleasure and gives me an assurance in this department which I certainly did not before possess. I have however experienced a prelude of that abominable idleness which every young man must experience in first starting, and though almost hardly a fraction of this listlessness has yet been my lot, yet the mere phantom of it I have had has brought me to regret— and it is only the prospective view which keeps my spirits up and prevents my cursing myself and everyone around me for advising this step. To guard against this I endeavour to cheat myself by forming schemes which I compel myself to believe must be accomplished, but whether this kind of self deceit can be kept up I shall see. But why, Tyndall, do I mention these things? When I sit down to write you I never know what I shall say as I should were I writing to others—there seems to be an understanding on my part to acquaint you with my inmost thoughts, and I even believe were I to fall in love (which will not be very unlikely now) I should tell you—whether or not I take no pains to restrain or conceal anything from you, and in your letters I naturally expect however unreasonable that you give me your counsel and advice. Let me now thank you for the readiness you have always acted towards me herein for I have some way overlooked this duty in the supposition you did not mind or care for them; but still it appears ungrateful on my part not to offer them though they be disdained. Now, what the deuce am I writing about—really thou wilt think me a little wrong somewhere, and that I am wandering— which is something I begin to fancy myself seeing my tastes and manners are so different from others, that I feel I am misunderstood, that I am placed in circumstances which somehow have a kind of spell or influence over me, that I wish for circumstances more exciting and interesting than these at present—that I feel in fact in a kind of nightmare bound by circumstances tightly enough and which I have not the power to break, and somehow dreaming at the same time of a more genial situation could I but free myself from the present. We are indeed curious mortals! Sometimes happy and sanguine, at others how depressed! how different we are and when one considers what a curious world are we surrounded by, how we can understand nothing, know nothing beyond the effects which certain things strike us as possessing but which if one wish to examine further elude us and leave us wondering—then that religion too with its host of bigotry which surrounds it, who can understand <u>that</u>, and who can reconcile all the inconsistencies which everywhere surround it and make and reduce it to a form and shape instead of the shadowy shapeless unsatisfactory thing it at present assumes. When we examine these things there is a something which would lead us to suppose that we are mere machines, gifted with faculties merely to torture us—and that we are entirely divested of power in guiding ourselves—in fact that we are guided and governed by circumstances and are the creatures of destiny. You laugh!— or smile at the remarks of the young surveyor—no doubt they are curious and flimsy to you, but I begin to entertain them as I can see no better conclusion from the premises we have to judge from. However, these are not subjects to interest you who have to my notions formed some ideas which if submitted to a thorough examination and compared with the standard from which I think you start no warranty can be found for them nor do I see that any hypothesis can be formed in accordance with that book and yet be rational and perfectly natural. If however I pursue reflections as these further and force them on your notice you will imagine that the writer had better have entered into any arrangement no matter what if it furnish plenty of employment than be as he is employed and have time to speculate over such matters as these. These are dangerous subjects, so to leave them.

I mentioned that I had received a letter for Hirst from Tidmarsh who it seems is following out his whim, though from the account I should fancy it has not answered so well as he had anticipated, and to me there appears notwithstanding his usual good-humoured style a certain cheerlessness and want of companionship one might easily suppose to be the case in such an out of the way land.[2]

I trust you are well and making that progress in your studies that you could wish, and hope that the experiments you undertake may be successful and that you are in good health. I close trusting to hear from you, and

Believe me, | Your friend, | James Craven.

Mr John Tyndall | Marburg.

Mrs Lewis—Lizzy Hebden[3] that was—has been pigging[4] and a son is the result. Eltershaw[5] has married a lady with some property and what is termed has made a good match of it.

Success in your Galvanic studies. | Amen.

RI MS JT/1/TYP/11/3547–3549
LT Transcript Only

---

1.   *Horton Street*: Hirst and Craven may have had their mail addressed to the Mechanics Institute which was on Horton Street. After his mother's death, Hirst lodged with the Wrights on Ferguson Street so this was not his lodging address (letters 0393 and 0329).

2.   *an out of the way land*: Tidmarsh had emigrated to South Australia.

3.   *Mrs Lewis—Lizzy Hebden*: probably Hebdon (the source here is an LT transcript), an earlier love interest of Tyndall. On leaving Preston in October 1848, he recollected a walk with 'Miss Hebdon 5 years ago' (Journal, JT/2/13b/ 389). Her parents were probably John and Alice Hebdon (see Volume 2).

4.   *pigging*: 1. farrowing (when used of pigs), or giving birth (for humans); 2. Huddling together, for example, sleeping two or more to a bed (*OED*). Craven, we suggest, intended a crude allusion to the second meaning.

5.   *Eltershaw*: not identified.

To William Francis   [3 or 4 February 1851][1]   0464

Dear Sir

I send you the promised memoir[2] within, I believe, the promised time—
It will speak for itself so it is needless for me to dwell upon it.

In the commencement I thought it very necessary to be clear in the description of the mode of experiment adopted. I hope you wont consider that I have degenerated into prolixity.

If any thoughts of this kind should strike you I not only give you full liberty to make any alterations you may think necessary but shall feel obliged to you for doing so.

The definition of the strength of a magnet which I have given is not perhaps the most philosophical but it has the merit of being practical; it sets the precise equivalent of magnetic strength before the reader—an equivalent which requires no abstraction to render it clear to the mind, for it is tangible to the eye.

I could have wished to have spent twice as much time over the matter and will assuredly return to it, but I intend to go to Berlin in 5 or 6 weeks and in the mean time must, according to promise, make an investigation on the old subject in company with prof. Knoblauch.[3]

This investigation I trust to send you before I leave Marburg.[4]

Prof. Magnus has been kind enough to promise to make room for me. I intend to work in Berlin for 3 or 4 months.

Would you when you next write have the goodness to let me know when the meeting of the British association will take place at Ipswich? If I can manage it I will be there. July 2nd/5

A thought struck me a few days ago in which I should like to have your council. Do you think a course of public lectures would be well received in London during the exhibition.[6] If you encouraged me I should not mind carrying over 300 of Bunsen's cells for the purpose of exhibiting those gorgeous electric experiments.

I shall be much obliged by your having the enclosed[7] posted for me.

As you see, I write to Frankland and have mentioned the memoir—if it be a convenience to you he will correct the proofs.[8]

I will post you a translation of the paper by Provostaye & Desains mentioned in one of my recent notes[9] within the coming week.[10]

Another Paper from Plücker has appeared in Poggendorff[11]—what do you think of it? Will you have it? —I have not yet had time to read it and dare not delay the memoir longer.

I have now only to repeat the request that if it be possible you will give the accompanying a place in your <u>March number</u>.[12]

There are two friends of mine engaged at a translation of it for Poggendorff,[13] but you will have it a month in advance.

I am dear sir | most sincerely yours | John Tyndall

Wm Francis Esqre | &c &c

I have one favour more to beg, and that is that you would get up 30 or 40 copies of the memoir for me in the same manner as you put up the copies of the second memoir on the magneto-optic properties of crystals,[14] and send a copy to each of the following addresses

Capt. Wynne, Royal Eng^rs | <u>Harrow</u>

George Edmondson Esqre | <u>Queenwood College</u>, Stockbridge, Hants

George Phillips Esqre | West Parade, Huddersfield

John S. Smith Esqre | Edith Cottages, North End, Fullham,

or office of Leigh Hunts <u>Journal</u> | 300 <u>Strand</u>

Wm Francis Esqre | &c &c

Dr Charles Mackay | Brecknock Crescent | <u>Camden Town</u>[15]

StBPL T&F, Authors' letters

---

1. *3 or 4 February 1851*: dated on the basis of letters 0465 and 0466 (4 February), both of which say that the memoir to the *Phil. Mag.* was posted in the same mail. This letter was therefore written either the same day or the previous day.

2. *promised memoir . . . promised time*: J. Tyndall, 'On the Laws of Magnetism', *Phil. Mag.* 1, no. 4 (April 1851), pp. 265–95. Tyndall promised the memoir on magnetism, at ever closer

time intervals, in letters 0459, 0461 and 0462. The processes of correction and publication can be followed in many letters over the following six weeks.

3. *an investigation . . . with prof. Knoblauch*: presumably the joint investigation referred to previously in letter 0461, n. 10.

4. *This investigation . . . leave Marburg*: there are no further references to this investigation. It seems that it came to nothing.

5. *July 2ⁿᵈ*: these words are inserted, as if Tyndall found the date himself after asking the question.

6. *lectures . . . during the exhibition*: the exhibition (letter 0399, n. 15), was to open in May. The lectures did not eventuate, but Tyndall's proposal initiated Francis's suggestion (letter 0470, n. 7) that Tyndall give an evening lecture at the British Association.

7. *the enclosed*: probably the letter to Frankland mentioned in the following sentence.

8. *proofs*: Tyndall had suggested that Frankland correct the proofs in letter 0462.

9. *the paper by Provostaye . . . recent notes*: the article by Provostaye and Desains is the first of those proposed in letter 0461 (n. 14); see that letter for the French source.

10. *within the coming week*: sent about five days later with letter 0467.

11. *Another Paper . . . in Poggendorff*: Plücker and Beer, 'Ueber die magnetischen Axen der Krystalle und ihre Beziehung zur Krystallform und den optischen Axen', *Poggend. Annal.* 82, no. 1 ([January] 1851), pp. 42–74. As always, Tyndall referred to this jointly authored paper by Plücker's name alone.

12. *March number*: the memoir was not published until April (see n. 2 above).

13. *two friends . . . translation of it for Poggendorff*: the translators are not identified, although Heinrich Debus, who was becoming closer to Tyndall at this time, is one possibility. The translation was published as Tyndall, 'Ueber die Gesetze des Magnetismus', *Poggend. Annal.* 83, no. 5 ([May] 1850), pp. 1–37.

14. *second memoir . . . of crystals*: see letter 0403, n. 2.

15. *or office . . . Town*: the second line of Smith's address is squeezed into the margin of the manuscript, and Mackay is an afterthought, his address being added below the second inscription to Francis.

To Michael Faraday 4 February 1851 0465

Marburg | Feb. 4ᵗʰ 1851

Dear Sir,

Your last kind letter[1] informed me that you were occupied with terrestrial magnetism. I had however read with deep interest previously that you had arrived at the probable origin of that hitherto enigmatical phenomenon—the variation of the needle.

During the last three or four months I have worked at the hem of the same great garment.[2] My belief in the living interest you feel in the progress of science encourages me to lay a brief abstract of my investigation before you. It relates however, not to terrestrial magnetism but to electro-magnetism.

The subject of the investigation embraces the following four propositions[3]:—

1. To determine the general relation of the strength of an electro-magnet and the mutual attraction of the magnet and a mass of soft iron when both are in contact.

2. A constant force, being opposed to the pull of the magnet being applied to the mass of soft iron, to determine the conditions of equilibrium between this force and magnetism, when the distance between the magnet and the force varies.

3. To determine the general relation between force and distance, that is to say, the law according to which the magnetic force decreases when the distance between the magnet and mass of soft iron is increased.

4. To determine the general relation between the strength of a magnet and the mutual attraction of the magnet and a mass of soft iron, when both are separated by a fixed distance.

The first proposition relates to the so called 'lifting power' of the magnet which, as you are well aware, has been the subject of manifold investigation. The results however heretofore obtained are incapable of being reduced to anything like law.

To avoid the causes of divergence complained of by previous experimenters a peculiar method of experiment has been adopted, and instead of irregular masses of iron I have made use of spheres. The coincidence of the results is truly surprising.

The reply to the first proposition is, <u>that the force with which the magnet and the sphere cling together is directly proportioned to the strength of the magnet</u>.

The 'strength of the magnet' is measured by the intensity of the current which circulates in the surrounding helix and the current was measured by means of a galvanometer of tangent In the investigation the tangent of the angle of deflection is taken as the measure of <u>the strength of the magnet</u>.

The reply to the 2nd proposition is, that when the distance between the magnet and the sphere varies, and a constant force opposed to the magnet is applied to the latter, to hold this force in equilibrium <u>the strength of the magnet must vary as the square root of the distance</u>.

I ought to mention that the 'distances' are very small—the unit of distance is $\frac{1}{1000}$ of an inch being the thickness of a leaf of foreign post paper; by placing a number of such leaves between the sphere and magnet the distance could be varied at pleasure.

The reply to the 3rd proposition is, <u>that the force varies inversely as the distance</u>.

You may perhaps find some little difficulty in separating the 2nd proposition from the 3rd This will vanish when you consider, that, in the former case a constant force (a weight) operated against the magnet, and the question was one between <u>magnetism and distance</u>:—in the latter case, the magnetism is preserved constant, and the question is one between <u>weight and distance</u>.

The 4th proposition embraces the rather celebrated law of Lenz and Jacobi[4]—who solved it by direct experiment. It can however be deduced <u>á priori</u> from the 2nd and 3rd proposition just noticed—I have submitted the deduction to experimental test and found the coincidence remarkably close.

The answer to the 4th proposition is, that <u>the attracting force is directly proportional to the square of the strength of the magnet</u>.

This latter law holds good when a distance of little more than $\frac{1}{1000}$ of an inch separates sphere and magnet. Is it not most singular that this small distance should so entirely change the nature of the law? In <u>contact</u>, as before remarked, the attracting force is proportional <u>to the strength of the magnet simply</u>.

A most remarkable analogy exists between some of the results established and the formulae which Poisson has developed for electrified balls.[5] I am not at all surprised that Prof. Barlow arrived at the notion, that magnetism is a surface phenomenon.[6] As far as I am able to judge at present the whole might be explained on this supposition.

A memoir containing an account of the investigation accompanies this letter to the office of the Philosophical Magazine. The memoir will, I trust, appear on the 1st March.[7]

When Science is a Republic as you say[8] it gains, and yet I dare affirm that no living man knows better than yourself how little benefit is to be derived in this way in comparison with that which results from the solitary communion of the individual with nature. There are people in the world who are very fond of what they call 'composite ideas'.[9] They imagine, if six men come together and talk on a matter, that more will be elicited than if one man held his tongue and simply <u>thought</u> over it. I must say that I have little faith in the proceeding, and if I needed an authority to confirm me in my scept[ic]ism I should without hesitation turn to Professor Faraday.

I remain dear Sir | Most faithfully yours, | John Tyndall.

Professor Faraday | &c &c

RI MS JT/TYP/12/4002–4005
Typescript Only

1. *last . . . letter*: letter 0453.
2. *hem of . . . garment*: a biblical allusion and Carlylean metaphor. The allusion to the 'hem' of the garment is a reference to a woman who was healed merely by touching 'the hem of his [Christ's] garment' (Matthew 9:20). Tyndall thereby expressed both modesty—his topic is small—and ambition—even outer edges can be powerful. The metaphor of nature as 'the garment' of God originated with Goethe who said, 'Nature is the living, visible garment of God' (*Faust*) and was explored by Carlyle in *Sartor Resartus*.
3. *The subject . . . four propositions*: this sentence and the four propositions appear (slightly altered) at the start of the published memoir (letter 0464, n. 2).
4. *law of Lenz and Jacobi*: from E. Lenz and M. Jacobi, 'Ueber die Gesetze der Elektromagnete', *Poggend. Annal.*, 47:6 (1839), pp. 225–66; was stated by Tyndall as: '*The attraction between two electro-magnets, or between an electro-magnet and a mass of soft iron, is proportional to the square of the strength of the magnetizing stream*' ('Laws of Magnetism' (letter 0464, n. 2), p. 289). Tyndall's interest in the law was made clear in this letter. See also letter 0458.
5. *formulae . . . for electrified balls*: Siméon-Denis Poisson (1781–1840) was a French mathematician who used Lagrange's potential function (originally applied to gravitation) 'to prove the formula for the force at the surface of a charged conductor, and to solve for the first time the charge distribution on two spherical conductors a given distance apart' (D. Millar, I. Millar, J. Millar, and M. Millar [eds.], *The Cambridge Dictionary of Scientists*, 2nd ed. [Cambridge, United Kingdom: Cambridge University Press, 2002], p. 292).
6. *Prof. Barlow*: In his 'On the Probable Electric Origin of all the Phenomena of Terrestrial Magnetism; With an Illustrative Experiment', *Phil. Trans.* 121 (1831), pp. 99–108, Barlow recounted his experiments in 1819 when 'a very remarkable fact was discovered; namely, that all the magnetic power of an iron sphere resides on its surface' (p. 101).
7. *on the 1st March*: Tyndall asked Francis for rapid publication, but it was not published until April. Tyndall appeared to agree with Francis's reasons for delaying publication in letter 0470.
8. *as you say*: letter 0453.
9. *what they call 'composite ideas'*: a concept characteristic of empirical philosophy in the tradition of Locke and Hume.

To William Thomson   4 February 1851   0466

Marburg 4th February 1851

Dear Sir.

I very gladly respond to the request contained in your last letter[1] which reached me immediately before my departure from England that is to write to you during the winter—the subject of my communication is not immediately connected with that which was brought before the British Association[2] nevertheless I believe it will interest you—setting aside all further preface I will at once plunge into it—I may remark that an account of the investigation will accompany this letter to the office of the philosophical Magazine.

The subject of the memoir[3] is embraced by the following 4 propositions—

1. 'To determine the general relation between the strength of an electro magnet and the mutual attraction of the magnet and a mass of soft iron when both are in contact.'

2. 'A constant force opposed to the pull of the magnet being applied to the mass of soft iron, to determine the conditions of equilibrium between this force and magnetism when the distance between the magnet and the mass varies.'

3. 'To determine the general relation between force and distance, that is to say, the law according to which the magnetic force decreases when the distance between the magnet and the mass of soft iron is increased.'

4. 'To determine the general relation between the strength of an electro magnet and the mutual attraction of the magnet and a mass of soft iron when both are separated by a fixed distance.'

The first proposition relates to the so called 'lifting power' of the magnet and as you are well aware has been the subject of manifold investigation. The results heretofore obtained have howev[er] defied reduction to a common law.

To avoid the causes of divergence complained of by previous experimenters I have applied a <u>sphere</u> of soft iron. This, combined with a peculiar mode of experiment has given results the coincidence of which is really surprising.

The reply to the 1st proposition is that <u>the force with which Magnet and sphere cling together is directly proportional to the strength of the magnet</u>.

The 'strength of the magnet' is expressed by the intensity of the stream which circulates in the surrounding helix—[4] This has been proved by Lenz and Jacobi[5]—The dimensions of the magnet render the recent objections of Müller[6] to this law inapplicable.

The reply to the 2nd proposition is, that when the distance between magnet and sphere varies and a constant force opposed to the magnet is applied to the sphere, to hold this force in equilibrium <u>the strength of the magnet must vary as the square root of the distance</u>. Calling the strength of the magnet m and the distance d we have the equation—m *[= n√d]* where n is a constant. From this we see, that, if *[the]* magnet have double the strength of another and if the latter exert a certain force at a certain distance the former will exert an equal force at four times the distance; & so on.[7]

I ought to mention that the distances I have chosen are very small: the unit of distance is $\frac{1}{1000}$ of an inch being the thickness of a sheet of foreign post paper by means of which the distances were regulated.

The reply to the 3rd proposition is that <u>the force varies inversely as the distance</u>.

You will perhaps find some little difficulty in separating the 2nd prop. from the 3rd This will vanish when you consider that in the former case a constant force (a weight) operated against the magnet and the question was one between <u>magnetism</u> and <u>distance</u>; in the latter case the magnetism is preserved constant and the question is one between <u>weight</u> and <u>distance</u>.

The fourth proposition embraces a law which in Germany is considered celebrated. It has been solved by direct experiment by Lenz and Jacobi. <It> follows, however *[simply]* as a corollary from the 2nd and 3rd propositions just noticed—The experimental proof and the deduction exhibit a surprising coincidence[8]

The answer to the 4th proposition is that <u>the attracting force is directly proportional to the square of the strength of the magnet</u>.

This latter law is true when an interval of little more than $\frac{1}{1000}$ of an inch separates sphere and magnet—Is it not singular that this small distance should so entirely change the nature of the law? In <u>contact</u>, as before remarked, the *[attracting]* force is <u>proportional to the strength of the magnet simply</u>—

You will doubtless detect a remarkable analogy between some of these results and the formulae which Poisson has developed for electrified balls. I am not at all astonished that Barlow arrived at the notion that magnetism was a <u>surface phenomenon</u>[9]—The memoir will I hope[10] appear on the 1st of march.

I remain dear Sir | very sincerely yours | John Tyndall
Professor W[m] Thomson &c &c

RI MS JT/1/T/1440
RI MS JT/1/TYP/5/1527–1529

---

1. *your last letter*: this letter (probably a reply to letter 0430) is missing.

2. *British Association*: Tyndall spoke 'On the Magneto-Optical Properties of Crystals'and Thomson spoke 'On the Theory of Magnetic Induction in Crystalline Substances', in which he anticipated reconciling his views with those of Tyndall and Knoblauch set out in their second joint 1850 paper (letter 0403, n. 2), see *Brit. Assoc. Rep. 1850*, p. 23.

3. *the memoir*: see letters 0464, n. 2.   The summary here is almost the same as the one sent to Faraday in letter 0466, which should be consulted for annotations.

4. *helix—*: The remainder of this paragraph differs from the parallel paragraph in letter 0465 to Faraday.

5. *proved by Lenz and Jacobi*: Lenz and Jacobi (letter 0465, n. 4), and M. Jacobi and E. Lenz, 'Ueber die Anziehung der Elektromagnete', *Poggend. Annal.* 47, no. 7 (1839), pp. 401–18.

6. *the recent objections of Müller*: J. Müller, 'Ueber die Magnetisirung von Eisenstäben durch den galvanischen Strom', *Pogg. Annalen* 79, no. 3 (1850), pp. 337–44. This paper was cited by Tyndall in his memoir on magnetism and also discussed in the first of his, 'Reports on the Progress of the Physical Sciences' (cited letter 0459, n. 1).

7. *Calling . . . so on*: the last two sentences of this paragraph are not in letter 0465 to Faraday.

8. *The fourth . . . surprising coincidence*: this entire paragraph is worded very differently to the parallel paragraph in letter 0466 to Faraday.

9. *Barlow . . . surface phenomenon*: see letter 0465, n. 6.

10. *I hope*: Tyndall inserted these words after writing the line, thus, in spite of his urging Francis to publish in March, he had doubts about such rapid publication.

To William Francis   [8 or 9 February 1851][1]   0467

Dear Sir

I send you the promised translation of the paper by Provostaye and Desains.[2] In the same number of the 'Annales' there is another memoir on the <u>reflection</u> of heat[3]—the result of this I have fastened to the end of the translation in a note—if you wish to have the thing more fully of course you have only to mention it.

I hope the paper on <u>magnetism</u>[4] <u>has reached you safe</u>—today I have been informed by the post office people that an error had occurred in the postage of it and they require an addition of 16½ Silbegroschen[5] to the amount already paid (1 Th. 22½ Sg.) this I will pay them, and I mention it to you lest you should be called upon to advance the same sum. Some such precaution is necessary as the officials here appear to be exceedingly stupid—I may remark that you ought never to pay anything for the communications posted here—they are invariably <u>prepaid</u>. I will now look into the last number of Poggendorff[6] and if I meet any thing interesting will send it to you.

Hoping to have the pleasure of hearing from you shortly.

I am dear Sir | most truly yours | John Tyndall

Wm Francis Esq

In page 270 of the 'Annales' there is a typographical error; the expression $25° \times \cos^2 85$ ought evidently to be $25 \times \cos^2 85°$—[7]

Would you have the goodness to append the following note[8] in some suitable place towards the end of my paper.

'The deviations with the leaves would not have shewn themselves at all had strong writing paper been used instead of the foreign post! The extreme thinness of the latter enables the imperfections of the surfases to pronounce themselves when magnet and ball are separated merely by a leaf or two'.

StBPL T&F, Authors' letters

1. *[8 or 9 February 1851]*: Tyndall recorded in his Journal (JT/2/13b/522) that he posted Provostaye and Desains, that is, this letter, on 9 February 1851. He may have begun the letter the previous day.

2. *promised translation . . . Desains*: the article translated is the first of those proposed by Tyndall in letter 0461 (n. 14), and promised within a week in letter 0464. The translation appeared as Frédéric Hervé de La Provostaye and Paul-Quentin Desains, 'On the Rotary Power which the Essence of Turpentine and Saccharine Solutions exercise on Heat', *Phil. Mag.* 1, no. 6 (June 1851), pp. 466–74.

3. *another memoir . . . in a note*: Tyndall had proposed translating this memoir also in letter 0461 (n. 14) but here he includes only an abstract at the end of the translation of the first paper (n. 2 above).

4. *the paper on magnetism*: the memoir on magnetism posted on 4 February (see letter 0464, n. 2).

5. *Silbegroschen*: see 'Note on Money'.

6. *last . . . Poggendorff*: as no. 10 and 11 of 1850 had been discussed in previous letters, this is presumably 1850 no. 12 or, perhaps, 1851 no. 1.

7. *typographical error . . . 85°—*: the error, a degree symbol in the wrong position, was not corrected in the *Phil. Mag.* translation.

8. *append . . . note*: this change, the first of many requested by Tyndall in his magnetism memoir, was not made.

To William Francis [14–15 February 1851][1] 0468

Dear Sir

May the gods grant you patience with me for I am a very troublesome fellow. I want you to make an alteration or two in the memoir on magnetism[2] for me. They are trifling but still I think worth writing about. In an early part of the memoir the method of experimenting is described, it is said in one place—'I sat myself before[3] the magnet, weighted the balance &c' dont you think that word 'sat' looks ill or sounds ill? if you do, pray be good enough to strike it out and set 'I took up my place before the magnet' or some such expression in its place. I have no copy of the memoir beside me and therefore cannot refer to pages. Under the head of <u>proposition 3 §.3</u> and immediately before the introduction of the tables the law is expressed in Italics. It is there said 'when a constant force opposed to the <u>magnet applied to the latter to hold this force in equilibrium &c</u>' it would be better instead of the last <*1 word missing*> expression to say 'to hold the ball in equilibrium &c'.[4]

Again immediately after the introduction of the table proving experimentally the law of Lenz and Jacobi[5]—I believe it is the <u>last table</u> in the memoir, a few considerations as to the general error of physicists are introduced.

at one place the following words occur 'I hardly think that Lenz and Jacobi meant to assume so much &c'—this private opinion of mine you will perhaps be good enough to expunge and introduce in place of the above and what comes immediately after the following words.—[6]

The Editor of | The Philosophical Magazine | Red Lion Court | Fleet Street | London[7]

StBPL T&F, Authors' letters

---

1. *[14–15 February 1851]*: The Marburg postmark is 15 February 1851; the letter may have been written the previous evening but, given the importance of the subject matter to Tyndall, he would not have delayed posting it.

2. *alteration in . . . memoir*: this is the second alteration requested (see letter 0467, n. 8). For the memoir see letter 0464, n. 2.

3. *sat myself before*: there is a heavy tick above 'before', apparently made by Francis to indicate he had made the correction. The amended phrase appears at p. 269.

4. *equilibrium &c'*: another heavy tick here, indicating that Francis made this correction (p. 275).

5. *law of Lenz and Jacobi*: see letter 0465, n. 4.

6. *expunge and introduce . . . the following words*: the offending passage was removed. The proposed insertion began 'Lenz and J' but Tyndall crossed that out and the remainder of the letter has not been found.

7. *The Editor . . . London*: address from envelope.

From John Haas [16–18 February 1851][1] 0469

Mein lieber Herr Tyndall,

Ich habe nur zwei minuten um diese wenige Seilen einem Briefe des Hrn. Edmondsons beizufügen. Ich erwartete alle Tage einen Brief von Ihnen; sonst hätte ich Ihnen längst schon geschrieben. Ich schreibe dies bloß meine herzlichsten Bitten denen des Hern. Edmondsons beizufugen daß Sie doch her-kommen so bald als nur möglich. Der Abgang des Hern. Galloway ändert sehr vieles und macht Ihnen alles alles viel leichter und angenehmer hier. O Glauben Sie nur jetzt werden *[Ihre schönen Tagen hier]*. ich sage Ihnen es stehe jetzt glaube ich schöne Tage vor Ihnen und vor ganz Queenwood. was mich selbst anbetrifft so verspreche ich Ihnen mein ganzes Schweizerherz und mein ganzes Schweizerhand ans Werk zu legen damit von jetzt an alles nach Wünschen gehe. vieles, vieles werde ich Ihnen zu sagen haben wenn Sie kommen; ich baue Schon die Schönsten Schlößer in die Luft, machen Sie doch daß Sie wahr werden, dadurch daß Sie kommen. Diese zwei Yahre haben mich manches Bitter*[n]* gelehrt, ich habe manches Dumme gethan, und manches Dumme angesehen. Das war aber auch der fieberhaften, ansteckenden Atmosphere zuzuschreiben in der wir oft lebten. Zeit einigen Monaten ist mir ein neues Luft über das Leben aufgegangen und Sie sollen mich bereit finden zu allem Großen, Guten, Edeln, und Wahren, kräftig handzubieten O dass ich etwas dazu vermöchte Ihr herkommen zu bestimmen, gewiß Sie werden es nicht bereuen aber Sie sollen kommen und sehen. Wie lange die Zeit sein wird bis wir eine Antwort von Ihnen haben! Die Schwierigkeiten die ich vor 4 Monaten vor Ihnen sah im Falle Sie herkämen sind alle jetzt verschwunden und glücklich hat sich alles gewendet. Die herzlichsten Grüße und Wünsche von Ihren treuen alten Freunde und, auf Wiedersehen!

John Haas.

My dear Herr Tyndall,

I have only two minutes to add these few lines to a letter from Herr Edmondson.[2] I was expecting a letter from you every day, otherwise I would have already written to you long ago. I am writing this merely to add my sincerest requests to those of Herr Edmondson that you will come here as soon as is only possible. The departure of Herr Galloway[3] changes a very great deal and makes everything, everything, much easier and more pleasant for you here. Oh, just think, now you are going *[to have lovely days]* here. I tell you, lovely days await you, I think, and the whole of Queenwood. As far as I myself am concerned now, I promise you that I shall put my whole Swiss heart and my whole Swiss hand to work so that from now on everything will go as you wish. I shall have much, much indeed to tell you when you come; I am already building the loveliest castles in the air, just make them come true by coming here. These past two years have taught me many a bitter thing, I have done many a silly thing, and seen many a silly thing. That, though, could also be attributed to the feverish, infectious atmosphere[4] in which we were often living.[5] For some months a new air has been rising over my life, and you should find me ready to lend a sturdy hand to everything great, good, noble, and true. [6]Oh, if only I could do something to ensure your coming here, I am sure you will not regret it but you should come and see. How long will it be until we have an answer from you! The difficulties which I saw awaiting you 4 months ago[7] in the event you came here have all disappeared now and everything has turned out happily. The sincerest greetings and wishes from your loyal old friend, and auf Wiedersehen!

John Haas.

RI MS JT/2/6/23
JT Transcript Only

1. *[16–18 February]*: Tyndall received this letter on 22 February (Journal, 23 February (JT/2/13b/522) enclosed with a letter from Edmondson. Given that letters from London usually took four days to reach Marburg we estimate five days for letters from Hampshire.

2. *letter from Herr Edmondson*: letter missing, but Tyndall received it on 22 February (ibid). Tyndall had discussed returning to Queenwood in August–September 1850 but had proposed to Edmondson that they defer negotiations for six months (letters 0434 and 0438).

3. *departure of Herr Galloway*: Robert Galloway (1822/3–1896) a chemist who wrote many text books, was born at Cartmel (Lancashire) and educated at the Grammar School there. His early career followed the same track as Frankland's. On leaving school, he was apprenticed to the same Lancaster druggist, and later studied under Hofmann at the Royal College of Chemistry, London. In 1848, when Frankland departed with Tyndall for Germany, he became chemistry teacher at Queenwood, and in 1851 left Queenwood for Putney College, where he became assistant to Playfair. He was succeeded at Queenwood by Debus. He was later Professor of Practical Chemistry at the Museum of Irish Industry in Dublin. (Obituary notice in *Journal of the Chemical Society, Transactions* 69 [1896], pp. 733–34, COPAC [for list of publications], and Russell, *Edward Frankland*, p. 61.)

4. *atmosphere*: Tyndall occasionally slipped into English when copying out Haas's letter. Here he wrote the English word rather than the German, atmosphäre.

5. *feverish . . . atmosphere in which we were often living*: there was much conflict at Queenwood when Tyndall and Haas had worked together there in 1848-9 (see Vol. 2 and letter 0492).

6. *true.*   : Tyndall left this and the later blank space in his transcription. It is likely that they were in the original.

7. *4 months ago*: Haas implied that he wrote to Tyndall four months previously; this letter is missing.

To William Francis [19 February 1851][1] 0470

Marburg Wednesday

Dear Sir

Instead of the term 'anchor' or 'Keeper' I propose the term <u>Submagnet</u>. It would be first introduced at A page 195[2] in the following manner.

'The apparatus used by Dub[3] was a steel yard from one end of which a bar of soft iron (<u>Submagnet</u>*) was suspended vertically, a weight being moved &c'.

Referring to the new term the following note might be introduced.

* This term, expressive of the mass of iron which is attracted by the magnet, and which, in its turn, attracts the magnet, is introduced here for the first time. The word 'Keeper' suggests an idea which has no reference to the present case. The word 'anchor', used in Germany, is also inapplicable. As the mass of soft iron, when induced, is itself a magnet, the term introduced above

seems to be appropriate. It suggests, not only the attractive property of the induced mass, but also its relation to the exciting magnet.

J.T.

The term <u>lifting power</u> is perhaps better than <u>carrying</u> power—at all events it is better known. Though it suggests the idea of the magnet pulling something <u>upwards</u>—on the whole I think 'lifting power' may be used with advantage.[4]

The term will have to be altered in my memoir[5] also.

To avoid the repetition of the term submagnet I have in many cases used the word bar, which I imagine is perfectly admissible. In every case where either term is to be used I have written it over the word <u>anchor</u> in the accompanying sheets.

Where the width of the line does not permit of the full word it may be contracted to <u>Submag.</u> or simply to <u>Subm.</u>

In Page 197 opposite a. the word <u>one</u> may be introduced instead of the words 'an anchor'—opposite b. the word 'anchor' may be entirely expunged—this will allow room for the term submagnet at C—[6]

If a better term strike you you will of course make use of it.

Many thanks for the trouble you have taken regarding my question—I <u>will</u> consider the observation you have made as regards the British association[7]—It would indeed be a great privilege.

I am quite content with your arrangement[8] as regards the memoir—indeed prefer it—the 'Report' and it, in the same number, would be too much of a good thing—it would pall the scientific appetite as both are on the same subject.

I will abide by your instructions as regards the translations, and all else that you have suggested shall be strictly attended to.

Accept my best thanks for your proposals as to letter of introduction. If I find them necessary I will at once write to you. I have been introduced to some of them already and Knoblauch will be with me for 5 weeks in Berlin and will break the ice in all quarters for me.

A fair Hessian[9] will pass through Marburg next Sunday on her way to England to unite her fortunes to those of Frankland. in other words Frankland will marry in a few days, Fraulein Sophie Fick, daughter of Oberbau Rath Fick, Cassel

The heading of the 'Reports' is as good as possible.[10]

I have not yet read the papers on the electricity of steam[11] but [have] nevertheless some ground for assuming that the doubt[12] you express is not ill founded. this however is but <u>hearsay</u>—I will read and judge for myself and report to you accordingly.

I have Schlagintweit's paper[13] nearly ready and will forward it tomorrow or next day to himself[14] with a request to attach the <u>French names</u> to the places of observations & send the thing forward to you—these will be more intelligible in England than the German names.

your letter reached me two or 3 hours ago. I will not delay the reply a single post.[15]

most sincerely yours | John Tyndall

StBPL T&F, Authors' letters

---

1. *[19 February 1851]*: problems with dating this letter have led us to query the *ODNB* date for Frankland's marriage (7 February 1851). Notices in contemporary newspapers agree in giving Frankland's wedding date as 27 February 1851 (for example, *Lancaster Gazette*, Saturday, 1 March 1851, p. 8). The 'next Sunday' when Sophie Fick (n. 10 below) passed through Marburg was probably 23 February and the Wednesday of writing 19 February. This date (rather than any other Wednesday) is consistent with the references in n. 7 below.

2. *page 195*: the page numbers throughout this letter are those of Tyndall's 'Reports on . . . Recent Researches on Electro-magnetism', sent to Francis with letter 0459, which Tyndall is here proof reading. Tyndall had previously raised questions about magnetic terminology in relation to translations (letter 0461, esp. n. 5) but without reference to Dub or his Report. At that stage he favoured 'anchor' and had not yet come up with 'submagnet'. The suggestions made here were followed, and the word 'anchor' not used in the published text.

3. *Dub*: article cited in letter 0457, n. 9.

4. *The term <u>lifting power</u> . . . with advantage*: both 'lifting power' and 'carrying power' appeared in the published article.

5. *term . . . my memoir*: his published memoir on magnetism (cited letter 0464, n. 2) mostly used 'lifting power' ('carrying power' only once).

6. *In Page 197 . . . at C*: these suggestions were followed.

7. *my question . . . British association*: In response to Tyndall's proposal to lecture in London during the Great Exhibition (letter 0464), Francis suggested that Tyndall give an evening lecture at the British Association meeting in Ipswich in July. Francis's letter has not been found but Tyndall recorded its arrival and its contents in his Journal (JT/2/13b/522 entry for 23 February covering 17–23 February); the earliest date of this reply is therefore 17 February. The possibility of such a lecture is later mentioned in letter 0485, Journal entry for 25 May, and letter 0491.

8. *content with your arrangement*: here Tyndall seemed to agree to his memoir on magnetism being published after his 'Report' on electromagnetism (letter 0459, n. 1).-

9. *fair Hessian*: Sophie Fick.

10. *heading*: Tyndall and Francis had previously discussed subtitles (letters 0460 and 0470) for his 'Reports on the Progress of Physics'.

11. *electricities of steam*: probably Reuben Phillips, 'On the Electricities of Steam', *Phil. Mag.*

36, no. 246 (suppl., 1850), pp. 50311 and 'On the Magnetism of Steam', *Phil. Mag.* 37, no. 250 (October 1850), pp. 283–88. See letter 0474.

12. *doubt*: this, along with other phrases ('your instructions', 'your proposals',) all refer to a letter or letters from Francis which have not been found, very likely to the recent letter identified in n. 7.

13. *Schlagintweit's paper*: Tyndall received a 'long memoir' on the 'Optical Phenomena of the Atmosphere of the Alps' (extracted from Hermann Schlagintweit and Adolph Schlagintweit, *Untersuchungen über die physicalische Geographie der Alpen* [Leipzig, 1850]) from Hermann Schlagintweit on 8 February 1851 and was translating it as requested by Francis (JT/2/13b/522). It was not published until 1852 (see letter 0589). See also n. 14 below.

14. *himself*: Hermann Schlagintweit (1826–82). Along with his brother Adolph (1829–57), he made his reputation from scientific studies in the European Alps in the late 1840s. Joined by their younger brother Robert (1833–85), they were commissioned by the East India Company to undertake scientific studies in central Asia from 1854–57, where Adolph was beheaded on suspicion of being a Chinese spy. See also letter 0474, n. 5 and 0498, n. 5.

15. *your letter . . . not delay a single post*: Francis's letter is missing. Tyndall's letter could not be delayed if he was to meet the publication deadline for the March issue of *Phil. Mag.*

From George Wynne 3 March 1851 0471

3 March 1851

My dear Tyndall

I hardly deserve a letter from you as I never answered your last, but your present one[1] does not admit of delay, I wish it did, for truth to say I do not happen to be well today, I will at once tell you the view I take of the question you put to me, you seek fame rather than money, or you would willingly defer the latter while you seek the former. Now if you accept the Quakers offer[2] I judge that you will be expected to give your whole energies as well as time to the task or at least so much of both as will leave you but jaded energies to pursue the abstract sciences. Even were the means at hand, it appears to me then that the labours of mind that you have for so many years been expending in your present pursuits will be in some if not in great part lost. You can plainly see to what my advice will tend, but there is yet a question to consider, have you the means of pursuing your studies that is of living without compromising your independence without involving yourself in debt if you have I say go on and prosper for I feel assured with your abilities and industry that you must succeed, providing always you forget not, 'what profiteth it a man if he gain the whole world and lose his own soul.'[3] The Queenwood offer does not appear to me to hold out any advantage except that of a present livelihood in fact of settling yourself down as a schoolmaster and perhaps rising some day to be the Principal of Queenwood.

I hope you will be able to read this very bad writing—I shall be most happy to make the acquaintance of your friend[4] & shew him the lions of Harrow and also to introduce him to M[r] Russell[5]—is there anything else I can make myself useful in, you are I am sure satisfied of my willingness.

I do not think these are the days when a man who has with ability & industry devoted himself to science and added to our stock of knowledge will be suffered to linger in obscurity or feel want, whenever it does happen something wrong I am sure may be traced in the man himself—Farewell for the present—M[rs] Wynne desires her kind remembrances, and with all our best wishes—

Believe me very truly yrs | Geo. Wynne
A Monsieur Tyndall, | Marburg[6]

RI MS JT/1/W/94
RI MS JT/1/TYP/5/1842a

1. *your last . . . present one*: letters missing, but comparison with letter 0472 shows that Tyndall had asked for advice about whether to accept the offer to return to teach at Queenwood College.
2. *the Quaker's offer*: George Edmondson, headmaster of Queenwood College, who had invited Tyndall to return to his old position teaching mathematics and surveying, was a Quaker. Tyndall recorded that the letter offering the position arrived on 22 February (letter 0469, n.1).
3. *what profiteth it . . . lose his own soul*: Wynne quotes, with some minor differences, the high call of Jesus to his disciples: 'For what is a man profited, if he shall gain the whole world, and lose his own soul?' (Matthew 16:26; similarly Mark 8:36).
4. *your friend*: unidentified, but perhaps the son of Schnackenberg, referred to in Journal entry for 23 February (JT/2/13b/522): 'I have been visited by Pfarrer [trans. Pastor] Schnackenberg and everything has been arranged as to the departure of his son for England'.
5. *M[r] Russell*: John Scott Russell.
6. *A . . . Marburg*: the address is not on the original letter but was transcribed by LT from the envelope. It suggests that Wynne expected the Germans to read French.

From William Francis 5 March 1851 0472

Brighton, | March 5/51

My dear Sir,

Your letter of the 28th ult.[1] fortunately reached me just previous to my leaving town, and although a few days' delay has unavoidably occurred, I hope that my answer will not be the last of the three.[2] I will at once proceed to answer your queries, and first of all that relating to the income likely to be derived from the Reports and Translations. Unfortunately, as I informed you at our last meeting,[3] the Magazine[4] is by no means a wealth-producing one, and cannot afford to pay more than £2 per sheet[5]—at all events for the present. As regards the quantity which I shall probably be able to dispose of, I fear it will be impossible to make room for more than one sheet per month—which is $^1/_5$th of the Magazine. This would only give you £24 per annum. I freely acknowledge that this is but poor remuneration for your labour; but for the present I find it impossible to give more. I shall certainly endeavour to make room for as much of copy as possible, and should the Scientific Memoirs be continued (which, however, is very doubtful) I would immediately call upon you to furnish me with Translations for that work. But taking all together, I do not think you ought to calculate upon deriving more than from £24 to £30 from the journal—upon this, however, you may safely reckon.

With respect to your accepting of the position at Queenwood College, I can readily imagine the difficulty in which you are placed from that in which I find myself on being called upon to advise you on the subject. As you correctly observe, good places are scarce; but it does appear to me that a person of ability who has devoted himself specially to the Physical Sciences would very soon meet with a desirable berth at one of our Medical Schools or other Colleges. I certainly should strongly advise against your going to Australia;[6] for that would seem to me to be like going out of the world as far as science is concerned, especially that branch to which you have devoted yourself. It is very probable that you might not succeed in finding a suitable place immediately upon your return to England, but I consider it very likely that after a short residence in London something might turn up. In the meantime you might perhaps manage to give a lecture or two at the Royal Institution, and possibly at the London Institution; and although I might not be able myself to furnish you with more occupation, I have not the least doubt I should be able to procure some literary occupation for you from other parties.

I regret that my present absence from town has prevented me from making some inquiries which might have enabled me to have given a more satisfactory answer. I fear I shall have contributed little to settle your doubts, but the question is so serious a one that I fear to <u>advise</u> one way or the other. Did I dare, it would give me the greatest pleasure to say: stop in Germany another year as you propose, and do not sacrifice those excellent opportunities you possess and which occur to so few persons—But then, on the other hand, there is the offer of, I will say, a comfortable position, and, I suppose, one safe to last, thus removing all anxiety as to the future. The choice is difficult, but should you decide upon remaining in Germany be assured that I shall keep a look-out for you here, and should anything present itself that I fancied would suit you, I would let you know at once and do what little lay in my power to serve your interest. I have felt greatly honoured by the confidence you have placed in me and which I shall endeavour to deserve by assisting you in any way that I may be able. Hoping that the other two letters expected may contain more and better materials for enabling you to come to a right conclusion,

I remain, dear Sir, | Yours very sincerely, | William Francis.
John Tyndall, Ph. D. | Marburg, | Hesse Cassel.[7]

RI MS JT/1/TYP/11/3573–3574
RI MS JT/5/16b/422
Transcripts Only

1. *Your letter of the 28th ult.*: letter missing.
2. *last of the three*: this implies that Tyndall had written to three people asking for advice on returning to Queenwood. He had received a reply (dated two days earlier) from George Wynne (letter 0471). The third person was George Edmondson who had not replied by 13 March (Tyndall's journal entry for 13 March, JT/2/13b/523).
3. *our last meeting*: before Tyndall left London in October 1850 (Journal, 3 October 1850 (JT/2/5/513)).
4. *the Magazine*: the *Philosophical Magazine*.
5. *one sheet*: that is, 16 pages in a quarto publication.
6. *going to Australia*: Tyndall must have mentioned emigration to Australia, although there is no such mention in any extant letters of this period. He had considered emigration to America in the early 1840s, and considered Australia again late in 1851 when a position was advertised at a new university in Sydney.
7. *John Tyndall . . . Cassel*: address probably from envelope.

To William Francis 8 March 1851 0473

Marburg 8th March 1851

Dear Sir.

Had I been aware of your inability to find a place for the memoir on magnetism until the April no. of the magazine[1] I should have spent more time upon it. —Not that I should have altered any of the facts but I would have modified the arrangement and improved the style. I have made an attempt to accomplish this in some measure with the imperfect means at present within my reach. I have not a copy of the memoir beside me[2] but I think my memory will help me to point out clearly the portions which might with benefit be modified—

1. Instead of Proposition 3 as contained in the memoir will you substitute that at α page 1.[3] The latter is more definite and reads better.

2. For the paragraph in the memoir which commences with the words 'Ten years ago Lenz and Jacobi occupied themselves with the solution of the first proposition &c ______ will you be good enough to substitute that at a. page 1.[4] You will at once see where the latter runs into the paragraph contained in the memoir—this is an important alteration—if I thought it would interest you I would tell you why—but I spare the space.

3. Instead of the long description of the <u>Taugentenbussole</u>[5] and the reference to Bunsens instrument I should feel obliged if you would substitute the two paragraphs which occur between b and b′—[6]

4. The definition of the strength of a magnet as contained in the memoir is unnecessarily prolix. Immediately after prop. 1. § 2. the paragraph at b′ might be introduced[7]—This paragraph is sufficient, and will render all that respecting the action of a magnet on a needle placed parallel and perpendicular to the magnetic meridian wholly unnecessary. In fact the development of the formula h = <u>H</u> tang α stands in almost every <u>Lehrbuch</u>[8] and is not necessary for the readers of your magazine. In this case also it is easy to see where the alterations must fit in—it will come between the end of prop 1. and the paragraph where Muller's name is introduced.

5. Immediately after <u>Table I.</u> in the memoir I should feel much obliged if you would introduce the passage at C. page 2.[9] The law expressed at the end of table I. will be thus pushed forward to the end of the third table*[10] [11]—where it may be expressed as at d. page 3. I have given a sketch of the arrangement as it stands so that I think mistake is hardly possible. The law must stand immediately <u>before</u> the paragraph which contains a description of the method pursued by Lenz and Jacobi with induced currents.[12]

6. In § 3. occurs the formula

    m=f(d)

    will you be good enough instead of the arrangement at present existing to adopt that at e. page 3. The law will then come in its proper place, that is <u>after</u> the table instead of before it as at present.[13]

7. Immediately after the reference to Dubs experiments[14] in §. 6. will you be good enough to introduce the remark at F. page 4. and 5. —It may be printed in the form of a note.[15]

8. In writing the last remark I made a little /spring/—that is to say there is something which requires modification <u>before</u> we reach the allusion

to Dub. Immediately after prop 4. §. 6 I have made use of an illustration which perhaps serves rather to confuse the subject in hand than to explain it. You would oblige me if you would omit all that about the melting point of fatty matter[16] and introduce simply what stands at G. Page 6.

9. §. 7. commences with the words—'Bodies capable of magnetization are divided into two classes &c'—A little further on the following words occur: 'At the commencement of this memoir we obtained a notion of the strength of a magnet by its action on a freely suspended magnetic needle.'[17] The change made by the paragraph b′ will render this reference inapplicable for the above words those at H.[18] page 7. may be substituted—A little further still it is said 'Were the case otherwise tang α would not be a correct measure of the magnetic strength'—These words for the above reason are also inappropriate—thier omission will in no way impair the sense.[19]

10. At the commencement I believe of (37) occur the words 'Whence then this singular result'? These words might be left out[20] and the words immediately following slightly altered to suit the omission.

11. In (36) I think it is said 'The following procedure occurred to me on observing the extreme tenuity of a gun cotton balloon' This remark appears to me to be altogether needless.[21]

12. The paragraph at the end where reference is made to the number of experiments ought to be entirely omitted—I mentioned this to you before.[22]

13. In the early part of §. 8. it is said—'We have thus arrived by fair deduction from established principles at the celebrated law of Lenz & Jacobi &c'—will you omit the word 'celebrated'?[23]—not that I wish to detract from the discovers but the word has no business there.

14. Where the former memoirs of Prof. Knoblauch & myself are mentioned—that is where the magnet is described—would you be good enough to refer to the vol and Page of the magazine which contains them.[24] and where towards the end of the memoir the pressing together of a plate of glass and a lens of weak curvature is mentioned it would be well to refer to Wildes memoir, a translation of which I sent you last November.[25] I have seen the translation but *<the>* magazine is not now in my possession.[26]

StBPL T&F, Authors' letters

1. *aware . . . spent more time*: in letter 0470 Tyndall had agreed to Francis's proposal that his memoir on magnetism would be published in the April issue for reasons of balance. His complaint therefore is not simply about the delay. It seems that Francis (in a missing letter) objected to the numerous changes requested by Tyndall, who is here making excuses, without acknowledging that the March publication date was his own wish.

2. *have not a copy . . . beside me*: although Francis had the only fair copy, Tyndall had made this from a draft (Journal, 23 January 1851, JT/2/13b/520). The level of detail here (including section numbers, proposition numbers and table numbers) suggests that he was relying on his draft and not on his memory.

3. *Proposition 3 . . . α page 1*: a list of corrections, at least 7 pages long, accompanied this letter and is referred to by page number and paragraph letter (α, a, b, etc). This list is missing. Where the changes can be identified the page number in the published memoir (cited letter 0464, n. 2) is given in the following notes.

4. *at a. page 1*: probably the last paragraph on p. 265.

5. *Taugentenbussole*: in the published text Tyndall does not use the German term but its English translation, 'tangent galvanometer', and variants.

6. *between b and b´*: probably the two paragraphs on pp. 267–68 under '3'.

7. *after prop. 1 § 2. the paragraph at b´ might be introduced*: this paragraph runs from p. 269 (bottom) to p. 270 (top).

8. *Lehrbuch*: textbook.

9. *C. page 2*: bottom of p. 271.

10. *[This is] not called Tab. III: if I recollect aright it has not a number*

11. *table**: Tyndall's footnote (above) was located at the bottom of his second sheet.

12. *The law . . . induced currents*: the law is in the second to last paragraph on p. 272.

13. *The law . . . at present*: the law and formula are in the upper half of p. 275.

14. *Dub's experiments*: which measured attraction between a bar and magnets of different thicknesses and at different distances (cited letter 0457, n. 9). Tyndall discussed the experiments in his report on magnetism (cited letter 0459, n. 1).

15. *in the form of a note*: note † on p. 284.

16. *the melting point of fatty matter*: p. 283.

17. *'At the commencement . . . magnetic needle'*: this passage remains in the published text (except that 'obtained a' is changed to 'arrived at the') at p. 287.

18. *H*: Tyndall has written this with parallel lines above and below the letter, though it appears simply as 'H' in the published translation.

19. *'Were the case . . . impair the sense'*: this passage also remains in the published text (although with 'strength' altered to 'power') at p. 287. Tyndall mis-spelled 'their'.

20. *These words might be left out*: this was done (p. 292).

21. *This remark . . . needless*: this sentence remains (p. 292).

22. *The paragraph . . . to you before*: Francis let this reference stand (p. 292). The earlier request was probably made in letter 0468, the latter part of which is missing.

23. *fair deduction . . . celebrated law*: without being so requested by Tyndall, Francis changed 'fair' to 'direct' and replaced 'celebrated' by 'well known' (p. 289).

24. *former memoirs . . . contain them*: see letters 0395, n. 22 and 0403, n. 2.

25. *Wilde's memoir*: a reference is inserted (at p. 293) but without mention of Wilde's name. For Wilde's paper see letter 0448, n. 8.-

26. *possession.*: the extant text ends here. The next fragment, letter 0474, could be a continuation of this letter (see letter 0474, n. 1), but although it is written at almost the same time and requested further changes to the memoir on magnetism, the changes requested do not follow the same numbering system and the manuscript is folded differently. We therefore number them as different letters.

To William Francis [8–10 March 1851][1] 0474

[2]For the sake of readers who have no time to dive into the thing it would perhaps be well to give the following summary at the end of the memoir.[3]

'The results arrived at in the present investigation may be summed up as follows:

1. The mutual attraction of an electro-magnet and a sphere of soft iron, <u>when <both> are in contact</u>, is directly proportional to the strength of the magnet, <or in> *[other]* words, to the intensity of the magnetizing current.

<2. When a> constant force opposed to the pull of the magnet is applied to the <sphere of> soft iron <and when> the distance between the magnet and <the> sphere varies; to hold this force in equilibrium, the strength of *[the]* magnet must vary as the square root of the distance.

3. When the strength of the magnet is maintained constant, the mutual attraction of the latter and a sphere of soft iron is inversely proportional to the distance between them.

Combining the last two propositions we obtain:

4. The mutual attraction of an electromagnet and a sphere of soft iron, <u>when both are separated by a fixed distance</u>, is directly proportional to the square of the magnetic strength, or in other words, to the square of the magnetizing current.

I am well aware of the annoyance which all this must cause you—the additional expense which the alterations will render necessary supposing the memoir to be already printed I shall be most happy to bear—and with regard to your share in the transaction what can I do but assert my readiness to do you a similar kindness should you ever require it at my hand—

most faithfully yours | John Tyndall

Owing to the stupid manner in which library matters are conducted here I have not yet been able to see a *<word missing>* magazine *[for]* the present year. This has delayed my remarks on Mr Phillips investigations.[4]

*<I was gra>*tified today by the receipt of a letter from Dr Schlagintweit[5] *<I had>* taken the liberty of making a few remarks on his memoir[6] which, were he over sensitive, might have displeased him—The reverse however is happily the case | J.T.

The Editor of | The Philosophical Magazine | Red Lion Court | Fleet Street | <u>London</u>[7]

Monday. —20 minutes ago I received your letter[8]—I shall not attempt to thank you for it—it is too precious to me to be held in equilibrium by thanks—Yours was the second reply—I await the third,[9] and shall let you know the upshot when all is decided. | J.T.

StBPL T&F, Authors' letters

1.	*[8-10 March 1851]*: dated by the postscript (Monday) and letter 0472. The Monday of the postscript, on which Tyndall received Francis's letter (0472) of 5 March, was almost certainly 10 March, and the main part of the letter probably written Saturday, 8 or Sunday, 9 March. Thus, this letter was written immediately after, or is a continuation of letter 0473, dated 8 March. But, although it was written at almost the same time and requested further changes to the memoir on magnetism, the changes requested do not follow the same numbering system (although the order of requests match the text of the memoir) and the paper is a different size and folded differently. As we cannot establish with certainty that 0474 is a continuation of 0473, we assign independent numbers to these letters.

2.	This entire manuscript is in very poor condition. It is yet another letter in which Tyndall requested changes to his memoir on magnetism, some guesses as to missing words can therefore be based on the published text.

3.	*the memoir*: Tyndall's memoir on magnetism (letter 0464, n. 2). An amended version of this summary is on p. 295. The most significant difference is the order of the four propositions. The order of the propositions here is the same as that in letters 0465 and 0466, although the wording is different, the wording of this letter being closer to that of the published paper. However, in the published version, the second and fourth propositions as listed here are switched, becoming the fourth and second propositions respectively. Letter 0475 is a fragment, listing Tyndall's corrections on proof reading the memoir. It is plausible that he requested a change in order when proof reading.

4.	*Mr Phillips investigations*: probably the papers on the electricity of steam in the *Phil. Mag.* alluded to in letter 0470 (n. 11). Tyndall had promised to read the papers for Francis, who doubted their merit.

5. *D[r] Schlagintweit*: Hermann Schlagintweit. See also letters 0470, nn. 13 and 14, and 0498, n. 5.

6. *remarks on his memoir*: shortly after 19 February (letter 0470) Tyndall had written to Hermann Schlagintweit asking him to give place names in French rather than German. He indicates here that he had also proffered some critical analysis.

7. *The Editor . . . London*: address from outside of the folded sheet.

8. *your letter*: Tyndall referred to letter 0472, in which Francis offered career advice.

9. *I await the third*: Tyndall had requested advice from three friends concerning the advisability of returning to Queenwood College, and had previously received a reply from George Wynne (letter 0471).

To William Francis   19 March 1851   0475

Marburg 19th March 1851

Dear Sir

I return you the proofs[1] and with them a thousand thanks. I can well forgive you if you wished me and my emendations[2] at the devil on finding your quiet enjoyment broken in upon—you were right in your anticipation—now that I see the thing before me I have not the courage to turn amputator. The majority of your readers will not perhaps find the matter a bit too plain. The description of the Taugentenbussole[3] may remain and so may the definition of the strength of a magnet. I was afraid that I had been too diffuse and popular upon these points, if so the error is on the right side I think the matter as it stands wont offend.

The paragraph in the 1st page commencing thus 'Ten years ago &c'[4]—you will be good enough to alter, and substitute that contained in the sheet of emendations.[5]

The 3rd proposition may remain in its present form.

The law at the commencement of page 272 will come in far better at the end of the table as I have altered it.[6]

The arrangement on P 275 may stand as it is.

[The] note at the bottom of page 285 would tell better if removed to the bottom <words missing> <to> refer to the star of the end of prop IV.[7]

<words missing> <ex>periments may be introduced into the text immediately after <words missing> a scrap containing a word or two of introduction in <words missing>

<*The allus*>ion to the melting point of fatty matter in page 282 you may *[exterminate or]* sp<*are*> as you please.[8]

The sheet of <*emendations*> will be only brauchbar[9] in two instances—1. for the alteration of the paragraph in the 1-2 page noted above. 2. for the introduction *[of]* <words missing> experiments—all the rest falls away. I have made several minor *[alterations]*, and introduced a few references which I trust you will contrive to *[introduce]*.

Finally I send you a leaf on which the results are summed up; the paragraph in page 294 commencing with the words 'In stating the case &c' may be suffered to *<words missing>* *[with]* the words 'The want here experienced it has been the object of the present inquiry to supply'.[10] *[There]* *<word missing>* the results *<words missing >* the affair *[1 word illeg]*!

Knoblauch read a portion of a letter for me a month ago in which referring *<words missing>* Schleg*<el>*[11] his father said—'Sie sehen aus als warm Sie in England sehr gefeiert wurden'.[12] you will find no particular genius in that paper which I have translated.[13]

Something similar to your suggestion about the Royal society[14] has often swum in my imagination But how can it be managed? I don't understand the thing, I know I can calculated [sic] on the interest of Scott Russel[15] for a brotherinlaw of his is one of my best friends—his address is Captain Wynne Royal Engineers Roxeth Harrow.[16] He sent a letter of introduction to Mr Russel after me to Edinburgh but it was unluckily mislaid in the assembly room of the association[17] and I did not get it until 3 weeks afterwards. I think Faraday would also interest himself for me—I have received two very kind letters from him. Captain Wynne is government inspector of Railways and is every day in the Railway Commiss^n office Whitehall He will right gladly respond to any enquiries you make regarding me. and I doubt not will act to the utmost of his power on any suggestion which your experience in such matters may prompt. Frankland had a valua*<ble>* friend in Playfair—I believe it was his agency—coupled of course *<with>* the value of the investigation—which obtained him *[a grant of £]* *<words missing>* 'the Royal' about a year ago.[18] In his last letter Faraday *<words missing>* he intended after some time to turn his attention to the infl*<uence of>* *<words missing>* arrangement upon Magnetism. This is a subject which *<words missing>* gladly pursue I have long worked at it and see high poss*<ibilities in>* this direction—Until within the last 4 years my life was devoted to Engineering and this accounts for my ignorance as to how matters are transacted in the Royal Society and *[such]* places. Lighten my darkness *<I>* beseech you[19]—I will do any thing you recommend.

most truly yours | John Tyndall

One question more and I have done. You say you will get some literary occupation for me.[20] Could this be done if I remain in Germany? When pressed to it I can write rhymes, romances, metaphysics, any thing at all but politics and lies! —The thing is yet pending as regards Queenwood—As soon as you can inform yourself on this head would you oblige me by sending me a line or two—without paying the postage?

You know Kohlraush, he is in Marburg at present; his experiments are looked upon in Germany as establishing the contact theory of *[galvanism]* a short digest of his labours would be highly interesting would it not?[21] —I will send *<1 word missing>* Clausius strictures when they appear.[22]

I should like very much *<to>* receive three or 4 copies both of the present memoir[23] and of the jet[24]—would you be good enough to send a few to the office of Leigh Hunts Journal, M[r] John Stores Smith sends me the Journal thro' Longmann, and he will be kind enough to forward copies at the same time.[25]

StBPL T&F, Authors' letters

1. *return you the proofs*: proofs of the memoir on magnetism (see letter 0464, n. 2). Francis's letter mentioning the despatch of the proofs has not been found, but see nn. 14 and 20 below.

2. *emendations*: Tyndall had sent a series of requests for changes to the memoir.

3. *Taugentenbussole*: see letter 0473, n. 5.

4. *Ten years ago &c*: the alteration was made for there is no such sentence on the first page.

5. *sheet of emendations*: the sheet which accompanied the letter has not been found; we assume it went to the typesetters.

6. *The law . . . have altered it*: the position was changed as requested and appeared two thirds down p. 272.

7. *[The] note . . . prop IV*: Tyndall may be alluding to the note (marked by a star) at the end of proposition 4 on p. 292; if so, it was moved as requested.

8. *you may . . . as you please*: the reference to fatty matter is in the published memoir (p. 282).

9. *brauchbar*: useful (German).

10. *'The want . . . to supply'*: this sentence was added to the end of the paragraph as requested (p. 295).

11. *Schleg<el>*: the allusion to translation suggests August Wilhelm Schlegel, whose translations of Shakespeare were widely admired, rather than his brother, Karl Wilhelm Friedrich Schlegel.

12. *'Sie sehen . . . wurden'*: 'You look as if you were celebrated very much in England' (German).

13. *that paper which I have translated*: not identified.

14. *suggestion about the Royal Society*: the discussion here suggests that Francis, in the missing letter (n. 1), raised the possibility that Tyndall could apply for a the Royal Society grant or be nominated for the FRS. The allusion to Frankland's grant (see n. 18 below), suggests the former.

15. *Scott Russel*: John Scott Russell.

16. *Roxeth*: Wynne lived at Roxeth House, Harrow.

17. *assembly room of the association*: that is, the administrative centre of the BAAS meeting at Edinburgh in August 1850.

18. *a grant . . . a year ago*: Frankland received his first grant from the RS in 1850 (Russell, *Edward Frankland*, p. 107).

19. *lighten . . . I beseech you*: a paraphrase from the (Anglican) Liturgy for Evening Prayer: 'Lighten our darkness, we beseech thee'.

20. *you say . . . literary occupation for me*: letter missing. This offer is stronger than that made by Francis in letter 0472.

21. *a short digest . . . would it not?*: it seems Francis agreed. Tyndall worked at translating Kohlrausch's memoirs in April and May; see Journal entries for 6 April (JT/2/13b/524), 30 April and 1 May (both JT/2/13b/541). Three papers by Kohlrausch formed the basis of a 'Reports' sent to Francis in early 1852 and published in the May 1852 number of the *Phil. Mag.* (see letter 0603, nn. 6 and 9). There are no allusions to Kohlrausch in other letters to Francis.

22. *Clausius . . . they appear*: probably an allusion to Clausius's ongoing dispute with Thomson. Tyndall sent two papers by Clausius to Francis on 2 April (letter 0476); in a letter to Francis (letter 0498, n. 4), sent in June, Clausius replied to Thomson.

23. *memoir*: his forthcoming memoir on magnetism (n .1).

24. *the jet*: see letter 0456, n. 1.

25. *I . . . time*: This second postscript is written, in a small hand, in the left margin of the page.

To William Francis   2 April 1851   0476

Marburg April 2—1851

Dear Sir,

I send you the paper of Clausius, and for the sake of completeness another short one from the same pen[1] which appeared in the same number of the Annalen—If you don't want the letter <u>dann Schadet es nichts</u>.[2] Should you find occasion to write to me within the coming 11 days your letter will reach me in Marburg.[3] Should you wish to write afterwards it will be best to address to the care of Prof. Knoblauch 23 Poststrasse Berlin.

most faithfully yours | John Tyndall.

StBPL T&F, Authors' letters

---

1. *paper of Clausius . . . another short one . . . pen* R. Clausius, 'Ueber das Verhalten des Dampfes bei der Ausdehnung unter verschiedenen Umständen', and 'Ueber den theoretischen Zusammenhang zweier empirisch aufgestellter Gesetze über die Spannung und die latente Wärme verschiedener Dämpfe', *Poggend. Annal.* 82, no. 2 (1851), pp. 263–73 and 274–79. The former was published as 'On the Deportment of Vapour during its Expansion under different circumstances', *Phil. Mag.* 1, no. 5 (May 1851), pp. 398–405. The 'short one', part of an ongoing exchange between Clausius and Thomson, was not printed in the *Phil. Mag.*

2.   *dann Schadet es nichts*: no harm done (German).

3.   *reach me in Marburg*: Tyndall left Marburg en route to Berlin on 18 April.

From Michael Faraday   19 April 1851[1]   0477

Hastings 19th April 1851

Dear Sir,

Whilst here resting for awhile I take the opportunity of thanking you for your letter of the 4th of February[2] and also for the copy of the paper in the phil. Magazine[3] which I have received. I had read the paper before and was very glad to have the development of your researches more at large than in your letter. Such papers as yours makes me feel more than ever the loss of memory I have sustained, for there is no reading them or at least retaining the argument under such a deficiency. Mathematical formulae more than any thing requires quickness and surety [in receiving and retaining the true value of the symbols used,][4] and whilst one has to look back at every moment to the beginning of a paper, to see what H or α or β mean there is no making way. Still though I cannot hold the whole tenor of reasoning in my mind at once I am fully able to appreciate the value of the results you arrive at, and it appears to me that they are exceedingly well established and of very great consequence. These elementary laws of action are of so much consequence in the development of the nature of a force which, like magnetism is as yet new to us.

My views with regard to the [cause of the][5] annual, diurnal, and other variations are not yet published though printed.[6] The next part of the philosophical transactions will contain them. I am very sorry I am not able to send you a copy from those allowed to me, but I have had so many applications from those who had some degree of right that they are all gone. I only hope that when you see the Transactions you may find reason to think favourably of my hypotheses. Time does not lessen my confidence in the view I have taken but I trust when relieved from my present duties and somewhat stronger in health to add experimental results regarding oxygen so that the mathematicians may be able to take it up.

As you say in the close of your letter I have far more confidence in the one man who works mentally and bodily at a matter than in the six who merely talk about it—and I therefore hope and am fully persuaded that you are working. Nature is our kindest friend and best critic (exciter?)[7] in experimental science if we only allow her intimations to fall unbiassed on our minds. nothing is so good as an experiment which whilst it sets an error right gives us a reward for our humility in being refreshed by an absolute advancement in knowledge[8]

I am my dear Sir | your very obliged and faithful Servant | M Faraday
Dᵣ J. Tyndall | &c &c

RI MS JT/2/6/52–3
RI MS JT/TYP/12/4128[9]

---

1. *19 April*: this letter, which Hirst forwarded from Marburg, took 9 days to reach Tyndall
    (see n. 9).

2. *your letter… February*: letter 0465 (in which Tyndall outlined his memoir on magnetism).

3. *copy … I have received*: presumably an offprint from the April issue of the *Phil. Mag.* in
    which Tyndall's memoir on magnetism was published (cited letter 0464, n. 2). Tyndall
    had asked Francis to send a copy to Faraday (see letter 0480).

4. *in receiving… symbols used*: there is an ellipsis in Tyndall's transcript; the passage is taken
    from the LT transcript (see n. 9 below).

5. *cause of the*: in LT transcript but not in JT transcript (see n. 9 below).

6. *my views… not yet published though printed*: appeared as 'ERE 26' and 'ERE 27', published
    later in 1851.

7. *(exciter?)*: Tyndall entered this option, showing that he was unsure of the reading. LT tran-
    scribed the word as 'exciter'.

8. *As you say… advancement in knowledge*: Tyndall took these remarks by Faraday as an enor-
    mous compliment. He copied the paragraph into his next letter to Hirst (0480). He cop-
    ied the entire letter into his journal (JT/2/6/52–3) on the day he received it (28 April).

9. *RI… 4128*: these two sources are both transcripts, the first by JT in his journal, the second
    by LT. Differences between them show that both were made from a missing original. We
    give priority to the JT version.

From George Edmondson   [c. 23 April 1851][1]   0478

My D.J.T.[2]

Thy[3] 2 letters of no date and of Ap. 19[4] are before me. I am glad to learn
that Dr D.[5] is likely to be with us so soon (sooner the better) as the young
man I now have[6] is very anxious to return to his own studies at the college
of Chemistry. But Jno.[7] I fear thou art too sanguine in thy anticipations as to
Dr D's readiness for duty. It seems scarcely likely he can take a class so soon. I
shall be glad if he be not obliged to suspend his lessons to study English, and
how shall we get on about lectures? Two per week fall to his share and the
difficulty is how are they to be supplied? I see difficulties in the way but am
not disposed to be deterred by them. The good Dr comes with a hearty good
will, and that is half the battle. It is better that we should have a clear understanding about salary now[8] than leave it till we meet. When I named £150[9] it

was with a perfect knowledge of my resources and that I could not go beyond this sum with propriety: I freely admit the truth of thy remark that thou art a stronger man now,[10] but alas! in the exchequer I am sorry to say that I am not: but this I will add, as my resources grow thou shalt share in the growth, and £30 or £50 shall cheerfully, gratefully be given as soon as I have it to give. Queenwood has great capabilities and by the united efforts of so many good men and true it will continue to grow in public esteem as it has done, and by a simultaneous effort I have no doubt we may make it take a higher position still. The arrangements seem so fixed that there is no chance of seeing thee before our vacation which takes place about the 12th of June. I have not yet seen my way clear as to a public examination, though I acknowledge the desirableness of it: I cannot accomplish the details of it yet, and though I highly approve of shewing the public what we can do, I doubt whether a prospective display would be so good as one shewing <u>results</u>—which I trust we shall be able to do in the course of time. 'Where there is the will &c'[11] and I feel persuaded and delighted to think that I shall have so many enthusiastic helpers in the good cause. I am my dear friend with feelings of esteem and respect

Thine sincerely

RI MS JT/1/HTYP/127-128
LT Transcript of JT Transcript Only

1. *[c. 23 April 1851]*: as Tyndall received this letter on 28 April (Journal entry, JT/2/6/52), the date of writing (estimated by postage times) was c. 23 April.

2. *D.J.T.*: abbreviation for 'Dear John Tyndall', probably Edmondson's efficient abbreviation of the plain Quaker form of address.

3. *Thy*: throughout this letter Edmondson uses the plain Quaker form of address, thy and thine, originally used for close friends and inferiors, rather than the polite plural forms, 'you' and 'your', originally used for superior persons.

4. *Thy 2 letters . . . Ap. 19*: both letters are missing. In his journal Tyndall recorded writing to Edmondson once in the week of 12–18 April, after receiving Edmondson's offer of £150, and again on 20 April (entries for 18 and 20 April, JT/2/13b/538). Thus, he does not record dates with scrupulous accuracy or, perhaps, he began the second letter on the 19th and finished it on the 20th.

5. *Dr D.*: that is, Heinrich Debus (Journal, 2 April, JT/2/13b/524).

6. *young man I now have*: Robert Galloway (see letter 0469, n. 3).

7. *Jno.*: an abbreviation for John.

8. *It is better . . . clear understanding now*: here Edmondson shifted to discussing Tyndall's position and salary.

9. *When I named £150*: in a missing letter which Tyndall received between 6 and 12 April (Journal entry of 12 April, JT/2/13b/538).

10. *£150 . . . stronger man now*: the amount proposed by Edmondson was the same as Tyndall
    had received in 1848–9 before his advanced studies. Tyndall must have asked for a higher
    salary, which Edmondson here declined.

11. *Where there is a will . . .*: Edmondson alludes to the proverb, 'where there is a will there is
    a way'.

To Thomas Archer Hirst   24 April 1851   0479

Beim Hrn. de Baux | Dorotheen Strasse 95. Berlin | 24ᵗʰ April 1851

My dear Tom.

I had a kind of presentiment as I delivered my chest at the railway station
that all would not be right—it was the first time I ever allowed my luggage to
escape my own vigilance and it will probably be the last—Here I have been
since last monday night not able to make use of the precious time which flies
by me on account of my chest not having come. I have been half a dozen
times at the Railway station to inquire about it and have just returned from
my last trip but have got no intelligence. will you go to the parcel expedition and enquire about it for me—The clerk—a little short fellow wrote the
'Frachtbrief'[1] himself and assured me that the chest would be here on monday
evening—it is now thursday—I had a good journey on monday did not feel
much tired. had a glorious sunset and a magnificent rainbow. had my pass
examined at the Railway station and was allowed to proceed. Climbed into
a Droschke[2] with two other travellers and drove to Töpfer's Hotel—Here I
stopped the night—had a capital bedroom and a sitting room adjacent—Up
next morning and set out in quest of lodgings, visited first of all those which
had been recommended to me but found them either occupied or unsuitable.
I examined about a score of them ranging from 6 to 14 Thalers a month and
towards evening decided on my present one for which I pay 8 Thalers. It is
clean and comfortable enough. My host is a painter, a nice frenchified little
man who has a rare predilection for painting war pieces[3] the actors in the
scene being however most sleepy looking rascals, appearing to set about their
work with the most mechanical deliberation possible—Berlin is a fine town
and were it not that dreams and shadows of my absent chest hover perpetually about my mind I dare say the impression of it would have been agree
able enough There are some fine monuments and fine buildings—yesterday
evening I enjoyed a walk to the post office very much, or rather a walk home
again—In crossing the Schlossplatz[4] I looked up at the clear silent sky with
its glancing stars, the figures and turrets relieved against the firmament—all
seemed chaste beautiful and inspiriting. There's one fine street in particular.
thro the centre runs an avenue or avenues of linden trees, the shade of which

is doubtless needed in summer. I have bought a number of tickets for the Cafe de Belvidere—5 for a thaler and each of these ensures me a dinner—today however I had a slice of blackbread and wurst and a quaff of Bavarian beer in my lodgings—this will doubtless be the case often. I sent a note to Knoblauch yesterday evening—he has been just with me and we have talked for half an hour together. he will return in a week or so and then give you the Tragheit's moment[5] of his magnet. Right opposite to the seat I occupy is a picture of two ladies. one drawing and the other leaning on her shoulder—they are like all M. de Baux paintings, colour enough—but no soul—they are flat stale and unprofitable—at my left is a lady intended to be very contrite—but she is too fat. there is no elevation in her sorrow, her grief is altogether a matter of blubber! There are 3 chairs in my room and a sopha. three tables on one of which my coffee is placed in the morning, on the other a decanter of water and a glass, and on the third I write—I attempted to do something at writing a short memoir for the British Association[6] yesterday but the noise of the town simmered so in my brain that I could produce only a disjointed affair— today it is better and doubtless after a little time the annoyance will totally disappear. I am well pleased with my lodging, they appear to be quiet unobtrusive people and procure whatever I require with sufficient alacrity. I miss my old tub in the morning very much and am obliged to content myself with a wash after your fashion. There is a *[fine]* stove in my room but what's the use of it when one cant have the pleasure of heating it—<u>The privy is vile</u>—it is the only thing to which I have any objection. I have remained very much within doors today, having merely walked to the railway station and back— The carrying of this letter to the post will afford me an additional half hour's occupation and if I meet a likely place I shall be tempted to step in and take a cup of chocolate—Knoblauch tells me every lady is expecting me[7]—I would prefer an *[undercurrent]* of life to this notoriety. I hate to do any thing when any thing is expected—What nobody expects is my own pure act and is not influenced by the probable opinion of others[8]—however I must content myself with circumstances and shall doubtless do so with sufficient comfort to myself. Remember me sweetly to all my Marburg friends—say to the walls of the old castle[9] that I have got a Daguerrotype portrait of them in my heart. to Spiegel*[s]* Lust and Augustin Ruhe[10] that I feel their foliage budding—I greet the church spires and the maidens—bless that little Sophy Bauer[11] for the last kind smile she gave me—bless them all, may they be happy and joyous as Westphalian larks—with their hearts full of everlasting melodies—Love to Noll, to Carl.[12] to Heinrich[13] if he be there. greet Dr Kohlrausch for me when you see him—I will send him a note when my brain settles. May god's grace attend you boy and keep your heart firm and faithful—Tyndall.

Herrn Thomas Hirst | Marburg | Kurhessen[14]

RI MS JT/1/T/538

___

1. *Frachtbrief*: bill of lading.

2. *Droschke*: hackney-cab.

3. *my host is a painter … war pieces*: Raymond de Baux (1785–1862) was a well-known Berlin painter, drawer, and lithographer. He painted portraits, especially war figures (for example, horses and soldiers on horses) and was widely known for his drawings of the Napoleonic Wars.

4. *Schlossplatz*: a large square in central Berlin, on an island formed by Spree River and canal. It is close to Dorotheenstrasse and adjacent to the Lustgarten.

5. *Trägheit moments*: inertial moments.

6. *a short memoir for the British Association*: probably 'On Air-bubbles Formed in Water' (abstract in *Brit. Ass. Rep.*, 1851 meeting, pp. 26–27). Tyndall's journal entry of 25 April (JT/2/13b/539) records that he finished the memoir that day.

7. *every lady is expecting me*: this implies that Tyndall had a reputation for charming ladies.

8. *What nobody expects … the probable opinion of others*: Tyndall alluded to the ethics of Fichte: only action that springs from inner choice, rather than any externally shaped need or impulse, is truly free.

9. *walls of the old castle*: Marburg or Landgrafen Castle (Landgrafenschloss), a fort built in the 11th century, and located on top of the Schlossberg. The walls were conspicuous from the town.

10. *Spiegel[s] Lust and Augustin Ruhe*: pleasant walking places on the outskirts of Marburg. There was a Spiegelslust area in green space on the north-eastern side of Marburg.

11. *Sophy Bauer*: not identified.

12. *Carl*: probably a fellow student, perhaps the 'Carl Schmitt/Schmidt' of letter 0565 (n. 16) and the 'Karl' of letter 0631 (n. 7).

13. *Heinrich*: Heinrich Debus.

14. *Herrn … Kurhessen*: Address on envelope. Postmarked Berlin 24 April, Marburg 25 April. Hirst noted on the envelope that he received it on 26 April. Tyndall wrote his own address in brief on the back of the envelope: '95 Dorotheen Strasse | B'.

To Thomas Archer Hirst   [28–29] April 1851   0480

29th April 1851

My dear Tom

   Your letter enclosing one from Faraday reached me today.[1] Prof. Knoblauch leaves for Marburg on Friday. I have not yet got to work so I cannot perhaps find a better time to write to you than just the present. Faraday's letter is a long one and a very friendly one. I gave directions to Francis to send him a copy of the last paper on magnetism,[2] this he says he has received but had read the paper in the magazine before he received it. With regard to the paper he says, 'I am able fully to appreciate the value of the results which you have arrived at and it appears to me that they are <u>exceedingly well established and of very great consequence</u>'. At the end he writes thus[3]:—'As you say at the close of your letter, I have far more confidence in the one man who works mentally and bodily at a matter than in six who merely talk about it; and I therefore hope and am fully persuaded that you are working. Nature is our kindest friend and best critic in experimental science if we only allow her intimations to fall unbiassed on our minds. Nothing is so good as an experiment which whilst it set an error right gives us a reward for our humility in being refreshed by an absolute advancement in knowledge'. This morning until 2 o'clock I have devoted to visiting, and have seen many of the great guns of science. I had an opportunity of making the celebrated experiment of du Bois Reymond, of exhibiting an electric current by the action of the muscles of my arm.[4] I have been with Dove, Magnus, Riess, and Poggendorff. Magnus' place[5] is out of order just now but by Wednesday he will arrange a spot for me to work in. Knoblauch and myself are to spend Wednesday evening with Poggendorff. The day after I wrote to you[6] my chest arrived—just in time sufficient to draw away coat and run to the 'Physikalische Gesellschafft'.[7] Du Bois Reymond is President and he delivered a short discourse on the decrease of force in a separated muscle. After him came a Lieutenant of infantry on the pendulum; at this time Knoblauch sidled over to me and said du Bois wishes you to deliver a 'Vortrag'[8]—Well, what was to be done? —I opened my eyes at first in astonishment, but as they were about to submit me to a vote of membership[9] I thought it would <u>look cowardly</u> to back out of it, so I scratched my forehead for 5 minutes, gathered up my thoughts and delivered a short discourse on that water affair.[10] They told me it was well done and quite fluent, but I felt myself tremendously fettered sometimes. This I think is the sum total of my life in Berlin since I wrote to you. I have been as yet to no place of amusement and have therefore nothing in that line to talk about, which of course you wont regret much. I enjoyed a speculative walk among the crowds yesterday evening very much. I managed to detach myself wholly from them and look at them as a beautiful picture swimming before my mind's eye. During this time I saw more deeply into Fichte's arguments[11] than ever I did before. Noll's apparent contentedness may be accounted for on the hypothesis that there is something <u>out</u> of him as well as that there is something <u>in</u> him—I speak only in the abstract. As regards myself I have only to say 'thank God I am that I am'.[12] Rest assured of it, Noll has his hours of shade as well as you, but he likes his work—he also likes to be liked by his patients and this is another inducement to attention. The summit of Noll's wishes is to be respectable country doctor, with a comfortable fire-side and a wife, and I honour him for the practical way in which he prepares himself for the realization of that ideal. Nevertheless there are men in the world with whom I feel more sympathy than with Noll.

A messenger has just brought me a letter from Mr Edmondson it runs as follows: —[...][13]

I will write to Debus[14] now and will send this off at the same time, the cost is all the same.

But the bowels, the bowels Tom! I have little faith in the injection method. I believe it is weakening even when it succeeds. I would strongly recommend application to Dixon in London.[15] I would put this earnest question to you: shall I write to him? If so, send me his address. I apprehend some mawkish indecisive reply to this question, the bare anticipation of which half enrages me. Send me his address. You will find it at the end of the preface of the Fallacies of the Faculty[16] which Noll has. | J.T.

RI MS JT/1/HTYP/127–128
LT Transcript Only

1. *Your letter . . . Faraday*: Hirst's letter is missing but Tyndall kept Faraday's letter (0477). He received both on 28 April (Journal, JT/2/6/52). Therefore this letter was started on 28 April. The 29 April date, is an LT date, and was probably the postmark date.

2. *the last paper on magnetism*: see letter 0464, n. 2.

3. *he writes thus*: the following extract is accurately quoted. Minor differences in punctuation may be errors in LT's transcription of this letter or in JT's transcription of Faraday's letter (on which 0477 is based).

4. *celebrated experiment . . . muscles of my arm*: see also letter 0489 to Faraday, which suggests Tyndall made this experiment again shortly before leaving Berlin.

5. *Magnus' place*: Magnus's private laboratory was one of the best-equipped in Europe.

6. *day after I wrote to you*: 25 April, the day after letter 0479 (24 April).

7. *Physikalische Gesellschafft*: the Physical Society of Berlin, f. 1845, is now the oldest scientific society in Germany.

8. *Vortrag*: lecture; give a lecture.

9. *vote of membership*: the election could not have taken place at this meeting for Tyndall did not hear until over a week later that he had been elected (Journal entry of 11 May, covering 7–11 May, JT/2/13b/543)

10. *that water affair*: either air-bubbles in water, the topic of a short paper finished that morning (letter 0479, n. 6), or his earlier investigation into water-jets (see letter 0456, n. 1).

11. *Fichte's arguments*: specific reference not identified. Tyndall read Fichte's popular ethical works (for example, see journal entries for 25 July and 13 August 1848 (JT/2/13a/371 and 374)).

12. *thank God I am that I am*: this expresses Tyndall's satisfaction that his ambitions, unlike

Noll's, were high. In an extreme assertion of his own free and independent identity Tyndall applied to himself the name by which God identified himself (Exodus 3:14: 'And God said unto Moses, I am that I am').

13. *letter from Mr Edmondson . . .* : letter 0478. The original letter is missing, but Tyndall copied it out in full in this letter to Hirst (which exists only in LT transcript form).

14. *write to Debus*: letter missing.

15. *Dixon in London*: Dickson (see n. 16 and letters 0482 and 0484).

16. *Fallacies of the Faculty*: Samuel Dickson, *The Principles of the Chrono-Thermal System of Medicine with the Fallacies of the Faculty, in a Series of Lectures* (London: Simpkin, Marshall, and Co., 1845). The Preface ends with Dickson's address: '28, Bolton Street, Piccadilly'.

To Thomas Archer Hirst   [1–4 May 1851][1]   0481

Dear Tom

Tomorrow morning I commence work so I cannot choose a better time than the present to write to you—I was in the midst of Magnus's apparatus today and choose what I wanted—he has placed a nice room at my disposal. So I am likely to be as comfortable in Berlin as a man could possibly wish to be—yesterday I dined with Knoblauch and some scientific friends of his—last night I spent at Poggendorff's till 12oC—Heinrich Rose and myself occupied the sofa—Dove, Magnus, Riess, Ehrenberg, and Knoblauch were also there—It was a most agreeable evening, and to me most gratifying. Sometimes I sat for half an hour a contented listener to the conversation—anon I found myself drawn into the whirlpool and enjoyed the gush of talk as much as the previous silence—verily Tom I never met with such kind people in my life—today I dined with Magnus—a good nature proceeds from these people which warms a man up like radiant heat. Knoblauch and myself accompanied Magnus to day to a sitting of the Academy[2] and saw how things are managed there—many celebrated men were present. among the rest Encke & old von Buch whom Emerson mentions[3] in one of his essays—Dove had some beautiful experiments to make[4] and he kindly came and led me over to where his instrument stood and shewed me the whole matter—there is something so hearty and unforced in all their kindness that it must be a mans own fault—his <u>subjectivity</u> must be cross grained if he does not feel pleased and comfortable—

Poor Smith[5] What will he do? —The intelligence shook me—These possible failures have always made me dread matrimony—but <u>He</u> tempers the wind to the shorn lamb—or perhaps better though not so beautiful, he hardens the lamb to bear the wind[6]—a wife, a wife, thats a terrible affair but I dare say the fellow has force enough bravely to meet the circumstance—and I cannot but think that London will offer opportunities to such as he, God speed him is all I can say—

Debus wrote to me[7] about Scheffer[8] I thought over the matter and laid the letter aside half intending not to write a single word in reply—To Mr Scheffer I am unable to give a word of advice negative or positive—I should be quite willing to assist him if it were in my power—but it is not—The circumstances which have led to Debus & Schnackenberg's going to England are altogether peculiar[9] and it would be unwise in Mr Scheffer to suffer their going to have the least influence upon him—In the present case he must just act as if there was no such person as John Tyndall in existence—These words may appear cold and hard but they are nevertheless true and perhaps also necessary.

Your arrangement[10] I believe cannot be altered for the better—I quite fall in with it—

I shall be anxious to hear the result of your bowel experiments,[11] may the Gods grant it a favourable issue.

This letter was commenced three days ago—it is now to be finished—I have yet to work, as yet I am merely preparing apparatus Last night I was at a meeting of the socalled Geographical Society[12]—about 200 met in a splendid room—2 or 3 lectures were delivered by travellers in various positions of the Globe—all the scientific men in Berlin were present—there was a supper afterwards which I dare say for the majority was the principal fact of the evening—I do not say that reproachfully, since the thing is by no means habitual, men of intellect may need a physical stirring up sometimes.

Knoblauch's father presented me with a ticket for the opera whither I go this evening There is to be something very particular I believe—how emphatically I feel the truth of Emersons assertion that gifts cannot make friends[13]—it needs a relatedness which all the world could not purchase—I feel myself as remote from Knoblauch as ever[14]—nay more so—for I see the opportunities which he has had here all his life long and what is the result—a reed shaken with the wind! —My stay in Berlin will I believe be chiefly precious to me on account of the glimpses into human character which I obtain here—In certain directions I feel my own strength gathering every day and I make a nearer approach to the practical belief that no man is to be dreaded except by my sufferance. If a man can smother egoism and rely upon himself it is wonderful what a little scrap of knowledge will carry him successfully through. this reliance imparts a certain abandonment and ease to a man which render his words when he speaks agreeable, for why it enables him coolly and unanxiously to look at the matter on which he speaks and he does not fear consequences. Only such a man can be properly polite—He defers to his fellow man but not through fear and therefore his deference is unencumbered, natural, and acceptable.

Knoblauch will carry this letter to you[15]—I dare say you will spend a portion of your time with him during the summer—doubtless you will find him willing to forward you in every way in his power—I would hear his physics again if I were you—and take up a subject for experiments in the long afternoons. As you have already said yourself <u>dive into it anywhere</u> for a readiness of manipulation acquired in the pursuit of any one subject is naturally available to others. In this matter you are yourself the most competent judge

I think to write more would be loss of time—good bye.

J Tyndall

RI MS JT/1/T/1012

---

1. *[1–4 May 1851]*: This letter was commenced three days before it was finished (see sixth paragraph and JT/2/13b/541) and was finished on 4 May (see nn. 12 and 15).

2. *the Academy*: the Berlin Academy of the Sciences or, during this period, the Königlichen Preußischen Akademie der Wissenschaften zu Berlin (Royal Prussian Academy of Sciences in Berlin).

3. *Encke . . . whom Emerson mentions*: Johann Franz Encke (1791–1865), Professor of Astronomy at the University of Berlin. Christian Leopold von Buch (1774–1853) was the pre-eminent geologist in Germany, and a friend of Alexander Humboldt since their student days together in Freiburg. Emerson mentioned von Buch while musing on man's role as an interpreter of nature in 'Uses of Great Men', one of a series of lectures on *Representative Men* (von Buch on p. 8).

4. *Dove . . . experiments*: it is unclear what these specific experiments were. Most of Dove's research was in meteorology.

5. *Smith*: John Stores Smith. The nature of his marital crisis is unclear.

6. *He tempers the wind . . . to bear the wind*: a proverb which had wide currency in the nineteenth century. It was most widely known from Sterne's *A Sentimental Journey* (1768) and was often mistakenly considered to be a biblical passage. Tyndall's reinterpretation suggests God strengthens the vulnerable rather than lessening their difficulties.

7. *Debus wrote to me*: letter missing.

8. *Scheffer*: not identified.

9. *circumstances . . . altogether peculiar*: the events alluded to have not been identified. They probably occurred in February for Tyndall recorded that he had been 'visited by Pfarrer [trans. Pastor] Schnackenberg and every thing has been arranged for the departure of his son for England' (Journal entry for 23 February 1851, JT/2/13b/522).

10. *Your arrangement*: unclear allusion. There are no extant letters from Hirst in the preceding month.

11. *your bowel experiments*: probably refers to the injections which, Hirst reported (letter 0482), were unsuccessful.

12. *Last night . . . Geographical Society*: the Geographical Society of Berlin (that is, the Gesellschaft für Erdkunde zu Berlin) met monthly, the first meeting after Tyndall's arrival

in Berlin was 3 May (*Monatsberichte über die Verhandlungen der Gesellschaft für Erdkunde zu Berlin* 13 [1851–52], pp. 399–400). Tyndall therefore finished writing this letter on 4 May.

13. *Emerson's . . . gifts cannot make friends*: see Emerson's essay on 'Gifts' published in *Essays: Second Series* (cited letter 0393, n. 3), pp. 104–9, esp. pp. 108–9. He expressed similar sentiments in 'Friendship', *Essays* (cited letter 0402, n. 18), pp. 191–219.

14. *myself as remote from Knoblauch as ever* : Tyndall often comments on his relations with Knoblauch in letters to Hirst.

15. *Knoblauch . . . to you*: LT noted 'Knoblauch, who conveyed this letter left Berlin on May 5/51' at the top of this letter.

From Thomas Archer Hirst   11 May 1851   0482

Marburg— | Sunday—May 11th/51

My dear John—

In a couple of hours I set off on my Sunday duty all alone, that is, a walk and dinner somewhere God knows where, and before starting I have time to have a word with thee instead of as usual having thee by my side. Debus sails from Rotterdam today if all be well Kolbe who only arrived here on Wednesday last tried to persuade him to remain some days longer to explain the state of the Laboratory but he would not and is off according to appointment. The Laboratory will not be open yet for a week Zwenger[1] & Kolbe divide it between them, I believe it will suffer much thereby in point of convenience as most assuredly it is doing by the delay—Marburg indeed by the absence of you all as well as other attendant circumstances[2] has suffered even more than we anticipated. There are few Students in all Scientific Branches. Gerling is rubbing on as miserably as ever with even fewer hearers. Stegmann for the first time at Marburg has only <u>one</u> Lecture & that the unimportant one (comparatively speaking) on Descriptive Geometry. For his Analytical Geometry im Raume[3] he had only 1 hearer who announced himself & it has been therefore abandoned. For me perhaps it is just as well—I hear his Descriptive Geometry 4 days per week and he gives me 5 private lessons. Knoblauch begins to-morrow morning with only a small number also. I shall hear him of course. I wished to put down my name but he would not allow it, and the kind little fellow placed everything at my disposal and requested me to take John Tyndall's position in every particular. Since your absence Marburg has received a curious addition in the shape of an Irishman with a wife, 3 small children[4] and a nurse. He comes from Dublin was recommended by Playfair to come here to Bunsen & to his astonishment found him not here.[5] He speaks next to no

German and is altogether a curious character. A long lanky body, with weak limbs and unless I am greatly mistaken weaker head. He appears a bit of an aristocrat who would learn Science in dilettante fashion, has some money at his disposal, & full of whims, has been already at Dublin & London colleges studying medicine & obtained a most suspicious title of Doctor; that is, if we take Mrs Simpson as authority who takes particular pains to give him his full title though it appears at the same time that he never practised. As far as regards Chemistry he asserts to have already studied it a few years ago under Prof. Graham but complains that the method of teaching was old fashioned and he wishes to begin anew at Qualitative Analysis & return to Dublin in October next (the 24th is the day already fixed to be again in Dublin) a Chemist!! You would have laughed however to see how he has pulled me out of my cosy nest the last week and for humanitys sake tramped over Marburg in search of Lodgings as their interpreter and the laughable blunders &c that my experience in household matters & interpreting capabilities have occasioned. How I have to sometimes smooth down but oftener put up in silence with the most insignificant complaints and outrageous expectations of the Frau Simpson who has brought with her her English notions of comfort &c & they even of the most artificial & dispensable nature. From the housekeeping department, the dessert service &c (/unfor[t]unately/ forgotten at Dublin) & these matters I hope I have freed myself by eliciting the services of the English Krantzchen[6] in her behalf & introducing her to Fraulein Spannenburg.[7] She will be an interesting though curious addition to their Society and as they will free me from certain undertakings for which I feel the greatest incapability, I am reconciled that their ideas of an English Matron should suffer (or otherwise) from such a representative. The chief failures of the Couple however I believe lie in these light headed artificialities; at bottom they are good natured I believe, but the anomaly of a Chemical Student in Simple Marburg with his Dublin drawing room characteristics & wife & family _is_ somewhat striking and not completely admirable. —She is shocked at German Sabbath Breaking and stand in open-mouthed horror when I was compelled to answer her enquiries about the doctrines, pew rents, &c at the Elizabeth Kirche[8] by a broad confession that for 6 or 7 months my head had never been in a gospel shop. She will relax her Sunday strictness however so far, to accommodate the Marburghers, as not to object to a drive for the Children & Doctors health after Divine service!! And now good bye to the good M^r & Mrs Simpson, my remaining time and paper must be otherwise employed—

<Handwritten letter ends; hereafter LT Transcript Only>

I am heartily glad that your Berlin prospects are so favourable, and that you met everywhere such kindness and willingness to please. In the common and usual aspect, your situation will not be without its advantages; though the manner in which you will make use of your environments will be of quite another nature. You say truly the presence of these men who surround you will serve you most and best as measures of your own strength; insight into your own relations towards the world and your fellow-men; in fact as one part of your apprenticeship in life, wherein we all have to learn to <u>use</u> the tools we find about us to the greatest advantage in order to qualify ourselves for taking our place on that more or less ideal platform, whereon the assembled students, as it were, receive the discipline and teachings of a higher Life—Philosophy. In one aspect John may not Life be considered as an immense College wherein are teachers and classes of every grade for the Human Student: wherein however his progress and rise depend materially and mostly on his own industry. The value of the whole institution reveals itself to him more as his own capability to appreciate increases. The lectures of the higher Professors therein are thrown away on the student whose place is lower. Emerson says somewhere we come into the world not only to act but to be acted on.[9] But <u>how</u> and after what manner we shall be acted on depends also upon our present action. I can recollect well how one after another 'the valuable' in the world, has changed to me, how it has taken the several forms of money, pleasure, society, literature, fame and God knows what. These are School-classes I have passed or may be passing through, and the continually increasing insight into the existence of a higher theory of what <u>is</u> valuable runs before, reconciles and rewards him who earnestly uses the tools and opportunities that now lie at his hand. I go to the Frauenberg;[10] Good bye till I return and have had a slice of black bread in the old Farm house there.

Here I am again after my walk and feel all the better for it; it rained rather heavily soon after I started. Mr and Mrs Simpson whom I met thought me, I dare say, a little insane; the former pressed me to dine with him in Ritter, but no, I persevered and have been rewarded with a sunshiny walk home. I read an article from January Searle on the way, which has reached me in the Truth Seeker, and which I will forward you, 'Scepticism and its Manifestations'.[11] It is a broad, manly, honest essay, and free from many of January's faults. Did I not know January I should satisfy myself with unqualified approval of the Essay and set down its author as a worthy man whom I would hail as Brother and whose presence in the world was cheering and satisfactory. As it is, another reflection will not withdraw itself, or hardly explain its presence, namely that there is in it no vertical deepening tendency. The man January by broad natural sympathies and (at some period of his life) active endeavour, has attained to a certain depth of vision, but at present his motion is for the most part lateral; I am almost afraid that he has too much of a notion that what insight he has already amassed is for the time being sufficient, if well spread and dispensed, to ensure him a certain literary status and accomplish a certain intended object. The materialities of the world have January too much in their grip at present to let him strip and use his boring-rods. Talking about him reminds me that only this morning a frank, honest-looking young fellow called on me and announced himself as Dr Huth of Wiesbaden,[12] and friend of Mr Phillips of Huddersfield.[13] He is a relation of Bunsen's, it appears, was a medical student here for 3 years, fell in love here and to-day has come, to get married to-morrow. His intended wife's name was strange to me and I have already forgotten it, I liked him well, there was an honest light in his eye, and a certain enthusiasm as he spoke of the good January's friendship that pleased me. In last week's Leader[14] is your 'Forester's Grave'[15]; it is a strange one for you, John, and as I take it, there is evident intellectual rust about the love machinery, the wheels don't revolve quite naturally but want a little more of the oil of actual Experience about their axles. You understand me no doubt—One would have no hesitation in saying that it would be difficult to catch the Weasel who wrote the 'Forester's Grave' asleep, and that for some time at least <u>he</u> had not been smitten or 'pulled far astray by bright eyes.

But about the bowels[16] I must not forget to report. The effect of the injection was as near 'nil' as is calculable. At the present moment they are far from being in natural action, though I am suffering under no crisis. Dr Dickson's advice I shall be glad to have. Noll has not got the book here, it is at home and packed up. Can you find the address any other way? When I have time I will write you as clear and concise[17]

RI MS JT/1/H/157
RI MS JT/1/HTYP/131–133

---

1. *Zwenger*: probably Konstantin Zwenger (1814–85), extraordinary professor in the Faculty of Medicine and lecturer in the Pharmaceutical Institute at Marburg, became Professor of Pharmaceutical Chemistry in 1852. He had studied chemistry and pharmacy under Liebig at Giessen. From this letter it appears that the chemistry laboratory served students of both chemistry and pharmaceutical chemistry.

2. *other attendant circumstances*: this allusion is unclear but the circumstances that led to Debus's leaving may be included (see letter 0481).

3. *im Raume*: literally 'in space'; perhaps he refers to analytical geometry in three dimensions.

4. *Irishman with a wife . . . and a nurse*: identified below as the Simpsons. The nurse is not further identified.

5. *Bunsen . . . not here*: at Easter, the end of the winter semester, Bunsen had moved to the University of Breslau to research with Kirchhoff (see letter 0458).

6. *Krantzchen*: correct spelling Kränzchen. This was an informal conversation club or circle at which members discussed English literature in English; dining, drinking tea, or eating may also have been a part of the meetings. It is mentioned in several letters between Tyndall and Hirst in 1851–2.   Men and women attended, but there were more female than male attendees. Compare with the mathematical Kränzchen (letter 0652).

7. *Fraulein Spannenburg*: not identified apart from this letter and 0516 (spelled 'Spannenberg'). She belonged to the English Kränzchen and had attended school near Geneva.

8. *the Elizabeth Kirche*: or Elisabethkirche, a church in central Marburg built, in the thirteenth century by the Order of Teutonic Knights, in honour of St. Elizabeth of Hungary.

9. *Emerson . . . to be acted on*: source not identified. This is an unusual sentiment for Emerson who, in *Representative Men,* praised Napoleon and Swedenborg because they were not acted upon by others.

10. *the Frauenberg*: see letter 0417, n. 21.

11. *'Scepticism and its Manifestations'*: *Truth-Seeker*, n.s. 2 (1850): 398–406; reprinted in J. Searle, *Essays* (cited 0399, n. 9), pp. 150–64.

12. *Dr Huth of Wiesbaden*: not identified beyond the information given here.

13. *Mr. Phillips of Huddersfield*: Hirst's friend, more commonly referred to by his pen-name, January Searle.

14. *Leader*: see letter 0398, n. 8.

15. *your 'Foresters Grave'*: [Tyndall], 'The Forester's Grave', *Leader*, 3 May 1851, pp. 420–21, appeared in the 'Portfolio' section for original writing. It is a romantic tale in which a man comes to terms with being rejected by a girl he loves, only for her then to fall in love with him. This is revealed when they meet, on an outing, at a forester's grave, shortly before he is to leave for America. She agrees to go with him. It is full of ponderous, pompous language. The forester's grave was a real place in the outskirts of Marburg (see letter 0508, n. 17).

16. *about the bowels*: Hirst's ongoing problem.

17. *concise*: the end of the letter is missing; the typescript ends here.

To Eilhard Mitscherlich   [13 May 1851][1]   0483

Tuesday Evening

Dear Professor.

I have just reached home and was not aware until the present moment of your kind invitation. I regret exceedingly that I am compelled to deny myself the pleasure of being present with you this evening—Last night I received a letter from the editor of the philosophical Magazine which demands my immediate attention[2] and thus throws an obstacle in the way of my enjoyment—Trusting that you will accept this as my apology and thanking you heartily for your past kindness

I remain | most respectfully yours | John Tyndall
Prof. Mitscherlich | &c &c

DM HS 897

---

1. *[13 May 1851]*: dated by letter 0484 to Hirst, which refers to this invitation as 'last tuesday'.
2. *letter . . . demands my immediate attention*: Francis's letter to Tyndall is missing but Tyndall noted in his Journal that he received 'a most kind letter' from Francis on 12 May (JT/2/13b/543). Whether this letter actually demanded Tyndall's immediate attention cannot be determined, but his account to Hirst (letter 0484) implies that he preferred to continue work on his investigation over sociability.

To Thomas Archer Hirst   17[–?] May 1851[1]   0484

Saturday night May 17th 1851

My Dear Tom

It is 11 oC and I ought to be going to bed instead of sitting down to write to you—well another week has closed down upon my Berlin life—I have been a pretty good boy this week, did not sleep very long in the morning and have stuck steadfast to my work until 7 and 8 in the evening—then home and busied myself with a report for the Phil. Mag.[2] I find Tom that hard work resembles teetotalism—it is easier for a drinker to abstain altogether than to half abstain—and the more I work the easier and pleasanter I find my work—at the tag end of a day for example during which I have worked well I find it easier to work an hour longer than at the tag end of a day in which I have worked ill—it puts one in mind of the parable of the 10 talents—'To him that hath to him shall be given, and from him that hath not from him shall be taken even that which he hath'[3]—I dont know whether you will understand me still I wont bother myself with further explanation—last tuesday eveng I came home at 7½ full of a project for a nights employment—I found an invitation from Prof. Mitscherlich before me—I was in no humour to accept, still he is a great man and I hesitated—I sat down finally and wrote a note declining the invitation[4]—I had no seal or my sealing wax was astray—I sent the note to my hostesses daughter begging of her to seal it for me. She stuck a wafer under it I have since learned and I doubt not stamped it with her thimble—thus the document reached the Hofrath[5] and formed likely a subject of speculation for him his wife and daughters—I felt annoyed, for I am sensitive on all these points—I would avoid above many things the imputation of being vulgar, it is a feeling which I remember existed strong in me when a boy, next day I felt in duty bound to pay the professor a visit—but why am I ranting thus? why merely to give you [another] one of those phases of character upon which you like to ponder—I know that you can see through the cobweb of an apparently trivial incident and detect the meaning beneath it—Well I disliked the idea of going in very much—but I felt that I ought to do so—At length I said to myself why do you suffer this feeling of annoyance to take hold of you, why do you shirk like a coward a task which you know to be your duty?—why not go forward and prove by the independence of your presence that you are not a vulgar man—My feeling [thus] reasoning became a kind of subdued counterpart to that of Teufelsdrock in the Rue de l'Enfer[6]—I felt eased in a moment from my annoyance and cried pshaw! at my previous cowardice—It was all myself—the thing assumed an importance merely because I permitted it to assume it, and the importance vanished the moment I was lord of myself. Rock crystal I believe Tom has surfaces of different texture. some are attacked by acid. others not—a man seems to be a kind of Rock crystal and he has surfaces which if he turns against the canker of the world the tooth of the latter will shew its blunt edge and base temper. Good—I have now done.

Thank you for your letter[7]—it was a sweet and precious one to me. of course boy the world's a school for the individual—but hang me if you draw me just now into your philosophic meshes—I cant afford it—and my eyesight in this direction is too dim just now to interest you with its revelations—still I cannot drive away that idea of the human <u>race</u> being an integral whole in the carrying forward of which the individual has perfect freedom within a certain circuit—the amplitude to which he can swing being never sufficient to disturb the higher designs which have respect to the race—Faraday says to me—'nothing is so good as an experiment which whilst it sets an error right &c.'[8] Cui bono? I exclaim—what good? Is it this paltry result that you dignify thus? —By heaven if it be to transact tricks of this kind that man is placed here life is not worth living for. Then creeps in that soothing whisper—'Know my son that thou art a link on which higher destinies hang. Know that thou in the faithful discharge of thy duty to day art forwarding the ulterior designs of the author of the universe—thou throbbest with his life and thy life must be given for the throb'[9] but I'm roaming again—

Your remarks on January[10] are just—I liked the article well though it revealed to me nothing which I did not already know—but there is a certain satisfaction in finding that January understands the matter so well. there is sometimes a looseness in his phraseology which I accustomed as I am to strict 'induction' would rather see banished—still I fully suscribe to those utterances of his which are apparently most illogical—January has had a very rich experience richer than most men, [so] it will be a pity if he now suffers himself through lazy inanity to become a castaway from nature.

The rust of intellect you say is about that article of mine[11]—now whether is the 'rust' my rust or your rust—it is a matter of contact so to speak—if the rust be in you it is practically all the same and your judgment has simply referred it to the wrong surface—I tell you boy I'm a very affectionate fellow and no 'old weasel'[12]—I'm sure I could love like a brick if occasion offered.

In a month I shall be leaving Berlin—Shortly afterwards I shall see Dickson personally—write me an account of your symptoms[13] in the mean time—

A few days ago I received a most kind letter from Francis[14]—he is one of the best fellows I know—He has obtained for me the office of Berlin correspondent to the literary Gazette[15]—to send articles and write Reviews at 10/6 a column—I wont undertake it for I cant—no time to spare—He has also proposed to Van Voorst the publisher[16] that I should write a treatise on physics for him, and the latter seems to like the idea—He is to see me in Ipswich[17] on the subject. I dont know whether I will undertake it—it costs so much time—He has sent copies of my memoir[18] to Wheatstone. Sabine and many others—on the whole his bearing towards me is most good natured.

I ask your pardon for this rigmarole—no I dont!—I forgot you and dropped into a silly conventionality. It is now midnight—good luck to you Tom. May your sleep be sound to night and—oh that my prayer could affect those bowels! I would wrestle with all heaven for the blessing[19]—Kind love to some of the ladies of the Kränzchen[20]—you may make the selection yourself—
Tyndall

RI MS JT/1/T/540

---

1. *17[-?] May 1851*: Tyndall started this letter late on 17 May, but it is a long letter and he enclosed it with letter 0485 of 23 May (see letter 0485, n. 2). All this suggests that he wrote it over a number of days.

2. *a report for the Phil. Mag.*: probably Tyndall, 'Reports on the Progress of the Physical Sciences', *Phil. Mag.* 2, no. 8 (July 1851), pp. 26–36, which Tyndall sent to Francis before the end of May (compare letter 0491). It discussed Dove, 'The Reversion-prism, and its application as ocular to the Terrestrial or Day-Telescope' and 'Description of several Prism-stereoscopes, and of a simple Mirror-stereoscope'; and Knoblauch, 'On the deportment of Crystalline bodies between the electric poles'.

3. *Parable of the 10 talents . . . he hath'*: there are very similar accounts in Matthew, Mark and Luke about a master rewarding his servants in proportion to their success, and taking from the failed one to give to the most successful. The Matthew 25:14–30 version is most frequently identified as the parable of the talents, but Tyndall's quote is much closer to Mark and Luke. 'For he that hath, to him shall be given: and he that hath not, from him shall be taken even that which he hath' (Mark 4:25). 'That to every one that hath shall be given, and from him that hath not, even that he hath shall be taken away from him' (Luke 19:26).

4. *wrote a note declining the invitation*: letter 0483. Tyndall received a second invitation from Mitscherlich on Wednesday 28 May and on that occasion accepted it (Journal, JT/2/13b/544).

5. *Hofrath*: (Hofrat in modern German) a court councillor or senior government official.

6.   *Teufelsdrock in the Rue de l'Enfer*: an allusion to Carlyle's *Sartor Resartus* (1836), Book II, chap. VII. Carlyle's protagonist, Diogenes Teufelsdröckh experienced an epiphany on the Rue Saint-Thomas de l'Enfer in Paris and refused to be defeated: 'Thus had the Everlasting No . . . pealed authoritatively through all the recesses of my Being, of my Me; and then was it that my whole Me stood up, in native God-created majesty, and recorded its Protest. . . . It is from this hour that I incline to date my spiritual new-birth, or Baphometic Fire-baptism; perhaps I directly thereupon began to be a man'.

7.   *your letter*: letter 0482.

8.   *'nothing is . . . right &c.'*: Tyndall quotes from letter 0477.

9.   *'Know my son . . . for the throb'*: not identified.

10. *Your remarks on January*: see letter 0482 for the discussion of January Searle and his article.

11. *that article of mine*: 'The Forester's Grave', which Hirst criticised in his previous letter (0482).

12. *'old weasel'*: Hirst had called the author of 'The Forester's Grave' a 'Weasel' in letter 0482.

13. *your symptoms*: Tyndall was encouraging him to consult Dr Dickson over his bowel problems.

14. *a most kind letter*: letter missing (see letter 0483, n. 2).

15. *Literary Gazette*: see letter 0501 where Tyndall alludes to doing work for the *Literary Gazette*.

16. *Van Voorst the publisher*: John Van Voorst (1804–98) was a London publisher who specialized in natural history books.

17. *in Ipswich*: Tyndall planned to attend the BAAS meeting in Ipswich in July 1851.

18. *my memoir*: the memoir on magnetism (cited letter 0464, n. 2).

19. *I would wrestle . . . the blessing*: Tyndall alludes to Jacob wrestling an angel (or God) and refusing to give up until he had received a blessing (Genesis 32: 24–30).

20. *the Kränzchen*: see letter 0482, n. 6.

To Thomas Archer Hirst   23 May [1851][1]   0485

May 23rd

Dear Tom

Read this one first and the other[2] afterwards. I received the enclosed[3] from Mr Edmondson yesterday—you see he has no ready money but requests me to draw on him at 3 months. Go to Bang and arrange this for me—draw on M[r] Edmondson at 3 months for £20 and have the money forwarded to me—I should say Bang wont have any objection—his interest will of course be secured. you may put your own name to the draft. —I shall be leaving here in about 3 weeks and I should like to have the money immediately as I will expend it all except what is necessary for the journey on instruments—Mr Edmondson appears willing to allow something on account of the Institution. I will purchase 10 or 15 pounds worth for <u>him</u>—I shall expect a reply to this immediately—letting me know whether Bang will promptly transact the affair—Another <*letter*> from Francis[4] reached me yesterday enclosing one from the treasurer of the British association—I am to lecture there[5]—terrible up-hill work for me just now. my investigation takes up from 8 AM to 7 P.M and the remaining time is all I have to get up *[three]* papers of my own[6]—3 papers belonging to others[7] which I am compelled to translate and <u>understand</u> and this formidable lecture—n'importe[8]—I shall get thro it—good bye my brother.

write quickly | Tyndall

not prepaid—the post office too far[9]

[An] | Herrn Stud | Thomas Hirst | beim Weissbinder Baum | Marburg | Kurhessen[10]

RI MS JT/1/T/902

---

1. *[1851]*: The context clearly dates this letter to May 1851.
2. *the other*: letter 0484. Hirst wrote '17 May 1851' on the envelope, indicating that he received both letters together. This later letter was more urgent hence Tyndall's instruction to read it first.
3. *the enclosed . . . Edmondson*: letter missing.
4. *letter from Francis*: letter missing.
5. *the treasurer . . . lecture there*: the letter was not from the treasurer (John Taylor) but, according to Tyndall's journal entry of 25 May (covering the week since 17 May, JT/2/13b/543), from John Phillips who was assistant secretary of the BA. He told Tyndall that there would be great interest in a demonstration of Du Bois's experiments. Tyndall replied to Phillips that he would be unable to perform them: 'but what can I do. Du Bois cannot come and I have no galvonometer' (Journal, 25 May 1851, JT/2/13b/543).
6. *three papers of my own*: 'On Diamagnetism and Magnecrystallic Action' and 'On Air-bubbles formed in Water', *Brit. Assoc. Rep. 1851*, pp. 15–18, 26–27). The third paper could be his demonstration on thermoelectricity (see letter 0501, n. 2).
7. *3 papers belonging to others*: possibly the previously unpublished papers by Dove (2) and Knoblauch (1) which Tyndall included in 'Reports on the Progress of the Physical Sciences' (see letter 0484, n. 2).
8. *n'importe*: never mind (French).
9. *not prepaid . . . too far*: postscript on back of envelope.
10. *Kurhessen*: address from envelope.

To Alexander Humboldt   [24 May 1851][1]   0486

Sir,

I am reluctant to leave Germany without seeing a man whom from my boyhood I have been taught to regard as the greatest genius now existing—I mean yourself. —Will you grant me the honour of an interview?[2]

I remain Sir | Your most obedient Servant | John Tyndall

RI MS JT/2/6/59
JT Transcript Only

---

1. *24 May 1851*: dated from Journal entry (25 May 1851, JT/2/6/59): 'Wrote the following note to Humboldt yesterday and sent it in accompanied by 3 memoirs'. Presumably, one was his recent memoir on magnetism (cited letter 0464, n. 2). Other candidates are the water-jet paper (see letter 0456, n. 1) and the memoirs written with Knoblauch (see letters 0395, n. 22 and 0403, n. 2).

2. *grant me . . . an interview*: Humboldt proposed Monday 26 May (letter 0487).

From Alexander Humboldt   [24 May 1851][1]   0487

Die scharfsinnigen und zugleich gründlichen Arbeiten von Herrn John Tyndall sind mir wohl bekannt, und es kann mir daher nur sehr erfreulich sein wenn er mich Montag um 2 Uhr mit seinem Besuch beehren wollte

Mit der innigsten Hochachtung
Ihr | Gehorsamster | A. V. Humboldt.
Sonnabend

RI MS JT/2/6/59
JT Transcript Only

The astute and at the same time thorough works of Mr John Tyndall are well-known to me, and it would therefore be a great pleasure if he honoured me with his visit on Monday at 2 o'clock.[2]

With high esteem
Your | most obedient | A. V. Humboldt.
Saturday

---

1. *[24 May 1851]*: Tyndall received this letter on Sunday 25 May 1851 and copied it into his Journal that day (JT/2//6/59). Humboldt wrote on Saturday (see bottom of letter), that is, 24 May.

2. *Monday*: Tyndall described the meeting in his journal entry for Monday, 26 May (JT/2/13b/544) and in letter 0489 (to Faraday).

From Thomas Archer Hirst   25 [and 31] May [1851][1]   0488

Marburg—May 25th | 1850

My dear John—

Another Sunday has come round in its due time, a small Havannah Cigar and an excellent chapter in Montaigne[2] have been simultaneously enjoyed The sunshine and the Gods have decreed as it appears to me that at 11 o'clock I must 'bundle my wallets and walk'[3] to Kirchhain[4] so that in the interval I take upon myself to say a word to thee more especially as a kind of presentiment that you very likely will be doing the same—I should have waited a day longer however & made this more of an answer (I may do so in posting yet) but a letter or two have come to hand—You have them enclosed. Debus in his note to me seems to have made the journey pretty well and is in a fair way of progressing and making himself useful and comfortable, so much will be off your mind. Jemmy Craven's letter to you[5] poor fellow is a queer production, nothing in it at all recommendable but this that he will not dissemble or pretend to what he does not feel. Two numbers of Leigh Hunts journal[6] have arrived for you along with 18 Copies of each of your last Investigations on the 'Water Jet'[7] and 'Laws of Magnetism'[8] from Francis—I have as yet made only one present, viz, a copy of each to Stegmann with whom I happened to be speaking about you at the time how the rest are to be disposed of you must let me know—and I will punctually execute.

In these few lines I find all business matters between us brought to a close a few more will exhaust the stock of news around me and then comes or perhaps will come as usual the paragraphs on our old subjects our worthy selves and experiences since we parted. —The Laboratory beautifully new cleaned painted and re-arranged has been open for a week, I work there from 9 to 1 A.M. but purely at Practical Chemistry. I hear no Lecture—Kolbe and I have not got into very good pulling harness yet, I mean with regard to my analysis & assistance therein I yet miss Debus & Bunsen very much, it will improve however every day. Ewart[9] is assistant, a good natured willing fellow enough but compared to Debus his ability to fill Laboratory duties far behindhand yet. There are about 14 students of whom 7 are working at Quantitative, the other beginners—Knoblauch is going a head a little more superficially perhaps than before though I have him at more advantage by far than last Session—He has only 5 lectures per week having abandoned the Theoretical Stunde[10] on Saturdays altogether. On Wednesday Evenings however he gives Wrightson, self and another a 'Theoretical stunde' to ourselves—Wrightson is yet 'muddy' and on still worse terms with Mathematical Formula[11]—& Knoblauchs bad hurried figures on the Black Board—His visits to me have increased of late & occur invariably in the muddiest phases of his muddiness namely after dinner. I treat him very candidly however, and he is forced to take it without offence—'My newspaper has just come M[r] Wrightson my Cigars you know where to find they are at your service as well as half the paper if you will sit down & use both you are welcome.' We thus sit silent & preserve the good feeling between us—At 3 O'Clock I have sometimes to add 'To a Student you are aware M[r] Wrightson that when the laws of the University & those of Hospitality come into contact the former must have the preference.[12] I have to go to Stegmann' And thus we rub on with the best understanding I deal also just as candidly & openly when he wishes any explanation & upon my word I think sometimes in point of trying to get information without appearing to need it, he is seeing its futility & attempting a less pretensious method. As to his Physical Investigations I see no sign of progress or even beginning—Davy & Dicks positions appear still more anomalous & sometimes unfortunate or rather incomprehensible—M[r] Simpson for a week at any rate has worked pretty diligently—His manners general appearance & German Language afford the students no small gigling opportunities The Postman entered with a letter from you[13] as I finished the last line but one— It stops my progress & probably will alter the following when read—I took it with me to Kirchain and read it two or three times through, it smelt of activity John, and to have such smells under my nose is more serviceable than a Scent bottle. Tomorrow I will make enquiries about the money and add a Postscript to this—The Bowels (hush! I dare not let them hear me) have gone a <u>little</u> by themselves for the last <u>3</u> days!! for a week or so before that they put me into one of the most unpleasant unnatural states of existence possible, that I would give much to be able to describe All I can say is it was a kind of dilirium—that is <u>I was conscious</u> that my senses were conveying to me sensations from outward objects different to what they ought & formerly did do—In contact with persons about me I felt it necessary to be careful if I would avoid revealing the inner abnormalism I felt—What do you think or can you conceive such a state? I scarcely can myself now that it is past. Tis now at any rate I am on something like Terra-Firma—although I dare not crow too much nor am I elated inasmuch as I find a few old infernal friends awaiting me on the Threshold I would send you an extract or two from the Diary but they are in general too what you call properly 'mawkish' & will as yet do nobody good.

Economy of Time. In that simple phrase lies an immense significance and I turned it over & over but confound it far too intellectually & too little practically. From my getting out of bed at 6 A.M to getting in again generally 11¼ to 11½ is 17 hours, and I daily ask what have I done with them? Not above 1 hour of those 17 is employed in eating <u>alone</u> & perhaps another 1½ hours we may reckon strolling about for health &c . . . Thus even reckoning accidentals of having to waste ½ an hour occasionally in '*/dummes zeug/*'[14] there is almost 15 hours per day <u>pretended</u> working i.e. self informing & improving time I protest that there must be some foul play in the matter some miscalculation at least, but no it is a positive fact, & the corollary from the fact is that I cut myself up & waste myself most miserably in fact as days pass they seem all chippings & only in a few years collecting can one see that a few solid blocks have by chance remained and become hoarded One thing it shews me clearly (many indeed but as you know John it is at bottom a matter entirely of individual insight and arrangement it would be useless to enumerate) that the advise of good people about us & their remarks generally on our actions <u>must be</u> futile & worth little to us—Mrs Simpson & her class smile benignly and say How pale Mr Hirst you are looking! I always see you standing by that old black upright desk when I pass—I'm certain you are working too hard &c &c—But good Mrs S. reckons by hours & minutes & as my calculations have long shewn there is no result as to positive work to be got out of them— alone. Moreover somebody else likely to know better than Mrs Simpson says—Tom Hirst not only does not work too hard but is properly considered actually '<u>lazy</u>'

Again with eating. Frau Baum[15] partly from pocket motives partly from better says every other day 'Och. Herr Hirst Sie essen gar zu wenig/, smächt es/ nicht gut dann oder fehlt etwas; sind Sie krank oder was ist es &c'[16] A Weiss Brod[17] sopped morning & evening by bits in my tea & coffee (lasting me 10 days) a basin of soup and a little 'obst'[18] at noon have for the last fortnight formed exclusively my meals & Frau Baum has not even sent up a Milch Brod or Zwieback[19] in the time. Frau Baum may well cry you eat too little but what say I—who know better—Tom many a time you are a <u>glutton</u>. Thus in matters of praise or blame or guidance must every man for the most part shut up himself from outward influence. But I must close or I may be getting mawkish there is a philosophy I feel at the end of all this—'Theres something in this world (or us?) amiss will be unravelled by and bye'[20] We'll help each other as far as possible to do it yet John, but not now—Good night! and a good days work to us both tomorrow (a PS not often added, but of the two <u>the</u> most important inasmuch as it reacts in no small degree on the former)

Saturday June 1st/51[21]

I have called every day during the week on Bang—There is no difficulty in the way of objection, but he has no ready money on hand—I am tired of waiting (& as you may be anxious) so dispatch this to day, and will send the

Money the moment I lay hands on it—He will charge 5 percent Interest for 3 Months & will give the Full exchange that appears in the Frankfurt paper on the day drawn—

In haste yours Affectionately | T A Hirst

RI MS JT/1/H/145

1. *[1851]*: Hirst wrote 1850 but the content of the letter clearly relates to other mid-1851 letters. See n. 21 for the postscript date.

2. *an excellent chapter in Montaigne*: according to Hirst's Journal (25 May 1851) he reread 'Of Virtue' (*Essays*, vol. 2, cited letter 0393, n. 12). Modelling his habits on the Bible reading of many devout Christian contemporaries, Hirst made a habit of reading a Montaigne essay on Sunday mornings.

3. *I must bundle . . . and walk*: from the poem, 'Fortuna', by Thomas Carlyle (*Critical and Miscellaneous Essays: Collected and Republished*, 3rd ed., vol. 2 [London: Chapman and Hall, 1847], p. 379, lines 19 and 20).

4. *Kirchhain*: a small town about 9 miles east (as the crow flies) of Marburg. Hirst was not sure of the spelling because he crossed out the second 'h' when he wrote it a second time.

5. *Jemmy Craven's letter to you*: letter missing.

6. *Two numbers of Leigh Hunts journal*: subtitle: 'a miscellany for the cultivation of the memorable, the progressive, and the beautiful', a weekly journal edited by Leigh Hunt, financed and published by John Stores Smith, which cost 1½ d and lasted only four months, from 7 December 1850 to 29 March 1851. Hunt blamed Smith for the failure of the journal.

7. *'Water Jet'*: cited letter 0456, n. 1.

8. *'Laws of Magnetism'*: see letter 0464, n. 2.

9. *Ewart*: see letter 0396 n. 6. In early 1850 Tyndall took Italian lessons from a 'Dr. Ewert' (JT/2/13b/481–3) who is perhaps the same person.

10. *Stunde*: class.

11. *Wrightson . . . mathematical formula*: the first of many allusions to Wrightson's inadequacies as a student.

12. *preference*: Hirst probably meant precedence. This letter shows that his punctuation, grammar, spelling and word usage were often incorrect.

13. *a letter from you*: letters 0485 and 0486, which were posted together. Clearly Hirst was answering 0486, but the allusion to Tyndall's 'activity' indicates that he was also answering letter 0485.

14. *'[dummes zeug]'*: nonsense.

15. *Frau Baum*: Hirst lodged with Weissbinder Baum and his wife in Marburg. Very likely they ran a student boarding house.

16. *'Och. Herr Hirst . . . ist es &c'*: 'Ooo, Herr Hirst, you're eating far too little, don't you like it then? or is something the matter; are you ill or something, is it &c'

17. *Weiss Brod*: white bread.

18. *'obst'*: fruit.

19. *Milch Brod or Zwieback*: milk bread or biscuit.

20. *'Theres something . . . by and bye'*: Hirst adapted and slightly misquoted from Tennyson's poem, 'The Miller's Daughter': 'There's somewhat in this world amiss | Shall be unriddled by and by' (first published in *Poems* (1833) lines 278, 4th stanza; these words were unchanged, but became lines 19–20 in the third stanza of the 1842 ed.).

21. *1ˢᵗ June /51*: 1 June was a Sunday. We assume Hirst was correct about Saturday, the day of the week, and therefore date the postscript as 31 May. Hirst's deletions show he was unsure of the date. After writing Saturday he initially wrote May, which he replaced by 'June 1st'. He also first wrote '4' for the year, before overwriting '51'. His Journal entry for 1 June is unambigious: 'Yesterday signed a three months' draft of Bangs . . . and sent Tyndall it'.

To Michael Faraday 26 May 1851 0489

95 Dorotheen-Strasse | Berlin | 26th. May 1851

Dear Sir,

Shall I thank you for your last encouraging letter?[1] By thus doing I should imply that a kind of equilibrium might be established between my thanks and your kindness—this cannot be done and I therefore refrain from making the attempt, appealing rather to my future actions to testify the effect which your inspiring words have had upon me.

I write now just to mention that I had the honour of an interview to-day with Humboldt. I introduced the topic of your recent investigations in terrestrial magnetism.[2] 'I have read them with astonishment' was his remark 'I do not imagine that very little variation can be thus accounted for but in the main he is correct' His last words to me were—'Tell Mr. Faraday that I am quite convinced of the validity of his hypothesis.' Dove has expressed the same opinion to me.

Could you not pay Berlin a visit? there is no place on earth where you would be more enthusiastically welcomed.[3] Your presence would call forth an exhibition of Hero-worship[4] with which even Thomas Carlyle himself would be satisfied.

I have been working for the last five weeks at diamagnetism. Prof. Magnus has been kind enough to place the necessary space and apparatus at my disposal—indeed I cannot speak too highly of the kindness of the men of science of Berlin generally.

My results I hope will interest you but I forbear mentioning them as they are not yet complete. It has been again my misfortune to arrive at conclusions very divergent from those of Prof. Plücker. A paper on the subject shall be ready for the British Association[5] at its next meeting.

Believe me dear Sir, | Most truly and respectfully Yours | John Tyndall
Prof. Faraday | etc. etc. etc.

Since writing the above I have spent a few hours with Du Bois-Raymond
and have succeeded completely in developing a current by muscular contraction[6]—I obtained a deflection of about 30°—right or left to the arm contracted. The experiment requires a delicate apparatus and some care. Du Bois'
multiplying Galvanometer contains 24.000 windings

RI MS JT/TYP/12/4006–4008
Transcript Only

1. *your … letter*: letter 0477 in which Faraday praised Tyndall's recent memoir on magnetism.
2. *your … terrestrial magnetism*: see letter 0477, n. 6.
3. *no place . . . enthusiastically welcomed*: these compliments were sincere on Tyndall's part.
    Although prone to exaggeration and even flattery, in the privacy of his Journal (30 April
    1851, JT/2/13b/541) Tyndall recounted the admiration with which his Berlin colleagues
    regarded Faraday.
4. *Hero-worship*: allusion to Carlyle's praise of heroes and advocacy of hero-worship in his
    On Heroes, Hero-worship and the Heroic in History (1840).
5. *A paper … British Association*: see letter 0484, n. 17
6. *developing a current by muscular contraction*: this was the second time Tyndall had conducted du Bois's experiment (see letter 0480).

To Thomas Archer Hirst   [30–31 May 1851][1]   0490

My Dear Tom.

I wrote to you precisely a week ago.[2]

My letter contained a portion of one for[3] M\u1d63 Edmondson in which he
requested me to draw upon him for 25 pounds at three months—

I requested you to go to Bang and draw upon M\u1d63 Edmondson for the sum
of 20 pounds at three months not thinking that Bang would have the slightest
objection to doing so—

I requested you for a speedy answer but no answer has yet reached me[4]—
If the letter has not reached you then go to Bang and do as I have requested.

Be good enough to let me know *[of]* his reply by return of post—I dont
want the money by return of post 10 days hence would be soon enough.

This delay is a dead loss to me—I intended to purchase nearly 20 pounds
worth of apparatus for Queenwood but now if I order them there is hardly
time to get them ready—They must be slurred[5]—perhaps not made at all.

Berlin is one wide scene of Hero worship[6] to day. The statue of Frederick the Great has been uncovered—it certain[7] was the grandest picture I ever witnessed.[8] The King princes Marshalls generals and about 20-000 troops Cannon roaring, bells ringing—men cheering[,] hats and handkerchiefs waving—

Good bye—write to me. | your affectionate Tyndall

Saturday evg.

Herrn Stud. Thomas Hirst | beim [Herrn] Weissbinder Baum | Marburg. Kurhessen[9]

Frei![10]

RI MS JT/1/T/541

---

1. *[30–31 May 1851]*: Tyndall dated his letter Saturday (at the end). The relevant Saturday was 31 May and the unveiling of the statue of Frederick the Great, alluded to in this letter, took place on that day (see n. 8), but see n. 2. Postmarks are 'Berlin | 31 5' and 'Marbug | 1 | 6'.

2. *wrote . . . precisely a week ago*: letter 0485, of 23 May 1851, is 8 days before the date of this letter. Tyndall's claim would be correct only if he began this letter one day (Friday 30th) and finished it the next day (Saturday 31st).

3. *for*: Tyndall meant 'from'.

4. *no answer has yet reached me*: Hirst sent his reply (letter 0488) the same day as Tyndall posted this letter.

5. *slurred*: done hurriedly (*OED*).

6. *Hero worship*: a Carlylean allusion.

7. *certain*: Tyndall meant 'certainly'. This is the second example of hasty writing in this short letter.

8. *The statue . . . ever witnessed*: Tyndall recorded the event, in some detail, in his journal (31 May, JT/2/13b/545).

9. *Herrn . . . Kurhessen*: address on envelope.

10. *Frei!*: trans. free, note on envelope alluding to his previous letter to Hirst (0485) which was not prepaid.

From William Francis   [1 or 2 June 1851][1]   0491

My dear Sir,

As you will no doubt be anxious to learn Mr Phillips's arrangements with respect to the lecture,[2] I forward this without delay. I hope to be able to write to you in a day or two from Hastings but have now merely time to thank you for the excellent report.[3]

Yours, etc. | W. Francis

Dr J. Tyndall | (Beim Herrn de Baux) | 95 Dorotheen Strasse, | Berlin[4]

[Enclosure] John Phillips to Francis　　1 June 1851　　0491encl

1 June, 1851 | London

My dear Sir,

I am much obliged by the information contained in the note from Dr Tyndall.[5] Since I wrote to you[6] the arrangements which were then in progress for the delivery of two Discourses in the evenings, have been completed, and we are to have Owen and Airy on those two evenings. It has been our custom for some time to provide for two such evenings, and no more; but I think at Ipswich a third discourse may very probably be desired, either in the General Meeting room, or in the largest of our Sectional rooms.

The best precedent for this kind of proceeding has been thus. A paper has been read to the <u>Section</u> in the morning, and repeated when of a suitable character to a larger audience in the evening. If Dr Tyndall were to follow this plan—if he were to read a paper on any selected portion of his subject in the morning, and be prepared to repeat it in the same or some other evening, I think it likely that the request would be made to him;[7] but having actually fixed the usual amount of evening business it is not desirable at present to make a positive arrangement for an additional evening.

Ever yours most truly, | John Phillips.

RI MS JT/1/TYP/11/3575
LT Transcript Only

---

1. *[1 or 2 June 1851]*: Phillips letter of 1 June, from one London address to another, could have arrived the same day or the following day. Francis forwarded it 'without delay', most likely 2 June, but perhaps 1 June.

2. *the lecture*: see prior proposals in letters 0464 and 0470.

3. *the excellent report*: probably Tyndall, 'Reports on the Progress of the Physical Sciences' (see letter 0484, n. 2).

4. *Dr . . . Berlin*: address presumed to be from envelope.

5. *note from Dr Tyndall*: letter missing, but presumably in response to the letter from Phillips (n. 6 below) received c. 25 May. Tyndall may have informed Phillips of his inability to reproduce the famous experiments of Du Bois, as he recorded in his journal. 'But what can I do. Du Bois cannot come and I have no galvanometer' (JT/2/13b/543).

6. *I wrote to you*: letter missing, but probably the letter which Francis forwarded to Tyndall, and which he recorded as receiving on, or shortly before, 25 May (JT/2/13b/543). Tyndall interpreted this letter as a firm invitation. Phillips mentioned 'Du Bois's experiments as likely to be of great interest' (Journal, ibid.).

7. *the request could be made to him*: this did not occur.

From Margaret Allen[1]   2 June 1851   0492

Stanley Terrace, [Preston,] June 2nd / 51

Dear Tyndall

 after a period of lengthened silence I almost feel culpable in disturbing the serenity which pervades your mind, nor would I do so, if I were not fully assured that you will be the gratified recipient of this letter and if the communication lightens one hour of (I am persuaded) your excessive labour I shall be amply repaid. I think I sent you a Preston paper & superscribed my beloved Robert's address, and if you were at all surprized at that, I fancy you will be a little more so presently. Robert is in the employ of Messrs Bridgens & Moody[2] Publishers of Maps & Lithographers; which M[r] Bridgens[3] is the Grandson of an English General, & the Son of an eminent engineer,[4] employed under the British government in the foreign service. And Robert is now with M[r] Bridgens in Charleston, 800 miles from their place of residence surveying that city where they expect to remain for two or three months, he speaks in high terms of the kindness of these gentlemen having resided in the family until their departure for the land of rice marshes & cotton groves, telling me that M[r] B. was taking a quantity of books which he told him were principally for Roberts use. Robert declares himself a changed man,[5] indeed I have faith in this for every sentence in his voluminous letters is confirmatory of the statement, but it is not only the relation of the matter which causes conviction in my mind, no, there is a secret concatenation in the mind which is more readily understood than I can explain to you here, you know, that words do not always convey power to bias the will . . . You will rejoice with me in the apparent success of my best wishes for our far distant one, and earnestly unite in the hope & prayer that his future career may be one of peace and prosperity. Life truly is intermingled with joy and sorrow, and they alone are happy who can enjoy its benefits in a becoming spirit, and, who are neither particularly elevated or depressed, by the circumstances which surround them. But I find contentment rather a difficult precept to practise; but still I do not despond. I have too many pleasurable excitements surrounding me ever to permit the evil one to establish that malady upon me. For instance, I fully feel the truth of that blessedness which he can enjoy who has found his work and who manfully exercises his ingenuity to accomplish that <u>life purpose</u> with the utmost facillity and dilligence in his power. And will you not agree with me that if I am not exactly a student in galvanism &c perhaps I may be a fellow worker otherwise. I cordially agree with the sentiments of your letter,[6] and although I have not as yet attained to the height of Transcendentalism[7] to which you appear to have reached, may I not with a more humble profession, be still making an advance in the scale of progression. I never feel more truly happy than when I act from a conscientious conviction of Duty and though I have much to perplex and annoy me from the scruplous foibles of those about me, yet for a moments apparent gratification I would not yield and thus forfeit that enjoyment of mind, such a course of action rewards. Every moment of life appears to me doubly precious, I cannot suffer idleness or the endeavour to kill time in useless engagements. Ah! life is a precious boon, I often sorrow to think how few there are who consider the great end of existence, and who endeavour to improve their moral condition by careful culture. I have been busily engaged in reading since my last. I have read Schlegels Dramatic Lit,[8] Goethe's Auto. & Travels,[9] a small work on the Origin & Progress of Language[10] and Emerson's Poems, &c. I shall never forget Goethe while I live; what brilliant aspirations together with profound and sublime thoughts are interspersed throughout those volumes, I imagine if I could peruse the work in its original language I should almost be transfixed with delight, how often I dwell on the concluding passages of the Auto. commencing 'Child Child no more, The [coursers] of Time, lashed as it were by invisible Spirits hurry on the light car of destiny'[11] &c perhaps you are equally familiar with the quotation, truly we know not the portion that is meted, but it is our privilege that we are able to take the reins in hand, and thus avoid by our own skill many threatening dangers.

And now after all this you will agree with me it is high time to conclude; what shall I say concerning the strange appearance my last letter presented, having four of her Majesty's portraits upon it?[12] You <u>must forgive</u> the mistake which originated with my sister. I was not able at the time to ramble so far as the Church Street post office.[13] I shall with the children be going to spend some time with Mr Allen's Father, and if you cannot write soon, will you address the letter according to the direction I have enclosed. You see I am venturing to trespass again on your precious time—I shall be in Blackburn on my return and trust to have the happiness of seeing you again. I must not omit to tell you that <u>Samuel</u>[14] is a Day Boarder at present at Tulketh Hall,[15] and I trust that after that he will be placed under M[r] Singleton's care, either as monthly or weekly pupil. he is a fine, intelligent lad, and I shall do all in my power that he may eventually reside at Spring Bank.[16] I trust you are well. Accept of my thanks for all your kindness, believing me

Ever Your affectionate | Maggie Allen.

Address | Mr Allen, | Newton Hamilton, | Co. Armagh. | For Mrs R.C.A.

RI MS JT/1/A/68

1. *Margaret (Maggie) Allen* (née Smith): the wife of Robert C. (Bob) Allen, a friend of Tyndall from his Irish Ordnance Survey days (see Volume 2). They married in 1844, and their first child, a daughter, was born in July 1845. At least two further children were born. As later letters (in Volume 4) show, Bob Allen had serious drinking problems and he may have departed for or been sent to America in order to reform himself. Three children (Charley, Sophy, and another daughter) died young, of consumption. Tyndall wrote friendly and supportive letters to Margaret Allen and occasionally gave money, for she was entirely dependent on her husband's parents for support. He helped her obtain a position in October 1854 and from that time she disappears from the correspondence.

2. *Messrs Brigdens & Moody . . . Lithographers*: not further identified.

3. *Mr Bridgens*: Henry F. Bridgens (c.1824–72), a surveyor, and map lithographer and publisher, was born in England. He emigrated to the United States before 1850 and settled in Philadelphia.

4. *Grandson . . . engineer*: both grandfather and father are unidentified.

5. *Robert declares himself a changed man*: the reform was only temporary (see Volume 4).

6. *your letter*: letter missing.

7. *Transcendentalism*: Margaret Allen's writing style and reading lists show that she shared Tyndall's introspection, and commitment to intellectual and moral self improvement.

8. *Schlegels Dramatic Lit*: Schlegel, *Lectures on Dramatic Art and Literature* (1809–11) was a classic of romantic literary criticism.

9. *Goethe's Auto. & Travels*: Johann Wolfgang von Goethe, *The Auto-biography of Goethe. Truth and Poetry: From my Own Life*, trans. John Oxenford and A. J. W. Morrison, 2 vols. first English ed. (London: H. G. Bohn, 1848 and 1849); volume 2, included *Letters from Switzerland and Travels in Italy*.

10. *Origin & Progress of Language*: probably George Smith, *The Origin and Progress of Language* (London, The Religious Tract Society, 1848), which was under 200 pp. long.

11. *'Child Child . . . of destiny'*: from the final paragraph of Goethe's autobiography, vol. 2, p. 168 (cited n. 9). The original reads: 'Child! child! no more! The coursers of time, lashed, as it were, by invisible spirits, hurry on the light car of our destiny'. Whether Allen wrote 'courses' or 'coursers' we cannot tell.

12. *four . . . portraits upon it*: possibly an allusion to the number of stamps on the letter.

13. *the Church Street post office*: not identified.

14. *Samuel*: not identified. Although the context suggests that Samuel is Margaret and Robert Allen's son, this is unlikely. The name does not match the names of children identified through letters (n. 1 above) and, as the first Allen child, a daughter, was born c. July 1845 (Volume 2, letter 0325) any son would be barely six years old at the time of this letter and therefore unlikely to be a day-boarder at any school.

15. *Tulketh Hall*: a Quaker school in Ashton-on-Ribble, Preston, Lancashire, started by George Edmondson in 1841, and run by Joseph Bray from the late 1840s.

16. *Spring Bank*: location of school started by Josiah Singleton.

To Thomas Archer Hirst   [5–6 June 1851][1]   0493

My Dear Tom,

I wish I had time to write a long letter to you, but I haven't[2]—The money reached me today quite correct—135 thalers—since I wrote to you[3] I received another letter from Edmondson[4] in which he expresses his consent to the purchase of £20 worth of apparatus for the college—I have written to him[5] to say that I will draw upon him for £30—will you therefore go to Bang and fill a second draft for £10 at three months?—

If the money reach me within the coming fortnight it will answer and if Bang can send me £5 English currency—ie Sovereigns and £5 German it will be all the more convenient for me—

Would to God that I could say to the bowels open! —I hate to see you hampered by such an impediment—but patience boy it will all work for your good—

When you see my paper[6] you will see that I have not been idle in Berlin. A good portion of my results however have been anticipated in a memoir from Bequerell[7] in the May number of the Annals der Chemie and Physique—He has not exhausted the matter however, and my method of experimenting is better than his—

Tom all your experiences are known to me—all barring the bowels—all I have to say is hope on hope ever—your work will not always be the dry and splintered affair which it now appears to be—your efforts lack a certain purchase which you will later discover—your present probation is unavoidable

—I had an interview with Humboldt a few days ago—I intended to put him a few home questions and thus extract something from his experience —but was defeated—He spoke straight forward and left me no opportunity. n'importe[8]—The same sky bends over him and me

as ever my brother | Tyndall

put down the postage of this The post office is a long way off—and there is a box quite near me.

RI MS JT/1/T/542

---

1. *[5–6 June 1851]*: Hirst wrote '7 June', the date on which he received the letter, on the first page. Letters usually took only one day, occasionally two, between Berlin and Marburg.

2. *I wish I had time … I haven't*: Tyndall's hand is here a hasty scrawl; he wrote foreign words and names inaccurately.

3. *I wrote to you*: see letters 0485 and 0490 in which Tyndall gave Hirst instructions regarding money.

4. *another letter from Edmondson*: letter missing.

5. *I have written to him*: letter missing.

6. *my paper*: Tyndall alluded to his paper 'On Diamagnetism and Magnecrystallic Action' (cited letter 0498, n. 6).

7. *a memoir from Bequerell*: Edmond Becquerel, 'De l'action du magnétisme sur tous les corps', *Annal. Chim. et Phys.* s. 3, 32 (1851), pp. 68–112, which discusses Plücker, Faraday, and diamagnetism.

8. *n'importe*: it doesn't matter.

From Heinrich Debus 9 June 1851 0494

Lieber John,

Die ersten Zeilen aus meinem Vaterland erhielt ich diesen Morgen; und sie waren von Deiner Hand. Ich habe diese wenigen Worte mehr als einmal gelesen. Ich kann nicht in wenigen Worten das ausdrücken was ich Dir sagen möchte, sondern ich hoffe es in 12 Tagen mündlich und mit mehr Worten thun zu können als ich auf dies kleine Papier schreiben kann. Wenn alle die sogenannten Philosophen und Menschen welche wie Hegel dencken, nur einmal einen wahren Freund erwerben würden, dann würden sie gewiss ausrufen dass ein Geist die Welt durchdringt, ein Geist der ohne Worte aus den Menschen redet, der die todte Materie zu unendlichen Wundern formt, und den Menschen mit dem Menschen verbindet. Zwischen uns John liegt ein Theil des grossen Weltmeeres; ich kenne nicht Dein Zimmer in Berlin; und doch bin ich oft darin. Goethe sagt dass es schön ist, die heiligen Räthsel des Weltalls zu lösen; ich sage aber dass es eben so schön ist zuweilen zu dem Menschen zugehn. Ich hätte gern mehr Wörter von Dir gehabt. Doch ich bin zufrieden damit, Dein kleiner Brief war gut.

Dein Heinrich Debus. | am IX. VI. 51

Dear John,

I received the first lines from my fatherland[1] this morning, and they were from your hand.[2] I have read these few words more than once. I cannot express in a few words what I would like to say to you, but I hope that I shall be able to do so in 12 days personally[3] and in more words than I can write on this small piece of paper. If all the so-called philosophers and men who think like Hegel would acquire a true friend just once, then they would surely exclaim that a spirit permeates the world, a spirit which speaks without words out of men, which forms dead matter into endless wonders, and connects man with man. Between us, John, lies part of the great ocean; I do not know your room in Berlin; and yet I am often in it. Goethe says that it is nice to solve the holy riddles of the cosmos;[4] I say, though, that it is just as nice to consider the human being every now and then. I would have liked to have had more words from you. But I am satisfied; your little letter was good.

Your Heinrich Debus | IX. VI. 51

RI MS JT/TYP/7/2402
LT Transcript Only

1. *first lines . . . fatherland*: Debus had just moved from Germany to take up a position at Queenwood.
2. *from your hand*: Tyndall's letter to Debus is missing. Debus used the informal Du form of address to Tyndall ('Deiner Hand').
3. *in 12 days personally*: Tyndall was about to return to Queenwood, but not as soon as Debus expected. He left Hamburg for England on 24 June (see letter 0497).
4. *to solve the holy riddles of the cosmos / die heiligen Räthsel des Weltalls zu lösen*: 'des Weltalls heilige Räthssel zu lösen': Alexander von Humboldt's quotation from Goethe's poem 'Die Metamorphose der Pflanzen' (1790), *Kosmos*, vol. 2, (Stuttgart und Tübingen: Cotta, 1847), p. 75. Debus's wording is closer to Humboldt than to Goethe.

From George Edmondson   9 June 1851   0495

Queenwood College | near Stockbridge |
Hampshire | June 9th[1] 1851

My Dear John Tyndall

Thy letter[2] dated 'Berlin Tuesday' has reached me to day, and at some risk of missing thee,[3] I hasten to remove from thy mind any impression of want of confidence, any expression of mine may have given rise to;—banish it I beg of thee:—I only wished to apprize thee that I am anxious to make Queenwood <u>self-supporting</u> and to do this I find it necessary to curtail expenses:[4] justice demands I should not run into debts, which I cannot promptly meet. The inconvenience of delay, is in my opinion, a degree of injustice to your creditor.[5] So far I hope I have set this matter right. I long to have thee on one side Debus on the other to become better acquainted with the good man, I feel there is good in him:—and Haas is another deserving worthy fellow. I shall be delighted to be of any service to thee at Ipswich,[6] though I am quite at a loss how it is to be done, I mean how I am able.

I am thy sincere Friend | Geo. Edmondson.

RI MS JT/1/HTYP/146
LT Transcript Only

———

1. *June 9th*: written shortly after Debus's arrival and dated the same day as letter 0494 from
    Debus to Tyndall.
2. *Thy letter*: letter written Tuesday, 3 June, is missing. On the Quaker use of 'thee' and 'thy'
    see letter 0478, n. 3.
3. *risk of missing thee*: Tyndall was about to leave Berlin but not quite as soon as Debus and
    Edmonson expected (see letter 0494, n. 3). He left Berlin for Hamburg on 22 June (see
    letter 0497).
4. *want of confidence . . . curtail expenses*: alludes to Tyndall's offer to purchase £20 worth of
    instruments for Queenwood and Tyndall's sensitivity to any criticism. Tyndall gave his
    version of the exchange in letter 0497 to Hirst.
5. *injustice to your creditor*: an expression of Quaker scrupulosity over debts.
6. *Ipswich*: the BAAS meeting at Ipswich in July (see letter 0484, n. 17).

From Thomas Archer Hirst   11 June 1851   0496

Marburg | June 11th / 51

My dear John—

Here goes for a business letter executed in two minutes—Herewith you
will receive 65 Thalers—The Exchange was as Before 6/23[1] per £:

Our account[2] stands on the opposite page—

Todays postage & future small items will straighten off the 21 ¾ Silber
Groschen[3] & make a clean-board of it—

Good bye & God Bless You | T.A. Hirst

No Prussian Money or English Gold was to be had.[4] These Hessian notes
will serve your purpose no doubt.—

RI MS JT/1/H/158

———

1. *6/23*: 6 Thalers 23 Silber Groshen, or slightly over 6 ¾ Th., was the German money
    received per pound. See 'Note on Money'.
2. *our account . . . page*: the account, which appears here below, sums up the financial dealings
    over the previous month and (third item in left hand column) the remains of an earlier
    exchange.
3. *21 ¾ Silber groshen*: the balance of 21.9 at the bottom of the right hand column in the table.
4. *No Prussian money . . . to be had*: see Tyndall's request in letter 0493.

	Received				Paid			
		Th[lrs]	Sg	–			Th[lrs]	Sg	–
1851									
May 31	1st Instalment from Bang	135–	10–	0	May 31st	Sent per Post	135–	0–	0
June 11	2nd " " "	67–	20–	0	June 11th	" " "	65–	0–	0
" "	Balance of your former draft on Edmondson	13–	25–	0	" "	Interest on Bang's two drafts– 3 Months 5 percent	2–	16–	3
					May 31st	Postage of Money to Berlin	0–	17–	0
					April 12	Paid you in Lieu of 10 Ths/ 25 Sg– in Bang's Hands	13–	0–	0
						Balance in Your favour to be carried to account	0–	21–	9
	Th[lrs]	216–	25–	0		Th[lrs]	216–	25–	0

To Thomas Archer Hirst　　　[14 June 1851][1]　　　0497

Berlin Saturday—

My Dear Tom.

Today I received 65 Thalers and the account[2]—the latter was a problem to me but I have got through it—

The steamboat that carries me from Germany starts from Hamburg on Tuesday week—I leave Berlin Tomorrow week and shall thus have one clear day in Hamburg.[3]

Give Kohlrausch a copy of each of the English memoirs[4] & greet him warmly for me. Knoblauch D°—Gerling D°—If you have any friend there to whom you would like to give a copy do so—to Wrightson for instance—send the rest to me—the parcel will cost about 10/SG/—Prof. Knoblauch will tell how to make them up—send them pretty quick—I shall want them in England—Kind wishes to Wrightson Davy and Dick—

I send you a letter which you sent me a day or two ago—it is from M[rs] Allen[5] and will interest you—

My next letter will probably be from England. Were you another than Tom Hirst should be giving you a parting wrd of advice—But Tom has better /monitors/ than I—one word however—<u>never despair Tom</u>! fight away boy it is only thus you will learn the length and breadth of your own capabilities—I have felt every one of your trials the bowels inclusive and am still alive.

and there are men who have battled through ten times as much as we both together—

I send you a scrap from Debus[6]—

I have worked constant here and have quarried out material for a good memoir[7]—you will hear of it in due time—

And now my pilgrimage in Germany is at an end—I can stick my Ebenezer here in the sands of Prussia and say with Jacob—'thus far have the gods helped me'[8]—

Goodbye Tom—give me some notice of your plans when you have any! I will write to you immediatly after the Ipswich affair[9]—Farewell my brother—
    your affectionate Tyndall

You may remember that passage in Edmondsons last note[10]—'I may trust you to lay out £20 judiciously'. To this I replied[11]—'of course you can trust me with the expenditure of £20—you must trust me with more than that— we must trust each other M[r] Edmondson—nothing more mischievous than a doubt could creep in between us—my ideal is that we should work together heart and hand and diligently trample out every suspicion'—to this he send me a prompt answer—I send it to you[12]—M[r] Edmondson is a man that was made to be ruled—If you suffer him to rule you he will set both himself and you out of order—

RI MS JT/1/T/983

---

1. *[14 June, 1851]*: the first Saturday after 11 June (see n. 2).

2. *Today I received ... the account*: letter 0496.

3. *steamboat ... leave Berlin ... Hamburg*: Tyndall expected to leave Berlin on Sunday, 22 June, spend Monday in Hamburg, and leave Hamburg by steamboat on Tuesday, 24 June.

4. *the English memoirs*: probably Tyndall's papers on the water-jet (see letter 0456, n. 1) and magnetism (see letter 0464, n. 2) published earlier that year in the *Phil. Mag.*

5. *letter ... from Mrs Allen*: letter 0492, which Hirst had forwarded from Marburg to Berlin (perhaps with letter 0496).

6. *scrap from Debus*: letter 0494.

7. *material for a good memoir*: probably the material for the paper on 'Diamagnetism'; see letter 0498, n. 6.

8. *stick my Ebenezer ... helped me'*: a biblical allusion. The prophet Samuel set up a stone to commemorate a successful battle: 'Then Samuel took a stone ... and called the name of it Ebenezer, saying, Hitherto hath the Lord helped us' (1 Samuel 7:12). In referring to Jacob, Tyndall seems to have muddled this event with a commemorative place name given by Jacob (Genesis 32:30).

9. *Ipswich affair*: the BAAS meeting at which Tyndall planned to give three papers and hoped to give an evening lecture (see letter 0484, n. 17).

10. *Edmondson's last note*: this missing letter was about the purchase of scientific instruments for Queenwood. Probably the letter mentioned by Tyndall in letter 0493 to Hirst.

11. *I replied*: letter missing.

12. *a prompt answer . . . to you*: letter 0495.

To William Francis   [19 June 1851][1]   0498

Berlin Thursday

Dear Sir.

It struck me yesterday that Dove and Knoblauch might like to see their names at the top of the page[2] where their papers are inserted—This is only a conjecture but I think it's a probable one—Would you therefore be kind enough to put their names there instead of mine—The title and the short introduction may remain as it is, only at the <u>top of the page</u> instead of 'Tyndall on the Phy. Sciences' put 'Dove on the reversion-prism' or something of the kind—

I start from Hamburg on Tuesday and hope to see you in two days afterwards.[3]

sincerely yours | J Tyndall

I saw Clausius yesterday eveng—he complains of Thomson having wronged him and feels himself compelled to reply. He will send you this shortly.[4]

Mr Adolph Schlagintweit is very anxious about some paper of his on the temperature of something[5]—I forget what—Its a question however whether your anxiety about the matter [mounts] so high—

auf Wiedersehen

I have a paper ready which will take up about 25 sides of the magazine.[6]

Dr Francis &c &c | Red Lion Court | Fleet Street | London[7]

StBPL T&F, Authors' letters

---

1. *[19 June 1851]*: the Thursday before Tyndall left Berlin (n. 3 below).

2. *names at the top of the page*: Tyndall included translations of two papers by Dove and one by Knoblauch in his 'Reports on the Progress of the Physical Sciences' (see letter 0484, n 2). Francis acknowledged both authors and translator; in the published text, even pages had Tyndall's name and odd pages the author's name.

3. *two days afterwards*: Tyndall expected to leave Hamburg on 24 June (letter 0497), and therefore expected to see Francis in London on 26 June. On arrival, however, he went to Queenwood immediately and did not see Francis until he returned to London on his way to the Ipswich BAAS meeting (letter 0501).

4. *Clausius will send you this shortly*: 'Reply to a Note from Mr. W. Thomson on the Effect of
    Fluid Friction, &c., appears in the June Number of the Philosophical Magazine', *Phil. Mag.*
    2, no. 9 (August 1851), pp. 139–42. Several of Clausius's texts published in the *Phil. Mag.*
    around this time reply to Thomson.

5. *M* *Adolph . . . of something*: possibly 'Untersuchungen über die Isogeothermen der Alpen',
    Poggend. Annal. 153, no. 7 (1849), pp. 305–56, or an extract from *Studies on the Physical
    Geography of the Alps* (Leipzig, 1850), as Hermann's article was. No translation appeared
    in the *Phil. Mag.* A few days later Tyndall met Adolf and his older brother Hermann. He
    liked Hermann, but judged Adolph to be 'very vain and very empty. When he speaks his
    lips are contracted to un-natural curves and his voices, which nature intended of course to
    be the instrument of thought is only used to make a noise. He seems to like to hear himself
    speak, regardless of whether what he says is to the purpose or not. He requested me to
    present his compliments to Faraday, to Sir Roderick Murchison, to Sir John Herschell,
    and other scientific aristocratic phenomena. Did I come into close contact with the young
    gentleman I should surely tell him a bit of my mind' (Journal, 22 June, JT/2/13b/548–9).
    See also letters 0470, nn. 13 and 14, and 0474, n. 5.

6. *paper ready . . . about 25 sides*: 'On Diamagnetism and Magnecrystallic Action', *Phil. Mag.*
    2, no. 10 (September 1851), pp. 165–88 (i.e., 24 pp.). He conducted the research for this
    paper while in Berlin; five days previously (letter 0498) he had told Hirst the material was
    'quarried out'. Perhaps he completed the writing before leaving Berlin. The material was
    the basis of his BAAS presentation of the same title (cited letter 0501, n. 3).

7. *D* *Francis . . . London*: address from envelope.

From Thomas Higginson[1]   24 June 1851   0499

Sibertswold[2] Vicarage | near Dover | June 24th 51

My dear Tyndall,

I only received your affectionate letter[3] a few days since, as since being
doubled I and my better half[4] have been travelling on the Continent, visiting
Belgium, Prussia, Germany and back by Paris

I hope now my dear old friend and companion that I have shown you the
way you will take the matrimonial leap Man was not made to live alone and a
good fellow like you was most certainly not. Only choose wisely and not a girl
that might be your daughter. My wife is only three years younger than myself.
I have known her for many years and am therefore well acquainted with her
character. If you go across the water this summer you might halt at Folkestone
and pay us a visit, when no doubt I shall be able to make you up a bed (not a
part of my own now) I can lend you all the night apparatus. I return to Shorncliff[5] on Friday next

Now as ever | Your attached[6] friend | Tom Hig.[7]

RI MS JT/1/TYP/2/600
LT Transcript Only

1. *Higginson*: Thomas Charles Higginson (c. 1821–98), an early companion of Tyndall in the Irish Ordnance Survey. Higginson had joined the Royal Artillery in 1842 and served in India (see Volume 1). In 1851 he was stationed at Shorncliffe and, as indicated here, recently married. In the next few years he travelled on army business in the United States before being appointed to superintend recruiting in Ireland (letters of 28 January 1853 and 6 July 1854, Vol. 4). Tyndall maintained contact over the years, writing letters and meeting on at least one occasion (in London, referred to in letter of 28 January 1853).
2. *Sibertswold*: Higginson was visiting Sibertswold, a village about 6 miles north-west of Dover. The letter implies he was living at Shorncliffe.
3. *your affectionate letter*: letter missing.
4. *being doubled . . . my better half*: allusions to his marriage (doubled) and wife (better half).
5. *Folkestone . . . Shorncliff*: Shorncliffe, a military camp 2 miles west of Folkestone.
6. *attached*: another allusion to his married state.
7. *Hig*: another use of a nickname among Tyndall's Irish friends.

From Anne Wynne   7 July [1851]¹   0500

Harrow. 7ᵗʰ July—

My dear Sir—Captain Wynne is from home & likely to remain from home to Thursday next, but has desired me to write to say to you that he hopes if you can you will let us see you here—

> believe me with every | wish for your success² | yours very truly A. Wynne
> Mr J. Tyndall, | Ipswich.³

RI MS JT/1/W/89
RI MS JT/1/TYP/5/1840a

1. *[1851]*: according to LT's note this was addressed to Tyndall at Ipswich, that is, the place of the BAAS meeting in 1851.
2. *your success*: success at the BAAS meeting.
3. *Mr J. Tyndall, Ipswich*: address from LT transcript; presumably taken from the envelope which is now missing.

To Thomas Archer Hirst   15 July 1851   0501

Queenwood July 15th 1851

My Dear Tom

I have just got my chaos of goods and chattels into something like order and ere the simmer of my brain has subsided I take up the pen to write to you—It will be a sad dislocated epistle no doubt as my thoughts are flying about like sparks from a crackling twig. From Hamburg to London occupied from 50 to 60 hours—I was one day sick and a dismal wet day it was—I was astonished to find the weather so beautiful on entering the Thames and more so to find that such weather had existed for a fortnight previous—Having got thro' the claws of the custom house officials I posted straight off to the waterloo Station and took my ticket for Queenwood—arrived safe—put some of my things in order and after 3 days stay started to Ipswich[1]—called upon Francis in passing through London and arranged that we should lodge together—got most comfortable lodgins but dear—paid my pound, got my ticket and thus found the doors of the British Association open to me—The wrong key note was struck at the very commencement—I had sent in the titles of 4 papers[2] with the probable time to be occupied in reading them, but the Secretary instead of sending them to the physical Section sent them to the chemical one—My name appeared in the list the first day and it was too late to transfer the paper; so that day passed without my having an opportunity of saying anything—Met Faraday in the street and a most friendly meeting it was—It was then thursday and he told me that the next day was all he could spend /at/ Ipswich—he expressed a strong wish to hear my paper on Diamagnetism[3] so it was arranged that it should come forward next day.—Then came the Prince[4] and his train of asinine flunkeys—He bothered the proceedings sadly—The president[5] and some of the chief men found themselves compelled to be near him and on this account many men before whom I should like to have read the paper were absent—many great guns were however present[6] It was towards the close of the day when I came on and the Section was already tired—at least I carried this feeling with me to the platform and this induced me to hurry over the paper more quickly than I otherwise should have done. It was well received and Faraday spoke at some length afterwards—the substance of what he said is contained in the Athenaeum[7] I sent you; but there is a stupid remark tacked on[8] which refers to a portion of the subject which is not contained in the abstract and therefore stands without meaning— Another blunder occurred on monday which caused me to be omitted in the physical list, and on Tuesday, the last day I got—not a paper read for I used no notes but a kind of a running lecture on 'Airbubbles formed in Water'[9] and illustrated by experiments—This was exceedingly well received—though towards the close of the day, and though the room at the commencement was thin, before I ended every seat was occupied—This has taught me a bit of experience and that is to trust myself more and my notes less—When you read a paper there is no danger of omission but it lacks the life and interest of an extempore discourse—The same day I managed to introduce a pretty experiment in thermoelectricity[10] and thus my connexion with the British Association ended—The two evening lectures were delivered, the one by Airy and the other by Owen I believe at the desire of both[11]—I reached London last Wednesday and remained there until Sunday afternoon, having some business to do for the Philosophical Mag. and for the Literary Gazette.[12] During the time I witnessed that pompous ceremony the Queens visit to the city—The streets along which she passed formed one great artery throbbing with the blood of all London—The illuminations were magnificent and the golden carriages must have shone like the holy of holies and highest heaven to the eyes of all flunkeys—but I can well excuse this bit of enthusiasm, far better than another bit got up by the Ipswichians to welcome prince Albert—arches of laurel were thrown across the streets and one of them bore in front the following words—'Albert the elevator of his Race' and behind 'Welcome patron of Science'—this piece of gratuitous valethood originating as it did with the quakers of the place stank in my nostrils[13]—what the devil had he done to elevate his race, or to patronize Science?—Many a decent fellow has conceived a disgust for another decent fellow purely as a reaction against the flatterers of the latter, and this flunkeyism of Ipswich has rather tended to make the prince despicable in my eyes N'importe[14] they have their reward—flunkeyism gets the loaves and fishes[15] and if you and I rebel against it it must be at the peril of the alimentary canal.

I like the way things are opening here. Though my arrival two days ago was accompanied by an unpleasant circumstance originating in Mr Edmondson's want of decision—it was with regard to a room—Debus had to vacate a room strongly against his will and which he had been given to understand was his but which they now say was intended for me—Debus is now content but the thing was very disagreeable—I find every proposal I make is acceded to and once acceded to prompt carrying out is the immediate consequence—we thus avoid hesitation and hope to continue to avoid it—Yates[16] is off, having got a situation in Worksop[17]—Even Debus during the short time which they have been together at Queenwood stuck his claws in him almost daily—He is a most incomprehensible fellow—Haas is in Switzerland—Debus and myself are here alone—I spent one day at the Great Exhibition[18]—It is an enormous affair—a decided hit in its way—the most pleasing part of the matter is to get upon one of the galleries and watch the variegated mass of life which swings to and fro underneath, and then the people are all so smiling and happy looking—it is really pleasant to behold. Is the Leader[19] defunct? I did not see its name in the usual place—I shall now work like a brick and live like a miser until I get myself out of debt[20]—A year will do it please the gods—How are you Tom? I wish to hear from you—I often think of you and those accursed bowels always draw their trail across my vision—Rest assured of it Tom your head has something to do with you abdomen—Give in a little at times boy—and if you think it a sin why lay it upon me—

Your affectionate Tyndall.

RI MS JT/1/T/543

1. *Ipswich*: Tyndall attended the BAAS meeting in Ipswich from Wednesday 2 July to Tuesday 8 July.
2. *4 papers*: Tyndall gave papers on diamagnetism (n. 3 below) and on air-bubbles in water (n. 9), and demonstrated an experiment on thermoelectricity (n. 10). We have not been able to identify the fourth paper.
3. *paper on Diamagnetism*: Tyndall, 'On Diamagnetism and Magnecrystallic Action', read on Friday 4 July. A lengthy abstract was published, *Brit. Assoc. Rep. 1851*, pp. 15–18. The paper was later published in full in the *Phil. Mag.* (letter 0498, n. 6).
4. *the Prince*: Prince Albert.
5. *The president*: George Biddell Airy, Astronomer Royal.
6. *great guns*: apart from Faraday, these are not identified here or in Tyndall's Journal.
7. *Faraday spoke . . . in the Athenaeum*: 'Twenty-First Meeting of the British Association for the Advancement of Science', *Athenaeum* 1237 (12 July 1851), pp. 745–57; see pp. 746–48 for the lengthy abstract of Tyndall's paper and Faraday's comments. (The abstract is identical to that in the *Brit. Ass. Rep. 1851*; see n. 3.)
8. *stupid remark tacked on*: probably: 'Prof. FARADAY felt prepared to admit that some of Dr. Tyndall's results seemed to promise an explanation of Plücker's perplexing results and conclusions; but for his own part he was anxious to keep his mind free from bias, to get well-established facts, and to free them as much as possible from all circumstances which could mark, or disguise, or mislead in the interpretation of them,—and such being his fixed determination and settled habit, he was rather at a loss to remember to what portion of his publications on the subject Dr. Tyndall referred when he considered him to have considered the facts now brought forward as improbable' (p. 748).
9. *'Airbubbles formed in Water'*: delivered on Tuesday, 8 July (letter 0485, n. 6).
10. *pretty experiment in thermoelectricity*: the experiment, which was conducted with the monothermic pile invented by Magnus, was recorded as following the paper on 'Diamagnetism' (*Brit. Assoc. Rep. 1851*, n. 3, p. 18).
11. *two evening lectures . . . at the desire of both*: Owen lectured 'On the distinction between Plants and Animals, and their Changes in Form' on Friday 4 July; Airy lectured 'On the Total Solar Eclipse of July 28, 1851', on Monday 7 July. Tyndall had hoped to deliver a lecture himself.

12. *the Literary Gazette*: see letter 0484, where Tyndall suggested he would not accept further work.

13. *the quakers of the place stank in my nostrils*: Tyndall was disgusted with the Quakers for forgetting their egalitarian principles. Traditionally, Quakers did not recognize distinctions of rank.

14. *N'importe*: never mind.

15. *they have their reward . . . loaves and fishes*: Tyndall uses biblical allusions to contrast high and base motives. 'They have their reward' is a quote from Jesus's criticism of hypocrites who do good to gain public approval (Matthew 6:2). Tyndall then alludes to the miracles of the loaves and fishes when, according to Mark 8:1-10 and Mathew 15:32-9, Jesus fed a hungry crowd of thousands with only a few loaves and small fish. Some, Tyndall implies, follow for the material benefits.

16. *Yates*: John Yeats.

17. *Worksop*: town in Nottinghamshire.

18. *the Great Exhibition*: had been opened by Queen Victoria on 1 May (see letter 0399, n. 15). Prince Albert was its most powerful promoter—the reason for the praise given him at the British Association meeting.

19. *the Leader*: a radical weekly newspaper (letter 0398 n. 8). It was not defunct.

20. *get myself out of debt*: Tyndall was in debt to Hirst (see letters 0393, 0395, and 0695).

From Elizabeth Steuart   16 July 1851   0502

Steuart's Lodge, | July 16th, 1851

Dear John,

I have been wishing much to hear from you since your return to England, feeling anxious to know the result of the part you have taken in the meeting at Ipswich,[1] of which, as yet, I have only seen a cursory account in the newspapers. I had hoped you might have been able to follow up your plan of taking a run-over here before settling down at Queenwood, but you cannot manage to do so at present, and I therefore send a line to beg you will, at your earliest convenience, let me have a prospectus of the terms etc. etc. of Queenwood College, and a few particulars from yourself with respect to the system of education pursued there: the friends of poor Dr Roche's two little boys[2] wish to place them at a respectable school where there is general instruction, and <u>moderate terms</u>, and I mentioned Queenwood to them. I should like to know what <u>religious</u> instruction is given, and by whom, for on account of the Master <u>not</u> being of the <u>Established Church</u>,[3] it is a matter of the <u>first</u> importance: please also to mention if the pupils are of a respectable class, such as would be put into learned professions: I am anxious about these poor little fellows, left to struggle through life as best they may, without the care and guidance of the parents who were so devoted to them. I think your being at Queenwood would be an inducement to their guardians to send them there, as you would naturally take an interest in their welfare and have a care over them. Mr Steuart unites in every kind wish for you, with

Yours very sincerely, | E. D. Steuart.

RI MS JT/1/TYP/10/3329
LT Transcript Only

1. *the meeting at Ipswich*: the BAAS meeting (see letter 0501 for details).
2. *Dr Roche*: Benjamin Roche, physician from Bagenalstown, County Carlow, Ireland (see letters 0074, n. 22 and 0100, n. 3). Local newspapers record his marriage to Elizabeth Noble in 1838, the birth of sons in June 1839 and December 1842, the birth of a daughter in April 1850, and Roche's death on 17/19 January 1851. Thus the two boys for whom Queenwood was considered were about 12 and 9 years old. This letter implies that Elizabeth Roche had also died. (For marriage and births of children, see *Belfast News-Letter*, 25 Sept 1838, *Freeman's Journal*, Sat 15 June 1839, *Dublin Evening Mail*, Mon 19 Dec 1842, p. 3, *Freeman's Journal*, Friday 5 April 1850).
3. *the master . . . Established Church*: Edmondson, who was a Quaker.

From Thomas Archer Hirst   20 July 1851   0503

Marburg. | July—20th 1851

My dear John—

Had your letter[1] not arrived to day it was already my intention to write to you so you will have a speedy answer. I read the Athenaeum[2] with great interest and liked your lecture well as far as one could judge from the Report; your Berlin experiments were very interesting and so far as they go they appear to me very convincing and simple, forming a beautiful corroboration of the theory that you had already deduced from your experiments with Crystals.[3] Now your time will be otherwise so much employed I suppose we shall hear less of your Scientific progress though I look forward to another memoir from you in the Phil. Mag:[4] With your indignation at the flunkeyism I heartily sympathise, though I would out of charity to the Prince[5] hope that is was also distasteful to him. One small act of Prof. Whewell which I learnt from the Leader[6] raised my choler a little. Stegmann and I had only the day before been speaking about him and his book on Mechanics,[7] but the day after I took particular delight in exposing his conduct here—I refer to his interrupting a paper on Sound as not being 'lively and interesting' enough for Princely ears. For which piece of flunkeyism—although it must necessarily be defeated in its purpose;—[8] I should take great pleasure in <u>pulling</u> Prof. Whewells ears. Thank Debus for his letter for me and shake his fist heartily for me also. I will write to him when I have something to say. How does he get on with English, and his Laboratory Pupils? You must continually report to me how you pull altogether I shall often wonder what transmutations you are effecting in that establishment now you have got it so much under you & so free from the Yates species[9]—Now let me gather for you all the sparing crumbs of Marburg News I possess, which without doubt you will agree with me is very little. The English Kränztchen[10] meet to-morrow evening in Fraulein Bückings[11] garden for the first time these many weeks. They have not forgotten you, but think that you have forgotten your promise to write to them!! Knoblauch is as usual, a harmless good natured little fellow, but as usual twittering timidly about the lamp of Physics—I am afraid Knoblauch as a Lecturer is <u>not</u> improving but otherwise. The childish delight in popgun experiments on the whole increases, and he seems more and more to be making his peace with the ruinous notion that <u>a</u> course of lectures on Physics, and not <u>the</u> best possible course, but a kind of sleight of hand skipping one, is all his hearers wish to have from him & consequently (so great is his respect & fear for their opinion) all that it is his duty to give. Kolbe has in some respects a similar tendency but of course not any thing approaching Knoblauch in extent he has too delicate a nose and idea of order for my taste, works also too little himself in the laboratory though for that he may have his own reasons. All I know is that the example of Bunsen and Debus working for themselves had the effect of stimulating more industry last Semester than I see this. Stegmann alone stands firm on his own legs and by comparison rises daily in my esteem. His lectures on Descriptive Geometry he has made very interesting throughout. I have spent a good deal of time with him altogether, and have gone over a pretty fair quantity of ground—A good part of Analytical Geometry (Plane) and all the introductory part of the Analysis as it stands in Burg,[12] we have gone through; now we are working away in Differential Calculus & hope before the close of the Semester to bring it near a Conclusion.—[13] Mathematics is and must be for some time yet my principal study, if I make an Examination it must furnish my dissertation. At Chemistry I spend from 3 to 4 hours per day—and have to leave experimental Physics for some future period—

I was surprised not long ago by Schmackenberg[14] stepping into my lodgings. I expected to hear of his having left England dissatisfied but was glad to find that he had only brought a student Dickenson in his charge to Germany & on the contrary liked England much. he has now returned as you will know, having received as I hope the dissertation sent you by Prof Hessel.

From across the way at this moment comes the well known cry 'Lebe hock!')[15] it is opportune for me; it comes from the Mouths of Marburg &

Giessen Professors who have at last succeeded in coming together—to eat & drink to the success of Science & for their own enjoyment. Yesterday morning Wrightson and I rose at 4 ½ AM. and walked over to Giessen[16] to hear Liebig—I dont know I was ever more pleased with a Chemical Lecture & Lecturer, he is a fine, able man—It was his first, on Organic Chemistry and lasted two hours, early in August he goes to England and consequently has much ground to get over He entered the room with an air that did not prepposses me much, I thought I could see in his bearing the evil effects of his celebrity. To a certain extent, though with considerable modification, I can say also that impression remains; it would take a man of immense strength to bear uninjured the praises and notoriety which such genius as Liebigs has gathered. This shews itself in many affect[at]ions[17] in manner, and a certain unmistakable expression in his countenance which says as plainly as words could, 'Listen to a great Chemist'. He sat down slowly in his chair before the table, looked all round in the calmest most self-possessed manner, for a period whose length was sufficient to make one notice it, & then commenced, with his elbows on the table and the fingers in emphatic motion, one of the clearest introductory lectures I ever listened to; without notes he continued giving to his subject that impressiveness which commands your attention, and leading you on without the least confusion or repetition deeper and firmer into his subject—Thus he treated the connection between Inorganic and organic Chemistry, and the several theoretical views of the changes of bodies in the former to others in the latter, & shewing also the influence of the Vital Principle which was now added to those forces considered already in Inorganic Chemistry—Soon he came to his experiments and, simple though they were, he put every process actually before them, not one necessary however insignificant would he pass over—Here however I should say if we compare him and Bunsen that in the Speaking explanatory part, Liebig is more clear and emphatic; but Bunsen more brilliant and convincing in experiment; it arises probably in a great measure from Bunsens being more self helping and Liebig more dictatorial to his assistant at his elbow—His lecture as I said which ought to have been an hour and a half long lasted more than two instead of closing at 12 ½ it was already past 1. still he continued, a rustling of feet, coughing and even hissing shewed that some students thought more of their dinner than even Liebig he turned to them with a sarcastic smile & said 'you know we have much to do and little time, you must not therefore begrudge it' for 5 or 10 minutes more out of pure malice as one could see by the corners of his eyes he continued, & then concluded. We followed him immediately but found him already at work in his private Laboratory. My opinion of him improved very soon, he shewed little or no assumption or pride as he shook us by the hand & spoke to us, his youngest daughter a nice little girl of ten years old jumped into the room, and playfully made him guess what colored cherries she had in her hand, as I watched him look and talk to her in return he rose another 20 per cent in my estimation and so on the whole I left him with an opinion that from the first to the last had been on the increase. But I have gossiped on too long I must 'shut up'. The Bowels John thank God, are immensely improved, as I anticipated the evil was but a mechanical one owing to my torpidity as Noll says, and at last I found a mechanical remedy which I need not explain but which with a careful attention as usual to my Diet has wonderfully improved my general Health—I remain even more than usual on this very account in-doors, & consequently (confound it) am weaker in body though better in bowels and clearer in head, consequently more capable of working. I will fetch up all loss of strength I hope in Switzerland in September. Good gracious this Semester has <u>flown</u> past. —And seen sadly too little done; although one has the satisfaction of knowing that we have been pretty constantly doing

But Good-Bye for a season to thee, John—We are often together I know & feel, although our interviews are no longer bodily ones—

T A Hirst

Post the enclosed[18] for me, it is of some importance so dont let it happen any misfortune or be forgotten:

RI MS JT/1/H/159
RI MS JT/1/HTYP/149–151

---

1. *your letter*: letter 0501.

2. *the Athenaeum*: Tyndall had sent Hirst a report of the paper he read at the BAAS in Ipswich earlier in the month (cited letter 0501, n. 7).

3. *experiments with crystals*: presumably those conducted with Knoblauch and published in the *Phil. Mag.* in 1850 (see letters 0395, n. 22 and 0403, n. 2).

4. *another memoir from you in the Phil. Mag:*: most likely Tyndall's paper on diamagnetism which was largely completed (see letter 0498, n. 6).

5. *the Prince*: Prince Albert.

6. *One small act . . . from the Leader*: according to the *Leader*, the Prince arrived while Rankine was reading a paper 'On the Velocity of Sound in bodies of limited dimension', 'but Dr. Whewell added another leaf to the court laurel he is weaving for his brow, by officiously interrupting the secretary, and requesting that a "more lively paper should be read to princely ears"!' ('Annual Meeting of the British Association', *Leader* [5 July 1851], p. 628.)

7. *book on Mechanics*: Whewell, *An Elementary Treatise on Mechanics* (Cambridge: Deighton & Sons, 1819; 7th ed. 1847).

8. —: this dash is very long, but cannot indicate a new paragraph as it is in the middle of the sentence.

9.   *free from the Yates species*: allusion to John Yeats.

10. *The English Kränztchen*: see letter 0482, n. 6.

11. *Fraulein Bückings*: not identified.

12. *in Burg*: probably Meno Burg, *Die geometrische Zeichnenkunst* (Berlin: Dunker und Humblot, 1822).

13. —: A very long dash here might indicate a paragraph break, but it doesn't correspond to a new topic (cf. n. 8).

14. *Schnackenberg*: Schnackenberg had left for England in May 1851 (letter 0481).

15. *'Lebe hock!'*: 'three cheers' or 'hurrah'.

16. *walked over to Giessen*: about 16 miles, which took them 5 to 6 hours.

17. *affections*: Hirst wrote affections; LT corrected it to affectations.

18. *the enclosed*: this may refer to the letter to Martin Kidd mentioned in letter 0504.

To Thomas Archer Hirst   29 July [1851]   0504

Queenwood 29th July

My Dear Tom

Your letter[1] pleased me exceedingly, your estimate of Liebig was cool, quiet, unexaggerated and just. I smell this, there is an internal odour[2] about the words which render external proof unnecessary—That was a pretty episode with the little girl. In short I believe you have measured him as accurately as most men will do.

In the last two numbers of the Athenaeum[3] there appears an advertisement regarding the University of Toronto. I believe you have a number which contains it, 6 professorships are vacant and amongst the number one for physics, for this I intend to become a candidate and have already written to Magnus and Poggendorff[4] for testimonials—A note accompanies this to Knoblauch[5] for the same object. It is not altogether certain that I shall propose for it, as I intend writing a long letter to Faraday[6] to day, laying my position here before him and requesting his advice. If there were sufficient grounds to hope that within a reasonable time some thing worth while would turn up in England I would remain. This he will probably know; at all events I will be guided by his advice. If I apply I should say it will take nearly a year from the present time ere the matter is decided, so that my removal from Queenwood—in case I should be successful—would not be too abrupt. Indeed I feel a strong reluctance to move from Queenwood so speedily. I would gladly remain a year or two if I thought my chances good at the end of that time; but it would be imprudent to permit an opportunity like the present to pass un-noticed.

Mr Edmonson as I often told you is a reed shaken with the wind. He is a man between 50 and 60 and an educator for nearly 40 years, but has never yet pierced beyond the skin of human nature. That undemonstrated force, alluded to by Emerson in his essay <on> character,[7] he lacks totally. He is a well cultivated cabbage plant, as Tom Carlyle would say, but he wants the elements which might enable him to turn to himself as an interpreter of what goes on in the heart of a sapling which nature intended for an oak.[8] He must, in short, be converted into a mere organ here. He is a kind of governmental fiction—like the queen of England—useful, perhaps, as a point of reference in certain cases, and that is all. His wife is a noble kind of woman.

Debus I like daily more and more, he is a strong-hearted upright fellow. Haas is good gold to the very heart but he lacks the iron and steel of Debus, the piercing power of observation—in a word the natural force.

My estimation of Stegmann never lessened. You see the man from the first and therefore cannot be disappointed with him. Stegmann belongs to the number who posses a power of character which is surprisingly rare. I am inclined to think that Liebig has a strong dash of this too, and his coolness and his almost scorn have this more perhaps as a basis than his scientific fame.

I wonder would Gerling give me a testimonial.[9] He has already given me one but a weak and good-for-nothing thing; indeed when he gave it he and I were not very warm. If he has read any papers of mine he would perhaps feel himself justified from this to speak a little more strongly. If you are on good terms with him I wish you would ask him. The more the better.

It is now approaching breakfast time. Martin Kidd[10] I expect has his letter[11] by this time. I will say no more. One word—I declare the intelligence about the bowels elevated me so that I was prompted to write a song of thanksgiving on the subject.

Yours | Tyndall

RI MS JT/1/HTYP/152
LT Transcript Only

---

1.   *Your letter*: letter 0503 (which clearly dates this letter to 1851).

2.   *odour*: Tyndall uses odour in the older sense of *sweet* smell (*OED*).

3.   *last two numbers of the Athenaeum*: *Athenaeum*, 19 July 1851, p. 761 and 26 July 1851, p. 793. The advertisement promised a salary of '350*l*. Halifax currency, per annum' as well as 'other emoluments arising from fees'. The deadline to submit testimonials was 19 November 1851.

4.   *written to Magnus and Poggendorff*: letters missing. Tyndall received a testimonial from Magnus on 16 August (letter 0512).

5.   *A note accompanies this*: Tyndall's note to Knoblauch is missing, but two weeks later he reported receiving a testimonial from Knoblauch (letter 0510).

6.   *letter to Faraday*: letter 0505, dated 30 July 1851.

7. *essay <on> character*: 'Character', in Emerson, *Essays* (cited letter 0393, n. 3), pp. 97–126.

8. *a well cultivated cabbage plant, as Tom Carlyle would say*: an allusion to Carlyle's *Sartor Resartus* (Book 2, Chap. 2, 'Idyllic'). Tyndall suggested that, as a 'cabbage plant', Edmondson was unable to relate to and educate those of his students whom 'nature intended for an oak'.

9. *would Gerling give me a testimonial*: see letters 0508 and 0510 for the responses of Hirst and Gerling.

10. *Martin Kidd*: not identified.

11. *his letter*: letter from Hirst to Kidd; see letter 0503, n. 19.

To Michael Faraday 30 July 1851 0505

Queenwood College, | Stockbridge, | Hants. | 30th. July 1851

Dear Sir,

It would have afforded me great gratification to have made these magnetic experiments[1] before you; but as the destinies seem disposed to deny me this pleasure, I take the liberty of forwarding you a few substances[2] which you perhaps will be good enough to suspend in the magnetic field when your leisure permits.

No.1.[3] contains two cylinders of bismuth; between the <u>flat poles</u> one sets axial, the other equatorial. Between the pointed poles when <u>near</u>, the former sets equatorial also, but on withdrawing the poles to a distance from each other it turns into the axial position. In this case the principal cleavage is transverse to the axis of the cylinder, as you can prove by trial with a knife; and I believe you will conclude that the cause of its setting equatorial between the near poles is due to the fact that the tendency of the mass to pass from stronger to weaker places of magnetic force overcomes the directive power of the crystal. If the mass were a sphere the powerful cleavage would always set equatorial, between both points and flat faces. This turning towards the axial line, as you are aware, is attributed by Plücker to a triumph of magnetism over diamagnetism; but I think you will see that it is on account of the more equable condition of the magnetic field permitting the directive action of the crystal to assert itself.

No.2. contains two bismuth plates; the magne-crystallic axis is perpendicular to the flat surface of each, but it will set equatorial. This has been brought about by compression between two plates of copper in a hard vice. Before compression the magne-crystallic axis set in both cases strongly axial.

<u>No.3.</u> contains a plate of compressed wax; it is diamagnetic, and the line of compression sets equatorial, thus causing the plate to set its length from pole to pole.

<u>No.4.</u> contains a crystal of carbonate of iron, and a model of the same, formed from the crystal when pounded into dust and sifted through linen. I think you will see that the deportments of both are so far similar as to be suggestive of a common origin. The dough from which the model is made was compressed in the direction of the optic axis.

<u>No.5.</u> contains a prism of carbonate of iron artificially made. suspended from the centre of one of its sides it will set axial—change the point of suspension 90° in the equatorial plane; it will set strongly equatorial. Between the near <u>points</u>, when they are not too strongly excited the prism suspended as in the latter case will set axial, between the distant points it will turn into the equatorial position. This is the phenomenon which Plücker attributed to the triumph of the optic axis force at a distance in the case of tourmaline, idocrase, and other magnetic crystals. In explanation of this I will merely say, that when the prism sets equatorial between the excited poles, the mass of dough from which it was taken was compressed in the horizontal line perpendicular to the axis of the prism—the line of compression sets axial and hence the prism itself equatorial. With regard to the turning round I believe you will see the analogy between this case and that of the bismuth cylinder, the one being <u>recession</u> and the other <u>approach</u>.

No.6. contains a bit of shale; suspended from one of its edges it sets axial, from the edge adjacent, it sets equatorial. This is similar to Plücker's first experiment with the plate of tourmaline. I had a much finer piece of shale to show this experiment but it has unfortunately got broken.

It is with considerable reluctance that I introduce any personal affair of mine to your notice, and yet I am induced to do so at present. I returned from Germany, as you are aware, a few weeks ago and entered a situation here. I am teacher in this college and when our work commences I expect to be occupied about 7 hours a day. My salary is £150 a year, rooms in the college and board; and I find the Principal and his family willing to do all they can to render my life agreeable. The institution is a private one, and with regard to its durability I am unable to utter a word. The boys whom I have to teach are young, and I shall be engaged with them at the lower mathematics, such as Euclid, Algebra perhaps to quadratic equations, and a little trigonometry, and elementary physics. I shall have to lecture twice a week, but to suit my audience must confine myself to elementary matters. There is danger in retrogression in such a position, and there is also the everlasting consciousness of want of permanency. I think this is a sufficient account of my present position.

In the last two numbers of the Athenaeum[4] I find an advertisement stating that in the university of Toronto there are 6 professorships now vacant at the annual salary of £350 Halifax currency;[5] one of these is the professorship of Natural Philosophy It occurred to me as I read the advertisement that I ought to make application for the post, and after turning this matter round in my mind I concluded that the wisest plan would be to solicit your advice, and this conclusion was confirmed by my friend Dr. Francis of the Philosophical Magazine. Nobody can be more sensible than I am how slender are the grounds upon which I can justify this liberty, and had I been left to my own unaided cogitations I should hardly have attempted it. But to return; if both choices were before me I should prefer remaining in this country to going to Canada—not that I lay stress upon locality any further than that I like to be near men of science. But I dont know my chances in England and hence I think it unwise to allow an opportunity like the present to escape me. To settle myself down at Queenwood, even granting it permanent, would be to sacrifice an object for which I have battled harder than anybody knows, and that is to approve myself a worker in science. Seven hours plus meal times and other contingencies, plus the time to depolarize the intellect[6] after having been engaged with other matters is a heavy subtraction from the day. I ask your counsel in this state of things—something doubting I must confess, for I know I have no right to expect it—I have already written to Magnus and Poggendorff[7] for testimonials so that if you advise the step I shall be ready to take it promptly.

I remain dear Sir | Most faithfully & respectfully yours | John Tyndall

Prof. Faraday F.R.S. | &c &c

RI MS JT/2/12/4009–4012

Transcript Only

---

1. *these magnetic experiments*: letters 0501, 0506 and 0507 all refer to experiments performed or described in presentations by Tyndall at the Ipswich BAAS meeting. The letters do not make clear what experiments Faraday missed and when Tyndall had performed or expected to perform them. Given Faraday's interests, the experiments must have related to his presentation 'On Diamagnetism and Magnecrystallic Action', a paper Faraday attended (letter 0501).

2. *forwarding you a few substances*: the substances described in the succeeding experiments, No. 1, two cylinders of bismuth and so on, to No. 6, the bit of shale.

3. *No. 1*: In the transcript, the six numbered paragraphs are formatted as hanging paragraphs but not consistently. We have made the formatting consistent.

4.	*last two numbers of the Athenaeum*: letter 0504, n. 3.
5.	*Halifax currency*: see 'Note on Money'.
6.	*depolarize the intellect*: Tyndall had learned to use physical metaphors for mental processes from his teacher, John Conwill (see Volume 1).
7.	*written to Magnus and Poggendorff*: letters missing. Tyndall mentioned the same requests in letter 0504.

# From Michael Faraday			1 August 1851			0506

Tynemouth, | 1 August, 1851

My dear Sir,

Your letter[1] finds me here ill of a quinsy[2] but now recovering, and though I cannot write much, I determined to answer you at once. In the first place, many thanks for the specimens which I shall find presently at home. I was very sorry not to see you make your experiments,[3] but hope to realize the profound results which interest me extremely. I want to have a very clear view of them.

But now for the Toronto matter. In such a case private relationships have much to do in deciding the matter, but if you are comparatively free from such considerations and have simply to balance your present power of doing good with that you might have at Toronto, then I think I should (in your place) choose the latter. I do not know much of the University, but I trust it is a place where a man of science and a true philosopher is required, and where in return such a man would be nourished and cherished in proportion to his desire to advance natural knowledge. I cannot doubt indeed that the University would desire the advancement of its pupils and also of knowledge itself. So I think that you would be exceedingly fit for the position, and I hope the position fit for you. If I had any power of choosing or recommending, I should aid your introduction into the place, both because I know what you have already done for science, and I heard how you could state your facts and treat your audience.

Now I do not, for I cannot, proffer you a certificate, because I have in every case refused for many years past to give on the application of candidates. Neither indeed have you asked me for one. Nevertheless I wish to say that, when I am asked about a candidate by those who have the choice of appointment, I never refuse to answer; and indeed, if my opinion could be useful and there was a need for it, you might use this letter[4] as a private letter, shewing it or any part of it to any whom it might concern.

And now you must excuse me from writing more, for my muscles are stiff and weak, and my head giddy.

Ever, my dear Dr. Tyndall, | Yours most truly, | M. Faraday.

Dr John Tyndall | &c &c &c | Queenwood College | Stockbridge | Hants—^{5}

RI MS JT/2/12/4129
Transcript Only
Faraday Correspondence, 4: 2452

1. *your letter*: letter 0505.
2. *quinsy*: a serious inflammation of the throat (*OED*). However, Faraday does not mention his throat; his symptons sound more like influenza.
3. *your experiments*: see letter 0505, n. 1.
4. *when I am asked about a candidate . . . I never refuse an answer; . . . you might use this letter*: Tyndall took up Faraday's offer and referred Airy to Faraday when asking him for a certificate (letter 0518). Airy (letter 0527) and Whewell (letter 0542) made similar responses. Tyndall mentioned both Faraday and Airy in his Toronto application (*Faraday Correspondence* 4:2452, n. 2) and cited Faraday in his application to Sydney (letter 0561).
5. *Dr. . . . Hants—*: This note is added by hand to the typescript. It is probably the address from the envelope.

To Michael Faraday   [2–3]^1 August 1851   0507

Aug. 1851

Dear Sir,

I have to return you my sincere thanks for your kind and valuable letter.^{2} I was already aware of your dislike to giving written testimonials and therefore took care not to ask you for anything of the kind—Indeed the manner of recommendation which you allude to would be more agreeable to me than that by certificate, and if the parties have an agent in this country (of which I am not yet aware) I doubt not it would be the most practical and effectual.

With regard to my paper at Ipswich there is one matter which I should like to call to your mind. During the reading of it I was haunted by the consciousness that the Section was already tired, and this caused me, I doubt not, to appear somewhat hurried.^{3} Had it occurred earlier in the day, as I hoped it would I think I should have satisfied you better.

I believe you will find the experiments^4 to be as I have described them. If you once turn the light of your intellect upon this matter I think you will not long halt between two opinions^5—that you will once more take the thought to your bosom which suggested that significant note to a passage in the Bakerian Lecture for 1849 'Perhaps these points may find their explanation hereafter on the action of contiguous particles'.^{6}

With best thanks, dear Sir | Most faithfully yours | John Tyndall
Prof. Faraday | &c &c

RI MS JT/2/12/4013
Transcript Only

1. *[2–3]*: probably written two days after Faraday's letter (0506) which was written two days
    after Tyndall's request (letter 0505). Given the importance of both Toronto and Faraday
    for Tyndall, he might have replied even more rapidly and 2 August should not be ruled out.
2. *your . . . letter*: letter 0506.
3. *my paper . . . hurried*: see letter 0501 on the circumstances which led to the rescheduling of
    Tyndall's paper, 'On Diamagnetism and Magnecrystallic Action'.
4. *the experiments*: this letter suggests that the experiments discussed in letters 0505 and
    0506 were those recounted in Tyndall's BAAS paper.
5. *will not long halt between two opinions*: 1 Kings 18:21: 'And Elijah came unto all the people,
    and said, How long halt ye between two opinions? if the Lord be God, follow him: but
    Baal, then follow him'.
6. *perhaps . . . action of contiguous particles*: Tyndall cites Faraday, 'ERE 22 (continued)', p. 30.

From Thomas Archer Hirst   3 August 1851   0508

Marburg—Sunday | Aug[st] 3rd 1851

My dear John—

I was glad yesterday to have a word from you[1] and am also glad to be in a
position even so early as to day to send you an answer I have made it a point
with me during this Semester inasmuch as I had forsaken his Lectures to
coddle[2] up little Gerling whenever an opportunity should occur; with this
intention I paid him a visit some time ago and took him a copy of both your
memoirs[3] whereon I wrote 'with the Author's kind regards' in good round
hand, this pleased the little Daddy Dogood[4] & as it has turned out was
remarkably lucky for this latter business—Yesterday evening there happened
also to be an English Kräntzchen[5] (at Pfeifer's garden[6] given by the Fraulein
von Tileman[7]) at which our old Friend was present. Knoblauch brought you
on to the table, or your representatives, in the shape of all your letters to him
since you left Marburg, whereby he proved that you still kept the Kräntzchen
in remembrance, and in <u>every</u> letter made mention of it and sent greetings
to its members. —This was again lucky to my purpose so that on our walk
home I tackled the old Boy[8] and mentioned your prospects with regard to the
University of Toronto. He immediately entered into a description of its situation, longitude and latitude, shewing an intimate acquaintance with the said

Toronto, told me there was an English Magnetical Observatory there, that Col. Sabine had at three distinct periods (dates after a little time also remembered) had forwarded him the results of some observations made there—&c. &c. &c. In short if I would go up this morning he would shew me these interesting documents as well as a map of North America, and at the same time I might pay a visit to his Frau who had often enquired after me. All this, to make a long tale shorter John I have done, I took the opportunity to ask also for a testimonial representing it as a project of my own and by no means as a <u>request</u> of yours[9]—though of course I told him how obliged you would feel for the influence of his name (which would be already known to the magnetical natives of Toronto and as a natural consequence esteemed!!) He jumped at the proposal wrote it immediately. we consulted over a draft copy of it, and tomorrow morning he is to send me down a fair copy for the inclosure with this. So far so good it will do you a <u>little</u> more service than blank paper perhaps and will it[10] also please him, thus to come before the inhabitants of Toronto. —All the family bid me greet you & should you go to America you are to give them one more call on your way! And now John having dispatched Gerlings share of the matter, for a word between ourselves, I look on it with a little jealousy. Toronto (according to Gerlings map of Upper Canada)[11] is situated on the shores of lake Ontario not far from the Falls of Niagara[12] now although it is under the same blue canopy of this universe of ours, yet I feel great reluctance to see the thread that binds you to good little England, who needs your services and more like them, so far expanded, True your Theory of the cause of the roar of the celebrated Falls[13] might become developed but even in a Scientific aspect you would be out of reach of many advantages which England can furnish. For a few years sojourn there there is not much objection but as a permanency do not let it enter your head. I have not forgotten that we have yet to work together (<u>bodily</u> as well as spiritually) and I have enough love of country in me to hope that in the end she also shall be the scene of our labours We are yet young and may wander through the world, it will help us perhaps to tools certain it is it will give us a discipline which we can eventually turn to account True also there is everywhere somewhat to be learnt, some good to do, to ourselves and those around us, space decreases daily as an obstacle to the influence of our labour and example. Toronto no more than Queenwood or Marburg can retain within its walls John Tyndall and his effect on Humanity, that I well know yet after all I hold it as due to the Land that gave us birth <u>if possible</u> to return her what extra advantages our presence may afford. You need not me to preach Patience to you, something and that before long <u>must</u> turn up in England, meanwhile you stand in readiness

Toronto will give you £350 per annum. 'Meine Ansicht nach'[14] you would reap no pecuniary

<Handwritten letter ends; hereafter LT Transcript Only>
advantage—in the long run certainly not—but rather a sacrifice. I rather anticipate and certainly hope that this also will be Faraday's opinion,[15] which of course must influence you most.

Have you seen Capt. Wynne or Dr Mackay since your return—remember me to the latter should you meet.

To-day has been a gorgeous day; in spite of the heat and Noll's advice, who cannot conceive the inducement I find in making walks alone, I have had a stroll to the Frauenberg.[16] The crops of rye are already reaped and in stacks, the <u>foresters grave</u>[17] is hid in its richest green foliage, and every beautiful branching valley tempts you to thread its fairy windings. I no more dine in the close room, but under an apple tree laden with branches as tempting as those in the Garden of Eden could have been, I have to-day swallowed my pancake and Sauer Milch.[18] That done, I spent an hour reading a poem of Schiller's, it pleased me so much I scribbled a literal prose translation. This evening I will put the same in my journal,[19] and as it will interest you perhaps to know what it was about, I will make this sheet my <u>rough</u> draft and leave you to smooth it for yourself, as I will in transcribing it in my journal. It is written in beautiful German hexameters, which would take a far cleverer translator than me to screw into unwieldy English, which is <u>not</u> well adapted thereto, as you know, and is entitled

The Genius.[20]

Thou asketh me 'Do I believe what the Wisdom Dealers would teach me, and what Youth so readily and confidently asserts; can Science alone lead me to true peace, only the beam of System support Happiness and Justice? Must I then still mistrust the search which softly teaches me the law which thou thyself, Nature, hast emplanted in my bosom until on the eternal page Philosophy sets her seal, and confines the soaring Soul in the formal vessel? Tell me! Thou hast pierced these depths, from the mouldy grave thou hast come back unsullied. Thou know'st what the crypts of the dark word preserve, whether among the Mummies the hope of the living still dwells. Must I wander it, that lonely path? I see a glimpse of it; wander it then I will, for it leads to Truth and Right.'

Friend, thou hast heard of the Golden Age? Have not the poets related simple and touching ridings of it, that Time, when Innocence yet wandered in Life, and Feeling preserved yet its maidenhood and chastity, when yet the same great law, that alike governs the sun in his course, and hidden in the egg moves the germinating point, yet the still law of Necessity, the eternal, impartial, also moved in free waves. the human heart, when the unerring understanding, true as the hands of the Dial, knew but the True and Eternal? Then was seen no Profaner, no Mocker; what the living enjoyed, was not at death expected; equally comprehensible to every heart was the eternal Law, equally hidden the source from whence it sprung.

But that happy Time is past! Imprudent Self-will has destroyed the God-like peace of trusty Nature, sacrificed Feeling is no more the voice of the Gods, and the Oracle is dumb in the confused breast. The listening spirit hears it now only in the stiller self and the Holy meaning thereof is clothed in the mystic word.

Here the Searcher prays the pure heart rises, and lost Nature gives him Wisdom in return. Hast thou, happy one, ne'er lost thy protecting Angel, never disobeyed the loving warnings of holy instinct; see'st thou yet Truth pure and bright in the chaste eye, in the childlike breast hearest thou yet its tones, in thy contented disposition has Doubt ne'er arisen, will it, art thou certain, be always thus silent, will the strife of sensations never require an Arbitrator, nor mere understanding becloud the simple heart?—

Oh then go forth in thy costly Unguilt, Science can teach thee nothing, but must learn from thee! That Law, which with honoured rod corrects the straying affects thee not! What thou doest, thy will, is law, and to all creatures goes forth from thee a Godlike command, what thou with holy hand formest, with holy lips speakest, will move almighty the astounded Understanding; thou only observest not the God that builds in thy bosom, thou knowest not the might of the Wand that bends all spirits to thee; simple goest thou and still through a conquered world.

From this you may derive a notion of how I passed the hour under the apple tree, but <u>not</u> of the poem itself unless it is already known to you. Remember me to Debus, Haas and Mr Edmondson and family.

Thine affectionately, | T.A. Hirst.

RI MS JT/1/H/160
RI MS JT/1/HTYP/153–155

------

1. *word from you*: letter 0504.
2. *coddle*: to treat indulgently.
3. *both your memoirs*: one was presumably Tyndall's memoir on magnetism published in the April issue of the *Phil. Mag.* (letter 0464, n. 2); the other is less certain. It may refer to his paper on water-jets (see letter 0456, n. 1) or to one of his 1850 papers with Knoblauch, probably the longer second paper (cited letter 0403, n. 2).
4. *little Daddy Dogood*: derogatory allusion to Gerling.
5. *Kräntzchen*: see letter 0482, n. 6.
6. *Pfeifer's garden*: a park in central Marburg.

7. *Fraulein von Tileman*: not identified.
8. *old Boy*: allusion to Gerling, which might be literal (he was in his early sixties), but also has a derogatory tone.
9. *I took the opportunity . . . <u>request</u> of yours*: Tyndall had asked Hirst to approach Gerling for a testimonial (letter 0504).
10. *will it*: 'it' is inserted above the line. Hirst probably meant to place it one word earlier; the phrase would then read, 'it will also please him'.
11. *Upper Canada*: between 1791 and 1841 (when they were merged into a single Province of Canada) British possessions in Canada were split into the colonies of Upper and Lower Canada. Upper Canada encompassed what is now the southern part of the Province of Ontario and the lands bordering Georgian Bay and Lake Superior.
12. *Niagara*: punctuation missing, an example of Hirst's habit of omitting punctuation marks that fell at the edge of the paper. This pattern of omission occurs throughout this paragraph.
13. *your Theory of the cause of the roar of the celebrated Falls*: refers to a passage in his water-jet article, (cited letter 0456, n. 1) where he suggested that the sounds of rippling streams and crashing waves was caused by the compression of airbubbles (as in a water-jet) rather than the impact of water against water. 'It is the same as regards waterfalls. Were Niagara continuous and without lateral vibration, it would be as silent as a cataract of ice' (pp. 110–11).
14. *Meine Ansicht nach*: in my view.
15. *Faraday's opinion*: given in letter 0506. Faraday did not agree with Hirst, offering contrary advice.
16. *the Frauenberg*: see letter 0417, n. 21.
17. *the <u>foresters grave</u>*: a few months previously, Tyndall had written a short story that situated a romantic meeting at this site near Marburg (see letter 0482, n. 15).
18. *Sauer Milch*: soured milk.
19. *I will put the same in my journal*: entry of 3 August 1851.
20. *The Genius*: 'Genius', originally published as 'Natur und Schule' in *Die Horen* (a monthly German literary journal edited by Schiller) 3, no. 9 (1795), pp. 89–93. Schiller made alterations (mostly removing passages) for later editions. Several English verse translations had been published, for example, in Edward Bulwer Lytton, *The Poems and Ballads of Schiller*, 2 vols (Edinburgh and London: William Blackwood and Sons, 1844; vol. 1, pp. 95–98). This poem should be distinguished from the much shorter poem by Schiller, 'Der Genius'.

From Hermann Knoblauch 11 August 1851 0509

Marburg 11 Aug. 51

My 'dear friend'

According to my promise I give you notice of our journey to England. I shall close my lectures on the 22th of this month and than depart the other day.[1]

My father (who is this moment with me in Marburg) and I have not yet decided whether we shall go by way of Ostende or Calais; but after all it seems sure that we shall come over to England on the 26th or near so.

The Hotel (von) <u>Morlay</u> (corner of Trafalgar Square and Strand, near Charing-Cross)[2] has been recommended to us. —As you were kind enough to offer to procure lodgings for us, we accept this favour, when it does not give you too much trouble. We want 2 bed-rooms (one of them for my father, not too small (<u>bequem und geräumig</u>[3]), and situated not too high in the house). Perhaps you could order them for us in the hotel above mentioned <u>from the 26th of August</u>.

We probably will remain there the whole time during our stay in London, but as we can not determine this before having seen the lodgings and knowing the price perhaps too expensive for a longer time pray order them only for a week or so, (if it be necessary to fix a time). (but we leave these pecuniary points to yourself.)[4]

The new Reostat has come to Marburg in these days, just in time before the beginning of the lecture on Electricity.

We should be very glad if you could arrange matters so as to meet us in London.

With my best wishes and kind regards of my Father and myself,

yours | most truly | Herm. Knoblauch.

A letter from you could reach me here with certainty only, when it comes until the 21th at farthest.

Dr John Tyndall, | Queenwood College, Stockbridge, | Hampshire, England[5]

RI MS JT/1/K/16

----

1. *than depart the other day*: Knoblauch's English was idiosyncratic.
2. *Hotel (von) Morlay*: Morley's Hotel occupied the whole eastern side of Trafalgar Square, except for the ground storey at the southern end which was occupied by the Post Office.
3. *bequem und geräumig*: comfortable and roomy.

4.   *(but we leave . . . to yourself)*: added in the margin.
5.   *Dr John Tyndall . . . England*: address from envelope.

To Thomas Archer Hirst   14 August 1851   0510

14th Aug. 1851

My Dear Tom

Be good enough to carry to the enclosed[1] to their respective addresses. That to Gerling is to thank him for the testimonial[2]—still a very dilute affair—Knoblauch has sent me a good one—a very good one—That to Hessel is requesting one, as I believe the man understands what I have done better than most physicists—He is a valuable old fellow, under his rough old bodily bark beats I believe a heart of oak—

Had I taken time to dwell upon it I might almost have arrived at the contents of your last letter[3] <u>a priori</u> God knows my readiness to respond to your wish and to wander the world with you; and if occasion ever demand it you shall not find me slow to gratify the feeling—But under present circumstances I think I am choosing the most prudent path. Faraday writes to me thus.[4] 'I trust it (Toronto) is a place where a man of science and a true philosopher is required and where in return such a man would be nourished and cherished in proportion to his desire to advance natural knowledge—so I think that you would be exceedingly fit for the position and I hope that the position fit for you'. Francis who at first rather opposed the idea of my going writes thus[5]—'I was much gratified by the perusal of Faraday's letter and entirely concur in his advice to you however much I shall regret your leaving England'. Now Tom what am I to do? —Dont you think that my decision is the best possible under the circumstances? and the decision once made dont you think it behoves me to set all levers at work and not to do the thing by halves—If I am successful I must manage to see you before I set out.

Many thanks for that odorous[6] passage from Schiller.[7] —I had never read it—but it is beautiful—It fell like dew upon my heart.

I was thinking of writing more but I wont[8]—good bye dear Tom | John you must take the note to Hessell yourself and <u>translate</u> it for him

Herrn Sind. Tho.s Hirst | <cannot illegible> beim Hrn Baum | Marburg | Hesse Cassel[9]

<u>frei</u>!

RI MS JT/1/T/544

1. *the enclosed*: at least two letters, to Gerling and to Hessell, were enclosed. Both are missing.

2. *Gerling . . . testimonial*: see letters 0504 and 0508.

3. *your last letter*: letter 0508.

4. *Faraday writes to me thus*: Tyndall quotes from letter 0506. He does not mark ellipsis in his quotation from Faraday. The quotation lacks punctuation when compared with letter 0506, but that was based on an LT transcription and LT often improved Tyndall's punctuation. Tyndall mistakenly inserts 'that' six words from the end (compare with the same quotation in letter 0512).

5. *Francis . . . writes thus*: letter missing.

6. *odorous*: sweet-scented (*OED*).

7. *passage from Schiller*: 'The Genius', a poem translated by Hirst in letter 0508.

8. *writing more but I wont*: Tyndall stopped when he ran out of space on the page, but he had not run out of paper for the back of the sheet, where he added the postscript.

9. *Herrn . . . Cassel*: address from the envelope, along with 'frei' (at the bottom left corner), to indicate postage paid.

From John Hall Gladstone 15 August 1851 0511

St. Thomas's Hospital | London | August 15th. 1851

My dear Dr. Tyndall,

You will readily believe that the intimate conversation we had on the board the steam-boat on the Orwell,[1] has not been forgotten by me. Indeed I have ever since purposed renewing the subject with you; but I was loath to commence writing, unless I had leisure quietly to sit down and consider the matter: and thus it has been deferred until now.

Need I tell you, my friend, my reasons for so doing? It is my custom to study the diversified aspects, under which the all-important questions of religion[2] present themselves to different inquiring earnest minds, and there are some peculiarities in your case, which I have never met with before, and which I do not precisely understand. If our intercourse should lead to a modification of any of my views so as to bring them nearer to truth, or if it should clear up any of those doubts with which your mind is oppressed, I believe we shall both rejoice, and esteem it very well worth the trouble of writing.

Our experience, you think, is similar; we evidently sympathize in our aspirations; and yet our belief is very different. Your earnest desire to know what is true, your wish to follow whatever is good, your readiness to converse about personal religion even with one who holds such widely different views from yourself, and your fellow-feeling with him in spite of them,—all tend to make me conclude that you must really have experienced that great change[3] which we observe not unfrequently in men of all sorts, and which the Bible calls 'being born again', 'new creation', 'newness of life', 'conversion'. Now this, wherever it exists, is a radical change affecting the motives and the will: it can certainly be produced only <u>ab-extra</u>,[4] and is ascribed by such as have felt it to the divine Spirit: no doubt it may take place in persons without much real knowledge, nay in spite of much error. Yet this change (as far as my observation and reading goes) usually takes a definite form, and is brought about by a full perception of certain truths. The poor soul feels (what he merely confessed before) that spiritual things are realities, that he has grievously sinned, and has given the Almighty just cause to be offended; moreover he feels incapable of leaving off sin, nor can he think of God with any complacency. This he feels more or less intensely as the case may be; and in all probability he tries various ineffectual means of obtaining peace of conscience. Then he finds that in the Bible God is represented as not really desirous of the death of the sinner, but that on the contrary He has in the person of Jesus Christ found a way, in which He can be at once just and a justifier. That the holy law having been magnified by the death of Christ as the sinner's surety, an atonement is already made, and pardon is already procured. The poor soul having been rendered willing to receive gratuitous forgiveness, believes this declaration, accepts salvation, and is happy. Now he seeks to obey the law, because he loves the Lawgiver, and acting from motives of gratitude and affection (aided too by a certain divine influence) he succeeds, more or less completely, in overcoming the evil, and following the good. But my friend, how far do you believe in this atonement—and in the statements of the Bible?

In respect to a revelation from God, I suppose you will agree[5] with me that we want it attested by miracles or some sort of superhuman proof; and that when we have found such an accredited revelation our highest wisdom is to put faith in its statements, merely employing our reason to ascertain the exact meaning of the divine declaration. Reason may very well decide where our faith is to be placed, but it cannot decide as to what is the subject matter of revealed truth. The conflicting and unsatisfactory theories of the Hegelian school of theologians might prove this to us, even if it were not almost self-evident a priori.

I have little doubt that the community of sentiments there is between you and me will lead you to second this attempt on my part to reopen to one another the feelings of our souls on these all-important topics. I therefore subscribe myself,

My dear Dr. Tyndall, | Yours most sincerely | J. H. Gladstone.

I take it for granted you are at Queen's College[6] now. I have neither space nor time to add anything scientific[7]

RI MS JT/1/TYP/1/400–401
LT Transcript Only

---

1. *the steam-boat on the Orwell*: the River Orwell runs through Ipswich. Tyndall and Gladstone spoke during an excursion organised as part of the Ipswich BAAS meeting.

2. *all-important questions of religion*: this letter and the subsequent correspondence illustrates how seriously Victorians treated matters of religion.

3. *that great change*: Gladstone, who held an evangelical form of Christian belief, alludes here to conversion or being 'born again'. His letter outlines the central doctrines of evangelical theology.

4. *ab-extra*: from outside or from without.

5. *I suppose you will agree*: this is one of the many points at which Tyndall disagreed with Gladstone. Tyndall used conventional religious language but invested it with very different, Carlylean meanings. Gladstone, and others, were often slow to recognise the extent of the differences.

6. *Queen's College*: Gladstone misremembered the name of Queenwood College.

7. *I take it ... scientific*: postscript written on the envelope.

To Emil du Bois-Reymond 17 August 1851 0512

Queenwood College Stockbridge | England | 17th Aug. 1851

My Dear DuBois.

I dare say you have already been informed of my intention to become a candidate for the professorship of Natural philosophy in the University of Toronto. I wrote to Professors Magnus and Poggendorff[1] some weeks ago for testimonials—yesterday morning's post brought me a most admirable 'Zeugniss'[2] from Magnus who is now in London—A few days ago I wrote to Professor Dove on the same subject and this post will carry letters to you and Professor Riess.[3] I have no doubt the support of your names will be of great value to me.

I have to beg of you (if you have not already done so) to look over the philosophical Magazine for July 1850[4] and over the various papers which have appeared in the Annalen[5] and see whether they will not justify you in sending me a testimonial. I do not want to dictate a single word to you, but it might be of service if you mentioned in the document the fact of my being elected a foreign member of the physical society.[6]

I wrote to Faraday[7] some time ago laying before him the nature of my situation here and asking his advice; he encourages me to make application for the post at Toronto—the following few lines are quoted from his letter[8]—'I trust it (Toronto) is a place where a man of science and a true philosopher is required and where in return such a man would be nourished and cherished in proportion to his desire to advance natural knowledge so I think that you would be exceedingly fit for the position and I hope the position fit for you'.—

I have nothing more to say, as I am a bad hand at writing complimentary letters. I will only add that I shall long remember the hours which I have spent in your society; with the best wishes for your happiness

believe me dear DuBois | most sincerely yours | John Tyndall
Remember me kindly to Dr Beetz[9] when you see him.[10]
Ihr Brief an mich muss natürlich <u>unfrankiert</u> sein.[11]

RI MS JT/1/T/397

1. *wrote to Professors Magnus and Poggendorff*: these letters are missing but are alluded to in letters 0505 and 0506.
2. *Zeugniss*: testimonial (German).
3. *wrote to Professor Dove . . . and Professor Riess*: letters missing. Riess replied in letter 0514 and du Bois-Reymond in letter 0515. Both sent testimonials.
4. *the philosophical Magazine for July 1850*: an article cited in letter 0403, n. 2.
5. *papers . . . Annalen*: Tyndall probably alludes to the two papers he co-authored with Knoblauch (cited letters 0392, n. 13 and 0458, n. 8), his paper on the water jet (letter 0458, n. 7), his memoir on magnetism (letter 0464, n. 13), and, possibly, his recent paper on diamagnetism, the translation of which was published as Tyndall, 'Ueber Diamagnetismus und magnekrystallinische Wirkung', *Poggend. Annal.* 83, no. 7 (1851), pp. 384–416.
6. *elected a foreign member of the physical society*: the Berlin Physical Society or Physikalische Gesellschaft. Tyndall had been elected in early May (letter 0480, n. 9).
7. *wrote to Faraday*: letter 0505.
8. *his letter*: letter 0506.
9. *Dr Beetz*: Wilhelm Beetz (1822–66) was at this time teaching physics at the secondary-level Cadet Training School in Berlin. In 1855 he moved to become Professor of Physics in the Berlin School of Artillery and Engineering, which trained officers for the military (Jungnickel & McCormmach, pp. 252–4)
10. *Remember me . . . him*: the postscripts were written on the back of the sheet.
11. *Ihr Brief . . . <u>unfrankirt</u> Sein*: 'Your letter to me must be <u>unfranked</u>, of course' (German).

To John Hall Gladstone   17 August 1851   0513

Queenwood | 17th. Aug. 1851

My dear Dr. Gladstone

I cannot pretend to reply to your friendly letter[1] which reached me an hour ago. My intellect is so to speak polarized[2] at present—directed upon objects which differ considerably from those treated of in your letter. Of course every scientific man feels himself clogged in this way at intervals, yet through all clogs and all hindrances your letter has reached my warmest sympathies. I thank you for it most heartily.

I believe undoubtingly that the religion of both of us is fundamentally the same[3]—that the same primitive material so to speak is cast into different moulds. There are some things in your form which I see I could not accept and I sincerely believe that the adoption of my form upon your part would fail to make you a better or a happier man.

I reverence the bible from my soul and have often drawn strength and encouragement from it. Here I am sure our experience will agree. I recognize your right to adopt the scheme of redemption touched upon in your letter; but I cannot say that I find the scheme suited to my case. —Well I know that the life of religion is not inseparably bound up with this scheme— nay I should even go so far as to say that religion may pronounce itself in its most exalted form in places and circumstances where the scheme was never heard of.

But you will ask me what is <u>my</u> scheme? Anything I have to say upon this subject must be referred to a future day when my mind is more attuned to the subject than it is at present—Well I know the difficulty of an answer; for the operations of God's spirit upon the heart of man are scarcely to be stated intellectually. This is not a solitary experience of my own—You know who has said 'The wind bloweth where it listeth and thou hearest the sound thereof but canst not tell whence it cometh or whither it goeth—so it is with every one who is born of the spirit'.[4]

Whenever I find myself in a condition favourable to the act I will write to you, and I shall ever be glad to hear from you. I do not want to modify your views, for if they elevate your heart and raise your life to pure and noble practices they do not want modification. I think our chief want is not a modification of our religious theories but strength and courage to act up to the knowledge we possess. This I believe is the only way of access to still higher knowledge—the practice of that which we possess. A religion founded upon logic is for the curious and argumentative, but not for the man who is in earnest about the matter. —I will now stop, excuse this unworthy acknowledgement of your truly friendly act and believe me

most faithfully yours | John Tyndall

Dr Gladstone.

RS MS/743/1/6
RI MS JT/1/TYP/1/402–403[5]

1. *your friendly letter*: letter 0511.
2. *My intellect is . . . polarized*: another use of a physical metaphor to describe a state of mind
    (see letter 0505, n. 6).
3. *religion . . . the same*: the continuing correspondence shows Gladstone puzzling over the
    extent to which they had the same beliefs or religion. Tyndall believed that religious words
    clothed a profound mystery. In using the same words, Gladstone invested them with more
    literal meanings in the tradition of a more orthodox Christian theology.
4. *'The wind . . . the spirit'*: the words of Jesus to Nicodemus, John 3:8. there are only minor
    variations between the KJV and Tyndalls' quotation: he omitted a few commas; he wrote
    'cometh or whither' instead of 'cometh, and whither', and 'one who is born' instead of 'one
    that is born'. But see n. 5.
5. *TYP/1/402-403*: the LT typescript of this letter reveals LT's editing processes (see 'Edi-
    torial Principles'). 'Hearest' was originally typed as 'heareth' and changed by hand; the
    revised quote is correct.

From Peter Riess   21 August 1851   0514

Verehrtester Herr

Es gereicht mir zu grosser Freude, Ihrem Wunsche durch das beiliegende
Zeugniss zu genügen, wenn ich auch leider mir davon wenig Erfolg verspre-
chen darf, da meine Stimme gerade in England nur ein sehr geringes Gewicht
haben kann. Meine früheren, wie ich glaube, besseren Arbeiten sind dort
völlig unbekannt geblieben, was zum Theil meiner eigenen Nachlässigkeit,
mich mit dortigen Gelehrten in Verbindung zu setzen, zugeschrieben werden kann. Vielleicht dass meine Electricitätslehre, an derer Herausgabe ich
jetzt ernstlich denke, Veranlassung wird, diesem Übelstande abzuhelfen. Ihr
gütiges Anerbieten, von meiner neusten Arbeit im 83ten Bande der Annalen
einen englischen Auszug zu geben nehme ich mit Dank an und bitte dabei
die beiden Punkte hervorzuheben, die mir die interessantesten des Aufsatzes
zu sein scheinen: die Abhängigkeit der verschiedenen Ströme von der Form
ihres Schliessungsbogens und den Schluss, den ich daraus auf die Richtung
der Ströme ziehe.

Mit dem herzlichen Wunsche, dass Sie recht bald Ihren nächsten Zweck
erreichen und dadurch Zeit gewinnen, die Wissenschaft ferner durch Ihre
Arbeiten zu bereichern bitte ich Sie, mir ein freundliches Andenken zu
bewahren. Frau und Kinder grüssen Sie bestens.

Ihr | ganz ergebener | P. Riess
Berlin 21 August 1851.

Most dear Sir

It gives me great pleasure to fulfil your wish[1] by means of the enclosed testimonial, even if I may unfortunately have little hope of success from it, since my voice carries but very limited weight in England. My earlier and, I believe, better works[2] have remained completely unknown there, which can be ascribed in part to my own negligence in contacting scholars there. Perhaps my theory on electricity, the publication of which I am now seriously considering, will provide an opportunity to remedy this unfortunate situation. I accept with thanks your kind offer[3] to publish an English excerpt from my newest work in the 83rd volume of the Annalen,[4] and ask in doing so that you emphasise the two points which seem to me to be the most interesting ones in the essay: the dependence of different currents on the shape of their closing arc, and the conclusion which I draw from this about the direction of the currents.

With the heartfelt wish that you will very soon achieve your next goal[5] and thereby gain time to enrich science further through your works, I ask you to keep me in fond memory. My wife and children send you their best regards.

Your | humble servant | P. Riess
Berlin, 21 August 1851.

R1 MS JT/1/R/25

1. *fulfil your wish*: the request from Tyndall to Riess is missing. It is alluded to in letter 0512.
2. *My earlier and, I believe, better works*: between 1829 and 1851, there are 41 papers by Riess in Poggendorff (including some co-authored papers), primarily in magnetism, terrestrial magnetism, and electricity.
3. *your kind offer*: presumably made in the same missing letter (n. 1) in which Tyndall requested a testimonial.
4. *to publish ... of the Annalen*: an abstract of two papers by Riess (under the single title, 'On Electric Currents of the First and Higher Orders') was published as Tyndall, 'Reports on the Progress of the Physical Sciences', *Phil. Mag.* 3, no. 17 (March 1853), pp. 173–85. The first paper had been published in 1850 (see letter 0457, n. 12.) and the translation sent to Francis in January 1851 (letter 0461, n. 7) but for unspecified reasons had not been published ('Reports', ibid., p. 174). The second paper was 'Ueber die elektrischen Ströme höherer Ordnung', *Poggend. Annal.* 83, no. 7 (1851), pp. 309–54.
5. *your next goal*: the Toronto position.

From Emil du Bois-Reymond 30 August 1851 0515

Verehrter Freund,
Ich sende Ihnen das verlangte Testimonial mit dem Wunsche, einmal dass

es seiner Form nach dem Zweck entsprechen möge, zweitens dass es wenigstens ein Milligramm in ihrer Wagschale austragen möge, woran ich mir zu zweifeln erlaube. Indessen fragt es sich ob nicht ein unbekannter Name bei dergleichen Gelegenheiten oft mehr imponirt als ein bekannter. Es thut mir übrigens sehr leid Sie nach Toronto gehen zu sehen. Da Sie hinwollen so müssen Sie wohl ihre Gründe haben und es mag auch am Ende für einen unternehmenden, reisegewohnten und in einer grösseren Weltanschauung lebenden Mann wie die Engländer sind nicht so arg sein als es uns armseligen Stubenhockern vorkommt. Ich habe im Athenaeum mit grossem Interesse ihr Auftreten in Ipswich verfolgt. Ihre Bemerkung über den Foucault'schen Versuch hat uns aber daraus nicht recht klar werden wollen. Plücker hat abermals Diamagnetica losgelassen, aber er scheint jetzt von seinem guten Engel verlassen. Er lässt den Sauerstoff dauernde Polarität annehmen u. *[d.]* m. Knoblauch will ja nach Paris und England gehen. Sie werden also wohl noch die Freude haben ihn zu sehen. Von Dove erhalten Sie in diesen Tagen auch das Verlangte. Verzeihen Sie meine Kürze aber ich bin einmal wieder von allen Seiten gehetzt. Auf Wiedersehen.

Ihr | E. du Bois-Reymond

P. S. Schicken Sie mir doch aus Toronto eine Familie Ursus americanus, die mit der gelben Schnauze; es sind niedliche Thiere.

Berlin, Carlstr. 25, 30. VIII. 51.

Esteemed friend,

I am sending you the testimonial you requested[1] with the wish, first, that by its form it might suit its purpose, and second, that it might tip the scales at least a milligram in your favour, something which I permit myself to doubt. The question is, however, whether an unknown name does not often impress more on such occasions than a known one does. I am very sorry, incidentally, to see you go to Toronto. Since you want to go there, you probably must have your reasons, and it may not, in the end, be so bad either for a man who is enterprising, used to travelling, and living in a broader view of the world—such as Englishmen are—as it seems to us miserable stay-at-homes. I have followed your appearance in Ipswich with great interest in the Athenaeum.[2] Your comment about Foucault's experiment,[3] though, did not leave us really in the clear from it. Plücker has let go of diamagnetism once again, but he now seems to have been abandoned by his ministering angel. He is letting oxygen acquire permanent polarity, among other things.[4] Knoblauch, of course, wants to go to Paris and England.[5] You are therefore probably still going to have the pleasure of seeing him. You will receive what you also requested from Dove[6] in the next few days. Pardon my brevity, but I am hounded from all sides once again. Goodbye.

Your | E. du Bois-Reymond

P. S. Do send me a family of Ursus americanus from Toronto, the ones with a yellow snout, they are nice animals.

Berlin, Carlstr. 25, 30. VIII. 51.

RI MS JT/1/D/142

---

1. *testimonial you requested*: in letter 0512.

2. *your appearance . . . in the Athenaeum*: a report of the BAAS which contained a lengthy abstract of Tyndall's paper on diamagnetism (cited letter 0501, n. 7); also see n. 3.

3. *your comment on Foucault's experiment*: following a paper by Professor Walker 'On the Influence of the Earth's Magnetism on the Pendulum of Foucault' (8 July) Tyndall suggested that 'some electric current, excited by the motion of the pendulum itself, may be the origin of the curious oscillation of the revolving motion of the pendulum as it approached the magnetic meridian' ('Twenty-First Meeting of the British Association for the Advancement of Science', *Athenaeum* 1238 [19 July 1851], pp. 776–87, on p. 776). Wartmann replied that he had tested for such a current but not detected one.

4. *among other things*: the German original looks like 'u.d.m.' but only 'u.a.m.' makes sense.

5. *Knoblauch . . . to Paris and England*: Knoblauch had written to Tyndall discussing his travel plans in letter 0509.

6. *what you requested from Dove*: a testimonial. Tyndall had mentioned writing to Dove in his last letter to du Bois-Reymond (letter 0512); the letter to Dove is missing.

From Thomas Archer Hirst   31 August 1851   0516

Marburg— | Aug^st 31st 1851

A few words dear John before I shoulder my knapsack and set off to Switzerland where I had always intended we should go together but fate would not have it so and I am satisfied. The Semester is closed and at the end of it I find myself better satisfied with the results than I was with the anticipation As we have many a time repeated gloomy forebodings only require to be moved under and are sure to be dissipated, they are ever of a darker hue than the realities. A weeks cold moist weather here has spurred me up and dissipated my inertia, by reminding me that the air on the Alps will be getting cool & that Summer (or Autumn) is flying fast. —As yet I have no settled route I shall wander my own wild way and trust all as usual to chance or providence, you will have a line from me now and then I dare say, if not you will know that I am among scenes familiar to you and as usual trying with more or less success as ever to interpret what they have to tell me.—

Gerard. Simpson & Co[1] advise me to remain in the outskirts of towns & watering places & not venture too much on the cold, lonely mountains, how much I value their advice you may guess. One object however I have before-hand, it appears from a conversation I had with Fraulein Spannenberg[2] that a relation of mine (a governess at a Moravian establishment) whom I have long lost sight of is in the neighbourhood of Geneva. I will seek her out if possible; I go provided with letters of introduction from Fraulein Spannenberg who was at school there, who is enthusiastic in her praises of the people and neighbourhood and who judging from her conversation has provided for me there a reception of the most hospitable nature—Knoblauch and his papa you will already have seen by this time remember me kindly to both, tell them if they should visit Bristol not to forget to make use of my Brother,[3] I gave Knoblauch his address and wrote to day to provide for my Brothers shewing them all in his power of the sights of Bristol, Clifton &c.—

Wrightson goes to England soon also, he will return to Marburg (according to his present intentions!!!) and make an examination here next Semester. His results at the Laboratory it appears are again = 0,[4] though one must give him credit for a pretty considerable stock of patience, I am inclined to adduce him as an example against your theory that Patience is synonymous with Genius. He continues doggedly to battle with Mathematics after dinner, often gives me his company at that time, under the professed objects of asking for some explanation, but I seldom have the honour of imparting it as by the time he reaches me I find he knows it already and his visit thus usually ends in his helping himself to a Cigar. He is a wonderful specimen of human nature and a complete puzzle to me in point of character. I never met his like. He and Davy have at last had an open rupture, and I have had to act partially as mediator though I could make a little of either. Davy it appears is articled to him (premium about £300) to learn Chemical Analysis!! Davy complains that on account of being left in the imperfect small Laboratory to himself, M^r W being always absent he has learnt actually nothing during the year, that owing to a difference (whose cause they yet both keep secret) between them which happened directly after coming here last year there has been a perpetual coolness & ill will between them which has rendered their living together for Davy no longer bearable At last Davy taking advantage of Wrightsons absence at Dusseldorf, whither he had accompanied Prosser[5] on his way to England, removed his boxes, &c & took up his lodgings wither Gerland, refusing to go back and alledging in his own justification that Wrightson had not fulfilled the conditions of the indentures & that the step was taken with his fathers knowledge. Wrightson somewhat frightened (the premium is not all paid yet) & perhaps also somewhat conscience-stricken asked me to try and come to some terms with him—I did so brought them together in my rooms. Davy is truly of a weak cast of character but scarcely more so than Wrightson himself, and who sees moreover that Wrightson is not well capable to teach much of

Chemical Analysis (he has in fact detected his inabilities in many points & seen that £300 is too great a price for all Wrightson knows about Chemistry) the result of the interview was that Davy would return on condition that he worked in the Laboratory next Semester and that they lived separately to which Wrightson consented. The relation between them is pitiable. Wrightson is totally unfit for the responsibility he has taken upon himself and Davy requires treating with a firmer hand if he is ever to be made anything of. He is for the most part of a harmless weak nature though this late difference has shewn me a secreted & hidden sullenness & obstinacy in his disposition. How it will end one cannot <u>predict</u>.

Smith[6] wrote me the other day. He is living precariously in London writing fugitive articles in the Leader Eliza Cook, Chambers[7] &c for an odd £1. I wish he had something of a more suitable nature to employ him. Poor Booth has been almost dead he sent me the enclosed interesting letter which you <u>must return me</u>. Of Phillips I have heard but little lately. Smith mentions that he has been in London & together with Caliban[8] had spent and Evening with him Smith speaks highly of Caliban who it appears has just had a fortune of £8000 left him, has given up his profession as Surgeon and is now in France—Has Professor Hessel sent your testimonial by Knoblauch? He said he would—

About Toronto John although I recognize the justice of your intentions I am by no means satisfied You know as yet too little of its nature. Faraday also does not it appears know what kind of a place it is, but only hopes that it and you will be fit for each other However you would do well to make every enquiry & preparation, if it turns out promising I myself will say go & God speed—and what is more <u>perhaps</u> follow thee. There is to be an immense railway made from Quebec to Halifax, if nothing else offers itself I can earn my living & do my duty thereon—

With patience all will order itself rightly at last, we will <u>force</u> nothing do what seems best to us at this present & leave the rest to gravitation. I have concluded to return here next Semester & probably at its close make an examination; as yet I can see no further (nor do I try) into futurity.—

Wrightson leaves here in about a week for England and returns again, his address is

Mr Francis Wrightson | Messrs Wrightson and Bell[9] | Birmingham.

I do not remember anything I can ask you to send me if you should do in the meantime you can send it to him—

Good-bye & God bless you— | Yours affectionately | T A Hirst

RI MS JT/1/H/161

[Enclosure]
Francis Booth to Thomas Archer Hirst 17 August 1851 0516encl

Birdcage, Skircoat.[10] | August 17, 1851

My dear Friend,

At last heartily do I rejoice to greet you once again. Some six months I think have elapsed since I last heard from you. I hope they have passed pleasantly and happily with you. I was about to add 'smoothly', but I bethought me that to you—and indeed to any man in a normal, healthy state, the smoothest path is rarely the pleasantest. The presence of difficulties is to be overcome—of objects to be attained that cannot be attained smoothly—is absolutely necessary to our enjoyment in the present sphere of existence; and I am of opinion that from the nature of things it must still continue to be so in <u>any</u> sphere of existence whatever. Were it to become otherwise we should lose those attributes that constitute us essentially what we are; we should cease to be <u>ourselves</u> in short, and each one become some other thing or being in whose future fate or condition <u>we</u> could have no interest whatever. Well I hope that you have been, and are still going on happily and prosperously, approaching nearer and nearer to the goal you have in view, and encountering no further difficulties by the way than may serve to give a pleasant stimulus to your energies. Where or what your goal may be, of course I know not, but I feel very confident that it is not for your own pleasure or profit merely that you are thus secluding yourself in Marburg, and pursuing a course of severe study there. —You have some higher object in view, and I shall look forward with interest for the next phases of development in your 'life history'. I hope that ere long you will emerge from your chrysalis state, so that I and the world may see what sort of butterfly you will make.

The months have passed sadly and drearily enough within me of late. I am now partly recovered from a dangerous illness that threatened to carry me down the stream

'whose current flows | Unto the land that no one knows'.[11]

It commenced with a severe cold to which I at first paid little attention. But soon a deep and violent cough settled on my lungs, accompanied by pains in the chest and shortness of breath. I was bathed in perspiration at nights. I grew weaker and visibly thinner every day. At last I could scarcely cross the room without assistance, and my breathings came short and quick after the slightest exertion. Finally I was confined to my bed, a victim to the doctor. I displayed all the symptoms of one in a consumption, and most of my friends gave me up for lost. I did not hear it from their lips, but I read it in their eyes. I myself seriously thought that I was soon to follow in the footsteps of my poor sister,[12] I remember wondering, one night as I lay in bed, whether in the next world I might still be permitted to trace those footsteps, and perhaps to overtake her! And if so, I thought how we should be happier to wander <u>together</u> through an immensity of desolation, than to be parted from each other and placed in <u>separate</u> heavens. But my thoughts seldom wandered in this direction. A deep despondency settled on my mind, but it was not occasioned by fear of death, or anxiety about my future. These things I must confess seldom occupied my mind. I could not bring myself into anything resembling that state of mind that religious people told me was necessary in my condition. The only sincere and earnest prayer I prayed—and <u>that</u> was sincere and earnest enough—was that I might recover and regain my health, and strength, for my mother's sake.[13] And I thought that if that Power to whom I prayed <u>could</u> be moved by prayer—if he indeed possesses any of those attributes which we look upon as forming the <u>divine</u> portion of ourselves,—if in short there could be any sympathy between Him and me, then my appeal could not be without effect. I may be superstitious, but I cannot help thinking now that my prayer <u>has</u> been answered, and that for her sake I have miraculously escaped the jaws of death.

But I must be brief, else I had many things to tell you, particularly about the unexpected illness, most of them, however, happily surmounted now. You see it has been by the strictest economy that Mother and I have been able to stay at the old home we are so loth to leave. In all our plans, however, I had never calculated on the possible failure of my health. When I became ill, and it was probable that I should have to leave my situation and be confined to a bed of sickness for an indefinite length of time, the gloomy prospect weighed heavily on my mind. I saw no way out of our difficulties, except by having recourse to the 'assistance' ('Charity') of friends, or relations, and thus sacrificing that valued independence which I had so long struggled to maintain. But I have no more room. The danger for the present is averted, I am almost recovered, miraculously as I think; at the same time I believe that cod liver oil has had something to do with my cure. Bottled porter too has been of great service. I am a firm believer in bottled porter.

I remain, dear sir, ever yours, | Francis Booth.

RI MS JT/1/TYP/11/3486–3487
LT Transcript Only

---

1.   *Gerard. Simpson & Co*: probably fellow students: Gerard, not identified; Simpson is Maxwell Simpson.

2.   *Fraulein Spannenberg*: not identified (see letter 0482).

3.   *my brother*: John Henry Hirst.

4. *his present intentions!!!) . . . again = 0*: Hirst often reported to Tyndall on Wrightson's incompetence in the laboratory and his social inadequacies (for example, see letter 0488, at n. 11).

5. *Prosser*: probably Richard Bithell Prosser (1838–1918), a young Englishman who had visited Marburg. About this time he enrolled in University College School in London.

6. *Smith*: John Stores Smith. A few months previously Hirst and Tyndall had learned of Smith's difficult circumstances (see letter 0481).

7. *the Leader Eliza Cook, Chambers*: three weekly papers. For the *Leader* see letters 0398, n. 8 and 0501, n. 19. *Eliza Cook's Journal* (1849–54) was a progressive weekly miscellany, edited by the poet Eliza Cook, and aimed primarily at women and members of the working classes (*ODNB*). *Chambers's Edinburgh Journal* was an inexpensive educational but entertaining weekly paper edited by the brothers William and Robert Chambers.

8. *Caliban*: see letter 0400, nn. 7 and 9.

9. *Wrightson and Bell*: publishers. This suggests that Wrightson's family was in publishing, and that both Wrightson and his pupil, Davy, were from Birmingham.

10. *Birdcage, Skircoat*: there is a Bird Cage Lane, off the Skircoat Road about 10 miles south of Halifax town.

11. *whose current flows . . . knows*: source not identified.

12. *my poor sister*: not identified. Booth implies that she had died of consumption.

13. *for my mother's sake*: references to Booth's mother imply that she was a widow and dependent on him for support.

To Margaret Ginty   [August 1851][1]   0517

Queenwood, Aug? 1851[2]

My dear Mrs Ginty,

I have been very busy indeed and continue to be so, this is my only reason for not writing. Yesterday morning before your letter[3] arrived, I think the thought of you was one of the first that entered my head and I muttered involuntarily to myself I must write to Mrs Ginty. Your letter has been a source of sincere satisfaction to me—I am very glad to hear that you are getting on so cheerily. That is a brave sentence: 'I am far more independent and comfortable than I was 9 months ago'. Our comfort and independence depend in a great measure upon ourselves—once we make up our mind, to face the circumstances of life valiantly and with courageous hearts,[4] difficulties which were previously deemed insurmountable dwindle into very sorry affairs—Keep to that Motto Peb[5] my girl—'don't get into debt'. The habitual tendency to get into debt is a snare of the devil.

I think I should like to see you just as well as you would like to see me—you know right well, you little minx, that I did like you a <u>little</u> <u>bit</u>,[6] otherwise you would never have been so impertinent as to mention it.

I hope that William[7] wont be long without a better salary not on your account for I know right well he gives you enough already. I shall see you in your new house sometime no doubt, but when is very uncertain—I must work like a negro and live like a miser for a long time to shake my own difficulties[8] from my shoulders—I shall get them underfoot at last I have no doubt.

You are not so lazy as you used to be are you not? I think if I were asked what is the greatest sin to which humanity is subject I should point to idleness.[9] The creator of man intended him to be an active being—Idleness in itself would not be so bad if it stopped there but it draws on other evils in its train. As long as a man works, his condition is hopeful and you could not give William a higher character than to say he works hard.

But now I must stop my sermon I have written more than I expected I could write—I used to be a good letter writer as Wm. can testify but I find myself getting daily more rusty in this respect. Give my love to William and take the same yourself.

Good bye my girl | your faithful friend | John Tyndall.

RI MS JT/1/TYP/11/3640
LT Transcript Only

---

1. *[August 1851]*: date assigned by LT (see n. 2). The allusion to financial difficulties and hard work (n. 4 below) supports a date in the period after Tyndall's return from Germany.

2. *Aug?1851*: this date is from LT's transcript. There is no indication of how she obtained either the month or the year.

3. *your letter*: letter missing.

4. *to face the circumstances of life valiantly and with courageous hearts*: a Carlylean sentiment.

5. *Peb*: a short (familiar) form of the name Margaret.

6. *I did like you a <u>little</u> <u>bit</u>*: since their Ordnance Survey days Tyndall had been a close friend of Margaret Ginty's husband, Bill (William). They had encouraged one another's love interests, even though sometimes interested in the same young woman. Margaret and William Ginty married in 1846. As reported in Volume 2 (biographical entry) Margaret was a notably attractive and congenial woman.

7. *William*: William Ginty, her husband.

8. *my own difficulties*: presumably a reference to his financial struggles and his debt. See letter 0501 (at n. 20).

9. *greatest sin . . . idleness*: another Carlylean emphasis.

To George Biddell Airy   1 September 1851   0518

Queenwood College | near Stockbridge Hants |
1st September 1851

Dear Sir

Circumstances have conspired to place me in the attitude of a petitioner to your kindness.[1]

I intend to become a candidate for the professorship of natural philosophy now vacant in the university of Toronto, and feeling assured as I do that a name so high as yours in the scientific world must have great influence in such a case I am induced to make an effort to secure your support.

I beg to lay before you the accompanying scientific memoirs;[2] should your leisure permit, and should your opinion of my labours justify the act on your part you would place me under a great obligation to you by granting me a testimonial. If you feel a desire to ascertain the opinion of a philosopher who has specially occupied himself with these subjects. I have much pleasure in referring you to Professor Faraday.[3]

I remain dear Sir | most respectfully yours | John Tyndall
Prof. Airy | &c &c

RGO MS.RGO 6/373.402–403

---

1. *Circumstances . . . to your kindness*: one of the set phrases Tyndall used when requesting testimonials from people whom he did not know (compare letters 0522 and 0523). See letter 0527 for Airy's reply.
2. *the accompanying scientific memoirs*: probably (compare letter 0524 from Joule) the English version of his first memoir with Knoblauch (cited letter 0403, n. 2), his water-jet paper (letter 0456, n. 1), the important memoir on magnetism (letter 0464, n. 2), and his most recent paper on diamagnetism (letter 0498, n. 6).
3. *referring you to Professor Faraday*: see letter 0506.

To William Thomson   1 September 1851   0519

Ardmillan | Sep 4[1]

Queenwood College | near Stockbridge |
Hampshire | September 1 / 51

Dear Sir,

You are already aware of my intention to become a Candidate for the professorship of Natural Philosophy in the University of Toronto, and my friend Dr. Francis has already informed me of the kind readiness with which you promised to support me—There are few indeed whose testimony on my behalf will be so gratifying to me as your own for I believe no Physicist of our day has penetrated the theoretic Mysteries of the subjects which have chiefly engrossed my attention so far as yourself.

I send you a copy of my last investigation which was made in Berlin and read before the Physical Section of the British Association at Ipswich.[2] I regretted your absence on the occasion very much—the paper I confess was in some measure framed with a view of meeting an anticipated onset on your part. I expect to have a short paper on diamagnetism ready for the October number of the Phil. Mag.[3] It is with some reluctance that I publish it so soon, but I am anxious to send it to Toronto with my other memoirs and therefore cannot delay its publication beyond the 1st of next month.

I have not yet had the pleasure of seeing the number of Philosophical Transactions which contains your Theory of Magnetism.[4] I heard of the investigation first in Berlin and afterwards in the address of Prof. Airey at Ipswich.[5] I shall endeavour to get into it if possible, though want of practice has rendered me rather rusty at the calculus.[6] I commenced Poisson's theory[7] 6 or 8 months ago, but never got through it; he writes with wonderful precision it is true, but he leaves many steps undeveloped. This made my reading so slow that I was forced to give it up for want of time.

It is hardly necessary to mention further the more specific object of this letter, but I shall consider it a very great favour if you would have the kindness to send me a testimonial.[8]

with best wishes | I am dear Sir | most faithfully Yours | John Tyndall.
Prof. Thomson | &c. &c.

RI MS JT/1/TYP/5/1530–1531
Transcript Only

1. *Ardmillan | Sep 4*: Thomson was not at home when the letter arrived and it was forwarded on to him (see his reply (letter 0528)). 'Ardmillan | Sep 4', added at the top of the letter, indicates the place and date of Thomson's receiving the letter. He probably refers to Ardmillan House, on the estate of the lawyer, James Crauford (later Lord Ardmillan). The estate was near the coast a few miles south of Girvan in Ayrshire.

2. *read . . . at Ipswich*: Tyndall, 'On Diamagnetism and Magnecrystallic Action', just published in full in the *Phil. Mag.* (letters 0501, n. 3 and 0498, n. 6).

3. *short paper . . . Phil. Mag*: the paper was published in November (see letter 0525, n. 2); thus Tyndall did not get it published as soon as he wished.

4. *your Theory of Magnetism*: Thomson, 'A Mathematical Theory of Magnetism' and 'A Mathematical Theory of Magnetism. Continuation of Part I' (read on 21 June 1849 and 20 June 1850), *Phil. Trans.* 141 (1851), pp. 243–68, 269–85.

5. *address of Prof. Airey*: Airy's presidential address to the BAAS delivered at Ipswich and published in the *Brit. Assoc. Rep. 1851*, pp. xxxix–liii. The address, which surveyed scientific progress over the preceding year, referred to Thomson: 'The first part of an elaborate mathematical theory of Magnetism, by Prof. Thomson, has been published' (p. xlv).

6. *rather rusty at the calculus*: Tyndall did not have the mathematical facility of Thomson (who had a mathematician father and had graduated second wrangler from Cambridge). Oliver Lodge later criticized Tyndall's mathematics: 'he certainly learned some mathematics . . . but he had not the genuine mathematical instinct' and elaborated on this weakness in his *Encyclopaedia Britannica* article' ('Tyndall, John', *Enc. Brit.*, 10th ed., vol. XXXIII, 1902). Ascribing Tyndall's mathematical limitations to innate incapacity rather than to education is a Victorian interpretation.

7. *commenced Poisson's theory*: Simon Denis Poisson (1781–1840) was a brilliant French mathematician who applied mathematics to many branches of physics including electricity and magnetism. Thomson had mentioned Poisson's theory of magnetism in discussion at the Edinburgh BAAS meeting (1850). Tyndall had asked Thomson for references (letters 0422) and Thomson responded (letter 0425) in August. It is not clear precisely what Tyndall was reading, but see Thomson's letter for his recommendations.

8. *send me a testimonial*: Thomson sent a testimonial with letter 0528.

From Robert Bunsen 1 September 1851 0520

Mein theurer Freund

Ich sende Ihnen das veränderte Zeugniss mit dem lebhaftesten Wunsche, dass es Ihren Zwecken dienlich sein möchte, wiewohl ich eine Art von Besorgniss kaum unterdrücken kann, ob es wohlgethan ist, dass Sie so fern von Ihrem engern Vaterlande eine wissenschaftliche Stätte suchen. Doch Sie werden gewiss Alles reiflich erwogen, und das Richtige gewählt haben. Gehen Sie wirklich nach Toronto, so erhalte ich zuvor wohl noch ein paar Zeilen von Ihrer Hand, denn ich nehme den herzlichsten und aufrichtigsten Antheil an Allem was Sie betrifft.

Ueber Debus Anstellung bei Ihnen habe ich mich herzlich gefreut. Er ist gewiss zu ängstlich, wenn er fürchtet, dass man ihm seine schnelle Abreise von Marburg dort verdacht habe. Ich glaube kaum, dass man ihm dort einen Schritt wird übeldenken können, der durch den Umstand geboten war. Ich bitte Sie ihn freundlichst von mir zu grüssen.

Was mich anbelangt so <lebe> ich hier sehr zufrieden. Die Preussische Regierung thut Alles um meinen Wunschen auf die bereitwilligste Art entgegenzukommen. <Mein> Laboratorium, welches reicher und schöner ausgestattet werden wird, wie vielleicht irgend eins in Deutschland, schreitet im Bau rasch vorwärts und ich denke es schon im nächsten Sommer beziehen zu können. Ich habe dabei auf eine Gas- u Wasserleitung bedacht genommen und gewinne so viel Raum dass ich über 40 treffliche Arbeitsplätze gebieten kann.

Ich bin so frei Ihnen einige Abdrücke meiner Arbeit über die vulkanische Gesteinsbildung mit der Bitte beizusenden, dieselben an ihre Adressen gelangen zu lassen. Da die Abhandlung innig mit der in den Schriften der Cavendish Society übersetzten zusammenhängt, so würde es mir sehr erwünscht sein, wenn auch diese Untersuchung dort oder in irgend einem andern englischen Journal übersetzt würde. Playfair oder Hoffmann, die ich freundlichst wie auch Frankland zu grüssen bitte, würden diese Angelegenheit gewiss gern vermitteln.

In aufrichtiger Freundschaft | Ihr | R. Bunsen

Breslau den 1^n Sept 1851.

My dear friend

I am sending you the amended testimonial[1] with the most eager wish that it might be useful to your purposes, although I can hardly suppress a sort of concern as to whether it is right that you are seeking a scientific home so far from your immediate native country. But you will certainly have considered everything carefully and made the right choice. If you really do go to Toronto, then I shall receive a few lines from your hand before you do, for I take the most heartfelt and most sincere interest in everything that concerns you.

I was heartily pleased about Debus finding employment with you. He is certainly being too apprehensive if he fears that people have held his sudden departure from Marburg against him there. I hardly believe that people there are going to be able to think badly of him taking a step which was offered by circumstance.[2] I ask you to give him my kindest regards.

As far as I am concerned, I am <living> very contentedly here. The Prussian government is doing everything to accommodate my wishes in the most eager way. <My> laboratory, which is going to be more richly and more nicely equipped than perhaps any in Germany, is making rapid strides forward in its construction, and I think that I shall be able to move into it as early as next summer. In doing so, I have taken into consideration the gas and water pipelines, and am getting so much room that I can have more than 40 excellent work areas at my disposal.

I am taking the liberty of sending you some offprints of my work on volcanic rock formation,[3] with the request that you forward them to their addresses. As the paper is closely connected with the one which was translated in the reports of the Cavendish Society,[4] I would much desire it if this investigation would also be translated[5] there or in some other English journal. Playfair or Hoffmann, to whom, as well as Frankland, I ask you to give my kindest regards, would certainly be happy to act as go-betweens in this matter.

In sincere friendship | Your | R. Bunsen

Breslau, 1st Sept. 1851.

RI MS JT/1/B/141

1. *amended testimonial*: there is no reference in letters or Tyndall's Journal which throws any light on the suggestion here that Bunsen had sent an earlier testimonial.
2. *by circumstance*: in letters 0481 and 0482 (n. 2) there are non-specific allusions to difficulties leading to Debus's departure.
3. *my work on volcanic rock formation*: Bunsen, sent this memoir by separate post. On receiving it Tyndall discovered it was very long (see letter 0526).
4. *translated in the reports of the Cavendish Society*: 'On the Intimate Connection existing between the Pseudo-Volcanic Phenomena of Iceland', in Thomas Graham, ed., *Chemical Reports and Memoirs* (London: the Cavendish Society, 1848), pp. 323–70.
5. *be translated*: Tyndall followed up this request with Francis (letters 0525 and 0526).

From Élie-François Wartmann   2 September 1851   0521

Monport, près Lausanne | 2 Sept. 1851

Mon cher Monsieur,

Votre lettre, datée du 2 Août, n'est arrivée à Genève que le 28. C'est ce qui vous expliquera le retard apparent de ma réponse.

Je vous adresse ci-joint, avec le plus grand plaisir, le certificat ou « testimonial » que vous me demandez. Je voudrais qu'il pût vous être utile et contribuer à vous faire obtenir la chaire que vous postulez. Mais il faudrait qu'il fût signé d'un homme plus connu dans la Science que je ne puis espérer l'être.

Recevez, mon cher Monsieur, l'assurance de mes meilleurs vœux pour vos succès. A Toronto, comme ailleurs, une belle carrière Scientifique s'ouvrira devant vous. Je ne doute pas que vous ne la parcouriez d'une manière honorable pour vous et profitable pour les autres. Toutes les fois que vous penserez que je puisse vous être de quelque utilité recourez amicalement à moi; et croyez à l'assurance de mes sentiments d'voués'

Elie Wartmann

Monport, near Lausanne | 2 Sept. 1851

My dear Sir,

Your letter[1] from August 2nd arrived in Geneva only on the 28th, which explains the apparent tardiness of my reply.

It is with the greatest pleasure that I attach the certificate or 'testimonial' which you asked for. I wish that it might be useful to you in that it might contribute to your getting the chair for which you have applied. However, it ought to have been signed by a man more renowned in science than I could hope to be.

Please accept, my dear sir, my best wishes for your success. In Toronto as elsewhere, you will surely have an excellent scientific career. I doubt not that you will continue in a way that brings honour to yourself and profit to others. Whenever you think that I may be of any assistance, appeal to me as to a friend;[2] and rest assured of my feelings for you.

Elie Wartmann

RI MS JT/1/W/4

---

1. *Your letter*: letter missing.
2. *a friend*: although these are formal expressions of friendship, they had a basis in a face-to-face meeting at the Ipswich BAAS meeting just two months previously (letter 0515, n. 3).

To Auguste Arthur de la Rive 4 September 1851 0522

Queenwood College | Stockbridge
Hampshire England. | 4th, Sept 1851

Dear Sir

Circumstances have conspired to place me in the attitude of a petitioner to your kindness.[1]

I intend to become a candidate for the professorship of physics now vacant in the university of Toronto, and as the support of your name cannot fail to have considerable influence in such a case I am induced to make an effort to secure that support.

You are already aware of my contributions to the philosophical magazine and to Poggendorff's Annalen.[2] In addition to the memoirs regarding crystals which you were kind enough to notice so favourably in the Bibliothèque Universelle[3] I beg to refer you to three others which appear in the February, April, and September *<word illegible>* of the Magazine.[4] I should have felt much pleasure in forwarding you the last mentioned by post but find the rate of postage so high as to amount to a prohibition. I trust however the magazine will reach you sufficiently early to enable you to read the memoir.

Should your leisure permit and should you deem my labours of sufficient value to justify the act on your part, you would place me under a great obligation to you by sending me a testimonial.

I shall forward my application to Toronto about the 1st of the next month. If therefore you grant me the honour of a reply any time within the present month[5] it will suit my purpose.

I remain Dear Sir | Most respectfully yours | John Tyndall

RI MS JT/1/TYP/1/1/333
LT Transcript Only

————

1. *Circumstances . . . your kindness*: compare with letters 0518 (n. 1) to Airy and 0523 to Plücker.
2. *contributions to . . . Annalen*: see letter 0512, n. 5.
3. *memoirs concerning crystals . . . Bibliothèque Universelle*: not identified.
4. *three others . . . the Magazine*: for the three memoirs in the *Phil. Mag.* see letter 0518, n. 2.
5. *within the present month*: De la Rive noted on the letter that he replied on 24 September and sent a certificate: 'Repondu le 24 7bre et envoyé un certificat' (see letters 0534 and 0538).

To Julius Plücker   4 September 1851   0523

Queenwood College | Stockbridge
Hampshire England | 4th September 1851

Dear Sir

Circumstances have conspired to place me in the attitude of a petitioner to your kindness.[1]

I intend to become a candidate for a professorship of physics now vacant in the University of Toronto, and as the testimony of one who ranks so high cannot fail to be of considerable influence on such I case, I am induced to make an effort to secure your support.

You know what I have done; it has been my misfortune to differ from you in some of your conclusions, but I believe this difference of opinion will have no influence upon my present application. I should have felt much pleasure in forwarding you a copy of my last investigation *[on]* 'Diamagnetism and magne-crystallic action'[2] which appears in the philosophical Magazine for the present month by post, but the rates of postage I find are so high as to amount to a prohibition.

Should your leisure permit and should you deem my labours of sufficient value to justify the act on your part you would place me under a great obligation to you by sending me a testimonial.[3]

with best wishes | I remain dear sir | most faithfully yours | John Tyndall
Prof Plücker

I shall forward my application to Toronto about the 1st of the next month. If therefore you grant me the honour of a reply any time within the present month it will suit my purpose.

J.T.

NRCC Plücker, Item #69 (Box 2/6/1)

1. *Circumstances . . . kindness*: compare letters 0518 and 0522 which have the same opening formula.
2. *investigation in 'Diamagnetism and magne-crystallic action'*: see letter 0498, n. 6. Tyndall also presented this paper at the BAAS meeting.
3. *a testimonial*: Plücker sent one (see letter 0549 to Francis).

From James P. Joule — 4 September 1851 — 0524

Acton Square Salford | Manchester Sep 4 1851

Dear Sir,

I have very great pleasure in enclosing a Testimonial[1] and sincerely hope you will succeed in obtaining a position which you are so qualified to fill.

Believe me | Dear Sir | Yours very truly | James P. Joule

Dr Tyndall | &c &c

I am much obliged for the papers which I have already seen in the Phil Mag[2] and studied with great interest[3]

RI MS JT/1/J/136

1. *enclosing a Testimonial*: the letter requesting a testimonial has not been found.
2. *the papers . . . Phil Mag*: see letter 0518, n. 2. Assuming that Tyndall sent the same papers with every testimonial request (to English speakers), this lets us know that they were all from the *Phil. Mag.*
3. *I am . . . great interest*: this postscript is written sideways, along the inner margin of the letter.

To William Francis — [6 September 1851][1] — 0525

Queenwood Saturday

My Dear Francis—

I just write to bespeak 10 or 12 pages (perhaps less) of the magazine for next month. I shall have a paper on Bismuth and its polarity,[2] which I think will at least establish the precise position of this subject at present—

Testimonials are coming in; an admirable one reached me this morning from Bunsen[3]—a capital one from Joule;[4] and rather a <u>mild</u> one from Wartmann[5] accompanied certainly by a very kind note—Du Bois[6] has sent me a most excellent one too—I have written to all those you mentioned[7] and expect answers in a day or two.

Bunsen mentions an article of his[8] which he wishes to get introduced into some English Journal and proposes that it might be done through Playfair or Hoffmann—Now I have reason to believe that the said article is of the highest interest, it is on the Composition of the Plutonic rocks of Iceland—would you not like to have it in the magazine? The memoir has not yet reached me but I expect it by tomorrow's post.

With kind greetings | believe me— | most truly yours | Tyndall

StBPL T&F, Authors' letters

1. *[6 September 1851]*: dated by references to testimonials: Bunsen, Wartmann and Joule had written on 1, 2 and 4 September, respectively (letters 0520, 0521 and 0524), which means this letter was written 5 September at the earliest. It was written before 11 September, when Tyndall received a letter (0527) from Airy and 13 September when he received a testimonial (letter 0528) from Thomson, as he would have mentioned these if he had received them already (receipt recorded in Tyndall's Journal JT/2/13b/549). The only Saturday between 5–11 September is 6 September.

2. *paper on Bismuth and its polarity*: Tyndall, 'On the Polarity of Bismuth, Including an Examination of the Magnetic Field', *Phil. Mag.* 2, no. 12 (November 1851), pp. 333–44.

3. *from Bunsen*: letter 0520.

4. *from Joule*: letter 0524.

5. *from Wartmann*: letter 0521.

6. *Du Bois*: letter 0515.

7. *all those you mentioned*: this letter from Francis is missing.

8. *an article of his*: cited letter 0526, n. 2.

To William Francis   [7 September 1851][1]   0526

Queenwood Sunday morng

Dear Francis.

Bunsen's Memoir[2] has just reached me—*[it]* is unfortunately rather long, taking up about 75 pages of the Annalen—What's to be done? if you don't make use of it I will send it off to Playfair and request him to have it translated in the reports of the Chemical Society.

Sincerely yours | J Tyndall

StBPL T&F, Authors' letters

1. *7 September 1851*: a Sunday, dated by reference to Bunsen's memoir. In letter 0525 (Saturday [6 September]) Tyndall said that he was expecting to receive Bunsen's memoir the following day.

2. *Bunsen's Memoir*: 'Ueber die Processe der vulkanischen Gesteinsbildungen Islands', *Poggend. Annal.* 83, no. 6 (1851), pp. 197–272. The length made editors reluctant to publish a translation. It was eventually published in 1853 in *Scientific Memoirs* (see letter 0666, n. 1).

From George Biddell Airy   10 September 1851   0527

Royal Observatory Greenwich | 1851 September 10

Sir,

It would give me pleasure to assist you in reference to the Toronto Professorship if it were in my power to do so.[1] Whether I can have the opportunity of doing it is not quite certain.

As a servant of the government, I am hardly competent to offer them a certificate in a matter which is not related to my own profession.

But I am prepared to respond to any inquiry which the government may address to me.[2]

You are therefore at liberty to refer the government to me: and I shall be happy if this gives me an opportunity of expressing the favourable opinion which I entertain of your qualification for the office.

I am, Sir, | Your obedient servant | G.B. Airy[3]
D[r] Tyndall[4]

RI MS JT/1/A/27

---

1. *assist . . . do so*: reply to letter 0518 in which Tyndall requested a testimonial. Tyndall replied in turn in letter 0532.
2. *respond to any enquiry. . . to me*: compare Faraday's response (letter 0506).
3. *Airy*: Tyndall copied this entire letter into his next letter to Francis (letter 0530).
4. *Tyndall*: this address form is located at the bottom of Airy's first page, rather than, as was then conventional, at the end of the letter.

From William Thomson   10 September 1851   0528

2 College, Glasgow | Sep 10, 1851[1]

Dear Sir

I enclose a short testimonial, according to your request, and I shall be very glad if you find it in any degree useful in promoting your present object. I see your letter is dated Sep. 1, and I regret the delay I have made in answering it which arose from my having been at a considerable distance from home when it was forwarded to me and only having returned today.[2]

I enclose also copies of some short papers on Electricity and Magnetism;[3] and by the same, or nearly the same post, you ought to receive some more extended papers on Magnetism, and on the Dynamical Theory of Heat,[4] all of which I shall be glad if you will accept.

If you read § 21 of my paper[5] (enclosed) entitled 'Remarks on the Forces &c.' you will perceive that the experiments described in the first part of your paper of this month[6] are particularly satisfactory to me. Indeed I have always felt very much inclined to believe that Plücker's 'loi générale' about magnetism decreasing less rapidly than diamagnetism was entirely a delusion, and I am still so inclined after reading your two last papers. Your experiments on the attraction of soft iron balls make it I think very improbable that Plücker has really observed the phenomenon indicated in § 23 of my 'Remarks &c.'[7]

I am very glad to see that, by your experiments[8] on the differences of force experienced by crystals of calcareous spar and other substances according to the directions in which they are turned with reference to the lines of force in a magnetic field, you have so amply confirmed the theory of magne-crystallic induction as suggested by Poisson, and by Faraday (2588), and verified experimentally by Faraday (2841)[9] for the single case of bismuth.

The only difference of opinion regarding the theory of magnetic induction which I am now aware to exist between us is regarding 'the influence of proximity' of which you speak.[10] Ever since May 1847 (See Cambridge & Dublin Math[1] Journal Vol II p. 235 § 12;[11] or British Association Report Swansea 1848 Physical Section p. 9[12]) I have been prepared to demonstrate that the effect of proximity among the particles of a diamagnetic powder is the reverse of what you assume it to be, but that it is so small as to be insensible in actual experiments. I think the very important experiments you describe in pages 19, 20, 21 of your last paper[13] demonstrate that the effects of compression which you observe are due to a molecular alteration of the substances, and they fully confirm the second of the conjectures which I threw out at Edinburgh[14] last year. I am quite ready to give up the first conjecture, the objection to it stated in p. 17 of your paper having occurred to myself as probably fatal to it, and your measurements (foot of p. 18)[15] being very decisive against it.

I hope before long to be able to write a short paper for the Philosophical Magazine,[16] explaining my views regarding form and proximity as affecting the bearing of single bodies or of groups, in a magnetic field.

I remain, Dear Sir, | Yours very faithfully | William Thomson
John Tyndall Esq

RI MS JT/1/T/10

1. *Sep 10, 1851*: Tyndall received this letter, along with letters from Francis and Dove (both missing), on 13 September (Journal, JT/2/13b/549).

2. *your request . . . returned today*: see letter 0519, n. 1.

3. *some short papers on Electricity and Magnetism*: only one of these papers can be identified with certainty, as it is mentioned below (n. 5). Thomson may also have sent a copy of his, 'On the Theory of Magnetic Induction in Crystalline and Non-Crystalline substances', *Phil. Mag.* 1, no. 2 (March 1851), pp. 177–86.

4. *more extended papers . . . Theory of Heat*: Thomson sent the magnetism papers that Tyndall told him he had not seen (letter 0519, n. 4). We have not identified the paper on the dynamical theory of heat.

5. *my paper*: Thomson, 'Remarks on the Forces experienced by inductively Magnetized Ferromagnetic or Diamagnetic Non-crystalline Substances', *Phil. Mag.* 37, no. 250 (October 1850), pp. 241–53. Thomson refers to pp. 250–52.

6. *your paper of this month*: Thompson alluded to Tyndall's paper 'On Diamagnetism' (cited 0498, n. 6), which Tyndall had sent him 10 days earlier (letter 0519, 1 Sept 1851), and which he had heard at the BA meeting. Tyndall discussed, and refuted, the general law that Thompson referred to here on pp. 165–7.

7. *the phenomena . . . my 'Remarks &c.'*: see n. 5; § 23 is on p. 253.

8. *by your experiments*: reported in 'On Diamagnetism' (see n. 6). Although there is no reference to Poisson in that paper, the detailed references to Faraday match Thomson's comments here.

9. *Faraday (2588), . . . (2841)*: the numbers refer to paragraphs in Faraday's series of articles: 'ERE 22 (continued)', p. 31 (§ 2588) ; 'ERE 26', p. 41 (§ 2841).

10. *difference of opinion . . . of which you speak*: in letter 0520 Tyndall told Thomson he had expected an 'onset' from him in response to his BAAS paper. Tyndall posits and provides evidence for 'the influence of proximity' on pp. 180–6 of his paper 'On Diamagnetism' (n. 6 above); Thompson quotes from p. 185.

11. *Cambridge . . . 12*: Thomson, 'On the forces experienced by small spheres under magnetic influence; and on some of the phenomena presented by diamagnetic substances', *Cambridge and Dublin Mathematical Journal*, 2, 1847, pp. 230–35.

12. *British Association . . . p. 9*: Thomson, 'On the Equilibrium of Magnetic or Diamagnetic Bodies of any Form, under the Influence of the Terrestrial Magnetic Force', *Brit. Assoc. Rep., 1848*, pp. 8–9.

13. *pages 19, 20, 21*: pages 183–5 of 'On Diamagnetism' as published in the *Phil. Mag.* (cited letter 0499, n. 6).

14. *Edinburgh*: at the Edinburgh BA meeting. Thomson gave a paper 'On the Theory of Magnetic Induction in Crystalline Substances' (*Brit. Assoc. Rep., 1850*, p. 23) in which he 'stated two principles, involving no physical hypothesis regarding the ultimate nature of magnetization'. The abstract only explains one—'the superposition of magnetic inductions'—which 'cannot be considered as fully established by experiment'. (The *Report* also noted that Thomson had discussed his research with Tyndall while at the meeting.) The paper was subsequently published in full in the *Phil. Mag.* (see n. 3).

15. *the objection to it stated in p. 17 . . . (foot of p. 18)*: 'On Diamagnetism', pp. 181 and 182 as published in the *Phil. Mag.* Tyndall quoted from, and replied to, Thomson on p. 181.

16. *short paper for the Philosophical Magazine*: this does not seem to have happened.

From James David Forbes   12 September 1851   0529

Craigieburn[1] | Moffat | 12 Sept 1851

My Dear Sir

I fear that you will think the enclosed Testimonial[2] rather meagre. But I make it a strict rule to confine myself in such cases to the limits of my personal knowledge of the facts.

I remain my dear Sir | Yours truly | James D. Forbes
John Tyndall Esq

RI MS JT/1/F/28

1. *Craigieburn*: Craigieburn House was about two miles from Moffat, a spa town 50 miles south of Edinburgh

2. *testimonial*: Tyndall's request to Forbes for a testimonial is missing.

To William Francis   [13 September 1851][1]   0530

Queenwood Saturday

My Dear Francis,

I was very glad to hear from you,[2] as I did not know but you had fallen in love, got swamped in the billows at Whitby, or done something else equally impudent with yourself but equally effective in introducing silence upon you. Accompanying this goes the one half of Beer's paper,[3] and the other half shall be in Red Lion Court almost by the time this reaches you. With regard to my own little paper[4] my object was to pack it off to Toronto along with the others and thus render the packet weightier. But if you find the slightest diffi- culty in making room for it by all manner of means exclude it—nay I should prefer this decision on your part for I could then make the matter much more perfect for the November number and you would perhaps be able to let me have a copy in time to send it off with the rest, as the Bursar's office is open to applications to the 19th Nov. Even here I would not press you, as it is almost a matter of indifference to me whether I send it at all or not. Now the case rests thus. If through my application you have been induced to make arrange- ments for the printing of the paper which you would find it inconvenient to alter then I will have it ready for you. If not then let it stand over until a more convenient season. I see the subject is fruitful and will give rise to many interesting speculations which are worth following up a little. Thomson wrote to me this morning and sent me a great number of papers,[5] among others his theory of magnetism. This is the most valuable part of his message to me. His testimonial is a closely clipped statement of facts—Joule's[6] is far better. I wrote to Airy and received the following reply.[7]

'It would give me pleasure to assist you in reference to the Toronto professorship if it were in my power to do so. Whether I can have the opportunity of doing it is not quite certain.

As a servant of the Government I am hardly competent to offer[8] them a certificate in a matter which is not related to my own profession. But I am prepared to respond to any enquiry which the government may address to me.

You are therefore at liberty to refer the government to me; and I shall be happy if this gives me an opportunity of expressing the favourable opinion which I entertain of your qualification for the office.' —

The matter however appears to be in the hands of the colonial government as I am not aware that the government at home takes an interest in the matter.

As yet I have received no reply from Whewell,[9] Grove,[10] Sabine[11] or Forbes.[12] I wrote to Liebig[13] also but I believe he has gone to Ireland with Reusch Rose & Magnus. I had Knoblauch down with me a few days ago. I trust he will be able to see you before he leaves this country.

with best wishes dear Francis | most faithfully yours | J Tyndall—

This morning a testimonial from Dove reached me[14]—I have a strong array from Germany but am as yet weak in England. This I hope to overcome in time, and one practical mode of overcoming it is the following up of your suggestion of coming to town now and then This I intend to do. | J.T.[15]

StBPL T&F, Authors' letters

---

1. *[13 September 1851]*: in this letter Tyndall thanked Francis for his letter, mentioned receiving Thomson's testimonial that morning and, in the postscript, mentioned receiving Dove's testimonial, also that morning. According to his journal these were all received on Saturday 13 September (JT/2/13b/549-50). Also, at the time of writing this letter he had not received Forbes's testimonial which, according to his journal, was received on 14 September (JT/2/13b/550). But, according to the same journal entry he wrote to Francis and sent half of Beer's paper that day. Rather than question the 13 September date, we assume Tyndall sent a second, missing letter with the second half of Beer's paper on 14 September.

2. *glad to hear from you*: letter missing, but Tyndall recorded receiving a letter from Francis on 13 September (Journal, ibid.).

3. *Beer's paper*: Beer, 'On the deduction of *Fresnel's* construction from the formulæ of *Cauchy* for the Motion of Light', *Phil. Mag.* 2, no. 11 (October 1851), pp. 297–303.

4. *my own little paper*: of 10–12 pages, see letter 0525, n. 2.

5.   *Thomson . . . papers*: see letter 0528.

6.   *Joule's*: see letter 0524.

7.   *I wrote to Airy . . . reply*: see letters 0518 and 0527. Tyndall quotes accurately from letter 0527, with only punctuation and capitalization differences.

8.   *offer*: at this point there is a page break and the two parts of the letter are separated in the Taylor and Francis archive.

9.   *Whewell*: request to Whewell is missing; for Whewell's reply see letter 0542.

10. *Grove*: request to Grove is missing; Grove eventually replied in letter 0547.

11. *Sabine*: request to Sabine is missing; Sabine replied in letter 0535.

12. *Forbes*: Forbes sent a testimonial with a letter written 12 September (0529). According to Tyndall's Journal (JT/2/13b/550), he received this letter on 14 September, that is, the day after writing this letter.

13. *I wrote to Liebig*: letter missing; it is not alluded to in other letters.

14. *testimonial from Dove reached me*: letter missing, but Tyndall recorded receiving it on 13 September (Journal, JT/2/13b/550).

15. *This morning . . . J.T.*: the entire postscript is written vertically across the second page of the letter.

To William Thomson   13 September 1851   0531

Queenwood 13th Sept 1851[1]

Dear Sir.

I am much obliged to you for the testimonial and still more for your valuable papers which I shall undoubtedly read with profit and pleasure.[2] I am sorry to say that during my last investigation in Berlin I was compelled so to devote myself to the experimental portion of the subject that the literature of it was comparatively neglected. Otherwise I should in writing my paper have referred to the close connection which subsists between the theoretic views advanced by you in the March number of the Philosophical Magazine[3] and my experiments—this neglect however I shall take care to redeem at some future time.

Your proof as to the influence of proximity would interest me extremely— As yet I have met with nothing which seriously militates against my notions on the subject. I am prepared at once to give them up if I see sufficient reason, but at present I have strong reasons for believing that the effect of proximity cannot be the reverse of what I have stated it to be.

I shall look with great interest to your promised paper in the Philosophical Magazine.[4]

Prof. Thomson. | &c. &c.

RI MS JT/1/TYP/5/1534
Transcript Only

———————

1. *13ᵗʰ Sept 1851*: Thomson wrote 'Sept 15', the date he received it, at the top of the letter.
2. *the testimonial and . . . papers*: Thomson had sent copies of memoirs and a long letter with the testimonial (see letter 0528).
3. *the theoretic views . . . Philosophical Magazine*: Thomson's articles are cited ibid, nn. 3 and 14.
4. *promised paper in the Philosophical Magazine*: ibid, n. 16.

To George Biddell Airy   13 [or 14]¹ September 1851   0532

Queenwood Stockbridge | 13ᵗʰ <u>Sept</u>. 1851

Sir.

I was induced to address you² because I knew that you were a physicist as well as an astronomer. I have now to return you my sincere thanks for the kind permission which you have granted me.³ Even should you never be referred to in connexion with the subject on which I have presumed to address you, it remains a fact full of encouragement and pleasure to me to know that I have been fortunate enough to secure the favourable opinion of an authority so exalted.

I am Sir | Your obliged & humble servant | John Tyndall
Prof. Airy | &c &c

RGO MS.RGO 6/373.413

———————

1. *13 [or 14]*: according to Tyndall's journal, he wrote this letter on 14 September (JT/2/13b/550).
2. *address you*: in letter 0518 Tyndall asked Airy for a testimonial.
3. *kind permission . . . granted me*: in letter 0527.

From Hermann Knoblauch   20 September 1851   0533

London, Morley's Hotel | Spt 20. 51

Prof. Tyndall, | Queenwood College.

Verehrtester Freund

Seit meiner Rückehr hierher habe ich mir täglich vorgenommen Ihnen zu schreiben und Ihnen sowie <u>Herrn Edmondson</u> und <u>seiner Familie</u> für die freundliche Aufnahme zu danken, welche mir in Queenwood-College zu Theil geworden ist. Ich rechne die Stunden, welche ich dort verlebt habe, <u>zu den schönsten</u> meines Aufenthalts in England. Queenwood wird mir <u>unvergesslich</u> sein! Wir denken am Dienstag von hier nach Edinburgh, Liverpool etc. abzureisen und etwa nach 8 Tagen wieder hier durchzupassiren, um unsre Pässe zur Rückehr nach dem Continent zu nehmen. Magnus und Rose sind eben von ihrer Reise nach Schottland u. Irland zurückgekehrt und bleiben noch einige Tage (unbestimmt wie lange) in London. Vielleicht gehen sie noch nach Brighton. (Auch Magnus bedauert es, dass Sie nach America gehen wollen.)

Frankland habe ich vergeblich in Putney aufgesucht, das er bereits seit 3 Wochen verlassen hatte. Wheatstone ist in der Isle of Whight. Dagegen habe ich Potter und durch ihn das University College so wie Brooke und seine sinnreichen selbstregistrirenden Instrumente gesehen.

Des Uebrigen haben wir hier noch so vieles kennen gelernt, dass es unmöglich wäre es mit 2 Worten anzuführen. London ist an Allem unendlich reich, nur eins habe ich vermisst: elegante physikalische Apparate. In dieser Beziehung können wir die Concurrenz in Deutschland aushalten. Bei Watkins u. Hill fand ich einige optische Instrumente von Interesse für mich, Darker will ich noch besuchen.

Leben Sie recht wohl. Grüssen Sie Debus herzlich von mir und empfehlen Sie mich Herrn, Frau u. Fräulein Edmondson. I shall ever remember the kindness and friendship, I have experienced in Queenwood-College with feelings of the sincerest gratitude and affection, more than it was perhaps expressed by me on parting in your language, foreign to me.

Mit den besten Wünschen für Ihr Wohl, und der Bitte von Zeit zu Zeit Nachricht von sich zu geben,

Ihr | treuer Freund. | Herm. Knoblauch.

Mein Vater grüsst Sie freundlichst und dankt *[Hrn.]* <u>Edmondson</u> für die Gastlichkeit mit der er uns <u>beide</u> bei sich aufnehmen wollte.

London, Morley's Hotel | Spt 20. 51

Prof. Tyndall, | Queenwood College.

Dearest Friend]

Since my return here I have been intending every day to write to you and to thank you as well as <u>Herr Edmondson</u> and <u>his family</u> for the kind welcome which was given to me in Queenwood College.[1] I count the hours I spent there <u>among the</u> <u>nicest</u> of my stay in England. I shall <u>never forget</u> Queenwood! We[2] are thinking of travelling from here on Tuesday to Edinburgh, Liverpool etc. and of passing through here again after about 8 days in order to collect our passports for the return journey to the continent. Magnus and Rose have just returned from their trip to Scotland and Ireland and will stay for a few more days (uncertain how long) in London. They may still go to Brighton. (Magnus is also sorry that you want to go to America.)[3]

I looked up Frankland in Putney, but to no avail;[4] he had already left there 3 weeks ago. Wheatstone is on the Isle of Wight. However, I did see Potter[5] and through him University College, as well as Brooke[6] and his ingenious self-recording instruments.[7]

Other than that, we have learnt so much here that it would be impossible to list it all in a couple of words. London is infinitely rich in everything, and there was just one thing I did not see: elegant physics apparatus. In this respect we can maintain the competition in Germany. I found some optical instruments of interest to me at Watkins and Hill's,[8] and I still want to visit Darker.[9]

Farewell. Give Debus my kind regards and give my compliments to Herr, Frau and Fräulein Edmondson. I shall ever remember the kindness and the friendship I have experienced in Queenwood-College with feelings of the sincerest gratitude and affection, more than it was perhaps expressed by me on parting in your language, foreign to me.

With the best wishes for your well-being, and the request that you give me news of yourself from time to time,

Your | faithful friend. | Herm. Knoblauch.

My father sends you his kindest regards and thanks *[Herr]* <u>Edmondson</u> for the hospitality with which he wanted to welcome us <u>both</u> at his place.

RI MS JT/1/K/17

---

1. *in Queenwood College*: according to Tyndall's journal, Knoblauch had arrived at Queenwood on Sunday 7 September and stayed one night (see entry of 9 November, JT/2/13b/549).

2. *We*: Knoblauch was travelling with his father.

3. *Magnus is also . . . America*: Knoblauch wrote this note in the margin beside the previous two sentences.

4. *Putney, but to no avail*: Edward and Sophie Frankland had stayed in Putney for the month of August in order to visit the Great Exhibition (Russell, *Edward Frankland*, pp. 186–7).

5. *Potter*: Richard Potter (1799–1886), natural philosopher. Born in Manchester; attended Queen's College, Cambridge (BA, 1838; MA, 1841); professor of natural philosophy and astronomy at University College, London 1841–42 and 1844–65 (*ODNB*).

6. *Brooke*: Charles Brooke (1804–79), surgeon and inventor of measuring instruments; FRS 1847; worked in London, at the Metropolitan Free Hospital and then Westminster Hospital (*ODNB*).

7. *self-recording instruments*: consisted of barometers, thermometers, psychrometers, and magnetometers that automatically registered variations using photography. This reduced

the number of assistants needed. They were employed at the observatories of Greenwich, Paris, Toronto, and Cambridge, MA (ibid., *ODNB*).

8. *Watkins and Hill's*: Watkins and Hills, an optical, mathematical and philosophical instrument maker, trading at 5 Charing Cross, London. See *ODNB* entry for Francis Watkins (1723–91) for details on the firm.
9. *Darker*: William H. Darker, optician and optical lapidary, and seller of scientific instruments, Paradise St, Lambeth, London.

## From Augustus de la Rive		24 September 1851		0534

Presinge. Sep. 24, 1851

Monsieur,

Je regrette de n'avoir pas répondu plus tôt à votre lettre du 4 Septembre; ce retard tient à des tristes circonstances dans lesquelles je me trouve et qui ne me laissent pas toujours toute ma liberté d'agir. Je ne connais pas la forme qu'il faut donner au <u>témoignage</u> que vous me demandez; j'espère que celui que je mets sous ce pli remplira votre désir; je l'ai fait avec d'autant plus de plaisir que je n'y ai exprimé que ce que je pense sans la moindre exagération, et j'ai été heureux d'avoir une occasion de vous le faire savoir et de vous dire avec quel véritable intérêt j'ai lu vos recherches.

Je désire vivement, Monsieur, que vous obteniez la place à laquelle vous aspirez quoiqu'elle doive vous éloigner terriblement de nous; mais j'espère qu'elle ne vous empêchera pas de continuer à vous livrer à vos importants et intéressants travaux.

Agréez, Monsieur, l'assurance de ma considération distinguée et de mes sentiments dévoués.

Aug. de la Rive.

Presinge.[1] September 24th, 1851

Sir,

I regret not having responded earlier to your letter of September 4th;[2] this delay is due to sad circumstances[3] in which I find myself, and which do not always allow me the freedom to act. I do not know the form that should be given to the <u>evidence</u> that you ask for; I hope that the one that I placed in this envelope will fulfill your wishes. I have done it with even more pleasure since I expressed my thoughts without the slightest exaggeration. And I was glad to have the opportunity to let you know my thoughts about you and tell you that I read your research with much genuine interest.

I truly hope, Sir, that you will obtain the position to which you aspire, even though it would take you so far away from us; but I hope that it will not prevent you from continuing to dedicate yourself to your important and interesting works.

Please accept, Sir, the assurance of my highest consideration and devoted sentiments.

Aug. de la Rive.

RI MS JT/1/TYP/1/334
LT Transcript Only

———————

1. *Presinge*: a village about 7 miles NE of Geneva on the French border.
2. *your letter ... 4th*: letter 0522, in which Tyndall requested a testimonial.
3. *sad circumstances*: not identified.

From Edward Sabine   27 September 1851   0535

Woolwich | Sept. 27/51

Dear Sir,

I have to apologise for the delay in replying to your note of the 1st of Sept.,[1] but I was desirous of reading over carefully the memoirs[2] which you were so obliging as to send for my perusal, and, having been much pressed by duties, I have been unable to command time to do so until now. I was moreover aware that a reply any time in this month would be sufficiently early for your digest.

Permit me to say that I could not read your writings without feeling some degree of regret at the prospect of the separation, more or less, which the distance of Toronto must occasion, from the intercourse which you have so happily cultivated with some of the most distinguished Physicists of Europe,[3] and which is so valuable as a stimulus. I trust however that you will find a residence in America not incompatible with very frequent visits to Europe.

I remain, Dear Sir | very sincerely Yours | Edward Sabine
Dr. Tyndall

RI MS JT/1/TYP/4/1295
LT Transcript Only

———————

1. *your note ... Sept*: letter missing.
2. *the memoirs*: presumably the same memoirs that Tyndall sent with other requests for testimonials (see letter 0518, n. 2).
3. *Permit ... Europe*: Tyndall quoted from Sabine's letter in letters to Francis (0536) and Hirst (0540). Notes to those letters identify the differences.

To William Francis   [29 September 1851][1]   0536

Queenwood monday—

My Dear Francis,

I send you two more testimonials one from Forbes[2] the other from Sabine[3]—the following note accompanied Sabine's.

[. . .][4]

Is it not kind?—I have written a reply to him[5] in the spirit which his beautiful letter aroused. told him of my writing to Faraday and the result.[6] Stated that though I should cheerfully go to Toronto if the honour of the lot fell upon me still I would willingly remain in my present position for a year or two, if a reasonable prospect awaited me of finding in these kingdoms the means of carrying on my investigations and of supplying my material wants.

I will send you either the whole or the greater part of my paper on bismuth polarity tomorrow.[7]

If Sir John Herschell's Treatise on Light, or Faraday's experimental researches on electricity, London 1839[8] should fall in your way <u>second hand</u> I should feel greatly obliged if you would purchase them for me.

Most faithfully yours, | Tyndall

I know that Herschell is to be had at Maynard's Earl's Court Leicester Sq.[9] price *[10/6]* if I mistake it not.[10]

StBPL T&F, Authors' letters

---

1. *[29 September 1851]*: the earliest possible Monday for this letter was 29 September 1851, for Tyndall had received a testimonial from Sabine posted on 27 September (letter 0535). Also, the allusion to his paper on bismuth (n. 8) indicates that this letter was written shortly before letter 0538, which was written on 30 September.

2. *from Forbes*: Forbes sent a testimonial with letter 0529 (which Tyndall received on 14 September). It is unusual for Tyndall to have been so slow forwarding a testimonial to Francis, but it was a modest testimonial. There are no extant letters from Tyndall to Francis between 13 September and this letter.

3. *from Sabine*: Sabine sent a testimonial with letter 0535.

4. *Dear Sir . . . Sabine*: Tyndall copied letter 0535 here. There are only minor differences in punctuation and capitalisation between the LT transcript (letter 0535) and the copy made by Tyndall; all such variations could be due to LT standardising punctuation in the typescript. For example, note that Tyndall uses D^r rather than Dear; from other letters we know that Sabine used such abbreviations.

5. *written a reply to him*: letter missing. The following account of the missing letter appears in similar words in letter 0540 to Hirst.

6. *writing to Faraday and the result*: see letters 0505 and 0506.

7. *my paper on bismuth polarity*: cited letter 0525, n. 2.   The bulk of the paper was sent with letter 0538. The last few pages of the paper were sent with letter 0539.

8. *Sir John Herschell's . . . on electricity*: Herschel's treatise on light was originally a long essay within the *Encyclopaedia Metropolitana*, vol. 4 (1827), pp. 341–586, and was republished in various collections in the following twenty years. Tyndall also requested the first of three volumes of journal articles by Faraday (*ERE*, 1).

9. *Maynard's*: Samuel Maynard, a bookseller who specialised in mathematical and philosophical books, traded at 8 Earl's Court.

10. *I know . . . it not*: this postscript was written vertically down the left hand edge of the last page of the letter.

From John Hall Gladstone 29 September [1851][1] 0537

Clevedon | Sep. 29th

My dear Dr. Tyndall,

I received your warm impromptu reply to my letter[2] in due course, and was much pleased by the friendly—brotherly—feeling which pervaded it. I thought it probable that I should receive another note explaining the scheme of redemption, in which you believe, as you promised again to write to me when you found your mind in a suitable condition. However as no such note has come, and as you say that you will always be glad to hear from me, I thought I would just remind you of it.

Although I cannot subscribe to every sentiment of your warm-hearted note, I can most thoroughly sympathize with the general feeling, and the earnest aspirations, and the kind wishes expressed in it. If indeed we be both 'born of the Spirit'[3] as you evidently feel confidant is the case, I do feel deeply interested in knowing how such subjective similarity can coexist with such great dis-similarity in our objective faith.[4] No doubt you have some positive belief, else you a sinner could not have that happiness in the contemplation of God, which you possess. No doubt too you seek for truth in this matter as in your more philosophical pursuits, seeking honestly to ascertain what it is that God has revealed—for certainly this subject lies far beyond the reach of our unaided reason.

Well I know, my dear friend, that all our conceptions of the sublime truths of religion must be extremely faulty, and my own theory is of course more or less incorrect.

I have been staying a while at Dr. Leeson's[5] country residence at Bonchurch;[6] now I am here in Somersetshire: on Wednesday our session at St. Thomas's Hosp. commences.

With Christian affection I remain | Most sincerely yours | J. H. Gladstone.

RI MS JT/1/TYP/1/404
LT Transcript Only

---

1. *[1851]*: the relationship of this letter to letters 0511 and 0513 between Gladstone and Tyndall dates this letter to 1851.

2. *your . . . reply to my letter*: letters 0513 (Tyndall's reply) and 0511 (Gladstone's letter).

3. *born of the spirit*: Gladstone is repeating Tyndall's quotation from John 3:8.

4. *great dissimilarity in our objective faith*: one of the many indications in this letter that Gladstone recognised, as he did not in letter 0511, the extent of the differences between his faith and Tyndall's.

5. *Dr. Leeson's*: Henry Beaumont Leeson (1803–72), FRS 1849; physician and lecturer on chemistry and forensic medicine (1840–52) at St Thomas's Hospital in London, where Gladstone was lecturer in chemistry.

6. *Bonchurch*: a village on the south coast of the Isle of Wight.

To William Francis   [30 September 1851][1]   0538

My Dear Francis.

I send you the paper upon the polarity of Bismuth;[2] with the exception of three or four pages which I shall probably forward to you tomorrow. I dont think the whole will occupy more than 12 pages of the magazine.

I send you a letter and testimonial from De la Rive,[3] both very valuable. Also a document from the president of the University of Toronto.[4] You will observe his remark regarding the authenticity of the documents. My notion was to forward the originals at the same time but I should certainly not like to lose them.

The translations will probably need authentication—and I know of nobody who can do this except yourself. Do you think you could manage it? —I think a testimonial from you also would be of service to me.

faithfully yours | J Tyndall

It would be well I think to print those French testimonials in French—dont you think so? My object is to get all out of my hands and on the way to Toronto by the 10th of October.[5]

13/6	Faraday[6]	vol I	1839
6/9		vol II	
1£[7]	sell for 27/6		

1. *[30 September 1851]*: in letter 0536 (29 September) Tyndall promised to send 'the whole or the greater part' of his paper on bismuth 'tomorrow', suggesting this letter was written on 30 September. This letter implies he had not yet finished the paper. The last few pages of the paper were sent the following day (1 October), with letter 0539.
2. *paper on the polarity of Bismuth*: cited letter 0525, n 2.
3. *letter and testimonial from De la Rive*: see letter 0522 for Tyndall's request to De la Rive and letter 0534 (24 September) for De la Rive's reply. This letter had not reached Queenwood by the time on 29 September that Tyndall wrote letter 0536.
4. *document . . . Toronto*: missing.
5. *October*: the end of this letter is missing.
6. *Faraday*: It is possible that the note on prices was written by Francis in the process of carrying out Tyndall's request, in letter 0536, that Francis buy him a second-hand copy of the first volume of Faraday's *ERE*. Although there is no direct evidence of Tyndall requesting the second volume, in letter 0569 he asked Francis to send both. The price difference may allude to the difference between second-hand and new prices.
7. *1£*: the precise total is £1.0.3d.

To William Francis [1 October 1851][1] 0539

My Dear Francis.

Gott sei dank die Arbeit ist jetzt fertig[2]—I send you the last pages of the paper[3]—I forgot to note the closing number of the paragraphs[4] and must beg of you to continue the numbers into the portion which I now send you—My brain is very tired and I will now give it two days rest, if it will take them which is a problem—

Sincerely yours | J Tyndall

StBPL T&F, Authors' letters

1. *[1 October 1851]*: This letter is dated by the allusion to Tyndall's paper on the polarity of bismuth (see n. 3). Tyndall posted the greater part of that paper on 29 September (letter 0536) and expected to send the remainder the following day. According to his journal (JT/2/13b/550), he finished the paper late at night on 30 September / 1 October and, according to letter 0540 (1–2 October) he posted the remainder of the paper that day. This letter was written on 1 October to accompany the paper.
2. *Gott . . . fertig*: Thank God the work is now finished (German).
3. *the paper*: Tyndall's paper on the polarity of bismuth (cited letter 0525, n. 2). While this letter does not identify the paper sent, the paper on bismuth was sent in two parts and in the journal entry in which Tyndall recorded sending the last pages of the paper on bismuth he wrote a similar comment on tiredness: 'my body was completely used up. I will now take two days rest—If I can'.

4.	*closing number of paragraphs*: that is the last paragraph number in the portion of the paper
	sent with letter 0538.

## To Thomas Archer Hirst	1–2 October 1851	0540

Queenwood Oct 1 1851. in the evening of a
stormy day. all alone, fire crackling before me.

My Dear Tom.[1]

I did not intentionally shut you out of my thoughts since your letter[2]
reached me, nevertheless you were not too often in them; sometimes in the
quiet evening I asked myself 'where is Tom now?' but you soon vanished like
a dissolving view in Swiss mist and I saw no more of you. Although however
not in the foreground of my thoughts, although I was as unconscious of you
as of my ear or my eye still were the question put directly to me I dare say I
should avow the same concern for you as for either. The fact is Tom I have
been working terribly hard.[3] I had three irons in the fire which demanded
constant attention—The school and lectures, the Toronto affair, and a little
investigation on 'diamagnetic polarity' the last of which I sent off today.[4] I
have been compelled to start before day in the morning and to work till late
at night, and the gods in their goodness have granted me barely strength to
finish the enquiry, did it endure much longer I should have been obliged to
yield for awhile, for never in my life did I lay down my pen with greater physical prostration than last night when the last work was written. Even this day's
quiet[5] has however done wonders for me. I have swung and jumped and forgotten every thing possible so that I find the machine[6] is again righting itself
and will be ready for another grind by and by. Poor Mrs Edmondson says to
her husband 'I'm concerned for that young man—he looks like an old one—
he will most certainly break down—in fact one needs only to look at him to
see that he is broken down already'—But they know how tough[7] I am—

The Toronto affair is going on in a proper manner. I am well supported.
As far as England, France, Germany, and Switzerland are concerned I have
strong testimony from the principle men in each, From Becquerel, Dela Rive,
Faraday, Sabine, Joule, and all the host of Berlin. Magnus, Dove, DuBois-Rey-
mond, Riess &c. Bunsen Frankenheim &c. &c. When I get that milk and
water little article which your small friend at the summit of the hill sent me[8]
beside some of these the contrast tickles one's fancy. Col. Sabine made the following remark in a note[9] which accompanied his certificate. —'Permit me to
say that I could not read your writings without feeling some degree of regret
at the separation which a residence in America will necessarily cause[10] from
the intercourse which you have so happily cultivated with some of the most

distinguished philosophers[11] of Europe.' I wrote to him back[12] and told him of my asking Faraday's opinion saying in conclusion—'I shall cheerfully go to Toronto if the honour of the lot fall on me, with the determination to do my best to sustain the credit of my department. Nevertheless I will say that I should willingly remain a year or two in my present position if at the end of that time a reasonable prospect awaited me of obtaining in these kingdoms the means of carrying out my investigations and of supplying my material wants'.

The rain is plashing outside, where are you now? Trudging along an alpine road or perhaps comfortably housed beside a Swiss stove. Knoblauch has told me Tom that you look miserable—I'm glad that you have looked miserable provided that the misery of your look is now gone—Remember Tom it is your duty to take care of that bodily instrument of yours. A man's value consists of his force and his capability of applying it. a vast deal of force may be silenced by an imperfect mechanism—I have no rule to give, you are bound to work;[13] but never forget that you are bound to work rationally—Remember also Tom that the principle of duty may take a morbid form, and become a devilish leech ever crying 'give', 'give'[14]—You must elevate yourself above this: you must be able to say to it sometimes 'no I won't give!'—A little daring rebellion of this kind has its value at times, though there are not many in the world who would bear to be told so.

I had Knoblauch down with me[15]—he stopped only one night. I expected his father also, but he did not come[16]—I must say it gave me great satisfaction to see him within the halls of Queenwood. He is now in Edinburgh but will probably return in a few days. You are aware I dare say that January[17] has closed a partnership with Tom Perkinton and has had to pay £300 for the privilege—Carter has taken Booth of Keithly[18] into partnership,[19] on what terms I know not.

Things looked very torpid when I came to Queenwood. there was no energy, no effort, the number of the pupils was low and altogether matters looked disheartening—There was[20] a number of farmers here who were like untamed colts, and the superintendence of this precious lot during the evening fell upon me. I took them together and explained my position to them got them all to promise that they would not enter a public house nor visit any of the neighbouring villages without permission. told them that they would ever find me their supporter in all rational demands which they wished to make of M[r] Edmondson, but also an obstinate opponent if circumstances called for it—For a few days things went on well, but the fellows had been accustomed to threatenings of M[r] Edmondson and to the nonfulfillment of these threatenings and doubtless imagined that they might get over me just as easily—One night a parcel of them came in in outrageous disorder—I followed them into their room—they seemed determined to shew me that they cared nothing about me, went on with their slang and their laughter as if they wished to provoke me—I listened to them for 2 hours without saying a word. They were at length silent when they saw me determined to see it out. By degrees they dropped off to bed the ringleader being the last to leave the room—I laid hold of him sat down and looking coolly into his face told him that if after that night I found him to cross the threshold of a public house, or commit even what might under circumstances be called a slight error I would resign my situation at Queenwood or he should be expelled. I broke the others up into parties next morning and took them in order, thus weakening the influence which mere numbers has with such fellows. told them the rules of our living together. 'If you cant abide by these' I said 'the most graceful exit you can make is to pack your trunks and leave immediately for most certainly if you infringe the rules I will do all that in me lies to have you expelled' of course this is the merest skeleton of my remarks—suffice it to say insubordination has been smashed to atoms; fellows which were the most goodfornothing scamps last half year are working—we stand on the most friendly terms and every thing wears the appearance of abiding good order.

The rain plashes Tom, where are you? Across waves and mountains I speak to you—I raise my eyes and right before me is a picture of Faraday, a noble countenance and withal a dash of sadness in it. M[r] Edmondson noticed this yesterday and said he could not understand why it should be so—I believe him—'To you it is given to know the mysteries of the Kingdom of heaven but to them it is not given'[21] He indulged in some eloquent wind about the beauty of the world &c and so I left him.

Debus made his first attempt at a lecture this evening and got through it well. He had a good many experiments which of course helped him. I never see the Leader[22] now and I am thinking of subscribing for it.

Morning of oct. 2nd

A watery looking morning: but up to the present time we have had fabulous weather. the most glorious sunshine day after day. A clear blue sky overhead and at night the firmament fretted by its star-spangles. If you have only had the same in Switzerland you will have been content with your journey. John Yeats met you![23]—well well how people knock against each other in this huge world. I shall now await your account of yourself, of your journey and your health—remember Tom what I have said about the latter. A few years will harden you and you may then go farther than you dare do at present. I have no more to say. —good bye Tom.

Your affectionate | Tyndall

RI MS JT/1/T/545

1. *Tom.*: In this letter it is particularly difficult to distinguish commas and stops. The apparent punctuation is often grammatically incorrect.

2. *your letter*: probably letter 0516, in which Hirst told Tyndall of his holiday plans.

3. *hard*: ten to twelve words are thoroughly scratched out here.

4. *investigation on 'diamagnetic polarity'*: see letter 0525, n. 2.

5. *this day's quiet*: the rest that Tyndall was taking after finishing the paper on diamagnetism.

6. *the machine*: that is, the bodily machine.

7. *know how tough*: Tyndall may have intended to write 'know not how tough'.

8. *milk and water little article . . . sent me*: probably refers to a bland testimonial, perhaps the one from Gerling, which Tyndall considered a 'very dilute affair' (see letter 0510).

9. *Col. Sabine . . . a note*: letter 0535. There are significant differences between the LT typescript of Sabine's letter and the quote here. See nn. 10 and 11 for details. There is no extant original MS, so it is possible that some differences were introduced by LT (or her clerk). It is unlikely, though, that she added phrases; more likely Tyndall omitted them.

10. *at the separation . . . necessarily cause*: the LT typescript of Sabine's letter reads: 'at the prospect of the separation, more or less, which the distance of Toronto must occasion'.

11. *philosophers*: in the LT typescript of the original letter this is 'Physicists'.

12. *I wrote to him back* : letter missing. Tyndall uses very similar words in his account to Francis (letter 0536).

13. *A man's value . . . bound to work*: Tyndall was here giving a metaphorical physical flavor to Carlyle.

14. *devilish leech . . . 'give'*: allusion to Carlyle, who often criticised the inner and outer demands to 'Give! Give!' He used the phrase combined with the leech metaphor of Jewish money-lenders (*Past and Present*, ch. 2.4).

15. *I had Knoblauch down with me*: Knoblauch spent the night of Sunday 7 September at Queenwood; see Journal, 9 September (JT/2/13b/549).

16. *he did not come*: both Knoblauch and his father had been invited (letter 0533), and Tyndall seems to have expected both.

17. *January*: January Searle, or G. S. Phillips.

18. *Booth of Keithly*: Tyndall may have meant Keighley, a town in West Yorkshire about 12 miles north of Halifax. Booth, a land surveyor, is not further identified.

19. *Carter . . . partnership*: Richard Carter, the surveyor who had employed both Tyndall and Hirst in Halifax.

20. *There was*: Tyndall originally wrote 'were' but corrected it to 'was'.

21. *To you . . . not given*: Mathew 13:11. It is the answer given by Jesus when asked by his disciples why he spoke to the crowd in parables. 'He answered and said unto them, Because it is given unto you to know the mysteries of the kingdom of heaven, but to them it is not given'. There is a very similar passage in Luke 8:10. Tyndall used the analogy of Edmondson with the crowd as a criticism of Edmondson.

22. *the Leader*: see letter 0398, n. 8.

23. *John Yeats met you!*: Hirst ran into Yeats, whom Tyndall disliked when teaching at Queenwood in 1847–48, in Lausanne during his walking tour in the Alps (Hirst, 'Journals', 19 September 1851). Tyndall's knowledge that Yeats and Hirst had met may have come from Hirst or Tyndall may have heard from a mutual friend. The rest of this postscript and Hirst's later letter (0553) suggest that Tyndall had yet to hear anything directly from Hirst about his trip.

From Lyon Playfair 3 October 1851 0541

London 3ᵈ Oct/51

Dear Sir

On my return to town I found your letter of the 17[1] enclosing a copy of Bunsen's paper.[2] Pray accept my best thanks for the trouble you have taken—
Believe me to be | Yours truly | Lyon Playfair
John Tyndall Esqʳ

RI MS JT/1/P/112

1. *your letter of the 17*: missing.
2. *copy of Bunsen's paper*: the paper sent by Bunsen (see letter 0520) was sent by Tyndall to Playfair because it was too long for the *Phil. Mag.* (see letters 0525 and 0526).

From William Whewell 9 October 1851 0542

Trinity Lodge Cambridge | Oct. 9 1851

Dear Sir,

I am so frequently called upon to give testimonials[1] to persons who have an official claim upon me and find it so difficult to do so to my satisfaction even in such cases that *[I]* must seek to be excused from the task in any other cases.

At the same time I may say that from the printed papers[2] which you have sent me it is apparent that you are well acquainted with the recent progress and present state of Natural Philosophy, and are occupied in promoting its further advances. If the expression of this opinion on my part is likely to be of any use to you you are at liberty to employ it
I am dear Sir | Your very faithful Servant | W. Whewell
Dr. Tyndall

RI MS JT/2/13b/ 716–7
LT Transcript of JT Transcript Only

---

1. *called upon to give testimonials*: Tyndall's request is missing. It was sent before letter 0530, in which he was waiting for a reply from Whewell; probably early in September when he sent out other requests (letters 0518, 0519, 0522, and 0523).
2. *the printed papers*: presumably the same as those sent with other requests for testimonials (letter 0518, n. 2).

To William Francis   [early-mid October 1851][1]   0543

I feel the value of remaining in England; the 'stimulus', as Sabine calls it,[2] caused by contact with men of science—I indulge in the hope of being able by hard work to climb to an average height as a man of science, but you know yourself how often such men are neglected in England and hence I imagine that it would be unwise to let the present opportun<ity> pass. —Give me a rational hope in England and I will yield up Toronto.

As ever Dear Francis yours | John Tyndall

StBPL T&F, Authors' letters

---

1. *[early-mid October 1851]*: This letter could have been written any time after 29 September when Tyndall first told Francis about the quoted letter by Sabine (letter 0535 and n. 2 below). The sentiment expressed suggests it was probably written soon after this date. 'The value of remaining in England' reads like a reply to a comment made by Francis but there are no extant letters from Francis until 21 October. The letter probably predates discussion of the Cork position in mid-late October. The manuscript is a fragment only. In content, it might follow from the next Francis letter (0544) in this sequence (which lacks a closing salutation), but the sheets of paper are of different sizes. All these considerations fall far short of certainty, and are merely the best guess based on available information.
2. *as Sabine calls it*: Tyndall alludes to letter 0535.

To William Francis   [c. 10 October 1851][1]   0544

The 'additions which accompany them'[2] are the memoirs[3] and my diploma—would you not think it well to send both? A couple of scraps from the Athenaeum and Literary Gazette, referring to the affair at Edinburg[4] last year might perhaps be sent also.—

I think the matter will be best arranged thus:—tomorrow you will send me the slips;[5] this will be sufficient for me. I will send you a copy of each of the four memoirs which I have in my possession together with the diploma; that upon the polarity of bismuth[6] and the testimonials may be added to the number; and thus if you would be kind enough to have them all packed up and posted for me the matter would be finally settled.—

If you think it as well to omit the diploma I shall be satisfied with your decision.

You also will be the best judge of the no of copies of testimonials—10, 12, 15 or 20 *[just]* as you please.

I think the matter is thus clear—you have a portion of the material; what remains over and above I will send you it remains then to tie them all together and send them adrift. —The name of the President[7] is you are aware John M'Caul.

Should I hear from Grove and Wheatstone[8] their communications may be sent afterwards.

How I should like to be beside you and Knoblauch[9] tomorrow—They are all teetollers down here,[10] otherwise I should pledge the health of both you in a bumper[11] I shall write to him to Morley's[12] tomorrow. It will reach him in time.

StBPL T&F, Authors' letters

---

1. *[c. 10 October 1851]*: This letter, like many other October letters, is difficult to date: it is incomplete (but see 0543, n. 1); many allusions are unclear; and many of the letters it can be related to are also undated. Tyndall implied that Francis will have printed copies of the bismuth paper, sent in MS only on 1 October. It appears to come before 0545 to Francis when their plans for forwarding the application to Toronto are more clearly formulated. Moreover, in both this letter and 0545 Tyndall hopes that he will receive a testimonial from Grove, which implies that both date before 13 October (when he would have received letter 0547 from Grove). We suggest c. 10 October.

2. *them*: presumably the testimonials.

3. *the memoirs*: the passage below implies 4 memoirs (not counting bismuth), presumably the same memoirs Tyndall had sent with requests for testimonials; see letter 0518, n. 2.

4. *affair at Edinburgh* the BAAS meeting in Edinburgh, when Tyndall defended his claims against many sceptics (see letter 0418).

5. *the slips*: proofs printed on long 'slips' of paper before the type is made into pages (*OED*).

6. *that upon the polarity of bismuth*: see letter 0525, n. 2. Tyndall had sent the manuscript to Francis in two parts, (see letters 0538 and 0539).

7. *President*: the president of the University of Toronto, to whom the applications were to be sent. Tyndall mentioned receiving information on the application procedure in letter 0538.

8. *Should I hear from Grove and Wheatstone*: Tyndall was hoping to receive testimonials from both.

9. *Knoblauch*: was touring around Britain in September–October.

10. *down here*: at Queenwood.

11. *bumper*: a cup or glass filled to the brim, or to overflowing; used particularly in drinking to health.

12. *Morley's*: Morley's Hotel, where Knoblauch was staying while in London (see letter 0509). It seems that Knoblauch was soon to leave. He had not arrived in Marburg on 19 October but Hirst was expecting him at any time (letter 0553) which supports a mid-October date for this letter.

To William Francis [12 October 1851][1] 0545

Queenwood Sunday mng.

My Dear Francis.

I am thoroughly satisfied with what you have done and see no plan better or more practical than that which you have proposed—The sending of one copy of the testimonials[2] renders the thing sure—in fact the others are rather a luxury than otherwise and if they arrive too late all is not lost.

In short I fully subscribe to your proposal and will thank you to carry it out for me.

I received the enclosed yesterday. perhaps it will be worth while to send it with the others[3]—should I hear from Wheatstone or Grove within the coming 10 days it will I think be judicious to post *[the]* letter at once Something less than a month seems sufficient to carry a letter to Toronto.[4]

I have got the number of Poggendorff which contains Riess's complete memoir[5] and will[6] complete the thing whenever you want it.

believe me dear Francis | most truly yours | J Tyndall

RDS 27/2

---

1. *[12 October 1851]*: dated by references to Toronto testimonials. Sundays in October were 5, 12, 19 and 26. It seems that Tyndall was replying to proposals of Francis (for example, on the number of testimonials, n. 2). He wrote while still hoping to get a testimonial from Grove and therefore before receiving letter 0547 (of 12 October). The enclosed may be letter 0542 (9 October) from Whewell (see n. 3), which would support a 12 October date.

2. *sending of one copy of the testimonials*: Tyndall had left the number of copies up to Francis's discretion (in letter 0544) and Tyndall seems here to agree that some larger number is not needed. Francis's reply (acknowledged in this letter) is missing.

3. *received the enclosed yesterday . . . with the others*: possibly the letter from Whewell which, as it was a statement of support but not a formal testimonial, might be described only

as 'perhaps' worthwhile to send. Whewell's letter written from Cambridge on 9 October could well have arrived in Queenwood on 11 October.

4. *within the coming ten days . . . a letter to Toronto*: the Toronto deadline for the reception of applications was 19 November 1851 (see letter 0504, n. 3). If Tyndall received the hoped-for testimonials by 22 October (10 days after 12 October) he proposed to post them separately, expecting that less than a month would be sufficient for them to reach Toronto by 19 November.

5. *Riess's complete memoir*: P. Riess, 'Ueber die electrischen Strüme höherer Ordnung' (cited letter 0514, n. 4). An abstract of this paper and an earlier related paper was published by Tyndall, but not until 1853 (ibid.).

6. *will*: the last word at the bottom of the second page. The remainder of the letter is written vertically at the left of the second page.

To John Hall Gladstone [12 October 1851][1] 0546

Queenwood. Sunday morng.

My Dear Dr Gladstone

I dont know what to say to you that will not be 'flat stale and unprofitable'. The day before your letter[2] reached me I had planned in my mind to send you a memoir[3] purely to shew that I had not forgotten you; but now that your letter has come it demands something else at my hands

It is perhaps a singular fact that I read the works of christian men, if they be experimental not doctrinal, with the deepest interest and I either see or fancy I see clear resemblances between their experiences and my own. I have been accustomed to watch the operations of my own mind and the fluctuations of faith and moral principle within me. This habit of mine has led me into many little secrets of my own nature which I find the greatest help in interpreting the experiences of others. I value the bible chiefly on this account; in the spirit of the woman of Samaria I may exclaim regarding it 'it is a book which tells me all things whatever I did'[4] But not only in the bible and in books of accredited genius do I take pleasure; I fancy I see profound wisdom at times in the writings of many whom the world term fanatics. I cannot help fancying at the same time however that the fanatic is wise without knowing it his wisdom consists in the faithful transcription of experiences the roots of which lie deep in human nature. In all this I feel pleasure, but when I come to talk about myself I feel as if every word I uttered was accompanied by a diminution of what I may term my spiritual strength—This is perhaps owing to want of practice, though I sometimes think it must have a deeper ground. It sometimes makes me appear uncommunicative and if I ever appear so to you I hope you will refer it to its proper cause. —You can, I dare say, fancy a man doing an upright act in secret and saying to himself 'nobody shall know that I have done it' and that such an act thus housed up in a man's own heart may contribute materially to his private strength and goodness whereas had he recited it this source of power would vanish—I do not mean to draw any parallel between my experiences and such an act further than the general analogy that the keeping of them to myself seems a kind of necessary fuel to me. When I contemplate the character of Jesus christ for example I find myself possessed of feelings which as far as I am concerned are better untold. There are I think two kinds of religion in the world; the religion of the head and the religion of the heart—The former is the religion we meet with in books of controversy; and 'schemes' of redemption as far as I have studied them contain a perilous infusion of the same—The writings of Saint Paul I am free to admit are strongly impregnated with the same kind of religion. A perfect illustration of the religion of the heart I find in the writings of Saint John—in the psalms of David and other portions of the bible, and these I confess are the portions which woo me most strongly—they woo me because they appeal to my own consciousness and they break the rigid barrier which any mere intellectual scheme would draw around me—They talk of faith and love both of which are beyond logic.[5] What you call 'objective faith' I call <u>form</u>—The intellect needs an image when it would contemplate the deeper workings of the soul—it throws a kind of drapery over essences just as the philosopher when he contemplates magnetism resorts to a 'fluid';[6] but the force is not the fluid, neither is the religion the form. I grant the right of men to represent their religious experiences by symbols—nay I see the necessity of this—But I am careful of making my symbol immortal; I feel the possibility of substituting another for it equally as good and hence I should be sorry to attempt to force my peculiar mode of presenting religious matters to my mind upon others. I am afraid that the various phases of our present christianity are so many symbols become rigid; and that the quarrels and dissensions of good men are to be traced to the unwitting substitution of the outward sign for the inward fact. I think you are tired of me by this time so I shall now stop—[7]most faithfully yours J. Tyndall.

RS MS/743/1/182

---

1. *[12 October 1851]*: Tyndall is replying to Gladstone's letter of 29 September (0537), thus this letter was written in October. The wording of the first paragraph suggests an earlier rather than a later date (5 or 12 rather than 19 or 26 October). Gladstone's reply (letter 0576, 28 November) suggests he received Tyndall's letter in mid-October. On the available evidence, 12 October is the most likely date, but we cannot rule out other Sundays in October.

2. *your letter*: letter 0537.

3. *send you a memoir*: possibly the paper on bismuth; Tyndall finished writing it and sent it to Francis around the time of receiving Gladstone's letter (see letters 0538 and 0539).

4. *the woman of Samaria . . . ever I did'*: reference to biblical story in which a woman said of Jesus he 'told me all things that ever I did' (John 4:29).

5. *They talk . . . beyond logic*: Tyndall crossed out two versions of this sentence. One version used the phrase 'defy intellectual investigation'.

6. *The intellect needs an image . . . 'fluid'*: Tyndall again uses a physical analogy to make a philosophical point: both religion and physics grasp truth through metaphors. His reference to 'drapery' is an allusion to the Carlylean conception of theology as a form clothing the inexpressible essence of religion.

7. *stop—*: because Tyndall was at the bottom of his sheet he did not have space to write his signature on a new line.

From William Robert Grove   12 October 1851   0547

London, | Oct. 12, 1851

Dear Sir,

I have been absent on the Continent for some time and only received your note[1] yesterday, which being enclosed in the copies of your papers was not forwarded to me. I read your first paper communicated to the British Ass^n [2] with much interest, and from that and from what I have heard of your remarks I have a most favorable opinion of your qualifications for a Professorship of Nat. Philosophy; but I should like to have an opportunity of reading your other papers and communicating with Mr Faraday (who is, I believe, personally acquainted with you) before I pledge myself more definitely.

Please let me know when the election takes place.[3]

I leave London in a day or two for Wales, and shall return in about a fortnight. If then I can be of service to you I shall be happy to do whatever I can and ought.

Faithfully yours, | W. R. Grove.

RI MS JT/1/TYP/1/468a
LT Transcript Only

---

1. *your note*: letter missing.

2. *your first paper . . . Ass^n*: Tyndall probably sent his paper, 'On Diamagnetism and Magnecrystallic Action', in the form that had been published in the *Phil. Mag.* (see letter 0498, n. 6).

3.   *when the election takes place*: Tyndall sent off the testimonials in mid-October; thus it is
      unlikely that Grove, according to the timetable set out here, could have sent a testimonial
      in time. Although Tyndall kept hoping for a testimonial from Grove (see letter 0554)
      there is no record of him receiving one.

To William Francis   [13 or 14 October 1851][1]   0548

My Dear Francis.

My object in requesting you to post these things[2] for me was to render
assurance doubly sure. We have a little cross post here in which difficulty often
arises as to my foreign correspondence. All this will be avoided by posting
in London and I shall therefore feel much obliged if you would have them
packed up as you propose.

When I made use of the number 15 or 20 with regard to the testimonials[3]
I referred exclusively to the number to be sent over—I shall thank you to have
100 copies at least printed for me—they may prove of great value on some
future occasion. I should like to send copies to Faraday and others and shall
therefore be glad to receive the remainder when they are ready.

The packet which I now send contains copies of the memoirs from the
2nd downwards,[4] and translations of some of these which appeared in Poggendorff.[5] The parcel which you were so kind as to send me a day or two ago
contained the translation of the last memoir[6]—I send you a number as you
may possibly decide on having it sent along with the others. Along with these
the parcel contains the English testimonials, the copy of Dr Percy's[7] and a
duplicate diploma—I shall leave it to you to decide whether the translation,
or the diploma should be sent or not, and shall rest entirely satisfied with your
decision whatever it may be.

I have read the testimonials carefully, and have made one or two slight
alterations[8]—I would propose that Sabines certificate should come between
that of Du Bois and Bunsen—The arrangement beyond this is I think the best
possible.

I omitted a few words in Magnus's testimonial which now stand upon the
margin—they are not very important and may be omitted if you like—

Dont you think that some remark like that which I have written at the
end might be introduced—A similar one in Percy's suggested the thing to me.

I shall gather up my thoughts this evening and write to the president[9]
—tomorrow I will send you the letter—Immediately afterwards I will write
to Sir Robert Kane.[10]

Most heartily do I thank you for all the trouble you have taken in this affair—I only wish that some opportunity may be afforded me of proving by my acts how I value your kindness.

with best wishes dear Francis | most sincerely yours | J Tyndall

My parcels which you send to me by mail be kind enough to direct to the <u>Dunbridge Station</u> of the S.W. Railway—Stockbridge is the post room town, but Dunbridge is the place for parcels[11]

RDS 27/1

1. *[13 or 14 October 1851]*: this letter is in the long sequence of letters concerning the sending of Tyndall's application to Toronto. References to the parcel(s) in this and in letters 0549 and 0550 seem to refer to the same parcel(s) and provide a means of dating this letter and the next. This letter was written to accompany a large parcel; the next letter was a brief note explaining that he saved money by dividing the parcel in two; the third letter, 15 October, mentioned that the parcel was posted 'yesterday'. Letter 0549 was therefore written on 14 October and this letter was posted on 14 October although perhaps written the previous evening.

2. *to post these things*: the application for the professorship at Toronto.

3. *the number . . . the testimonials*: Tyndall previously discussed the number of testimonials Francis was to print and send in letters 0544 and 0545.

4. *the memoirs from the 2nd downwards*: that is the four (identified in letter 0518, n. 2) sent out when requesting testimonials; his first, 1850 co-authored paper with Knoblauch is again omitted.

5. *translations . . . in Poggendorff*: the translations of Tyndall's papers published in *Poggend. Annal.* are listed in letter 0512, n. 5.

6. *The parcel . . . containing the translation of the last paper*: Tyndall had given a manuscript copy of his diamagnetism memoir to Poggendorff (for translation and publication) before leaving Berlin in June (JT/2/13b/548). Presumably, the package which Francis had forwarded contained copies of the paper, and had been sent care of Francis by Poggendorff.

7. *Dr Percy*: John Percy (1817–89), an English metallurgist and Edinburgh M.D., had been FRS since 1847. Earlier in 1851 he had been appointed metallurgist at the museum of the Geological Survey and lecturer in metallurgy at the newly-established Government School of Mines in London.

8. *I have read . . . slight alterations*: Tyndall had read the proofs ('slips' referred to in letter 0544) of the printed English and French testimonials and, as the names here indicate, translations of the German testimonials. From his comments it would appear that translation could shift over into emphasis and interpretation.

9. *the president*: of the University of Toronto, John M'Caul (letter 0544).

10. *I will write to Sir Robert Kane*: regarding a potential position at Cork. Francis must have told Tyndall about the position in one of the missing letters (compare later letters 0551 and 0556).

11. *My parcels . . . parcels*: as this postscript makes clear, letter post and parcel post were independent systems. That Tyndall needed to inform Francis of the local offices suggests that Francis had not previously sent parcels to Tyndall.

To William Francis [14 October 1851][1] 0549

Dear Francis.

By breaking the parcel[2] into two I save a penny which is a bit of wisdom I have alighted on by the purest accident. I should feel obliged if you print those French testimonials[3] in <u>French</u> for me. Plücker[4] you see has proved a trump[5]

J. Tyndall

StBPL T&F, Authors' letters

1. *[14 October 1851]*: see n. 2 below and letter 0548, n. 1.
2. *the parcel*: the parcel containing copies of Tyndall's memoirs (see letter 0548). The reference to the parcel, the present tense ('I save a penny'), and the small size of this scrap of paper suggests that this letter was enclosed with one of the parts of the original parcel.
3. *French testimonials*: testimonials from Wartmann (received by 6 September), de la Rive (sent 24 September and received by 1 October; see letters 0522 and 0534), Becquerel (received by 1 October) were probably in French. Tyndall had previously asked Francis whether translation was advisable (letter 0538).
4. *Plücker*: Tyndall requested a testimonial in letter 0523. Plücker's reply, which is missing, must have arrived only recently.
5. *trump*: a person of surpassing excellence; a first-rate fellow (OED).

To William Francis 15 October 1851 0550

Queenwood 15th Oct. 1851

Dear Francis.

I went to the station yesterday morning and secured the parcel[1] containing all mentioned in Mr Gyde's note.[2] I am much obliged to you for all and now proceed forthwith to Wood's paper.[3]

I scarcely know what to say to it—There is manifest cleverness evinced throughout but the subject appears to be hastily handled—If you have patience I will run through a portion of it and dot down the impressions that occur to me.

1. The spelling and grammar will require a little attention. There are some pencils marks, by whom made I know not, but they are too faint to be read.

2. In (14) I would say '— instead of a cube of iron we substitute a cube of ice at 0°F.' The cube of iron was not at 0°F but at 0°C.

3. What is 'motion abstractedly considered'? (15) his argument here is against heat being the result of motion has, I conceive, no force; it appears to rest upon a misconception.

4[4]. That is a sweeping conclusion at the end of (15). The idea of equilibrium implies a balance of forces. If two bodies are in equilibrium true they remain so until acted on by a third; this destroys the equilibrium—of what? Of forces to be sure; of what forces? 'not attraction and repulsion saith D[r] Woods'[5]; of what then? Subtract these and what remains to be equilibrated?

5. He says in (16) the greater the amount of matter in a given space the more difficult it is to take its particles asunder. Now water contains more matter than ice bulk for bulk, mercury than steel but are the particles of the solid more easily taken asunder than those of the liquid?

6. I'll leave (19) and (20) to be dealt with by the author of the Vestiges who imagines the sun to have at one time filled the entire space between here and the uttermost planet.[6]

7. We might fasten an argument on almost every paragraph. With regard to (24) it might be urged that 1 vol. of carbon vapour unites without consideration with 1 vol of oxygen, the result

<*up to here the source is* RDS 27/4; *the continuation is from* StBPL T&F, Authors' letters>

being two volumes of carbonic oxide. The combination is accompanied by heat—but there is no approach of atoms—whence comes the heat then?

<8>. If heat by developed by the explosion of chloride of Nitrogen which is certainly exceedingly probable his observations in (26) would fail. That's a rich illustration in (27) 'boiling water will scald &c'[7]—

I think I may stop here as what I have written will impart to you my notion of the paper. It is marked by cleverness but not by the strictness which scientific matters require. *[His]* mind is wandering about an important subject which I cannot think he has yet mastered. *[His]* paper may serve to agitate an important question, but my notion is that such articles ought to be introduced sparingly. *[This]* remark applies, I imagine, with still more force to the paper of Reuben Phillips.[8] Such theories are more interesting to the finder than to any body else. This clipping and patching up of an old hypothesis is thankless work. [Walloston's][9] theory is at best a make-shift and the filing of it up will not add much to our intellectual comfort. Such notions as those entertained by M[r] Phillips may be of a certain practical value to a man in his private investigations but I doubt the wisdom of giving them publicity—you must however take all I say on this subject with great caution—I have seen so many pretty theories which proved pure brain-bubbles that I have conceived a certain prejudice against them.

At the end of (16) of my last paper[10] stands the letter B; 'the magnetic sphere will move to the pole B'—could you change this into <u>A</u> for me—If the copies are already struck off it might be mentioned as an erratum. The letter d' is omitted in fig. I; but this is of little consequence.

most faithfully yours J. Tyndall

When next you have occasion to write to me be good enough to send me Sir John Herschell's address[11]

RDS 27/4 and StBPL T&F, Authors' letters

---

1. *the parcel*: presumably the parcel of testimonials and associated documents discussed in letters 0548 (see n. 1 of that letter) and 0549 to Francis.

2. *all . . . M[r] Gyde's note*: the note from Charles Gyde, Francis's office manager, is missing. According to letter 0548, Tyndall sent some additional items which had not been requested.

3. *Wood's paper*: published as Thomas Woods, 'On the Heat of Chemical Combination', *Phil. Mag.* 3, no. 15 (January 1852), pp. 43–53. It was framed as a letter to the editor and dated December 1851; presumably this is the date that Woods finalised and submitted the paper. See letter 0551 for a second paper by Woods.

4. *4*: Tyndall ceased indenting each point when he turned to the back of his sheet of paper.

5. *repulsion saith D[r] Woods'*: Tyndall's quotation mark is in the wrong place; it should follow 'repulsion'.

6. *author of Vestiges . . . planet*: the author of the anonymous *Vestiges of the Natural History of Creation* (1844) was still unknown in 1851, but he was judged by most scientific men to be ill-informed, as Tyndall implied here.

7. *'boiling water will scald &c'*: cf. passage in section 29, p. 52 of the published paper (see n. 3): 'If we take a certain quantity of steam and of ice, one will scald, the other freeze . . .'.

8. *the paper of Reuben Phillips*: Phillips published papers in the September 1851 and January 1852 numbers of the *Phil. Mag.* The allusion to Wollaston indicates that Tyndall is here discussing the forthcoming paper ('On Frictional Electricity', *Phil. Mag.* 3, no. 15 [January 1852], pp. 36–43), which discusses Wollaston's theory, as expounded by Faraday

(pp. 39–40). Francis must have seen an early copy of the paper or been informed of its subject.

9.   *Walloston*: Tyndall wrote Walloston or Wolloston. He was referring to William Hyde Wollaston (1737–1815) and his theory about frictional electricity.

10. *my last paper*: Tyndall refers to his paper on the polarity of bismuth, sent to Francis with letters 0538 and 0539, and published in the November number of *Phil. Mag.* (cited letter 0525, n. 2). Both changes requested here were made (the end of section 16 is on p. 339 of the published paper; figure 1 is on p. 338).

11. *Sir John Herschell's address*: this postscript is inserted vertically in the left margin of the first page (that is, of the RDS ms). The reason for requesting Herschel's address is unknown. The applications for the Sydney position were to be posted to Herschel, but Tyndall did not know about this position until 3 November (letter 0558).

From William Francis   [15 October 1851]¹   0551

Richard Taylor's Printing Office, |
Red Lion Court, Fleet Street | Wednesday

Dear Tyndall,

A few words before starting on an expedition to Hampton Court with Willhelm Rose. Mr Taylor wrote yesterday to Sir Robert Kane respecting the vacancy at Cork requesting early information in order if possible to keep you here in the Old World. The following is a copy of the letter that you may know what I have been after and blame me if what I have done does not meet with your approval

'Dear Sir Robert

When Prof. Magnus was here a few days ago, I heard from him that there was a likelihood of a vacancy in the Chair of Natural Philosophy at Cork. I have since been informed by Dr. Ronalds² that Prof. Shaw³ has resigned. I am very desirous to obtain correct information on this subject for my friend Dr Tyndall who is at present a candidate for the Professorship of Natural Philosophy at Toronto.

His testimonials are of the highest order—but I am sorry to think that so very able a man should be lost to our country:—a feeling which is entertained by Faraday, Sabine, Magnus, De la Rive and others. Should there be an opening at Cork and the least probability of his success I should advise him to delay sending in his testimonials to Toronto &c. &c'.

You need not write to Kane therefore unless you have already done so as I have no doubt Mr Taylor will soon receive an answer.⁴

Your parcel⁵ came safely to hand this morning and I will attend to your instructions and post the parcel immediately on receiving the letter for the President.[6]

When you have a little leisure I should like you to go over a second paper I have received from Dr Woods[7] and which I think will prove interesting to you.

I have received a long paper from Thomson[8] this morning
Yours very sincerely | W. Francis.
Dr. Tyndall.

RI MS JT/1/TYP/12/3981
LT Transcript Only

1.   *[15 October 1851]*: reference to the parcel (which 'came to hand this morning') and to Tyndall's intention to write to Kane date this letter to shortly after letters 0548 and 0549 [13–14 October]. Given the urgency of the application and the efficiency of the Victorian parcel delivery and train services, we date this letter 15 October rather than 22 October.

2.   *Dr. Ronalds*: Edmund Ronalds.

3.   *Prof. Shaw*: George Ferdinand Shaw (1821–99), first Professor of Natural Philosophy (1849–55) at Galway, who continued to hold a position as tutor at Trinity College, Dublin. In 1855 he returned to TCD and in 1856 obtained MA, LLB and LLD qualifications. In his later career he combined journalism and academic positions.

4.   *You need . . . an answer*: Tyndall had indicated his intention to write to Kane in letter 0548. His journal entry for 30 October (JT/2/13b/550) indicates that he wrote to Kane himself, either before receiving Francis's advice in this letter or despite it.

5.   *Your parcel*: the parcel discussed in letters 0548–50.

6.   *the President*: as for letter 0548, n. 9. Tyndall had promised to send Francis this letter the following day, that is, on 15 October. Here, Francis is expecting the letter imminently.

7.   *second paper from Dr Woods*: not identified.

8.   *paper from Thomson*: probably 'On the Mechanical Theory of Electrolysis', *Phil. Mag.* 2, no. 13 (December 1851), pp. 429–44.

From Dionysius Lardner 18 October 1851 0552

Paris, | 18 Oct. /51

Dear Sir,

I have not seen the Philosophical Magazine, and don't know where it could be referred to here. You could send me the numbers by the usual conveyance from London,[1] and on reading them I shall be happy to do what you desire.[2]

I remain, dear Sir, | Yours very truly, | Dion. Lardner.
J. Tyndall, Esq.

JT MS JT/5/16b/431
LT Transcript Only

---

1.   *send . . . by the usual conveyance from London*: Tyndall referred the request to Francis (see letter 0554).

2.   *on reading them . . . what you desire*: it seems that Tyndall had asked for a testimonial and Lardner here agrees to write one after reading his published memoirs. However, when Tyndall asked Francis for advice (letter 0554) he did not admit he had already made the request.

From Thomas Archer Hirst   19 October 1851   0553

To John Tyndall. | Marburg— |
Oct 19th 1851 | i.e. <u>Sunday</u> Morn[g] [1]

Dear John

My Swiss tour is ended, I arrived (after a months trip) in Marburg three weeks ago,[2] stronger in body and soul than when I left it. Knoblauch I am afraid has frightened you[3] and exaggerated a little my then melancholy aspect, though truly I was very weak. I account for it thus, when we parted at Cassel I was attacked by one of my most stubborn bowel cases, took a fearful amount of Senna & Pills,[4] which brought me down to a pitiful state of weakness; just at the time too I caught a cold which during the whole semester never left me but renewed its attacks in various shapes, throat, chest nose &c &c. Warm though it was, I was ever shivering, & rose every morning with fingers and toes benumbed; it was a miserable time that I can tell you, I felt my intellect <u>cringing</u> before its task, and retired often conquered from a Mathematical problem that I was conscious of being able to unravel A curious symptom now shewed itself after standing one morning for two or three hours at my desk I felt a curious sensation about my legs and on examination found that from ancle to calf were swollen double their usual size, the skin stretched as if it only required to be pricked and a stream of water would issue forth; somewhat startled I sent for Noll, he pulled a serious face & with unusual earnestness made me shut up my books, take an immense amount of medicine & lie all day on the sofa, in which comfortable situation I remained a week, if I attempted to change the legs into a vertical position they began immediately to swell.—[5]

He told me after it was a kind of dropsy which rather startled me, it never left me completely during the whole semester & from the reaction of the medicine I never fully recovered; my intellect never resumed its vigour completely, and self mortified at its failure, I got into a kind of morbid and abnormal state of mind and body. Thus my miserable appearance, which people <u>would</u> attribute to my studiousness, & thus that the morbidity was increased by the contradiction which the consciousness of self-incapacity offered to the opinion of the world It was a miserable time that John but it contained its own lesson which I shall not soon forget, my Sunday walks were pure tasks, Nature was dead to me[6] wherever I turned. My thoughts turned ever inwards, I shunned all company which might have withdrawn my conscience from its self prey, & thus nourished the fiend that never ceased gnawing inwardly. I see now my error I stuck doggedly to my work, but with a feeling of revenge instead of love, and naturally success or satisfaction never followed. The Semester closed however I shut up my books pocketed Schiller,[7] strapped on my knapsack & away I started for Switzerland, determined if possible to use it at any rate to get bodily strength. I did not at once shake off my tormentor however, a few days toil even on my pleasure tour were allotted me but fresh air, exercise, & a <u>sleeping</u> intellect had at last their effect. Light at last shone on me, I could spring from my bed with a cheerful heart, and feel myself not entirely an outcast from Nature, but truly a relation if yet a somewhat distant one. The month passed by, into it were crammed experiences, pleasures, trials and scenes which I do not expect yet to unravel but from which God willing I shall yet draw lasting value—I am now strong and hearty John and during the last three weeks have done more work than perhaps the whole of last semester (that is efficient work) There now you have the whole matter, you say truly I must be my own physician, (Noll and his pills together with their sister Senna are strangers to me, whose acquaintance I hope never to renew) but you preface your advice with an account of your own toil[8] which puts all my efforts in the shade; I could not help smiling as I read of your working night and day & then in the next line your advise[9] to me not to do so—But I must gallop over my tour for you. I did think of writing a small account for the Leader. I partly completed it but I burnt it,[10] what I did and saw & felt I have yet to digest myself before I attempt to offer it to others; my Diary is pretty full however, though as I said its contents are confused & useless to all but myself (and perhaps you) some day we will con it over together, and through Times telescope with adjusted focus we may measure it and learn its true bearings upon me & my life history.—

I started with bad weather, in Frankfurt Heidelberg & Basel got several wettings through—On a rainy Sunday Morning I started in low spirits with the diligence from Basel to Zurich, during

<Handwritten letter ends; hereafter LT Transcript Only>

the day, however, it cleared up, the sun shone brightly as we reached that beautiful valley of the Aare at Brugg, and for a fortnight through the whole of Berner Oberland he scarcely ever hid himself behind a cloud. The valley of the Aar will perhaps remain longer than all the other scenes I saw, in my memory. The rapid river leaped and laughed in windings the most fascinating (reitzend[11]), clouds chased each other over the green rich meadows studded with clean white Swiss cottages; with their overhanging thatch and climbing vines they seemed to me the ideal of comfort and cheerful industry, as simple and natural as the daisies in the surrounding meadows. From what appeared a mysterious inland fairy kingdom, rounded hills clothed to their summit with rich green foliage came and peeped coyly on the pastoral vale, and by their side were one or two higher, sterner beauties with their summits veiled in silver clouds. I <u>saw</u>[12] that valley, John, and, like the Ancient Mariner[13] who blest the fishes and serpents of the sea, I felt my heart beat towards the scene. And immediately, returning health of soul, like a ray of sunshine, lit up that inner kingdom so long o'erclouded with the mists of selfishness and indifference. Next morning found me sailing on Lake Zurich, a clear blue sky o'erhead reflected still clearer and bluer in Zurich's fair waters. At the village of Horgen I landed, and here verily commenced my tour, and tramped over the hills to Zug. At every turn of the road scenes new and fresh to me presented themselves, too quickly almost to be enjoyed. I felt as if squandering and revelling in riches, and longed to stop and wander up but one of those fairy valleys; that is an old trick of Nature's, however, were it not for this ever near but never grasped enjoyment, we should cease to hunt her over Alpine hills and dales. That night I sat with my cigar on the forsaken pier at Zug, and in the rippleless water saw a strange and to me a wondrous sight—namely a sunset. But I could stop and prattle over every little scene and send you a volume. I must be briefer. Next morning I followed the lake's edge, towards Goldau. At noon as usual I threw off my knapsack on the roadside, to rest and smoke half a cigar, and pull out my note book, or Schiller, just as I felt inclined. These midday resting places, on my whole journey, are stamped most indelibly on my memory. In this one I thought such and such a thought, in another I read such and such a poem, in a third made a translation of a poem—to all which the place and scene is now a beautiful frame. Today, on the opposite shore was a brave old hill with rough, shaggy slopes, covered with alternate scaurs and firs. Three times successively I watched him doff his cloudy cap as if he were greeting Mother Nature from his height; the old fellow seemed alive to me, and out of respect and sympathy I doffed my hat to him, and then—looking in my map to see if he had a name, and if not, to give him one, I found that I had been saluting the world-famed Rigi. It was not long before I made closer acquaintance with him. He gave me such a sweating, toiling up his sides as I never had before, and when after losing my way I arrived just at sunset on his summit, my eyes opened to a scene that surpassed all I had before seen in paintings, and thought imaginary. I slept on the summit, and next morning saw if anything a more wondrous sun-rise; description would be useless, we will <u>talk</u> about it some day.

From the Rigi I went to Lucerne. As an incident there I may mention that I left my old hat, for whom I had such affection and which you so often insulted. With some sorrow and scruples of conscience I interred him in a small hat shop there, and in his stead brought away with me a small Swiss 'Jim Crow',[14] From Lucerne after sailing over the Vierwaldstätter See, I landed at Altdorf and trudged along up that memorable St Gothard's street. My intended route was in such delightful uncertainty that here on the summit I tossed up whether I would go into Italy or turn off through Oberland. The thaler was flung in the air and fell down a 'tail' (? has it one) which decided for the Berner Oberland, so I turned aside to Realp, passed over the Furca to the Rhone Glacier, forward to Grimsel, that magnificent land of barrenness; followed the vale from there to Meyeringen, passing through all zones of cultivation, from Grimsel where no blade of grass and nothing but moss will grow, down to the fertile lovely meadows of Meyeringen. I slept that night Brientz and next morning in a small oared boat went forward to Interlaken. It is a delightful spot, but in my humour was too artificial. I sought and found here one of the snuggest and quietest of little Gasthauses[15] that I met with in my journey; in my journal is an effecting page[16] devoted to my separation from the snug little bedroom, with its sofa, and small table, where I spent most of a whole day writing up my journal and resting awhile to digest what I had so hastily swallowed for the last 3 or 4 days. Here I wandered up Lauterbrunnen valley, and made my acquaintance with your favourites, the Jungfrau and Monk. The Staubbach is one of my dedicated resting places. From Interlaken I went to Thun, from there to Berne, and from Berne forward to Neuchatel; there I sought and found out my relation.[17] It was strange there, in a land where all were strangers to me, to find among a crowd of girls one who on the instant knew me and welcomed me, although we had never seen each other before. Here rainy weather again overtook me and compelled me to use diligences for the longer stages. I followed Lake Neuchatel to its extremity, Yverdon, where Pestalozzi[18] had his establishment, and from there across the country to Lausanne on Lake Geneva. It was early in the morning, between 4 and 5, when I reached Lausanne, having travelled overnight in the diligence. None but the inmates of too great a Gasthaus were stirring, and I found my way to an eminence above the town, and there made my acquaintance with Lake Geneva. I did not wonder at Byron, Rousseau, Voltaire and Goldsmith choosing it as their abiding place. A more heavenly spot than its upper extremity from Lausanne to Villeneuve could not be imagined. Every village and scene here is immortalized by some or other of the litterati that have dwelt there. Chillons Castle[19] stands as picturesquely, the placid waters kissing its feet as lovingly, as when Byron saw it, yet just behind it now is the immense Hotel de Byron, which I must confess does not improve it. From Lausanne to Geneva I went by steamer; this part of the lake is flat, and compared with its upper extremity uninteresting—at least, so guide books say, but I had often to differ from my respected friend, Mr Murray;[20] and as jackal or Lion's provider[21] can well dispense with his services. Among the English tourists I met, he is held as a God, and not to see and praise the tit-bits of scenery he mentions is held as highest blasphemy.

From Geneva I went forward through Savoy to Chamonix. For three days I stopped at the foot of Mont Blanc, snowed-up; in the valley it was 5 or 6 inches thick, and during my stay I did not ascend a single mountain. I enjoyed my three days, nevertheless, for it was a grand and impressive sight. Snow everywhere, and thick snowy clouds often silvered by the sun, clung to the huge mountain steeps, and hid their summits. Up in a mysterious cloud region I could hear avalanches thundering unceasingly; and as a cloud for a moment was driven aside I could see huge peaks where I had thought were clouds alone. The third day the weather cleared a little, and I journeyed onward; as I reached the upper extremity of Chamonix valley and was about to leave it for good, I turned round and for the first time saw the summit of Mont Blanc. I found that here are two worlds, often separated from each other by a stratum of cloud which we inhabitants of the lower regions call the sky. I had now pierced this cloudy stratum, the lower world was hid from me, and I stood alone among the giant mountain tops. Overhead the <u>veritable</u> blue sky, clear as crystal, but the air cold and cutting. This part of my journey, from Chamonix over the Tête Noir to Martigny, in the valley of the Rhone, was by far the wildest and most magnificent I had seen. I learnt here what a precipice was, for often the path led over abysses that made me dizzy to look at; a false step and I should have been dashed to atoms. This was the extremity of my tour. I had thought to have pierced as far as St Bernard, and brought back a dog as souvenir and companion, but the weather again threatened; and (a plague on it!) this conscience of mine began to point towards Marburg and my work. It seems a wonder to me now that I went back at all, what should hold me? I had money, and none but my own will to command me; why not have gone on roaming freely over the hill and dale, and instead of a close garret, breathing the bracing air on the mountain tops? So I asked myself once or twice after my return, and before I got reconciled to my work once more. It did not last long, however, and I shall look back on my journey all the more cheerfully, that it was relinquished when duty called.

From Martigny I followed the Rhone again to Lake Geneva, and made my nearer acquaintance with this, its more beautiful part. As I climbed the hills on my homeward road to Berne, it as usual looked lovelier than ever as I was about to leave it. I never felt tempted more to return than now when I was bidding good-bye perhaps for ever. If anyone had told me that a mere scene could have had such hold on me I should have laughed at them. To all such feelings I had thought my Marburg coat of indifference had rendered me proof. Doggedly at last I had to turn my back and tramp; yet as if to shake my resolve a turn of the road when I was miles distant would shew me a corner of her blue waters glistening in the sun. I passed through Freiberg again to Berne, from there past Hoffwynl (I don't know how to spell it[22]—I mean where Yeats and Haas came from), Soleure and over the Jura hills, back to Basel.

Thus ended my journey Tyndall, poor in anecdotal incident, rich and valuable to me, however. It forms part of my funded property, the interest of which I look forward to with more pleasure than to the drafts from Messrs Charlesworth & Co., Bankers, Huddersfield.[23] Yes, in Lausanne I met Mr John Yeats[24]—he would have passed me by, my moustache having metamorphosed me, but I called out to him; we did not remain with each other, our orbits unluckily (?)[25] only crossed. He described to me the magnificent tour he had in view, and spoke of making it pay its own expenses, but how (!) I could not gather. One little incident you will know to be characteristic of John Yeats.

<u>Yeats</u>. Have you read ___'s works? (God forgive me, I have forgotten the name already)

<u>Tom</u>. No.

<u>Yeats</u>. No!!!!! Why, he is the Walter Scott of Switzerland!!!

<u>Tom</u>. Indeed.

<u>Yeats</u>. I called on him the other day.

<u>Tom</u>. So!

<u>Yeats</u> I am translating one of his novels which he gave me.

Here he brings Tom a book, taking particular pains to display the page on which was transcribed 'John Yeats, from the Author.', which Tom, however, to his disappointment would not see.

I am afraid from your letter[26] there is little chance of your losing the Professorship of Toronto, that is, if impartiality rule the decision. You have a more valuable list of testimonials than many will be able to present. Knoblauch is not yet returned to Marburg, nor Wrightson either; Davy and I are the only Englishmen at present honouring Marburg. Remember me to Debus; I was glad to hear of his being able to lecture, he must have made good use of his time, I would not like to undertake one in German.

Enclosed I send a few translations;[27] read them, and if you think them worth anything, or what you think the best, send forward to the Leader. The Diver[28] will be too long I fancy. Perhaps your friend the Preston Chronicle would be glad in the scarcity of the news to find a place for it. If you find blunders, correct them, and recollect, only those you think worth it you must send to the Leader[29]—the others burn.

And now John, goodbye. I have written you a long letter, but far shorter than I could wish. Let me hear of thee soon.

T.A. Hirst

Phillips in a letter he sends me by Dr Huth[30] mentions you, and desires me to give you a sound kicking out of love. He wants to know if you are in Marburg yet—tell him.

RI MS JT/1/H/162
JT/1/HTYP/162-167

---

1. *[To John Tyndall . . . Morn^g]*: these lines are a later insertion (in different ink, but in Hirst's hand). It seems that Tyndall returned the letter to Hirst and that he identified it—as he identified letters received and kept.

2. *My Swiss tour . . . three weeks ago*: Hirst had spent almost a month travelling in the Swiss Alps, departing Marburg on 3 September and arriving back there on 30 September. His journal records the tour in detail.

3. *Knoblauch . . . frightened you*: Knoblauch travelled to England in late August and visited Tyndall at Queenwood on 7–8 September; see letter 0533, n. 1. He was expected back in Marburg imminently.

4. *bowel cases . . . Senna and Pills*: an ongoing medical problem for Hirst. Senna remains a treatment for constipation.

5. —: we interpret Hirst's ambiguous punctuation here as a paragraph break.

6. *Nature was dead to me*: here, as at later points in the letter, Hirst expresses romantic, Carlylean ideas about the importance of being in communication with Nature.

7. *Schiller*: Hirst took a volume of Schiller's poems on his tour, reading and translating poems as he travelled; see allusion to 'The Diver' later in the letter.

8. *you preface . . . own toil*: Hirst refers to Tyndall's letter 0540.

9. *advise*: Hirst altered 'advising' to 'advise'; he probably intended 'advice'.

10. *a small account . . . burn it*: there is a similar piece, written with publication in mind, entered at the end of Hirst's Journal, ff. 855–56 (undated).

11. *reitzend*: probably Hirst meant 'reizend', meaning charming or lovely (German).

12. *saw*: in underlining the word, Hirst emphasized that his seeing was a transcendental or Carlylean kind of seeing, a seeing of the essence.

13. *the Ancient Mariner*: an allusion to 'The Rime of the Ancient Mariner', a poem by Samuel

Taylor Coleridge (first published 1798), a story of disaster befalling a ship when a seaman shoots an albatross, and of the seaman's efforts to expiate his guilt.

14. *small Swiss 'Jim Crow'*: Hirst recorded buying the 'broad-brimmed' hat on 10 September, while in Lucerne. The name alludes to a style of hat made fashionable by popular black impersonators.

15. *Gasthouses*: inns or hotels (German).

16. *in my journal is an effecting page*: 16 September.

17. *found out my relation*: Hirst had been informed by a friend in Marburg that a relation of his (Miss Shawe) worked as a governess near Geneva (see letter 0516, in which he discusses his holiday plans). Hirst met her at the Moravian Institute in Montmirail and enjoyed their discussion of 'mutual friends and relations in England' ('Journals', 18 September).

18. *Pestalozzi*: Johann Heinrich Pestalozzi (1746–1825), a Swiss educational reformer who opened an institute at Yverdon in 1805, which attracted visitors and pupils from across Europe and served as a model of child centred education until the mid-1820s.

19. *Chillons Castle*: a medieval fortress built on an island in Lake Geneva. While visiting the castle in 1816, Lord Byron wrote a poem, 'The Prisoner of Chillon', about the sufferings of a Genovese monk who was held captive in the castle from 1532 to 1536.

20. *Mr Murray*: John Murray began publishing guidebooks for travellers in 1836; by the 1850s they had achieved immense popularity.

21. *as jackal or Lion's provider*: Hirst alluded to the belief that jackals alert lions to prey, or actually hunt prey for lions. The allusion was often used figuratively, for example by Byron (*Don Juan*, ix, 27) and Carlyle (*Sartor Resartus*: Book 1, Chapter 3).

22. *Hoffwynl . . . how to spell it*: the correct spelling is Hofwyl. It was a school founded by Philipp Emanuel von Fellenberg (1771–1844), initially in collaboration with Pestalozzi (see n. 18) in 1799. At first, it combined manual training, agricultural and academic instruction, and moral training, and was intended for poor children. Von Fellenberg later added a Classical institute for middle-class children and, by the time of his death, the school attracted pupils from across Europe. Queenwood had connections to this educational tradition, hinted at here by the reference to Haas and Yeats.

23. *Messrs Charlesworth & Co., Bankers Huddersfield*: presumably the bankers from whom Hirst received regular payments (from the inheritance received from his mother).

24. *in Lausanne I met Mr John Yeats*: Tyndall had mentioned this encounter in his last letter to Hirst (see letter 0540). Yeats was a tutor at Queenwood during Tyndall's first stint at the school.

25. *unluckily (?)*: the question mark is probably an ironical comment on 'unluckily', for Hirst knew that Tyndall did not like Yeats.

26. *your letter*: as per n. 8; Tyndall had listed some of the testimonials he had received.

27. *a few translations*: apart from 'The Diver' (following note), Hirst translated a number of poems by Schiller: 'To Emma', 'The Words of Faith' (a slightly inaccurate rendering of 'The Words of Belief'), 'The Flowers', 'The Outcast' and 'The Philosophers' over 5–19 October

(Journal, 12 and 19 October). He copied the translations into his Journal (ff. 775–9 of the typescript [in the middle of the entry for 26 October].) It is likely that he sent all of them to Tyndall with this letter. Tyndall gave his opinion on Hirst's translations in letter 0568.

28. *The Diver*: a poem, 'Der Taucher', written by Schiller in 1797.

29. *the Leader*: see letter 0398, n. 8.

30. *Dr Huth*: see letter 0482 (at n. 12).

To William Francis   [21 October 1851][1]   0554

Tuesday.

My Dear Francis,

This morning I received the enclosed from D[r] Lardner. It is[2] worth while to have a testimonial from him. could you manage to enclose a memoir or two if I send them to you in your packet of next month?[3] —I don't know the 'usual mode of conveyance' to which he refers.

Before I received your last note[4] I had sent a copy of the bismuth paper to Faraday[5] and another to Grove[6] (from whom I expect a certificate daily[7]) these are the only two that I have sent away, and had I received your note in time even <u>they</u> should have been kept back.

most truly yours | J. Tyndall

StBPL T&F, Authors' letters

---

1. *[21 October 1851]*: Tyndall had just received letter 0552, dated 18 October 1851, from Lardner from Paris. 'Tuesday' is therefore probably 21 rather than 28 October, that is 3 days by post rather than 10 from Paris to Queenwood. But see n. 7 below.

2. *It is*: Tyndall probably intended 'If it is' or 'Is it'.

3. *enclose . . . your packet of next month*: Tyndall is avoiding expensive foreign postage by asking Francis to enclose his mail in a larger Taylor and Francis parcel to Paris. The allusion to 'next month' implies that Tyndall sent his translations for *Phil. Mag.* to Francis in a regular monthly posting; other letters show that additional postings were sometimes made.

4. *your last note*: letter missing. Francis had apparently advised against sending out early copies of the memoir on bismuth (which was published in the November issue of *Phil. Mag.*).

5. *the bismuth paper to Faraday*: cited letter 0525, n. 2; the accompanying letter is missing.

6. *to Grove*: letter missing. Tyndall possibly sent the paper on bismuth to Grove with his reply to letter 0547.

7. *expect a certificate daily*: see letter 0547. Grove had not promised the testimonial as quickly as Tyndall expected it—unless this letter dates from 28 October which is unlikely (see n. 1).

From Michael Faraday 21 October 1851 0555

Lee Road, | 21 Oct. 1851

My dear Sir,

Many thanks for your note[1] and the paper[2] and the testimonials.[3] I hope you will obtain your desire at Toronto. I have read the paper briefly, but must do so again: at present I am writing, or rather copying, a paper on <u>lines of force</u>, which touches the point of <u>Polarity</u>, so that it is only hereafter I shall be able to collate all together. I propose giving the paper in to the Royal Society today.[4]

Ever truly yours, | M. Faraday.
Dr. J. Tyndall, | &c. &c. &c.

RI MS JT/TYP/12/4130
Typescript Only

---

1. *your note*: missing.
2. *the paper*: Tyndall's paper on the polarity of bismuth (cited letter 0525, n. 2), which he told Francis he had sent to Faraday (see letter 0554).
3. *the testimonials*: only a week previously (letter 0548) Tyndall had asked Francis to print additional copies of the testimonials. They were printed, despatched to Tyndall by Francis, a set sent to Faraday, and Faraday replied, all within seven days.
4. *paper on <u>lines of force</u> . . . today*: Faraday, 'ERE 28', was dated 9 October 1851, received 22 October, read 27 November and 11 December.

To William Francis [20–31 October 1851][1] 0556

[Queenwood][2]

My Dear Francis,

The thought occurred to me last night that in such cases as that of D[r] Woods[3] it would be well to write to him, stating an objection or two to which his views appeared liable. He might be thus induced to reconsider his subject, or if the objections arose from a misapprehension of his meaning he would have an opportunity of setting us right. In this way you would retain his good will, and if any thing were in him might bring it out, at the same time preventing the magazine from being the vehicle of inconsiderate theories. If you fall in with this idea send me his paper again and I will select one or two points which appear to be vulnerable, make a few remarks upon them, send the remarks to you and you could enclose them with a short note from yourself.

Many thanks for Kane's letter[4] n'importe.[5] in späteren Jahren werden wir hoffentlich doch ein Glass Bier und eine Cigarre zusammen haben.

Leb'wohl | Dein Treuer Tyndall[6]

I shall have Riess's paper[7] ready by the time you mention.

About a year ago Regnault made an investigation[8] on the thermo-currents of Bismuth and Antimony—it appeared I think in the Comptes Rendus, or perhaps in *[the Annales]*. If you could find it I should be very glad if you would send me the 'heft'[9] containing it when you send me Magnus's paper.[10] Another paper from Svanberg[11] appeared in the Comptes Rendus—this I should also like to have if possible.

StBPL T&F, Authors' letters

1. *[20–31 October 1851]*: The sequence of letters from Taylor to Kane (copied by Francis to Tyndall in letter 0551 of 15 October), and then Kane to Taylor, which was passed from Taylor to Francis and finally forwarded from Francis to Tyndall (as acknowledged here) suggests that this letter is after 20 October. Tyndall copied Kane's reply into his Journal on 30 October (n. 4), setting that as the latest possible date for this letter. Similarly, the reference to Regnault's paper implies that this letter predates letter 0558 (3 November) by at least a few days (see n. 8 below). Thus this letter was written between 20 and 31 October.

2. *[Queenwood]*: Queenwood embossing on the paper.

3. *Dr Woods*: Francis had asked Tyndall to read through two papers by Woods and Tyndall had already sent Francis a list of objections to the first of those papers (letters 0550 and 0551). The plan Tyndall suggested here was adopted as he reported receiving answers from Woods in letter 0563.

4. *Kane's letter*: Robert Kane to Taylor, which Francis forwarded to Tyndall (no covering letter found). Kane explained that the professorship at Cork was not vacant because the incumbent had changed his mind and was no longer resigning. Tyndall copied Kane's letter into his Journal (30 October, JT/2/13b/550). Kane's letter was a response to Taylor's letter (copied in letter 0551).

5. *n'importe*: never mind (French).

6. *in späteren . . . Treuer Tyndall*: In later years, we shall hopefully have a glass of beer and a cigar together. Live well | Yours faithfully Tyndall (German).

7. *Riess's paper*: cited letter 0514, n. 4. In letter 0545 Tyndall informed Francis that he was ready to translate it whenever required. This letter implies that Francis had replied (letter missing), giving a deadline.

8. *Regnault made an investigation*: Tyndall misremembered. He had seen a translated article in the *Phil. Mag.*, not an article in a French journal (see letter 0558, n. 7).

9. *heft*: German for number (of a periodical).

10. *Magnus's paper*: Heinrich Gustav Magnus, 'Ueber thermoelektrische Ströme', *Poggend. Annal.* 83, no. 8 (1851), pp. 469–504. This was translated and condensed as 'On

Thermo-electric Currents' and included alongside discussions of three other papers in Tyndall's February 1852 'Reports on the Progress of the Physical Sciences' (see letter 0581, n. 20).

11. *paper from Svanberg*: Lars Fredrik Svanberg, 'Expériences sur le pouvoir thermo-électrique du bismuth et de l'antimonie cristallisés', *Comptes Rendus* 31 (1850), pp. 250–52. Tyndall included an extract alongside the paper by Magnus (see n. 10).

To William Francis [1–2 November 1851][1] 0557

receive[2] the papers.

I entirely agree with your remark as regards Lardner[3]—It is a thousand pities that he made that false step[4]—He is an able man. There is a symmetry so to speak in his manner of thinking which I have seldom seen surpassed. As I have written to him[5] I should certainly like to send him a memoir or two; though I shall make no use of his testimonial.[6]

The end of the month[7] will answer for Magnus &c.[8] Just let me know when you purpose publishing the report. You will scarcely get it out before the 1st of January. The report of Riess's paper[9] I will send you some day of the present week.

I have been requested by a Society in Romsey to give them a lecture and have half engaged to the same at Southampton. The subject I have chosen is optics illustrated by the electric light.[10] I will send you your number of Poggendorff in a few days.

most faithfully yours | J. Tyndall

RDS 27/3

---

1. *[1–2 November 1851]*: This letter dates from before, perhaps some days before, the Romsey lecture which took place on Tuesday 11 November (*Hampshire Advertiser & Salisbury Guardian*, 15 November 1851, p. 5, under 'Recent events in Romsey'). Comparison of allusions to the lecture and the Magnus paper between this letter and the following one lead us to place this letter, somewhat tentatively, before letter 0558 (3 November).

2. *receive*: this letter is a fragment.

3. *your remark . . . Lardner*: the letter, which would be the letter Francis wrote in response to Tyndall's letter 0554, is missing.

4. *that false step*: Lardner had eloped with a married woman in 1840. Her husband followed them to Paris and subjected Lardner to a flogging, but his wife refused to return to him. They were divorced in 1841 and she married Lardner in 1846. The scandal effectively ended Lardner's career in England and he lived in Paris until his death.

5. *I have written to him*: letter missing.

6. *use of his testimonial*: Tyndall had raised the possibility of obtaining a testimonial from
    Lardner in letter 0554 to Francis. This letter reveals that Tyndall had already requested a
    testimonial from Lardner but that Francis had advised that Lardner's reputation made a
    testimonial from him of doubtful value.

7. *end of the month*: the month of November. Tyndall intends to send his Magnus summary
    and translation in time for 1 January publication, which seemed to him the earliest possi-
    ble publication date; it was not published until February (letter 0556, n. 10).

8. *Magnus &c*: see letter 0556, n. 10 for references.

9. *Riess's paper*: cited letter 0514, n. 4.

10. *requested by a Society … electric light*: the lecture was delivered at the request of the Mutual
    Improvement Society of Romsey. As this was before the invention of the incandescent
    light bulb, electric light was arc lighting.

To William Francis   3 November 1851   0558

Queenwood College 3rd November 1851

My Dear Francis.

I have read the Sidney paper which you were good enough to send me[1]
carefully through. The spirit in which the matter is conceived is most gratifying; it is catholic,[2] as the ruling spirit of such an institution ought to be, and
I doubt not if men be found to carry out with energy the design so liberally
expressed the undertaking will prosper. One thing I did not like at first, and
that is the distinction which is made in point of salary between mathematics
and experimental philosophy. On the continent, as you know, there is no such
distinction,—if any it is the other way, and it seems strange that in England,
the birthplace of Inductive philosophy such a difference should appear. However the authorities know their own affairs best, they make their offer, and of
course he who does not like it need not apply. There is a kind of set off however in the fact that the physicist will have three fourths of the fees, and also
in the expense which his necessary apparatus entails.

Another circumstance recommends the thing to me. In all probability
the situation will permit of more time being devoted to private investigations than that at Toronto. The success of such an undertaking must of course
chiefly depend upon local effort: but the position occupied by its professors
in their various departments must also greatly influence its power and usefulness. Shutting out the interest which the professor himself must feel in such
an arrangement the prosperity of the institution would be best consulted by
combining public instruction with private research. The confidence thus created, and the example of industry thus set are inspiring and often of far more
value than the direct communication of knowledge.

So far as I can judge from the paper the amount of chemical knowledge required is such as need [not] deter one from undertaking it.[3] I have worked in Bunsen's laboratory both qualitative and quantitative for a year, and have heard his excellent lectures two or three times over, so that in this respect I apprehend little difficulty would arise. The sum granted for apparatus is also an inducement. Were the lot to fall upon me I should make a quick tour through Germany, visit Giessen, Marburg, Göttingen and Berlin; inspect the various cabinets, and note down the improvements recently introduced in the construction of apparatus. With the knowledge thus acquired I conceive the money might be laid out to the best advantage.

All this is of course supposing a case that may never occur; but I think it right to let you know all I have thought about the matter. The distance is certainly great, but the opportunity of lending a hand in the establishment of an institution so liberally based is also tempting. In fine,[4] if you do not advise me to the contrary I will become a candidate.

Yours dear Francis | Most sincerely | John Tyndall

Directly I get this lecture off my hands[5] I will commence Magnus's paper; it shall be ready for you in time for the January number.[6] You have referred me to the exact thing that I required in the Mag. for 1850. I knew I had seen a memoir from Regnault[7] somewhere, and the one you mention is it.

I received the enclosed from Buff yesterday[8]—put it in the bag with the rest—I will write off to Weber and Reich[9] this week.[10]

StBPL T&F, Authors' letters

---

1. *good enough to send me*: letter missing. Francis had informed Tyndall (Journal, 4 November, JT/2/13b/550) that a university was being founded in Sydney and that applications for professorships were to be made to Herschel and Airy.

2. *catholic*: inclusive.

3. *need [not] deter one from undertaking it*: the advertised position was in 'Experimental Philosophy and Chemistry'; Tyndall felt he knew enough chemistry to be a serious applicant.

4. *fine*: to conclude or sum up (*OED*).

5. *this lecture*: Tyndall was preparing a lecture to deliver at Romsey on Tuesday 11 November (see letters 0557, n.1 and 0568, and Journal, 4 November, JT/2/13b/550).

6. *Magnus's paper . . . the January number*: Tyndall was translating and condensing a paper for publication in the *Phil. Mag.* (see letter 0556, n. 10 for original publication). It was published not in January, as expected, but in the February issue (letter 0581, n. 20).

7. *memoir from Regnault*: Tyndall had asked Francis (letter 0556) to send him a copy of the paper, either because he wanted to translate it or because (as it seems here) he wanted to read it again. Francis had informed him (in a missing letter) that he had published a translation ('On the Measurements of Temperatures by Thermo-electric Currents', *Phil. Mag.*

36, no. 245 [June 1850], pp. 409–20). Tyndall discussed Regnault's results, in light of the investigation reported in Magnus's paper (n. 6), in a section entitled 'Application of the results of M. Magnus to the solution of certain difficulties encountered by M. Regnault' in his February 1852 'Reports on the Progress of the Physical Sciences' (cited letter 0581, n. 20).

8. *the enclosed from Buff yesterday*: letter missing. Heinrich Buff (1805–78) was a German physicist and chemist. He had studied at Göttingen and worked with Bunsen in Kassel before being appointed professor physics at the University of Giessen in 1837.

9. *Reich*: Ferdinand Reich (1799–1882), German chemist and professor of physics at the Frieberg University of Mining and Technology.

10. *I received . . . this week*: the first postscript filled the bottom of Tyndall's sheet of paper. This second postscript is written vertically in the left margin of the letter.

From Edward Sabine   6 November 1851   0559

Woolwich Nov. 6/51

Dear Sir;

I rec^d at the Royal Society today a copy of your paper on the Polarity of Bismuth[1] for which I am very much obliged to you, and acknowledge it at once as I may be a day or two before I am able to read it.

I take it for granted that with such testimonials[2] your prospect of obtaining the Professorship at Toronto is secure. I confess I do feel no inconsiderable regret that the mother country is about to be deprived of two such persons as yourself and Mr Huxley[3]—May this loss be compensated by the gain to yourselves and to the Colony in which you propose to make your residence!

Would not the interests of all be advanced by your becoming a Fellow of the Royal Society before you go,[4] & looking to that Society as the Channel of your future communications? Should this appear to you as it does to me, I should be very happy to put matters in train for your election

Believe me, my dear Sir | Sincerely yours | Edward Sabine.
Dr Tyndall.[5]

RI MS JT/1/S/4

---

1. *your paper on the Polarity of Bismuth*: see letter 0525, n. 2.

2. *your testimonials*: Tyndall must have sent a set of the printed testimonials along with the bismuth paper to Sabine.

3. *Mr Huxley*: Huxley was also applying for a position at the University of Toronto.

4. *becoming a Fellow . . . before you go*: Tyndall found this offer so signficant that he copied the

letter into his Journal (JT/2/6/73–4), copied it out in letters to close friends, and saved the original (after loaning it to the Steuarts).

5.   *Dr Tyndall*: Sabine noted the addressee's name at the bottom of the first page (which was in this case at the end of the first paragraph) rather than at the end of the letter. This was a more formal and, perhaps also, an old-fashioned practice.

From William Thomson   7 November 1851   0560

2 College, Glasgow | Nov 7, 1851

Dear Sir

I am much obliged to you for the copy of your interesting paper on the Polarity of Bismuth, and your Testimonials[1] which I have read with pleasure.

In the passage you have quoted from my paper of last March,[2] the word 'more' is wrong and ought to be 'less'. You will find the mistake corrected in the separate copy of my paper which I sent you,[3] and in the 'Errata' of the Volume of the Magazine containing the paper (concluded I presume last July)[4] When so corrected, the whole passage you quote is consistent with itself, and is verified by your experimental results. I need scarcely add that the mistake was merely a slip of the pen, as the concluding part of my sentence lays down the principle in correct terms, which are exactly contrary to the 'more' in the first part of the sentence. The mistake was first pointed out to me by Prof. Stokes[5] last April, and I immediately sent an intimation of it to the Editors for the 'Errata'.

I regret greatly that it has occasioned you so much perplexity regarding my meaning.

I remain, Dear Sir, | Your's very truly | William Thomson

RI MS JT/1/T/11

---

1.   *paper … and your Testimonials*: as he had to Sabine (letter 0559, at n. 2) Tyndall had sent copies of the testimonials with the bismuth paper.

2.   *quoted from my paper of last March*: Thomson, 'On the theory of magnetic induction in crystalline and non-crystalline substances', *Phil. Mag.* 1, no. 2 (March 1851), pp. 177–86 (a paper read at the 1850 BA meeting). Tyndall quoted from p. 183 of Thomson's paper in a footnote to his bismuth paper (cited 0525, n. 2) on pp. 333–34.

3.   *copy of my paper which I sent you*: see letter 0528.

4.   *'Errata … last July)*: printed in the front matter, *Phil. Mag.* 1 (1851), p. viii. Communications from both Thomson and Tyndall clarifying the issue were subsequently published in the *Phil. Mag.* (see letters 0567, 0569, and 0570).

5. *pointed out by Professor Stokes*: the contrast between the meticulous Stokes and the enthusiastic Thomson is well brought out in *The Correspondence between Sir George Gabriel Stokes and Sir William Thomson, Baron Kelvin of Largs*, ed. David B Wilson (2 vols; New York: Cambridge University Press, 1990).

To John Frederick William Herschel   8 November 1851   0561

Queenwood College near Stockbridge Hants |
8th November 1851

Sir,

I beg leave to offer myself as a candidate for the Professorship of Experimental Philosophy and Chemistry in the University College Sidney.

I have the honour to submit to you the accompanying printed copies of Testimonials. They have been duly authenticated in Guild Hall before one of the Magistrates of the City of London. If the original French and German documents be required I shall be happy to produce them.

In addition to these I beg to lay before you a copy of a private letter from Professor Faraday[1] in which the absence of a certificate from him is accounted for.

To the five scientific memoirs[2] which accompany this I beg also to refer, respecting my qualifications as a physicist. With regard to Chemistry I have to state that I have practically pursued both qualitative and quantitative analysis in the laboratory of Professor Bunsen of Marburg (now of Breslau) and have attended his excellent lectures during two or three successive half years. My chief attention, however, has been directed to Experimental Philosophy, and I have made my chemical studies accessory and subordinate to this.

I have the honour to be | Sir, | Your most Obedient Servant | John Tyndall
Sir John F.W. Herschel Bart.[3]

RS HS 17.381

---

1. *private letter from Professor Faraday*: letter 0506.
2. *five scientific memoirs*: the four papers sent with requests for testimonials (cited letter 0518, n. 1) plus the recent paper on the polarity of bismuth (letter 0525, n. 1).
3. *Sir John . . . Bart.*: in this very formal letter Tyndall placed the addressee's name at the bottom of the first page, not at the end of the letter.

# To John Frederick William Herschel	8 November 1851	0562

Queenwood College Stockbridge Hants |
8th November 1851

Sir,

A sentence which occurs at the 329th page of your Discourse on the Study of Natural Philosophy induces me to hope that those of the enclosed memoirs which have reference to the 'magnetism of crystalline bodies' will interest you.[1]

I am afraid however that the theoretic notions as to the atomic mechanism of Iceland spar expressed at page 25 of the paper on the 'Magneto-optic properties of Crystals'[2] will not meet you approbation, as they differ from the hypothesis made use of by yourself in the 989th article of your treatise on Light.[3] The reasoning at this place alluded to will not, however, be materially affected if we substitute spheroidal for rhomboidal molecules. We shall only have to suppose that the distance from end to end of the larger diameters, across the space which separates two adjacent molecules, to be greater than the distance between the ends of the short diameters. This view of course implies that the spheroids are not in absolute contact.

I ought to mention that the Testimonials were collected with a view of making application for the Professorship of Natural Philosophy in the University of Toronto; they have been authenticated by D[r] William Francis Editor of the Philosophical Magazine.[4]

I remain Sir, | Your most obedient Servant | John Tyndall
Sir John F.W. Herschel Bart.[5]

RS HS 17.382

1.	*A sentence . . . interest you*: 'The magnetism of crystallized bodies . . . has not hitherto been at all examined, but would probably afford very curious results', *Preliminary Discourse on the Study of Natural Philosophy* (London: Longman, Rees, Orme, Brown, & Green, and John Taylor, 1831; first published as part of Lardner's *Cabinet Cyclopaedia*), end of paragraph 367, on p. 329. Of the five papers Tyndall sent to Herschel with the preceding letter (0561), all, apart from the paper on the 'Water-Jet', discuss the 'magnetism of crystalline bodies' (most notably iron and bismuth). This letter seems to have been a more informal letter enclosed with letter 0561 and the larger package of papers.

2.	*paper . . . Crystals*: cited letter 0403, n. 2.

3.	*your treatise on Light*: six weeks previously Tyndall had asked Francis to find him a copy of the treatise (letter 0536); the 989th article was on p. 537 (cited ibid, n. 9).

4.	*have been authenticated . . . Magazine*: Tyndall refers to two different processes of authentication. Here he refers to authentication by Francis that the translations of the German letters were reliable (compare letter 0538). The French ones had not been translated. In the previous letter (0561) he refers, we presume, to authentication that the printed testimonials matched the manuscripts.

5.	*Sir John . . . Bart.*: As in the previous letter, Tyndall placed the addressee's name at the bottom of the first page.

## To William Francis	[9 November 1851][1]	0563

My Dear Francis,

I am in receipt of Wood's answers.[2] his reading is evidently extensive and his conceptions ingenious—he calls to my mind the case of Walsh in Cork so admirably described by Mr Boole.[3] not that he is any thing like so verkehrt[4] as Walsh but I cannot help thinking that he is 'kicking against the pricks'[5]—His answers don't convince me—but more of that anon—I will write to you again on the subject in the course of the week.[6] I may be hasty in thus expressing myself and shall be happy to retract if deeper study of the subject shews me my error.

I will translate Wartmanns paper[7] this week.

I send you some British Association proofs[8]—would you be good enough to hand them to the proper party.

Col. Sabine writes to me as follows.

[. . .]![9]

I hardly know how to reply to this!

I sent off the application regarding Sidney last night.[10]

I received a short letter from Thomson this morning[11] which I shall beg of you to publish in [this months] number.[12] It wont occupy more than half a page—

most truly yours Tyndall

StBPL T&F, Authors' letters

1.	*[9 November 1851]*: written one day after Tyndall sent the application for Sydney, which was 8 November (see n. 10). This date is consistent with the allusions to other letters (e.g. from Sabine and Thomson, see nn. 9 and 11).

2.	*Wood's answers*: Francis had consulted Tyndall about two papers by Woods, and Tyndall suggested (letter 0556) that Francis ask Woods to respond to some objections. Letters from Tyndall to Francis, Francis to Woods, and Woods back to Tyndall via Francis, went forwards and backwards in approximately two weeks. Tyndall's initial criticisms were expressed in letter 0550, but there is no record of what he sent to Woods.

3. *Walsh . . . Boole*: John Walsh (1786–1847) was an eccentric mathematician from Cork, whose biography was written by George Boole (1815–64), English mathematician and logician, who was Professor of Mathematics at Queen's College Cork from 1849.

4. *verkehrt*: wrong.

5. *'kicking against the pricks'*: 'to kick against the pricks' was a proverb used in Acts 26:14. Tyndall implied that Woods, like Saul the persecutor of early Christians, was bringing trouble to himself.

6. *I will write . . . of the week*: Six days later Tyndall sent Francis a further question for Woods (letter 0567).

7. *Wartmanns paper*: Élie Wartmann, 'Note sur la polarisation de la chaleur atmosphérique', *Bibliothèque Universelle Archives des Sciences Physiques et Naturelles* 18 (October 1851), pp. 89–95.

8. *I send you . . . proofs*: possibly the proofs for the abstracts which would appear in the *Report* of the Ipswich BA meeting. Tyndall was saving postage by sending them via Francis.

9. *[. . .]!*: Tyndall here copied out letter 0559 from Sabine. He copied fairly accurately. Minor changes include: spelling out some of Sabine's abbreviations, altering capitalisation and, the biggest change, omitting 'at Toronto' after 'professorship'. The exclamation mark, which Tyndall inserted after Sabine's signature, indicates his pleasure and surprise.

10. *I sent . . . last night*: see letter 0561.

11. *I received . . . this morning*: letter 0560.

12. *beg of you to publish . . . number*: Tyndall wanted the confusion created by Thomson's error (see letter 0560) to be clarified as soon as possible.

To Edward Sabine   [9 November 1851][1]   0564

Dear Sir,

I have spent some time reflecting on what I ought to say in reply to your note[2] of the 6th which reached me this morning, but I feel that the best that I could utter would be but a weak and defective acknowledgment of the obligation which I owe to you. I have had indistinct visions from time to time latterly of an investigation which might be found worthy to place in the hands of yourself or M[r] Faraday and communicated to the Royal Society, but even as the result of such an investigation I did not dare to promise myself the high reward which you have proposed. I regarded the distinction as a kind of mountain summit which I was determined by patient climbing one day to attain, and if I should be elected Fellow of the Royal Society I shall undoubtedly consider it an honour paid in advance; it will place me in the position of a debtor to science which debt it shall be the religious duty of my life honorably to liquidate.

John Tyndall[3]—

RI MS JT/2/6/74–75
JT Transcript Only

---

1. *[9 November 1851]*: this is a reply to letter 0559. In his journal entry for 9 November (JT/2/6/74–75) Tyndall recorded receiving Sabine's letter that morning (3 days from London to Queenwood was an unusually slow post). He wrote this reply the same day and copied both Sabine's letter and his reply into his journal.

2. *your note*: letter 0559.

3. *John Tyndall*: The original would have included conventional and formal subscription, such as, 'I remain, dear Sir . . .', but Tyndall omitted these details when he copied the letter into his journal.

From Thomas Archer Hirst   9[-10][1] November 1851   0565

Marburg, | Nov 9th 1851

My dear John,

Thanks for the Literary Gazette containing review of Carlyles Life of Sterling,[2] for which it was I have no doubt sent; the following number containing the second notice[3] I have also been expecting, if you have it by you it would be acceptable. But better than the review Wrightson who returned to Marburg last week brought me according to order <u>the</u> book itself, which for the last week has filled with unusual interest my few hours devoted to such reading. For many reasons I hold this book among the costliest of my collection, and among Carlyles it ranks among, if it be not <u>the</u> most valued, inasmuch as through it, more directly than through his others, I can make the acquaintance of Carlyle himself—his homelier attributes, and the manly honest and hearty love to his friend that is so often apt to be hid by a rough exterior.

This morning to breakfast I read the last chapter and as is ever the case with the best books I read, it induced me to seize my pen and in the absence of your bodily self, attempt through a scribble to tell you how and why it has so pleased me. Thus my letter.

Of Sterling himself, even in so far as his own writing could inform me, I know next to nothing. The former Biography by Archdeacon Hare[4] whose shortcomings and misinterpretations this present work of Carlyle's was mainly undertaken to remedy I never saw, consequently how far this object has been attained I am for the most part unable to judge. One only of Sterling's writings fell into my hands, as you will remember, and that was his essay on Carlyle;[5] fresh, as if it were but yesterday I remember how it pleased me, and to day as I finish Carlyle's judgement of Sterling, the two appear to me beautifully complimentary to each other, opening up by the mutual bearing of their subjects, a field at once rich in matter of thought and full of a valuable significance. If I had time, and were able, I could linger with profit here, and learn much from these honest records of each other that two friends, bound together by manly sympathies, like and yet unlike, have left to us. Carlyle the massive, strong and energetic is himself seen in more vivid outline by the taper which he holds up to his brilliant, genial yet weaker and less persever-ant Friend. In Sterling's essay Sterling and not Carlyle was measured, the for-mer appears as the unit employed to gage the latter, so far as it goes the task too is well performed, but one feels that said task is not completed in some directions, the outlines of the thing measured begin to fail in distinctness, until at last on the chart we come to the regions marked 'unexplored'. Far otherwise is it with Carlyles Life of Sterling, yet fully substantiating the opin-ion just expressed, complimentary, as I said thereto. The thing measured here is an Island not a Continent. the infinite ocean surrounds it, yet its shores are marked out in clear and defined outline. With an admirable pride our guide leads us through the riches of his possession, lingers with fondness in the beautiful flowery meadows, shews us the reclaimed lands that have with labour been tilled and arrowed,[6] and on which the corn is springing; nor does he forget the moorlands and barren places which <u>would</u> have also been tilled and ploughed had not the husbandman been called away.

Sterling was a brave, hard striving fellow brother; of visible success mak-ing itself apparent to the world there was little, yet the struggle was not the less noble, and one feels thankful to the loving hand that has wrested such struggles from their obscurity, and held them up as example, encouragement or caution to the obscure ones that are now struggling. 'I have remarked', says Carlyle, and the same remark I find almost verbatim in a page of my late Swiss Journal, 'that a true delineation of the smallest man and his scene of pilgrimage through life, is capable of interesting the greatest man; that all men are to an unspeakable degree brothers, each man's life a strange emblem of every man's; and that Human Portraits, faithfully drawn, are of all pictures the welcomest on Human walls'.[7] As he also says, Sterling's life is the more significant for us, as an emblem for his own and our time, on account of his intense susceptibil-ity to the impressions it had to offer, 'he fashioned himself eagerly by whatso-ever of noble presented itself; participating ardently in the world's battle, and suffering deeply in its bewilderments'.[8] The History of Sterling's search for a religion, is handled ably. I know nothing on the same subject treated with a hand so masterly, or offering to the doubter, who feels and labours under the same want that Sterling did, so significant a lesson. With an intellect keen and searching he had unmasked the hypocrisy and emptiness that lurks beneath existing formal creeds. The destructive part of his labour, the pulling down; he accomplished, as it is not difficult for any of us to do, bravely; but the new building on its ruins! That is the difficulty to him and us, and as I said Sterling's nature was such as to make his essay thereon more than usually instructive.

With aspirations abhorrent of all <u>Vacuum</u>, with a soul that had sworn fealty to Truth and hatred to hypocrisy, yet with a want of a self-sufficing principle of reverence in him, he felt deeper than most the barrenness of the field he had fallowed. <u>He</u> with unusual sincerity and earnestness tried to requicken the old forms (since it must have a form) with his 'reason' to accept what his 'understanding' had falsified, and we see the result. At a crisis in his life he hastily grasped at the ecclesiastical anchor, became a Church of England Parson, prayed as earnestly as most could do that said Church would be his Spiritual Mother,—she would not answer him. Coleridge it appears had great influence on him in this respect. He was the originator of this scheme of reconciliation with the old defeated universalities. There is a chapter devoted to Coleridge[9]—A chapter of as trenchant character sketching as I remember ever to have read. In some respects though with a different result it has great similarity to Emersons picture of Socrates,[10] which in its way is a masterpiece. —Coleridge was one of the most brilliant surprising talkers that ever lived, and by the young thinking souls of the day was looked up to as a prophet, with a strange fascination they hung upon his eloquent words as the[y] rolled unceasingly from him, as if <u>there</u> was a living spring of inspiration Yet there was <u>one</u> amongst them too sturdy for Coleridges' eloquence to warp. I can see him, the future exploder of 'shams' and piercer of 'windbags' standing amongst that group, though in a retired corner of the same, listening with serious face to the words as they issued unceasingly forth, and at the end when all around him with looks of admiration pronounced it as the height of eloquence & wisdom, he with disappointed look turns silently away, his firm lips but opening to grumble forth a 'transcendent moonshine'.[11] Yes, Coleridge here is turned literally inside out; the young Carlyle continued to attend these literary reunions around the great Lion, returned ever dissapointed, puzzled at first whether to deem it inspiration or 'flat infatuated imbecility',[12] began at last to harbour treasonable opinions with regard to it, and at last came to the conclusion that 'it was talk not flowing any whither like a river, but spreading every whither in inextricable currents and regurgitations like a lake or sea; terribly deficient in definite goal or aim, nay often in logical intelligibility; <u>what</u> you were to believe or do, on any earthly or heavenly thing, obstinately refusing to appear from it. So that, most times, you felt logically lost, swamped near to drowning in this tide of ingenious vocables, spreading out boundless as if to submerge the world'.[13] Coleridge little thought how one of his hearers was <u>not</u> fascinated, but silently measuring him all this time, one who at some future day (1851) would give the world the result of said measurement:

'To the man himself Nature had given, in high measure, the seeds of noble endowment; and to unfold it had been forbidden him. A subtle lynx-eyed intellect, tremulous pious susceptibility to all good and all beautiful; truly a ray of empyrean light; - but embedded in such weak laxity of character, in such indolences and esuriences[14] as had made strange work with it. Once more, the tragic story of a high endowment with an insufficient will—an eye to discern the divineness of Heaven's splendours and lightnings, the insatiable wish to revel in their godlike radiances and brilliancies; but no heart to front the /seathing/ terrors of them, which is the first condition of your conquering an abiding place there . . . . For pain, danger, necessity, slavish harneshed toil, and other highly disagreeable behests of destiny, shall in no wise be shirked by any brightest mortal that will approve himself loyal to his mission in this world'.[15]

So far I had got when Carl Schmitt[16] came to pull me out for a walk, and thus put me out of my 'gang';[17] however, the above has served my purpose, for which alone it was undertaken you have perhaps, or at any rate must read the book for yourself, and /thus/ render what I or another reviewer have to say quite dispensable. Write soon, John, and tell me how the Toronto affair progresses and what you are at present doing with yourself. All is once more in working order here, and taking all together I never felt myself better able to meet it. The English Kräntzchen[18] has begun again, the last meeting was at Fraulein Baumbach's; Knoblauch was there, and in his element—relating his English journey. They make good progress in English and none more so than Fraulein Baumbach,[19] who at every such Aesthetic Tea-drinking raises herself in my estimation, she is unassuming, unaffected yet at bottom an able as well as amiable girl, were it not for Mathematics, and seeing her so seldom, one cannot tell what might happen: If I had nothing else to do I think sometimes there might be a possibility of even me, slow blooded as I am, falling in love. —But occupation is the greatest enemy womenkind have. Were it not for that, John Tyndall would before this have been spliced. 'Nicht Wahr'!![20] In absence of a wife, however, it may be some consolation to thee that thou hast
    an affectionate friend | Tom Hirst.

Monday evening—11 P.M. Just found out that I have forgotten to post my letters today—Ball in Ritter,[21] music as in days of yore comes wafted on the still breeze across the Ketzerback.[22] The lamp up at the Castle[23] glistening as usual, Moon almost at full tips the /Accacia/ leaves (now covered with snow!) with silver. I wonder if you're looking at the same moon now. I believe you are. | Lebe—wohl.[24]

RI MS JT/1/H/163

1. *9[-10] November 1851*: the postscript was written on Monday evening, that is, the following day.

2. *Thanks . . . Sterling*: [anon], 'The Life of John Sterling', *Literary Gazette*, 1813 (18 October 1851), pp. 701–3. It is unclear when Tyndall sent this review to Hirst. The last extant letter to Hirst is 0540, 1–2 October; since then Hirst had sent letter 0553, 19 October. Neither refers to Carlyle's volume.

3. *The following . . . be acceptable*: ibid., 1814 (25 October 1851), pp. 721–23. Tyndall did not think the second part merited sending to Hirst (see letter 0568).

4. *Biography by Archdeacon Hare*: Julius Hare, 'Sketch of the Author's Life', *Essays and Tales, by John Sterling*, vol. 1 (London: John W. Parker, 1848), pp. i–ccxxxii.

5. *his essay on Carlyle*: Sterling, 'On the Writings of Thomas Carlyle', first published in the *Westminster Review* 33, no. 1 (October 1839), pp. 1–68 and later in Hare, ibid, pp. 252–81.

6. *arrowed*: Hirst may have meant harrowed, or he may have been using Yorkshire dialect, which often dropped the h.

7. *'I have remarked . . . human walls.'*: Hirst quotes (accurately) from the final paragraph of the introduction to Carlyle's *Life of Sterling* (London: Chapman and Hall, 1851), p. 10.

8. *'he fashioned . . . its bewilderments.'*: from the conclusion, ibid, p. 343.

9. *a chapter devoted to Coleridge*: ibid, chapter 8, pp. 69–80.

10. *Emerson's picture of Socrates*: Hirst probably refers to the portrayal of Socrates in Emerson's lecture on Plato, published in *Representative Men*.

11. *'transcendent moonshine'*: Carlyle referred to Coleridge's philosophy as 'moonshine' on several occasions in the *Life of Sterling*; he twice described it as 'transcendental moonshine' (pp. 81 and 126).

12. *'flat infatuated imbecility'*: Hirst misquoted the passage that reads 'getting infatuated into flat imbecility' (ibid, p. 77).

13. *'it was talk . . . submerge the world.'*: ibid, p. 72.

14. *esuriences*: appetites (*OED*).

15. *'To the man . . . in this world.'*: *Life of Sterling*, pp. 78–79 (Hirst quoted accurately except that he wrote 'harneshed' for 'harnessed', and there are minor differences in punctuation).

16. *Carl Schmitt*: possibly the same person as Carl and Karl in letters 0479 and 0631.

17. *'gang'*: meaning unclear; if a German word was intended, 'gang' could mean 'course'.

18. *The English Kräntzchen*: see letter 0482, n. 6.

19. *Fraulein Baumbach*: Tyndall offered advice on Hirst's romantic interest in letter 0568.

20. *Nicht Wahr*: Isn't that so!

21. *Ball in Ritter*: there was a Rittersaal or Knight's Hall in Marburg Castle, which may have been a venue for balls.

22. *the Ketzerback*: the Ketzerbach is a main road in central Marburg; the Elizabeth Kirche stands at one end.

23. *lamp up at the Castle*: the castle was conspicuous above the city (see letter 0479, n. 9).

24. *Lebe—wohl*: farewell.

From
John Frederick William Herschel   10 November 1851   0566

Harley Street | Nov 10 / 51

Sir

I have received your packet of Testimonials and letters[1] to which all due attention will be paid—as well as the Memoirs accompanying them. At present of course I can only acknowledge their receipt

I am Sir | yours truly | J F W Herschel

J. Tyndall Esq

RI MS JT/1/H/89

---

1.   *your packet . . . letters*: Tyndall's application for the professorship of Experimental Philosophy and Chemistry at Sydney, sent with letter 0561. Herschel was one of those responsible for appointing the professors.

To William Francis   [15 November 1851][1]   0567

Queenwood Saturday—

My Dear Francis.

I send you a question <for> Dr Woods[2] if you like to send it to him. I have confined myself to one point which I imagine he cannot get over without involving himself in manifest absurdity.

I send you also a note[3] which you will oblige me much by publishing in the forthcoming number of the magazine. I am engaged with Magnus[4]—the investigation admits of being greatly drawn together without the slightest injury to its real merits.

I got the following note from Sir John Herschel a few days ago.

[. . .][5]

As you are not in a hurry for Elie's paper[6] I will delay it a little—that is if you do not give me instructions to the contrary.

I am my Dear Francis | most truly yours | John Tyndall

If you see any advantage in it I have no objection <to> your publishing all about Woods's affair as he suggests.[7] I am perfectly indifferent about it. It would however be a matter of amusement rather than of instruction.[8]

StBPL T&F, Authors' letters

1. *[15 November 1851]*: this was written 'a few days' after Tyndall had received letter 0566 (10 November) from Herschel, making it likely this letter was written on Saturday 15 November. As letter 0569 (also a Saturday in November, probably 22 November), postdates this letter, 15 November seems certain.

2. *question <for> Dr Woods*: Tyndall had promised to write again about Woods (letter 0563) and continues here the questioning method proposed in letter 0556.

3. *a note*: in response to letter 0560, Tyndall was seeking public clarification. Given that both his and Thomson's papers had been published in the *Phil. Mag.* this was the appropriate forum. His 'note' (which is missing), probably added his comments to a paraphrase of Thomson's letter. As events unfolded, only the last paragraph of the note was published (see letters 0569 and 0570).

4. *engaged with Magnus*: Tyndall was translating and—this letter suggests—condensing a paper by Magnus for publication in the *Phil. Mag.* (see letter 0556, n. 10).

5. . . . : Tyndall here copied out Herschel's letter (0566) acknowledging receipt of his Sydney application. The only differences are in capitalisation and punctuation (Tyndall uses fewer capitals and omits one dash), and in the signature (he expanded J to John).

6. *you are not in a hurry for Elie's paper*: refers to a paper by Élie Wartmann that Tyndall was translating for the *Phil. Mag.* (see letter 0563, n. 7). The letter from Francis about deadlines is missing (it would be dated 10–14 November).

7. *as he suggests*: later letters (0581) indicate that Woods had suggested that Tyndall's objections be printed alongside his paper; this was not done (see 0589).

8. *If you see . . . instruction*: postscript written vertically in the left hand margin of the first page of the letter.

To Thomas Archer Hirst   15[–16][1] November 1851   0568

Queenwood Nov. 15/ 51

My Dear Tom.

It is excessively cold this morning and I have no fire. My finger ends are cold and, what is worse, sore, being blistered by citric acid which I contracted a few nights ago in Romsey.[2] I lectured there on light to a full house and attentive audience, and illustrated the lecture by the electric light from 50 of Bunsen's cells.[3] I saw Kossuth[4] yesterday, he appeared a wearied man; bowed down either by the officious hospitality of the English nation which scarcely permits him time to sleep, or else by cares and sorrows of another kind. I received your bundle of translations.[5] I am glad you did not send them to Clayton[6] for they are very defective in many respects! now dont bristle up—I tell thee the plain truth. I read your letter first as a matter of course; there you were, free, fresh, and vigorous; I turned then to your translations and could see you through them—you were a machine, a stone mason miserably hacking up your solid material, and rounding off their corners to make them fit. They had undergone no smelting process, they were as I have said solid—you wrought with words not with the ideas to which words owe their symmetry and life, and consequently the poetic edifice instead of having its beams 'laid in music'[7] was a defective earthly production. Through some of the pieces however I can discern the soul of Schiller shining—a beautiful soul, warm with celestial radiances and such pieces shall receive all due attention. That fine piece of the Diver[8] you have not rendered with half the force which it is capable of and the last verse if I mistake not you have absolutely misunderstood, and thus deprived it of all its cadence of woe. I speak from memory, having no Schiller by me. I began to think; this boy is turning clod over his mathematics, but then I recollected your letter, and then again luckily came your letter to Debus, which fell like a shower of Eau de Cologne upon my hot and angry temples, cooling me and soothing me by the assurance that the gold of Ophir[9] had not yet become dim. To drop figures Tom, you must take care, great care, of your spelling—any thing you have doubt of refer to your dictionary about; this is perhaps dry work but it is essential, and <u>now</u> is the best time to look to it. You will find it 10 times as dry by and by. Again with regard to the poetry either make your rhyme good or make no rhyme at all, that is confine yourself to blank verse; you appear to have no ear for rhyme, you are worse than Emerson, which makes me believe that your proficiency at the bass fiddle &c was all a sham and that if the gods had put a fiddle in my hands when I was young I should have beaten you hollow. I just lay my hands by utter chance and without choice on the first bit of your translation that I meet—what is it—those beautiful 'words of faith'; by the way, not <u>your</u> words for if one could not see through these there would be little beauty in the affair. 'they echo' does not rhyme with 'they grow', 'terror' does not rhyme with 'ever', 'short' by no means rhymes with 'thought', and so of many others—What is worth doing at all is worth doing well and therefore I say if you rhyme see that you do it properly, if not give it up altogether, and turn to blank verse, or better to good fresh vigorous prose for here you are untrammeled. I will not imply by the latter sentence that rhyme is a trammel, I hardly believe it to be the case with some, the words come more freely with a *[jingle]* than otherwise, as sheep follow a bell.

Poor A.S.[10] I have given you a terrible thrashing—oh that I could see you rolling your great eyes over the above—an active fancy must be a substitute for the reality. Now and then however I picture a smile rippling across those broad features which proves that the feeling *[evoked]* by John with his pale face & cold fingers blistered by acid is not all of anger.

I have many things to say to you Tom regarding my doings and prospects, some of which you will not like to hear, and which I therefore hardly like to communicate. We must look bravely at destiny however although she does threaten our warm hopes and early visions—There is a prospect of my going to Australia; indeed I have become a candidate for the professorship of Experimental Philosophy in the new University which is about to be established at Sidney.[11] The choice of Professors rest with Sir John Herschell, Professor Airy and two others. The salary is £300 per annum three fourths of the students' fees which amount to £6 each per year £100 a year for the house and £160 for passage money—They also place £500 at the disposal of the physicist for the purchase of apparatus. The affair is liberally conceived; all sectarian distinctions are rigidly banished from it.[12] The University is to open her arms to all without reference to creeds. This is the last *[move]* Tom; what do you think of it?

This matter will soon be decided, perhaps before Christmas. If so I will take a scrip[13] in my hand and face the road to Marburg, purposing to see thee before I go.

This is Sunday morning; today at half-past ten I meet my friend D[r] Fox[14] get astride a horse and ride through the country for 3 hours.—

The ride is over and a magnificent one it was. On starting poured down a glass of warm sherry the sun shone clearly but there was a keen cutting wind; for a mile slowly then at it—you would have *[passed]* your eyes in admiration boy could you behold my horsemanship. Once in passing at full gallop under a beech tree a branch caught me and almost took the crown of my head away—my hat went! —At intervals the D[r] dropped in to see his patients and I drew into a sunny corner and there enjoyed the literal Sabbath stillness of the landscape. In one place a splendid old English Mansion stood before me, with its spreading lawns and lowering woods—the last beauty of autumn appeared to have been reserved for the especial glorification of the day. The beech was not all dismantled but reared its hills of red foliage over the surface of the wood; the taper birch yellow as gold peeped through here and there, while the naked twigs and branches mingled like a softening mist with the scene. I blessed the gods that England has such homes, but alas the other pole of the magnet stood hardly by—a squalid village, filled with an ignorant toil-ground peasantry—I have never seen a more decrepid[15] peasantry than that of Hampshire. We *[have]* got a parson here who receives £900 a year and for this he preaches an occasional Sunday and spends his weekdays seeking partridges in the Queenwood turnips[16] I have seen him at it, and were I near him I should have most assuredly told him he might be better employed. And I have heard that fellow imprudently open his mouth in the pulpit and preach to the crushed wretches beneath him of the manifold mercies of god and of the necessity of gratitude for the blessings which he pours on the children of men. However the ride ended, through turnips, over bad stubble, along smooth pastures, through winding wood walks over hills, down slopes back

again to Broughton at 3 where a plentiful leg of choice mutton and a generous glass of sherry awaited us; talked two hours 'across the walnuts and the
wine'[17]—and then home. Had you seen me after the ride Tom you would have
pronounced me positively handsome, the lilly and the rose so struggled for
mastery upon my cheek—I'm afraid however my nose shared in the general
blush, which some people might deem a detraction from my beauty. Many
a time during the day; especially during [these] dreamy quarters of an hour
which I enjoyed in the sun alone while the D[r.] was visiting, did you mingle
with my thoughts—I thought of the English Kränzchen[18] too and of other
things connected therewith. You speak of a certain fair girl;[19] as far as I know
[here] she is all you describe her to be; In fact I once felt the sunny side of my
own heart turning towards her; and [it] required a trifle of philosophy backed
by that never failing remedy, honest work, to keep me from feeling towards
her more tenderly than was prudent. Her mother is a proud woman, it has
sometimes been a problem with me how she managed to educate such a girl. It
is perhaps well that she was so, as a counter-pride on my part gave me a kind of
purchase which enabled me to shake the entire matter away from me like dew
drops from the Lion's mane. I have no advice to offer; if she were younger I
might even encourage a thought of the kind on your part. All I know and can
testify is that should you be 'crossed' your work will prove a perennial source
of recovery to you. I would balance myself by means of it; preserve my soul
from waste and foolish hazard;—in a word Tom I would not have you stake
too much upon such a cast.[20]

Now then, love, tenderness, poetry, and all such bottled moonshine
avaunt![21] and welcome the cold heights of science, twinkled over by the everlasting stars! —I must say that the reception which my labours have met in
England is sufficient to satisfy a vainer fellow than I am. I receive a note from
Col. Sabine, Treasurer of the Royal Society, a few days ago of which the following is a copy.

[…][22]

I must now shake you off; but not without a word about your last letter.[23]
It reached me since the commencement of this; the allusion to Fraulein V.B.
will have apprized you that it has come to hand. It tells me a good deal about
Sterling which I could never learn through the brains of that muddy reviewer.
I did not send you the next number[24] because it was flat, stale, and unprofit-
able. The fellow began[25]

I wont write anymore but merely say that the letter was manna to my hungry soul.[26]

Tyndall

RI MS JT/1/T/546

1.	*[–16]*: this letter took at least two days to write. Tyndall began on 15 November and about half way through notes that it is now Sunday morning (16 November). Given the length, he may not have finished it on Sunday.

2.	*a few nights ago in Romsey*: 11 November (letter 0557, n. 1).

3.	*Bunsen's cells*: primary cells (or batteries), as devised by Robert Bunsen, consisted of a zinc cathode immersed in dilute sulphuric acid and a carbon anode immersed in concentrated nitric acid. The electrolytes were separated by a porous pot. Each cell gave an e.m.f. of about 1.9 volts.

4.	*Kossuth*: Louis Kossuth (1802–94), Hungarian nationalist leader in the 1848–49 revolution. See Journal, 14 November (JT/2/13b/551), for a fuller account of his encounter with Kossuth.

5.	*your bundle of translations*: sent with letter 0553.

6.	*Clayton*: Joseph Clayton, of the *Leader*. Hirst had asked Tyndall to forward to the *Leader* any of his translations that were worth publishing (ibid.).

7.	*'laid in music'*: a phrase from a poem written in praise of the immensely famous opera singer, Jenny Lind: 'O world within, that lies between our souls and God's above | Whose beams are laid in music and whose dome is arched in Love' (Samuel H. Lloyd, 'Jenny Lind' [1849], published, for example, in *Glimpses of the Spirit-Land. Addresses, Sonnets, and Other Poems* [New York: 'for private distribution', 1867], pp. 57–58).

8.	*the Diver*: cited letter 0553, n. 28.

9.	*gold of Ophir*: Ophir was a port or region mentioned in the Bible, that was famed for its wealth. King Solomon and the Tyrian king sent an expedition to Ophir that brought back large amounts of gold, precious stones, and 'algum wood' (for example, 1 Kings 9:26–28).

10.	*A.S.*: Archer Stanley, a pseudonym occasionally used by Hirst.

11.	*the professorship . . . at Sidney*: see letters 0558 and 0561, and journal entry for 4 November (JT/2/13b/550).

12.	*sectarian distinctions . . . banished*: Tyndall emphasizes the difference from England where, at this time, Oxford and Cambridge colleges were restricted to students willing to affirm Anglican doctrine and conform to Anglican forms of worship. King's College, London, was more 'liberal' or free of religious tests for students, but required its professors to be Anglican.

13.	*scrip*: (archaic meaning) a small bag, pouch or wallet carried, typically, by a pilgrim, shepherd or beggar (*OED*).

14.	*Dr Fox*: a doctor from the near-by village of Broughton. Six of his sons were Queenwood pupils (Brock, 'Queenwood', p. 16).

15.	*decrepid*: a common alternative spelling of decrepit until the early twentieth century (*OED*).

16.	*seeking partridges . . . turnips*: hunting, a past-time of the wealthy, across Queenwood farm land.

17.	*'across the walnuts and the wine'*: a quote from Tennyson, 'The Miller's Daughter', first published in *Poems* (1833). Hirst quotes from the greatly altered, later edition (vol. 1; London: Edward Moxon, 1842, pp. 102–14, last line of the 4th stanza).

18. *English Kränzchen*: see letter 0482, n. 6. Hirst had mentioned it in his last letter (0565).

19. *a certain fair girl*: Fraulein von Baumbach (Fraulein V.B., below), had attracted Hirst's interest.

20. *such a cast*: such a chance, that is, depending on the cast of a die or dice.

21. *avaunt*: archaic term for begone! (*OED*)

22. *[. . .]*: here he copied the entirety of letter 0559.

23. *your last letter*: letter 0565. This letter began as a reply to letter 0553, but, in the course of writing it Tyndall received 0565.

24. *the next number*: Tyndall had sent Hirst an issue of the *Literary Gazette* containing the first part of a review of Carlyle's *Life of Sterling*; Hirst had asked Tyndall to send the issue containing the second part (see letter 0565).

25. *The fellow began*: at this point Tyndall reached the end of the sheet and, it seems, decided not to start another sheet. The farewell is written down the left-hand margin of the last page.

26. *manna to my hungry soul*: the phrase is from the famous hymn, 'How sweet the name of Jesus sounds' (written by the converted slave-trader, John Newton) and is still widely used to indicate spiritual or emotional sustenance. 'Manna' was bread that came down from heaven to feed the people of Israel (see Exodus 16:14–15 and 16:31–5). Tyndall emphasized that Hirst's second letter cheered his inner being.

To William Francis [22 November 1851][1] 0569

Queenwood Saturday

My Dear Francis.

I send you Riess's paper,[2] Thomson's note and a title for the former. I quite chime in with your suggestion of tacking my last paragraph as an addenda to Thomson's note;[3] it will be quite sufficient. Many thanks for your information regarding the 'Royal'[4]—£20 is an important reduction—I will endeavour to have the matter to deferred a little as it would be inconvenient to me to cash out just at present.

I hope Hunts Physics[5] is not a large one, and that it will give me occasion to praise instead of to find fault. Reuben's remark[6] is certainly amusing A man who knows the true nature of a scientific investigation would have never made such a remark—One may expect crudities from such a man but there are a million chances to one against his producing

<the two manuscripts are separated in the Taylor and Francis archive but clearly fit together>
any thing really valuable.

I am engaged at Magnus[7] and will have it ready for you at the beginning of next month. I will send you Wartmann's paper[8] about the same time. The following is a copy of a note *[rec'd]* from Sir John Herschell acknowledging the receipt of my application,[9]

Sir.

I have received your packet of Testimonials and letters, to which all due attention will be paid as well as to the memoirs which accompany them. At present of course I can only acknowledge their receipt.

I am &— | John FW Herschell

[We] shall soon know the issue of the matter; probably before christmas.

most faithfully dear Francis yours | John Tyndall.

'Elastic—slightly elastic!' —There is *[some]* masonry[10] here—There is! <u>after</u> the Red Lion dinner I could understand it but beforehand it is incomprehensible—perhaps they <u>were</u> tacked on afterwards eh?[11]—

Could you arrange to send me Faraday's two volumes of 'Researches in Electricity'[12] at the end of the month? I have still another favour to beg of you—I mentioned to you[13] that Poggendorff was[14] getting up a volume of biograp<hical> notices of people who have occupied themselves with the exact sciences—Do you know <an>y English work or works which contain such biographies? If such should occur to you I should <be> <mu>ch obliged if you would note *[them]* down and send the titles to me when you have occasion to *[write]*.

StBPL T&F, Authors' letters

1. *[22 November 1851]*: Many allusions show this letter to be closely related to, and written after, 0567 (15 November). Tyndall saying he expects Magnus and Wartmann to be done by the beginning of the next month rules out a December date and makes Saturday, 29 November unlikely. See also the allusion to Royal Society costs (discussed n. 4 below and letter 0572, n. 2).

2. *Riess's paper*: see letter 0514, n. 4 for original publication. Tyndall abstracted two papers by Riess under his 'Reports on the Progress of the Physical Sciences', *Phil. Mag.* 3, no. 17 (March 1853), pp. 173–85; subtitle 'On Electric Currents of the First and Higher Orders'. Tyndall had planned to send the paper earlier (letter 0557). We have no explanation for the delay.

3. *Thomson's note . . . my last paragraph . . .* : the letter containing Francis's suggestion is missing. Thomson's note was probably a letter which Thomson wrote directly to the *Phil. Mag.* (not letter 0560 to Tyndall), which Francis then sent to Tyndall after receiving his request in letter 0567, with the proposal that, rather than publish the letter Tyndall had sent him, he publish Thomson's short letter followed by the one paragraph from Tyndall's original letter. This arrangement was adopted (see letter 0570).

4. *the 'Royal' . . . deferred a little*: Tyndall is here concerned about the costs of being elected FRS, for which Sabine had offered to nominate him (letter 0559). Tyndall had informed

Francis of the offer on 9 November (letter 0563). It seems that, in a missing letter or letters, Francis mentioned expense, because (as indicated in letter 0572) concern about cost of membership led Tyndall to write to Sabine.

5. *Hunts Physics*: Hunt's *Elementary Physics: an Introduction to the Study of Natural Philosophy* (London: Reeve and Benham, 1851) was nearly 500 pp. long. Presumably Francis had asked Tyndall to review it (in a missing letter). In his review, Tyndall was strongly critical of the volume (letter 0588).

6. *Reuben's remark*: Tyndall alludes to Reuben Phillips, who had recently written a paper for publication in the *Phil. Mag.* (published in the January 1852 number). Phillips's paper is dated 19 November but Tyndall had commented on a version in letter 0550, 15 October (cited n. 8). Whether the remark was in a version of the paper or in a letter to Francis cannot be determined.

7. *engaged at Magnus*: Tyndall was translating and condensing a paper by Magnus (cited letter 0556, n. 10). He sent it to Francis with letter 0581 (see n. 20).

8. *Wartmann's paper*: the translation was eventually published as Élie Wartmann, 'On the Polarization of Atmospheric Heat', *Phil. Mag.* 3, no. 16 (February 1852), pp. 108–11. No extant letter indicates when it was sent to Francis. Original cited in letter 0563, n. 7.

9. *note . . . my application*: his application for a professorship in Sydney. Tyndall copied letter 0566, from Herschel, for the second time (previously copied in letter 0567) to Francis, making similar minor errors in copying; we can only assume that he had forgotten he had done so.

10. *masonry*: perhaps Tyndall alludes to the semi-mystical, ritual practices of freemasonry with which the Red Lions had some parallels.

11. *Elastic . . . eh?*: the Red Lions was a dining club, initiated at the BAAS meeting in 1839, as a protest against the expense and pomposity of the Association's formal dinners; it had a 'London pride' (f. 1844). The meetings became ritualistic and rambunctious: for example, a lion skin for the chairman's chair and roaring in place of applause (Hannah Gay and John W. Gay, 'Brothers in Science: Science and Fraternal Culture in Nineteenth-Century Britain,' *History of Science* 35 [1997]: 425–53). We cannot identify the meeting attended by Tyndall: he did not attend the London dinner regularly until he lived in London; perhaps he attended during the Ipswich BAAS meeting (but his journal is empty for that period).

12. *send me . . . Electricity'*: for earlier discussion of these volumes see letters 0536 and 0538 n. 6.. Francis sent them as requested (see letter 0581).

13. *mentioned to you*: letter missing.

14. *was*: the first part of the postscript is at the bottom of the last page of the letter, however Tyndall ran out of room and from here onwards it is written vertically in the left hand margin of the last page of the letter.

To the Editors of the *Phil. Mag.*   [22][1] November 1851   0570

I have only to say that the facts are precisely what they are here stated to be. Previous to writing the remarks in question, I looked to the Errata, but not it seems with sufficient attention, for Professor Thomson's correction escaped me. Not only do our results agree in principle, but the same substance and form of substance which Professor Thomson had referred to in illustration of his theory was unwittingly examined by me in Berlin, and the exact result which he had theoretically predicted arrived at by way of experiment.

Quoted in the editor's note to William Thomson, 'Magnecrystallic Property of Calcareous Spar' (letter to editors), *Phil. Mag.*, 2:14 (December 1851), pp. 574–75.

---

1. *[22]*: this note, an addendum to a letter by Thomson, was the last paragraph of a letter Tyndall had written on or shortly before 15 November (see letter 0567). Francis suggested (letter missing, but Tyndall agreed in letter 0569) that it would be sufficient to publish the last paragraph. Thus, although the letter was written earlier, its published form was decided only on 22 November.

  Thomson's longer letter was an explanation of how an error of transcription in his published paper had led to apparent disagreement between his and Tyndall's results. The published letter from Thomson was longer and clearer than his original letter to Tyndall (letter 0560) although it is also dated 7 November, thus Thomson wrote his explanation to the *Phil. Mag.* on the same day that he wrote to Tyndall.

From Edward Sabine   24 November 1851   0571

Woolwich Nov. 24./51[1]

My dear Sir

Will you be so kind as to give your correct designation, such as it must be entered in your certificate as a candidate for the R.S.—

Name (including Christian names)
Rank
Profession   } Title or Designation
Qualification, Author of —
Usual Place of Residence
Distinguished for his
Acquaintance with the science of —

M[r] Faraday will sign from personal knowledge, and there must be two other signatures from <u>personal</u> knowledge,[2] who must be, of course, Fellows of the Society. Can you kindly tell me, who you would wish me to apply to for this purpose; I shall find no difficulty in obtaining three other signatures from <u>general</u> knowledge, from persons who have read your papers: six signatures is the number required.

The titles of your publications must be specified: will you be so good as to write them down for me.

Sincerely yours | Edward Sabine.

D[r] Tyndall.[3]

RI MS JT/1/S/6

---

1. *Woolwich Nov. 24./51*: this letter was enclosed with letter 0572.

2. *two other … personal knowledge*: Tyndall chose J. J. Sylvester and T. H. Huxley (Journal, 27 November and 25 December 1851, JT/2/13b/552 and 556) but the approaches were made by Sabine or Francis, rather than directly. For example, Sylvester's letter agreeing to supply a signature was passed on to Tyndall by Francis and Tyndall initially passed on his thanks to Huxley through Francis (letters 0581 and 0590).

3. *D[r] Tyndall*: as in the earlier letter (0559) from Sabine, the addressee's name was written at the bottom of the first page rather than at the end of the letter.

From Edward Sabine   24 November 1851   0572

Woolwich Nov. 24./51:

Dear Sir

I had written the enclosed,[1] when your note[2] arrived; and I accordingly suspend any further steps.

The entrance fee is £10—the annual subscript[n] £4, commutable at any time by a payment of £40 by those who have contributed a memoir printed in the Transactions—The Phil Trans. are given gratis to Fellows: it is in fact their subscriptions which pay the expense of the Transactions.

Sincerely yours | Edward Sabine

There is but one election in the year—i.e. in June, but all candidates must be declared before the 1st of March.

RI MS JT/1/S/5

---

1. *the enclosed*: letter 0571.

2. *your note*: missing. From letter 0560 (22 November) it appears that Tyndall was concerned

at the cost of the FRS. He must have written to Sabine almost immediately, on either 22 or 23 November, suggesting delaying the nomination. His concerns, however, were rapidly allayed and when he received this pair of letters from Sabine he continued the nomination process (Journal, 27 November, JT/2/13b/552).

From George Wynne   26 November 1851   0573

Board of Trade | Whitehall | 26 Nov. 1851

My dear Tyndall,

I do not deserve that you should write to me for I am acting shamefully in leaving your letter[1] unanswered—but I have been very busy for some time back, and my journies to and out of town daily[2] occupy so much time that I may say I am chasing time all the rest of the day. A vain pursuit you will say—I read with real pleasure the testimonials you sent me and Mrs Wynne read them out to Mr Russell[3] who was greatly pleased with them and his expression was that of regret at a probability of your being withdrawn from the country—I should greatly wish to see you established here, but with the easy communication that now exists with America I do not look upon a residence there as the banishment that others do, there would be a great field of usefulness before you and you would have time and means to prosecute the studies you have here engaged in. I would much prefer seeing you established in this country but if that does not offer I would not regret the other—Were I to find myself placed in a position similar to yours, I would seek at the highest source[4] to have my way made plain before me, and when such information is sought in faith it is seldom sought in vain—perhaps you will think this foolishness, but if you do you will I feel assured respect the motive which makes me speak on a subject on which I feel strongly but to which I am not in the habit of alluding except with those that I know well.

I have at the present time for no more—pray keep me au fait[5] as to your movements and prospects and believe me to be
    most truly yours | Geo. Wynne

RI MS JT/1/TYP/5/1843
LT Transcript Only

---

1. *your letter*: missing.
2. *my journies . . . daily*: Wynne lived at Harrow (see letter 0577). There was a train link from Harrow to Euston station (distance approx. 12 miles), but he then had to get about 2 miles south across London to Whitehall.
3. *Mr Russell*: John Scott Russell, Wynne's brother-in-law.

4. *highest source*: the Christian God. As the letter goes on, it becomes clear that Wynne expects divine guidance will be given when requested.

5. *au fait*: fully informed.

From Josiah Singleton 27 November 1851 0574

Spring Bank, | Over Darwen, |
Lancs. | Nov. 27, 1851

Son[1] Tyndall,

Hearken—and be sure thou pay'st[2] attention now,
For I'm a loving father, but a rebel son art thou;[3]
By far too fond of star-dust and too much attached to rags
and tatters that are hardly worth the gathering into bags.
Nor know I how to purge thee, John, of all the atrocious stubble
thou gatherest and heapest up with such a world of trouble.
And when I think upon it, John, my heart, alas! doth ache,
that thou should'st fill thy belly with the husks that swine[4] should take.
But patience, John, I know indeed that thou art far above
my little skill to mend thee, or my feeble power to move.
So I try to take the wise advice that Jesse's son[5] has given,
and fret not, John, when naughty men desert the road to Heaven.
Still, still one truth remains, dear John, when all is said and done,
That I'm thy loving father, and tho' naughty, thou'rt my son.
And the wish to see thee once again before I pass away,
springs warmer in my bosom, John, with each returning day.
So come at Christmas, John, my boy, and with my best I'll treat thee,
and now and then, as chance occurs, most probably may beat thee.
For sure, son John, it's always safe in such a way to greet thee,
And thrash away with conscience clear whenever one may meet thee.
For if thou'rt not in mischief now, it matters little, John,
Thou art plotting to begin it soon, or else it's just been done.
But notwithstanding, come, old boy, we'll kill the fatted calf,[6]
And if we do not dance we'll sing, and many a time we'll laugh.
And on the fattest of the land we'll live, son John, you'll see,
For I've two Irishmen[7] to kill and jolly ones they be.
And though no puddings black[8] you'll find, nor such unchristian food,
No doubt we'll please your carnal mind with what will be as good.
  So come, old fellow, and we'll try if 'Tris'[9] we can't persuade
to join us and to help the fun with his most potent aid.
And if we don't kick up a shine and merry merry be,

I think that you'll acknowledge, John, that sadly changed are we.
 And furthermore, to M.Haas my kind remembrance bear;
And tell him that we wish he'd prove our welcome and our fare.
It is but little that we have, we're neither great nor gay.
But never fear we'll manage, John, to keep the wolf away.
And if an English welcome warm can make him feel at home,
He shall not wish himself away nor sorrow that he's come.
 And now I do not know, my son, that further much remains,
Than to wish you may have pleasure from this sample of my pains.[10]
We're as we were when last I wrote, in all essential things,
And though Christmas takes away some boys, yet others Christmas brings.
We've got some colds, and chilblains some, and some are coughing badly,
And some are working very hard, and some are idling sadly.
And some grow fat but I grow thin, and some are restless ever;
And some are dull and thick i' th' skull, and some are very clever.
And little wife and little son are both of them but ailing,
For Mistress coughs most dreadfully[11] and baby's often wailing.
But as I've fetched the doctor I suppose they'll soon get well,
Although it needs a 'prophet', John, the very day to tell.
 And so you see of ups and downs we get a decent share,
And like our neighbours we are made to taste of 'neighbours fare.'
But more I need not tell you; it's a waste of pains and time,
To prolong a tedious twice-told tale and turn it into rhyme.
For a 'Martyr' and a 'Prophet' as you've been for many a year,
will see as well at Queenwood as a common man would here.
And now farewell again, my son, your mother sends her love,
And hopes you'll mind the good advice I've given you above:
And thinks as I'm so busy now 'tis time I made an end.
So believe me yours, | J. Singleton, | Your father and your friend.
Dr John Tyndall, | Queenwood College, | Stockbridge, Hants.
P.S. Write soon—say Yes,[12] and may your soul repose in bliss. But if you
don't, I wouldn't be where you'll be sure to go,[13] you'll see.

RI MS JT/1/TYP/11/3861–3862
LT Transcript Only

1. *Son*: Singleton had been a good friend of Tyndall and Frankland when they taught at Queenwood in 1847–48. They had called Singleton 'father' and, as this letter demonstrates, he called them his sons, in an affectionate and humorous style.

2. *thou pay'st*: for 'you pay'. Unlike the rest of his family, Singleton was not a Quaker, but he uses Quaker forms of expression here (see letter 0478, n. 3).

3. *Harken . . . now, For . . . thou*: this entire letter is written in rhyming couplets and we present it here in verse format. The LT transcript is not in verse format.
4. *fill thy belly . . . swine should take*: a near-quotation from the New Testament parable of prodigal son who was so hungry that he ate pig food (Luke 15:11–32; Singleton draws on 15:16: 'And he would fain have filled his belly with the husks that the swine did eat: and no man gave unto him'). Singleton's style is shaped by the language and cadences of the King James version of the Bible.
5. *Jesse's son*: Jesus Christ in Christian tradition.
6. *kill the fatted calf*: another allusion to the parable of the prodigal son. When the son returned, his wealthy father 'killed the fatted calf' (Luke 15:23) and held a feast.
7. *two Irishmen*: this seems to be an allusion to pigs. Pig farming was an important export industry in nineteenth-century Ireland, but we can find no source which supports this meaning.
8. *no puddings black*: some Christians followed Jewish food prohibitions and did not eat black pudding because it was made with blood.
9. *Tris*: nickname for Edward Frankland, given after he discussed alchemy and Hermes Trismegistus in a lecture on the history of chemistry.
10. *sample of my pains*: allusion to writing the letter in rhyming couplets.
11. *coughs most dreadfully* (and above, *coughing badly*): when consumption/tuberculosis was a common cause of death, allusions to bad coughing could be ominous.
12. *Yes*: this may be a continued rhyme, although with shorter lines, compare Yes/bliss and be/see.
13. *where you'll . . . go*: a joking allusion to hell, in contrast to the bliss of the previous sentence.

To Michael Faraday　　　28 November 1851　　　0575

Queenwood College | Stockbridge, |
Hants. | 28th Nov. 1851

Dear Sir,

From a letter received yesterday from Col. Sabine[1] I conclude that you are already aware of his kind intentions towards me In that letter he tells me that your signature will be one of those from personal knowledge required by the Royal Society previous to the admission of a new member. For this allow me to return you my sincerest thanks; I do most fervently hope that the great kindness of which I have of late been the recipient from you and others will bear its proper fruit in keeping me faithful to my work.[2] Few have had greater encouragement than I have had, and the greater will be my condemnation if it be not turned to suitable account.[3] I owe you much—more perhaps than you are aware of. However sweet the instances of personal good will with which you have favoured me may be,—and sweet they are beyond a doubt—you have been my unconscious benefactor in other ways to an extent which, as far as the cultivation of my proper manhood is concerned outweighs those instances of private friendship. To me you have been a preacher of toil, courage, and humility—a preacher in the highest sense of the term,[4] demonstrating by the silent power of example the possibilities which are open to the honest worker. This has been your relation to me and may it long continue so.

I remain dear Sir, | Your faithful and obliged servant | John Tyndall

Professor Faraday | etc. etc.

RI MS JT/1/TYP/12/4014

LT Transcript Only

1. *letter . . . from Col. Sabine*: letter 0571, 24 November. Most letters from London reached Queenwood the following day but Sabine's letter took 3 days to reach Tyndall.

2. *bear its proper fruit . . . faithful to my work*: this letter is full of biblical and Carlylean metaphors and concepts. Bearing fruit and being faithful are Christian metaphors; work and toil are Carlylean.

3. *greater encouragement . . . greater will be my condemnation*: Tyndall alludes to the statement in Luke 12:48, 'unto whomsoever much is given, of him shall much be required'.

4. *a preacher in the highest sense of the term*: that is the inner, almost transcendental, meaning; not the mundane sense.

From John Hall Gladstone   28 November 1851   0576

Stockwell | Nov. 28th. /51

My dear Dr. Tyndall,

So pleased was I with the spirit of your last letter[1] that I was near replying to it immediately in order to assure you more fully than heretofore of my deep-felt sympathy with your feelings; but more pressing engagements interfered, and now above a month has past

I like much your remarks about the religion of the head, and the religion of the heart. How strikingly do we often see the two displayed in the characters of different Christians! And I think I can admire your cast of mind, akin to that of the apostle John, the more because I feel myself to resemble rather the apostle Paul. As far as a man's own self is concerned the religion of the heart is of course the all-important; but if Christianity is to be studied and taught, we must have also the religion of the head. God has given us the means of attaining both. He has given us the Bible, where, in a series of narratives and miraculously attested declarations, certain <u>doctrines</u> (undiscoverable by mere human reason) are revealed; and he has given us examples of renewed men, both in the Bible and out of it, and the Holy Spirit, to teach us the <u>sentiments</u> of religion. The religion of the head and the religion of the heart ought surely to be combined: in fact I doubt whether either can exist in any high degree of perfection without the other. Religious knowledge of itself is apt, as you see very clearly, to run into formality and become petrified; but religious feeling of itself is equally liable to become fantastic and unstable. Love is certainly what God requires of us; but I think the more thoroughly we know him, and believe what he has said of himself, the more we shall love him. It is our beloved Saviour himself who said—'This is life eternal, that they should know thee, the only true God, and Jesus Christ whom thou hast sent'.[2]

Pray excuse me, my Christian brother,[3] if I have seemed here to sermonise; I could not write what I wanted so well in another manner. And these are certainly important considerations, especially (it appears to me) in the present day, when, perhaps from far too exclusive an attention having been paid to the 'form', thinking men are now disposed to make light of scientific theology, and attempt to arrive at truth merely through 'sentiment'. I fear there is a secret mistrust either of God himself, or of the evidence of revelation, at the bottom of this: at any rate in dealing with our fellow creatures the genuineness of that affection would be questioned which should lead a man to avoid examining the lineaments and character and wishes of the beloved object, and confine his regards to the ideal of his imagination.

Well, it is delightful to trace the workings of the same divine power on minds differently constituted: this is part of the charm of Christian fellowship. All the Lord's children have a genuine resemblance amidst their specific differences: and you and I can both love and enjoy and glorify Him, and feel ourselves even whilst in the land of our pilgrimage 'complete in him';[4] and 'what we know not now we shall know hereafter'.[5]

As to scientific news I suppose you hear what is going on, even though in so outlandish a place as Queenwood. The English savants seem to have spent the summer in gazing at the Exhibition,[6] and to have little in the way of discovery to announce. However something will I dare say be forthcoming. The Chemical Society meets now in very smart rooms of its own, and has had large presents to its museum.[7] Liebig has now got to work again at Giessen, the better I believe for his English visit.

If I did not know that it takes much shorter time to read a letter than to write it, I should fear I had thoroughly tired you, but believe me My friend
Yours sincerely | J. H. Gladstone.
Dr. Tyndall.

RI MS JT/1/TYP/1/408-409
LT Transcript Only

1. *your last letter*: letter 0546. This letter is the fifth letter in which Gladstone and Tyndall discuss their religious beliefs.
2. '*This is . . . hast sent*': John 17:3.
3. *my Christian brother*: here and throughout this letter, Gladstone identifies Tyndall's Carlylean religion as within the Christian tradition.
4. '*complete in him*': Colossians 2:10, 'him' being Jesus Christ. Jesus was a moral teacher in Tyndall's religion, but at the heart of Gladstone's devotional practice.
5. '*what we . . . hereafter*': John 13:7: 'Jesus answered and said unto him, What I do thou knowest not now; but thou shalt know hereafter'. Gladstone was pushing Tyndall to make more definite statements, rather than to rest satisfied in his 'religion of the heart'.
6. *the Exhibition*: the Great Exhibition.
7. *The Chemical Society . . . its museum*: the Chemical Society had recently relocated from 142 Strand to 5 Cavendish Square, where it rented 2 rooms (that were part of the premises of the Polytechnic Institution) for £120 a year (*The Jubilee of the Chemical Society of London* [London, 1896], pp. 156–57).

To Elizabeth Steuart   29 November 1851   0577

Queenwood | 29th. Nov. 1851

Dear Madam,

It is a long time since I had the pleasure of hearing from you.[1] As I appear to be now approaching a new phase in my career, in accordance with my usual custom I write to you upon the subject.[2]

The accompanying list of Testimonials will throw some light upon my movements. I am a candidate for the Professorship of Natural Philosophy in the University of Toronto, and indeed also in a newly established University in another Colony.[3] My chances of success are good.

It would have been better if I could have established myself in England, but of this I see no immediate prospect and it would therefore be culpable on my part to allow the present opportunities to escape me.

My testimonials as you see[4] embrace men of various European countries, and I believe I may safely say that they embrace the best in each. I have worked hard for the last 3 years but the reward is more than I ever dared to promise myself. Only a few days ago I received a note from Col. Sabine[5] Treasurer of the Royal Society proposing to put matters in train for my election as a Fellow of the Society—I send you the note and will thank you to enclose it when next you favour me with a letter.

I often receive letters from my valued friend Captain Wynne I know no man for whom I have a deeper esteem; certainly his bearing towards me is more like that of an elder brother than otherwise; nor can I speak too highly of the kindness of Mrs Wynne—I have spent some most gratifying hours with them at Harrow. I send you Capt. Wynne's last letter[6] to me, and will thank you to send it to me again when you write to me.

I took up a volume of the poet Burns's letters[7] a few days ago and could not help drawing a parallel between his case and my own—I derived both instruction and warning from the reading of a few of them; my pursuits have a more cool and chastening influence than his had, and I believe my natural temperament is also more sober, so that I am thus comparatively protected from the rock upon which he, poor fellow, split.[8]

I shall now bid you good bye for the present, and with every kind wish for the continued happiness of yourself and Captain Steuart

I am dear Madam | most faithfully yours | John Tyndall.

RI MS JT/1/TYP/10/3330
LT Transcript Only

1. *long time . . . from you*: the last extant letter from Steuart to Tyndall was five months earlier (letter 0502, 16 July 1851).

2. *usual custom . . . the subject*: Steuart was an early patron of Tyndall's.

3. *University in another Colony*: a university in Sydney, New South Wales.

4. *My testimonials as you see*: Tyndall was enclosing a copy, as letter 0586 indicates, not just a list of names.

5. *note from Col. Sabine*: Tyndall enclosed his highly valued letter 0559.

6. *Capt. Wynne's last letter*: letter 0573.

7. *a volume of the poet Burns's letters*: the edition which Tyndall was reading cannot be determined. Many single volume editions of *The Poetical Works of Robert Burns* were published in the nineteenth century.

8. *the rock upon which he, poor fellow, split*: possibly an allusion to his promiscuity or, more likely, his reputed alcoholism, to which his early biographers often attributed his early death (*ODNB*).

From Anne Edmondson   [30 November 1851][1]   0578

Dear John Tyndall,

My boy[2] having failed as a messenger of peace from me to thee,[3] I will try a second time through the medium of my pen; not with a view to enter into explanations as before but to induce thee to let the affair[4] be considered settled between us. —If I understand right I have offended thee by interfering to remedy a mistake into which I thought thou had inadvertently fallen, uncourteousness towards myself followed on thy part.

Let both be consigned to oblivion, not that of forgetfulness, but of <u>forgiveness</u> and let us meet as if no misunderstanding had occurred[5]

Thy sincere friend Anne Edmondson.

RI MS JT/2/6/80
JT Transcript Only

1. *[30 November 1851]*: the source for this letter is Tyndall's journal (JT/2/6/80), copied in on 30 November, the morning it was written and received.
2. *My boy*: her son, George William.
3. *Dear John Tyndall . . . thee*: see letter 0478, nn. 2 and 3, on Quaker usage.
4. *the affair*: this occurred on the evening of 27 November (Tyndall had avoided the Edmondson's since). It is explained, from Tyndall's perspective, in the following letter.
5. *Let both be consigned . . . no misunderstanding had occurred*: after this sentence, and at the bottom left of the page in his journal, Tyndall added 'most magnanimous!' in rather bold handwriting.

To Anne Edmondson   [30 November 1851][1]   0579

Dear Madam,

The note[2] with which you have favoured me is doubtless the natural result of your manner of viewing the subject; but an object may change its aspect by changing the point of view, and hence the only way of forming a correct judgment seems to be to walk round the object and view it upon all sides. Might I beg your kind attention to the following statement of the manner in which I have acted throughout this affair; it will be interspersed here and there with my present impressions as to the said manner of action.

On thursday I went into the dining room, Mr Wright[3] was at the fire, he turned towards me and said 'you have not invited Battersby, he also is a farmer'.[4] I thanked him for reminding me of it and indeed it gave me the most lively gratification to be able to shew a kindness to so good a boy. I believe I threw this feeling into my invitation, for joy beamed from the youth's countenance, and certainly found a reflexion in my own breast. I heard nothing more of the matter until 7 oC. that evening when I came down stairs in good spirits and prepared to enter upon the provided amusements with cheerfulness, when I noticed Battersby undressed. I asked him the cause, and he replied that George William[5] had been sent down to say that he was not to go. The tears rose to the boy's eyes as he spoke to me; I told him that the fault was mine, that I had mistaken your wishes, and expressed less sorrow at the circumstance than I really felt. I have been accustomed to regard boys of his years as beings possessed of feelings, and Battersby peculiarly so from natural temperament, and I must confess that at that particular moment the prohibition appeared to me to savour of something like cruelty. I went up stairs to Mr Edmondson and learned from him that you had so arranged the matter. I did not then decide upon my own course of conduct, but the more I thought of the matter the less sympathy I felt with the festivity, and at the end I knew it would be a mockery for me to join in it, for I could neither enjoy it myself nor contribute to the enjoyment of others. This, and not any sudden act of anger upon my part, was the cause of my absenting myself. George William came to me at 9 oC. The substance of his communication so far as I recollect it was that Battersby did not belong to N° 9, that you meant to restrict the invitation to N° 9 and that furthermore you had invited a German Lady[6] and expected when you did so that you would have somebody to speak to her. I should have been very glad indeed to have entertained your guest, but under the circumstances I felt my inability to do so in a proper manner, and therefore thought it better not to make the attempt.

I believed, and still believe, that it would have been not only kind but politic to have admitted Battersby; it must also be borne in mind that I was totally ignorant of the circumstance that the boys were to have a feast also, and did not learn until the afternoon of the next day. True Mr Edmondson said to me that Battersby was intended to come in with another party, but his communication was so brief that I actually understood him to mean that the boy would be admitted at a later hour in the evening. Had George William been entrusted with the simple message that Battersby had permission to come I should have gladly accompanied him; as it was, I should have considered enjoyment or gaiety a reproach to me. Even had the true state of the matter been explained to me in time, I might have contrived to let Battersby down softly and thus spare him a painful disappointment; but nothing of the kind was done and I acted upon the knowledge which I possessed.

I would submit it to any court of etiquette in the world whether in thus deporting myself I acted an uncourteous part. Sure I am that every man who still retains so much of his boyhood as to enable him to enter into and sympathize with the feelings of a boy will acquit me of any such charge—sure I am that every mother who pictures her boy far from home and similarly circumstanced to Battersby will pronounce me 'not guilty'. Were the case simply as stated in your note—did I feel it to be so—that I had acted uncourteously towards you for correcting an error into which I had fallen, I should undoubtedly cover my head with the ashes of repentance.[7] But whatever you may think of me now while your feelings are alive on the subject, I fearlessly appeal to your future self to say whether my act has been one of rudeness to yourself or of sympathy with the wounded feelings of a most superior boy.

Furthermore, my dear Madam, I apprehend that we should differ in our definitions of courtesy. My courtesy must be a free act; if it be <u>demanded</u> there is a law of my constitution which compels me to refuse it. Courtesy is a sweet and graceful flower, I love it and acknowledge its high significance; in fact my respect for it is too high to permit of my substituting a blossom shaped by the scissors in its stead. With some people a demand for outward respect may have a certain value, as forming a barrier against vulgar impertinence, indeed I have been compelled to resort to it in certain cases myself. But there are vast distinctions to be made here. Allow the man of natural courtesy and gentleness his freedom, trust him with his liberty, and his courtesy will shine through every act of his life; but once appear to demand this from him, and at the risque of appearing rude he will reject the moral serfdom implied by such a demand. I dont know any subject which requires nicer powers of discrimination than this, and it has been my lot to witness many mournful mistakes made in connexion with it

I know you are unaccustomed to such writing as this, and I am also unaccustomed to it—yet I feel no difficulty in writing. I believe you will blame me for having written it; but as I before stated, I appeal to your future, not to your present self, for a verdict. And what is implied in this appeal? There is an honest free admiration of you in many respects implied in it—the admiration of a man who while he expresses his belief in your natural nobleness and intrinsic soundness of heart as not afraid to tell you that he firmly dissents from many of your acquired notions.

I remain dear Madam | your obedient Servant | John Tyndall

RI MS JT/2/6/81–85
JT Transcript Only

1. *[30 November 1851]*: from Tyndall's journal (JT/2/6/81).

2. *the note*: in response to Anne Edmondson's brief note (letter 0578) Tyndall wrote this lengthy, self-righteous justification of his actions. He also recorded the incident in his journal on the day it occurred (27 November 1851, JT/2/13b/552) and recorded further thoughts in a letter (0592) to Hirst. Clearly, it bothered him greatly. This version, from Tyndall's journal, may be either a draft for the letter sent or a fair copy.

3. *Mr Wright*: probably Richard Pearl Wright, a fellow teacher

4. *Battersby … a farmer*: that is, one of the older pupils, who studied the theory and practice of farming. Not otherwise identified.

5. *George William*: the Edmondsons' son, often used as a messenger.

6. *a German Lady*: not identified.

7. *cover … repentance*: an allusion to an ancient practice, recorded in the Bible, which was adopted by early Christians and enshrined in the celebration of Ash Wednesday.

From Thomas Archer Hirst   30 November 1851   0580

Marburg, Nov. 30th, 1851

My dear John—

It is Sunday Evening (7-30) outside, a miserable night, cold, dirty splashy, rainy and snowy all at once, inside however it is as warm and cheery as my black companion the stove can make it, (you might consider it too warm) I did wash myself all over when I got out of bed and also pulled on my breeches and 'Schlafrock',[1] but the day was not tempting enough to make me further decorate myself with the appendages of silk handkerchief, clean collar etc. etc. I intended to solve a somewhat stiff Integral first thing and then spend the rest of the day in preparing a packet of letters for England, aforesaid Integral however would not solve as readily as I anticipated, it was obstinate nettled me, and I determined to sit it out. I did so and hatched, or was delivered of, a solution at about 4 <u>P.M.</u> Then came Mr Wrightson, stuck fast in a puddle, in the shape of a simple equation with three unknown terms,[2] he had kicked about in his puddle & made it terribly muddy, & it took us until 6 P.M. before we had emptied it and got a sight of the bottom.

Thus has the day flown and the English Packet must wait another week. By a strange coincidence I read at Breakfast this morning the following sentence in Montaigne[3] which reminded me to answer your last[4] and thank you for one of the most prized letters you ever sent me. Hear my old friend 'A man had need have long Ears to hear himself frankly censured.[5] There are few who can bear to hear it without being nettled, and those who hazard the undertaking it to us, manifest a singular effect of Friendship; for 'tis to love sincerely indeed, to attempt to hurt and offend us for our own good. I think it rude to censure a man whose ill qualities are more than his good ones'.[6] Such would have been my reply had Montaigne not helped me to it, and so I can conscientiously use it. Amongst the mass of the usual smooth penned epistles that reach me yours was a reality, not flat and stale, but fresh if somewhat pungent. I will not deny that at first I was somewhat 'nettled' as Montaigne has it, and if you had said as much to me face to face should perhaps have shewed that I was nettled & answered accordingly. But all the stinging properties of my nettle vanished at a second perusal when I felt that the various indictments against me were but too just, and in my heart I thanked God that he had given me a friend who <u>dare</u> tell me my failures & to whom I could willingly listen. —I had many misgivings that the cement and plaster work you discovered in my poor translations were but too visible, I felt myself unable at the time I sent you them to decide whether they had a value or not, hence I sent them to you instead of to Clayton well knowing that the rubbish might stand a chance of being detected and detained at Queenwood. Keep them there John. They have served their purpose, and profited me far more than if they had appeared in the Leaders columns.[7] I do not grudge a moment that was spent on them. The Blunder in the Diver[8] is inexcusable, and yet, strange to say I accused Schiller of a lame conclusion to a beautiful ballad without detecting (stupid) that 'Den Jungling' in the last line

Den Jungling bringt keines wieder.[9]

was accusative and not nominative, and the lame conclusion my fault and not his. —Much as I deserve kicking for this I will not plead guilty to others of your allegations Namely about the musical ear. I believe there to be as great an excess possible in good (syllabic) Rhythm as in bad. It is well known that far from harmonious chords (viz exact thirds & fifths &c) constituting harmony, or music, that discords are absolutely necessary. I plead not for any particular instance I know well that 'short' & 'thought' 'echo' and 'they grew' as far as syllables are concerned do not rhyme, but I question whether if I had it to do again I would go far in search for better, they do not convey a sensation of pain to the most sensitive ear when the whole forgoing & following chords are struck symultaneously, like I have often heard you do on my piano you have struck upon two isolated keys, inharmonious when so thumped on alone but capable of administering to true harmony when in their proper combination. I grant to you that if one uses rhyme one ought to do it well, but would humbly add thereto that

Tweedledum tweedledum fiddledidee | Although passible rhythm may <u>not</u> music be.

To the insinuation about the bass fiddle[10] also I protest firmly. God made you sensitive to a musical <u>thought</u>, for which be thankful, he also gave you an ear for the music of the Auditorium. Knoblauchs Acoustics[11] to wit, but he knew you would desecrate a bass fiddle and he <u>therefore</u> kept it out of your reach. After this slight retort on my part, which is not so much an exoneration of myself as a protest against the ability of my judge, I again pull in my horns & confess you have once more pricked me in a soft place, viz my spelling. Your advice is good and shall be looked to, I had already found it necessary often to use my dictionary & what you say proves to me that I err not only in doubtful words for which I always use my dictionary but that the habit has so far got hold of me that I pass over errors unquestioned. It shall be looked to. [12]

RI MS JT/1/H/164

1. *Schlafrock*: dressing gown.

2. *Mr Wrightson . . . unknown terms*: Hirst often referred to Wrightson's incompetence in mathematics and the lack of progress in his research.

3. *Montaigne*: Hirst often read Montaigne on Sundays (see letter 0488, n. 2).

4. *your last*: letter 0568.

5. *frankly censured*: Tyndall had criticised Hirst's translations of Schiller.

6. *'A man . . . good ones'*: from 'Of Experience', Montaigne, *Essays*, vol. III, essay XII (cited letter 0393, n. 12). The quote is fairly accurate, the major difference being that Hirst used 'bear' rather than 'endure'.

7. *to Clayton . . . the Leaders columns*: Hirst refers to poems he had translated. He wanted any publishable translations to go to John Clayton of the *Leader* but Tyndall did not consider them worthy of publication (see letters 0553 and 0568).

8. *the Diver*: a poem by Schiller. Tyndall had claimed that an error of translation had led Hirst to misinterpret the poem's dénouement (letter 0568).

9. *Den Jungling . . . wieder*: 'Not one [wave] brings the young man back'. (The poem is about the drowning of a young man, hence the allusion to the waves of the sea.)

10. *this insinuation about the bass fiddle*: Tyndall had insinuated that, given the opportunity, he would have been a better player than Hirst, whose ear for rhythm was so poor (letter 0568).

11. *Auditorium Knoblauch's Acoustics*: not identified.

12. *It shall be looked to.—*: LT added a note, 'Letter ends thus'. We consider that Hirst held this letter for a week and enclosed it with letter 0587.

To William Francis   [30 November–2 December 1851][1]   0581

Queenwood Sunday night!—and |
monday morning long before day dawn![2]

My Dear Francis,

your parcel with all the *[recounted]* contacts has reached me safely. thank you for Faraday.[3] thank you for the magazine and thank you for the trouble you have taken about the Signature. To Mr Huxley pray present my best thanks.[4]

I have just read Donovan[5] through—and while it is fresh in my mind I will set down what I think about it—I can now fancy some of the difficulties of editorial life—the impossibility to please all parties and still preserve the high standing of the magazine. Mr Donovan's paper would be perused with great interest by the general reader, his style is exceedingly perspicuous and attractive. Besides he appears to be an old contributor; he has also won a prize at the Royal <*Irish Ac*>ademy;[6] and I must say that I feel a certain shudder in pronouncing an adverse judgement on such a man. But what can I do—you have asked me for my opinion and of course expect me to give a conscientious one. I believe as regards the advancement of science that M^r D's paper is of no value. Every practical man will look askance at such memoirs; It is very easy to imagine matter for such, though perhaps not so easy to throw them into the beautiful language which Mr Donovan appears to have at command—were there a dearth of matter—had you not your drawers already crammed with contributors *<I should>* say at once 'insert it'; but I certainly think *<your>* space might be more valuably filled—all he has advanced could be said in one fourth of the words which he has used about it. He intends to publish more—would it not be well to request him to send in the whole? you might then propose a condensation of the affair and I am inclined to think you could reduce it as Brahma's press[7] reduces a pack of cotton—it is very porous! The term electric fluid is a kind of symbol by which we tie certain phenomena together, a kind of scaffolding for thought, to be removed when the solid masonry has been put together. My notion is that any attempt to parcel the fluid out into elementary constituents arises from an ignorance of the true aid which it renders to science. M[r] Donovan appears to be not aware of the fact that Melloni has by means of the thermo-electric pile proved that the moon's beams contain heat.[8]

Why does man seek for causes? *<words missing>* result of some unifying principle in human nature,—this principle has ever shewn itself and ever will shew itself, and it shews itself in M[r] Donovan though he regrets it in theory, and would not hesitate to assign a distinct cause for every fact, thus forming as many causes as facts. Nearly the whole of his objections against 'identity' are answered by the assumption of different degrees of tension in expansive force in one and the same fluid. My advice would be to get all he has to say upon the subject said, and then propose this kind of smelting process; otherwise the thing is likely to be *<words missing>* of 'linked sweetness long drawn out'[9]

With regard to D[r] Woods I believe the fairest way of acting would be to print the whole matter as he proposes. If you tell me when you would like to have it I will send you a fair copy of the objections.[10] I do not think that he has at all answered them. In fact he just argues as if I, not he, had proposed a theory. Even supposing that no single fact which I have adduced rests upon a certain foundation, the fact of their being generally received places the onus of positive disproof on D[r] Woods, before he can ground his theory; but I apprehend that he will find some of the facts stubborn ones.

I have had some sweet imaginings over Reuben's paper[11]—did you notice the hand? a woman's,—perhaps his wife, perhaps his daughter, feeling a pride perhaps in her husbands or her father's attainments as she transcribed his speculations Well Reuben appears to have a good deal of ingenuity *<1-2 words missing>* think on the whole his paper may be inserted when he modifies the following fundamental article of a portion of his scientific creed. In page 2 he says 'Suppose a particle to be electrified by having positive electricity developed on one end and its equivalent of negative electricity on the other end, then however intense this polarisation may be the neutrality of the mass is perfect, because [of] the equality of the two opposite electric forces and the minuteness and independence of the particles'.[12] *<1–2 words missing>* minute and independent particles *<so>* endowed must take up a certain determinate arrangement dependent on the position of /their/ poles[13] Now this is exactly the state in which the molecules of a magnet are supposed to be. The magnet is not polarised as a whole, for break it in two and you have two perfect magnets, continue the subdivision down to the molecules and you will find them polarised. But the particles by no means neutralise each other so as to render the outward action of the magnet zero. In fact it has been mathematically demonstrated by Poisson, that a molecular polarization such as that assumed by M[r] Phillips produces precisely the same effect as if free magnetism were distributed like an electric layer over the <u>surface</u> of the mass. I should undoubtedly expect a volume of hydrogen to affect the electrimeter if *<its>* particles were in a state of 'intense polarization' as assumed by M[r] Phillips.

Furthermore it would be difficult to assign 'independance' to particles polarized as M[r] P. supposes the particles of hydrogen to be; they ought to unite, the same force and disposition of force being present among themselves alone, as when chlorine is introduced among them.[14]

But in these speculations we should soon get into a region where no safe footing *<2-3 words missing>* my part I imagine that M[r] Phillips's un*<doubted>* ingenuity might be better expended than on mending these old theories— His remarks on abrasion appear to be of value; but his theoretic views of the polarity of atoms although they may be very useful to himself and may help him materially to a certain extent as keys to a certain number of phenomena do not appear likely to contribute much to the advancement of Science.

One word more as to M[r] Donovan. The magazine has a commercial as well as a scientific aspect. I believe the paper would interest many who do not look far beneath the surface of things. Now I would make its insertion or noninsertion to depend upon the number of this class of readers which you may suppose the magazine to possess.

most truly yours | John Tyndall

D[r] Francis | ——[15]

over.[16]

Would you stick M[r] Huxley's address on the enclosed[17] and send it to him for me? —Yours containing the note from M[r] Sylvester reached me to day[18]— all my difficulties are now surmounted—thanks to you for that same—[19]

I send you a 'report'[20] and am not ashamed to say that it is a good one—I spent a good deal of time at it but have the satisfaction of believing that I now know as much of the real secrets of the matter as anybody that has written upon it. When I have time at disposal I hope to be able to bore a little deeper than any of them have yet done.

Woods's private note[21] accompanies this—

StBPL T&F, Authors' letters

1. *[30 November–2 December 1851]*: The allusions to Huxley, Sylvester and signatures refer to the RS nomination certificate and clearly date this letter to late November-early December 1851. Thus the Sunday and Monday at the head of the letter indicate Sunday, 30 November and Monday, 1 December. The allusions to Reuben Phillips and the 'Report' (see n. 20) are also consistent with this date. After completing the letter very early on Monday morning Tyndall received a letter from Francis containing a note from Sylvester (see nn. 18 and 19) which he responded to in the postscript, which most likely dates from later on Monday the 1st, or possibly Tuesday the 2nd. Moreover, the Huxley letter which he enclosed was dated 2 December. Tyndall usually posted expeditiously but perhaps he delayed posting the letter in order to finish the 'Report' which he sent with it (see n. 20).

2. *before day dawn*: this is a particularly hasty letter, written with a scratchy nib, in a poor hand, with little attention to punctuation and many deletions. All this may be explained by Tyndall's writing in haste with lack of sleep. Many holes in the paper contribute to the uncertain readings.

3. *thank you for Faraday*: probably refers to the 2 volumes by Faraday that Tyndall had requested in letters 0536 and 0569.

4. *thank you … my best thanks*: Francis had approached Huxley and Sylvester, asking if they would sign Tyndall's certificate for the RS. Although Tyndall here asks Francis to pass on his thanks to Huxley, he decided to write directly (see the postscript below and letter 0582).

5. *Donovan*: a series of 8 papers entitled 'On the Supposed Identity of the Agent Concerned in the Phenomena of Ordinary Electricity, Voltaic Electricity, Electromagnetism, Magneto-electricity, and Thermo-electricity' were published in the *Phil. Mag.* in 1852. Presumably the paper under discussion here was the first of those, subtitled 'On the Constitution of the Electric Fluid', *Phil. Mag.* 3, no. 16 (February 1852), pp. 117–27. It would seem that Tyndall's reservations about Donovan's work were outweighed by other considerations.

6. *won a prize at the Royal* <Irish Ac>*ademy*: Michael Donovan had won a prize at the Royal Irish Academy in 1815 for his *Essay on the Origin, Progress, and Present State of Galvanism* (Dublin: Hodges and M'Arthur, 1816).

7. *Brahma's press*: the hydraulic press developed by Joseph Bramah.

8. *Melloni . . . contain heat*: Macedonio Melloni made the first sensible measurements of lunar heat in 1846, by placing a thermopile at the focus of a telescope.

9. *linked sweetness … out*: from Milton's *L'Allegro* (1631), line 140.

10. *With regard to Dr Woods . . . fair copy of the objections*: Tyndall's objections to the paper had previously been passed on to Woods, whose replies had not satisfied Tyndall. Tyndall expressed his willingness to provide a copy of his objections, as Woods had proposed, for publication in conjunction with the paper (for further discussion see letter 0589).

11. *Reuben's paper*: cited 0550, n. 8.

12. *'Suppose … the particles.'*: on p. 36 of the published article.

13. <1-2 words missing> . . . *words*: there is a large X written in the centre across these three lines at the bottom of the page of the manuscript. Whether written by Tyndall, Francis or someone else is unclear.

14. *Furthermore . . . among them*: the objections Tyndall outlined in this letter were passed on to Phillips. In the published article (cited letter 0550, n. 8), immediately following the quoted passage, Phillips responded to Tyndall's objection in a lengthy footnote (on pp. 36–37).

15. —: faint squiggles here are Tyndall's allusions to the '&c &c ' that usually stood for an address at the foot of letters.

16. *over*: written in the bottom right corner of the page to warn Francis to turn the page for the postscripts.

17. *the enclosed*: presumably letter 0582. In the interval between writing the body of the letter and writing the postscript, Tyndall had decided to write directly to Huxley, rather than confine himself to sending thanks through Francis.

18. *yours containing the note . . . me today*: Francis's letter is missing but the enclosed note (dated 29 November) survives. In it Sylvester assured Francis: 'I shall have great pleasure in placing my name to the certificate of a gentleman for the Royal Society so well qualified as our friend Dr. Tyndall'. (RI MS JT/1/TYP/4/1506, 29 Nov. 1851). If each post took 24 hours (and within London the post could be faster) and Francis forwarded Sylvester's note immediately, then Tyndall would have received it on 1 December; if any delays in posting, then 2 December.

19. *Would . . . same—*: on our interpretation (n. 1) the following postscript was written at least a day after the rest of the letter.

20. *a 'report'*: Tyndall, 'Reports on the Progress of the Physical Sciences', *Phil. Mag.* 3, no. 16 (February 1852), pp. 81–92. The report discussed papers by Magnus (cited letter 0556, n. 10) and, more briefly, Svanberg (cited letter 0556, n. 11), Regnault (cited letter 0558, n. 7), and Rudolph Franz ('Untersuchungen über thermo—elektrische Ströme', *Poggend. Annal.* 83, no. 7 [1851], pp. 374–83). Tyndall met the deadline that he had set for himself (in letter 0569).

21. *Woods's private note*: probably a response to Tyndall's objections.

To Thomas Henry Huxley   [2 December 1851]   0582

[Queenwood, Stockbridge, Hants. | 2nd Dec. 1851.][1]

Dear Sir,

It would be unbecoming in me to permit your kindness in consenting to Sign my certificate for the Royal Society to pass unacknowledged.[2] Most sincerely do I thank you for it; and shall endeavour as far as Science is concerned to prevent your name from being disgraced by the association. I suppose by this time your papers are under the eyes of the authorities in Toronto. Whether it be my privilege to accompany you or not, I wish you success with all my heart.

believe me dear Sir | yours most faithfully | John Tyndall
H. Huxley[3] Esq[re] | &c. &c. &c.

IC HP 1.1

1. *[Queenwood . . . 1851]*: this address and date are on the typescript copy (HP 8.4) but not the manuscript letter (HP 1.1). They were probably assigned by Huxley's son and biographer, Leonard Huxley. Given that Tyndall enclosed this letter with letter 0581 to Francis (probably no later than 2 December) and wrote a parallel thank you letter to Sylvester on 2 December (letter 0583), the 2 December is probably correct.

2. *consenting . . . unacknowledged*: before writing to Huxley directly, Tyndall had asked Francis to thank Huxley on his behalf, but then decided to write directly (see letter 0581).

3. *H. Huxley*: Tyndall missed Huxley's first initial, which suggests that he did not know Huxley well, and sent the letter via Francis (letter 0581), which suggests that he did not have Huxley's address.

To James Joseph Sylvester   2 December 1851   0583

Queenwood, Stockbridge, Hants | 2nd December 1851

Dear Sir.

Permit me to return you my sincere thanks for so kindly consenting to sign my certificate for the Royal Society.[1] I was much in doubt whether you would consider our brief acquaintance a sufficient warrant for this act of kindness upon your part. I only trust that the future will in some measure justify the generous confidence which the act implies

believe me dear Sir | most faithfully yours | John Tyndall
J.J. Sylvester Esq[re] | &c &c &c.

StJCL JJS 3/7/183

1. *Permit me . . . Royal Society*: Tyndall was replying to the news, passed on by Francis (letter missing but see letter 0581, n. 18) that Sylvester was willing to sign Tyndall's FRS nomination certificate.

From Thomas Henry Huxley 4 December [1851] 0584

4 Upper York Place | S Johns Wood | Dec[r] 4th

My dear Sir

I was greatly rejoiced to find I could be of service to you[1] in any way—and I only regret, for your sake, that my name is not a more weighty one—Your election I should think can be a matter of no doubt—

As to Toronto I confess I am not very anxious about it—Sydney would have been far more to my taste[2] and I confess I envy you, what as I hear, is the very good chance*[s]* you have of going there—

It used to be our head quarters in the 'Rattlesnake'[3] and my home for three months in the year—Should you go, I should be very happy if you like to give you letters to some of my friends

Greatly as I wish we had been destined to do our work together, I cannot but offer the most hearty wishes for your success in <u>Sydney</u>

Ever | yours very faithfully | Thomas H. Huxley

John Tyndall Esqr.

IC HP 9.1-2

1. *rejoiced . . . of service to you*: Huxley had offered to sign Tyndall's FRS nomination certificate. Here he replies to Tyndall's thanks (letter 0582) within one day of receiving it.
2. *Sydney . . . my taste*: Henrietta Heathorn, Huxley's fiancée, was in Sydney.
3. *'Rattlesnake'*: the ship on which Huxley had voyaged in Australian waters.

From Edward Sabine 5 December 1851 0585

Woolwich. Dec 5. / 51

My dear Sir

I have rec[d] yours of the 3[d][1] and have put your certificate in form and will send it to the different persons for signature. You need give yourself no further trouble about it.

Sinc[y] yours | Edward Sabine

D[r] Tyndall.

RI MS JT/1/S/7

1. *yours of the 3[d]*: letter missing. Tyndall's letter (sent after receiving the information from Francis, as implied in letter 0581) presumably informed Sabine that Sylvester and Huxley would sign the nomination certificate.

From Elizabeth Steuart   5 December 1851   0586

Steuart's Lodge | December 5th. 51

Dear John,

I am sorry for having <u>appeared</u> unkind in being so long without writing to you,[1] but feel that you will excuse me when I tell you how much distress of mind I have suffered for the last two months on account of Mr Steuart's health, which has been affected in an alarming manner, and I need scarcely mention that this subject engrossed <u>all</u> my thoughts and attention, otherwise <u>you</u> should have had an ample share of both. I am most thankful to say he is now very much better, and able to go out as usual, but strictly prohibited all sorts of exertion, in the way of hunting &c. the physicians fearing that such might bring on another attack of this serious illness. He has suffered a great deal, and I trust will be careful of himself for the future—Your friends Mr and Mrs Wynne called here some time since and told me of your intention to seek a situation in the far 'west', and I heard it with <u>regret</u>, for your <u>success</u> puts an end to my hopes and wishes concerning you being realized, so far as regards England, <u>at present</u>, but I cannot willingly relinquish the prospect so long indulged, though distant it may be, of your being established <u>there</u> amongst the scientific elite of that enlightened land.

I feel quite obliged to you for sending me the enclosed letters,[2] which are most gratifying: Mr Wynne's is that of a Christian who knows where to seek for support in time of need:[3] may <u>you</u>, dear John, be directed to <u>that</u> fountain whence flow rivers of living water[4]—I enclose a letter to Mr C. Magrath,[5] which might be of service to you: he is a son of the clergyman who lived in Bagnalstown[6] some years ago, and I understand an influential person in Toronto: perhaps you will approve of writing to him at once, and sending Mr Singleton's letter[7]—I keep the copy of your testimonials, which I should like to <u>have</u>, if you can <u>spare</u> it, if not, it shall be returned—I hope you will write soon again, and let me know every thing which occurs relative to your welfare. Mr Steuart sends you his kind regards and wishes, united with those

of your sincere friend | E.D. Steuart.

RI MS JT/1/TYP/10/3331
LT Transcript Only

---

1. *so long without writing to you*: the last extant letter from her to Tyndall is letter 0502. In letter 0577, Tyndall noted her lengthy silence.

2. *sending me the enclosed letters*: Tyndall had sent letters from Sabine (letter 0559) and Wynne (letter 0573), as well as a set of the printed testimonials (as mentioned here, Steuart kept the testimonials, suggesting that Tyndall had sent printed copies) with letter 0577.

3. *Mr Wynne's . . . time of need*: Wynne had suggested Tyndall pray for guidance on his career
    path (letter 0573).

4. *fountain . . . living water*: alludes to Jesus's saying in John 7:38: 'He that believeth on me,
    as the scripture hath said, out of his belly shall flow rivers of living water'. Steuart wanted
    Tyndall to see Jesus as a source of life or living water.

5. *C. Magrath*: Charles Magrath (c.1809–84) the third son of Rev. James Magrath, emigrated
    with his family from Ireland to Canada in 1827. Charles returned to Ireland in 1844–46 to
    take care of family properties there and, presumably, met the Steuarts during this time. He
    became a law student in 1847 and practised as a lawyer in Toronto from that year. He was
    Secretary and Bursar of the University of Toronto from its foundation. ⸱

6. *Bagnalstown*: also known as Bagenalstown, a small town on the river Barrow about 2 miles
    south of Leighlin Bridge.

7. *Mr Singleton's letter*: the letter, written about Tyndall rather than to him, is missing.

From Thomas Archer Hirst   7[–8] December 1851[1]   0587

Sunday Evening. Dec. 7th | Marburg—

I found myself inwardly rejoicing this morning that there was a prospect
of our soon talking face to face,[2] it was but a gleam of sunshine however, for
a cloud soon hunted it over the meadow land of my thoughts when I remembered it would be our last meeting probably for a long period. That Australian
Cloud, John looks gloomy to me in spite of my effort to believe that there is
the sun still shining behind it. The point of intersection of our Earthly orbits
is once more at hand however and I hail it; for such to me are the mile posts
in my life journey, since we last met I have lived enjoyed and suffered, and the
brief moment allowed us seems almost too fleeting to chalk off the old score
and begin another reckoning. I thought of innumerable things that would
have to be canvassed & interchanged, yet probably when you get here I shall
find myself dumb.

Sabine's letter[3] was a cheering one to me but it would have been more so
if instead of deploring Englands loss by your departure he had held out some
prospect of a worthy situation in England—To that last straw I yet cling—
The position you have undoubtedly gained in the Scientific World ought and
would inevitably reward you with a suitable situation. Backed by the influence of such men as have sent you testimonials you would not have long to
wait for preferment I feel sure. However if the Gods have told <u>you</u> to go to
Toronto or Australia then I too will say <u>go</u> and may Gods blessing follow you
(and if it seem to Him good bring you back again) I will not utter a syllable
of complaint. Of the two I prefer the Australian project, it appears a liberally
conceived one,[4] and you would remain still in closer connexion with England
than if you went to Toronto which I cannot help picturing to myself as deep

in the back woods & primeval forests of the world inhabited chiefly by Buffalos and red Indians—The thought struck me too that without your intervention destiny may be about once more to throw you and Jack Tidmarsh together. You will find in him if I mistake not a different man to the one who bore that name in Halifax. Such freaks of Destiny preach to us a doctrine of anti-despondency. After years sojourning together Jack and you seemed at last separated for good, but it seems it was not so to be. Poor little Knoblauch who has been to see me to day, is also sorry that you are going. but principally because you are going before next April when he will be married & could have entertained you; as a householder. The little fellow is head over heels in Love, works continually with a Photographic picture of his Angel under his nose and I solemnly believe when our backs are turned caresses it and falls down in worship before his fair Idol. His first word to me on his return from England was that he had 'verlobt'[5] himself, and in all our meetings since she has been mentioned. When you come he will probably be in Berlin. I look forward to every post now to bring me news that you will be with us soon. Last Xmas eve we spent together over a glass of rum and there seems a probability of our doing so again, with a far different prospect ahead of us however.—

The V. B. affair[6] you have regarded more seriously far than I have yet or probably shall allow myself. I have felt the same counter-pride that you mention, in the Mothers presence and the very feeling has occasioned me to despise all attempts to undermine it and ingratiate myself in her good opinion. Even to the Fraulein herself it has caused me to show less attention than to her fellow members of the Kränztchen[7] so that a stranger to myself would fancy that far from esteeming her the most I had almost shewn her neglect.—

Much as I might value her favour I would not bend an atom to gain it, so that if she be ever won by me it will be through almost a studied bluntness of demeanour towards both her and her Mother. Such being my determination therefore you will believe with me that no nearer relations are ever likely to ensue.—

I received the Weekly Dispatch[8] you sent me and read with interest the article on 'Sterling Coin'[9] it was worth a hundred Literary Gazettes[10] it is true, and was permeated with a honest respect for his author as well as sympathy with that authors teaching. A caution against spiritual pride might however not be thrown away on the writer. He was somewhat dogmatic in his anti-dogmatism; he regarded the 'Life of Sterling' in this light merely which is far from being its only or even its most valuable aspect. The watchman on the Katzerback[11] is just blowing his melodious horn which reminds me that it is bed time so Good night John. Say good night to Debus also for me & that you may sleep and live well Is my last wish before jumping into my own sheets. (or between my own beddings)

T A Hirst

<Handwritten letter ends; hereafter LT Transcript Only>

You can do me a favour, John. If you have a £5 note by you (not otherwise) enclose it in an envelope directed <u>by some other person than yourself</u>, to | F. Booth, Bird Cage, Halifax. | I have reason to believe the little fellow is in want of it and too proud to ask. He has been ill for some time, and I fear is consumptive.[12] If possible also do not post it at Queenwood, but send it, say, to Ginty to post in Manchester. I want to avoid all suspicion of its having come from us and the Queenwood postmark would reveal it.

Monday morning, | Dec. 8/51.

Dr Tyndall, | Queenwood College, | Stockbridge, Hampshire, England.

RI MS JT/1/H/489
RI MS JT/1/HTYP/176–177

1. *7[–8] December 1851*: a postscript is dated 8 December 1851. This letter may have followed on from letter 0580, which would explain the missing salutation. Tyndall considered this letter so important that he copied it into his journal.

2. *prospect . . . face to face*: in his last letter (0568) Tyndall had promised to visit Marburg before taking up an appointment in Toronto or Sydney, should he be appointed as his mentors expected.

3. *Sabine's letter*: letter 0559. Tyndall had copied it into his last letter to Hirst (0568).

4. *liberally conceived one*: see Tyndall's description of the new university (letter 0568).

5. *'verlobt'*: engaged (German), that is, engaged to be married.

6. *The V.B. affair*: this refers to Hirst's affection for Fraulein Baumbach, discussed in letters 0565 and 0568.

7. *Kränztchen*: see letter 0482, n. 6.

8. *the Weekly Dispatch*: a weekly newspaper founded in 1801.

9. *article on 'Sterling Coin'*: not identified.

10. *a hundred Literary Gazettes*: Tyndall had previously sent Hirst a copy of the *Literary Gazette* with a review of Carlyle's *Life of Sterling* (letter 0565, n. 2).

11. *watchman on the Katzerback*: Ketzerbach (see letter 0565, n. 22 and 0592).

12. *I have reason . . . consumptive*: Hirst had mentioned Booth's troubles (and passed on a letter from Booth) in letter 0516.

To William Francis   [7 or 14 December 1851][1]   0588

Queenwood Sunday

My Dear Francis,

I send the review of Hunt's work[2] and should have rewritten it and perhaps added a polish here and there were it not for the reflection that you would never print it[3]—If he is a near friend of yours you will be angry with me—I have hit him hard but not half so hard as he deserves—I shall now leave the matter in your hands—do unto it what seemeth to thee good.[4]

Dein[5] | Tyndall

StBPL T&F, Authors' letters

1. *[7 or 14 December 1851]*: there are four undated letters that we have assigned to mid-December. The second, 0589 is clearly an answer to both this letter and letter 0581 of 30 November. This letter dates from after 22 November when Tyndall noted that he was to review Hunt's book (letter 0569), and would not have been written on the same Sunday as letter 0581 (30 November–2 December) to Francis. It dates before 26 December when Tyndall revised the review (n. 3). The likely Sundays are 7 and 14 December, although 21 is possible.
2. *Hunt's work*: cited letter 0569, n. 5.
3. *send the review . . . would never print it*: when Francis accepted the review (see letter 0589), Tyndall revised it (Journal, 26 December, JT/2/13b/556).
4. *do unto . . . thee good*: a biblical allusion expressing submission to higher authority: 'And he said, "It is the Lord. Let Him do what seemeth to Him good"' (1 Samuel 3:18).
5. *Dein*: Yours (German).

From William Francis   [mid-December 1851][1]   0589

Red Lion Court. | Dec. 1850

My dear Tyndall,

I have been passing the last few days at Brighton in order to escape a little of the London fog[2] which will account, I hope satisfactorily, for my long silence; and now I find I have so much to write about that I hardly know with what to commence. You talk of the difficulties of editorial life,[3] but it has also its pleasures. To the Magazine I owe much for making me acquainted with many friends and many a pleasant letter, such for instance as accompanied the return of Donovan and Phillips' MSS.[4] It is fortunate for you that you are destined to emigrate, otherwise I should certainly request you to allow your name to figure on the wrapper of the Magazine that you might partake not only of the trouble. With respect to Donovan's paper[5] I quite agree with you as to its value and the absurdity of attempting to assign a distinct cause for every fact; but as you truly observe the Magazine has a commercial as well as a scientific aspect and in order to print the good papers it is sometimes necessary to give one of little or no value. I have therefore decided upon printing it especially as I found that course would be most agreeable to Mr Taylor's feelings from his being an old correspondent and moreover likely to interest many who do not plunge deep into the mysteries of science and amuse some that do.

Woods also will be printed.[6] In writing to him I informed him that the objections would probably appear in the next following Number[7] but I leave this for you to determine and will therefore furnish you with a duplicate proof of his paper as soon as it is in type. You will then be able to put your objections into proper form and give them either as a separate paper or if you wish to preserve the incog. append them to his paper in the form of an editorial note. This morning I received a note from him complaining of Thomson who in his Mechanical Theory of Electrolysis 2nd paragraph[8] states as one of his grounds of argument that decomposition causes the absorption of as much heat as combination produces without alluding to Dr Wood's paper. He asks how Prof. Thomson found this out; if he proved it himself he ought to say how—I shall advise him to put the question to Prof. Thomson himself.

In stating the objection you made to Reuben Phillips' view[9] on molecular polarization I availed myself of the opportunity of telling him what you mentioned with respect to the application of his ingenuity in the hope it might do him some good—He took it kindly but declined to modify his views—I however send you his note and as soon as the foot-note alluded to is in type will furnish you with that also that you may see what answer he makes[10] to the objection.

And now a few words with respect to Hunt. The review I think will do exceedingly well and I return you many thanks for it—I had more than a suspicion when I sent you the book that you would find it a most miserable production; and from the opinions I have heard expressed by others I think you have been pretty merciful.[11] I had much rather he had not written the book for to me it is most disagreeable to have to print any thing unfavourable to an author; but in the present case I consider it my duty to print the review for it is scandalous that a person so unfit should have been appointed by Government to teach Mechanics and Physics to the rising generation of miners—I have long thought the author to be a conceited person and do not feel any inclination to spare him. Mr Hunt should stick to the Poetry of Science[12] and not expose himself to be more than laughed at. From his Introductory Lecture[13] you would fancy him to be a most wonderful man; although people do say that the glass covering at Kew is a miserable failure.[14]

With this I inclose proofs of your Report on Magnus[15] and also a portion of Schlagintweit—It is absolutely requisite that I should make some attempt to get rid of the latter, otherwise it will become stale. I therefore propose, subject to your approval, to bring it out, at least a part, this month;[16] and to leave Magnus over to form the first Article in the February Number, but the contrary can be done just as easily if you consider it advisable or if it should be most agreeable to you that Magnus should appear without delay—Pray let me know your wishes on the subject. In Magnus's paper there are several references to figures to which queries are attached. Figs 1,2, and 9 of the orig. I suspect you do not consider necessary and I hope it may be so, as those referred to by you will just fill an Octavo plate—I shall not place them in the Engraver's hands until I learn from you whether my supposition be correct.

I also inclose copies of the Introductory Lectures delivered at the School of Mines[17] thinking they might interest you—Should you in perusing them find anything worth noticing for the Magazine I should be glad to devote a page or two to the subject. Forbes and Playfair's[18] seem to be good but I can't say that I altogether approve of the way they run down the study of the Classics—They appear to me just as one sided as the people at Oxford and Cambridge, and although I consider the study of Natural History most excellent, I do not exactly see that it is necessary to understand Zoology to be a miner. It should be as in Germany. Teach both at the High Schools. Undoubtedly there are too many people even among the educated classes who are totally ignorant of the very first principles of the Physical and Chemical Sciences, but that is no reason for casting a slur upon the cultivation of other departments of knowledge.

But enough, I fear before you have half got through this scrawl your patience will be sadly tried and you will set me down as a great bore. Yours most sincerely. William Francis.

RI MS JT/1/TYP/12/3978–3979
LT Transcript Only

1. *[mid-December 1851]*: the LT date, 'Dec. 1850', is incorrect. This letter replies to both letter 0581 (30 November–2 December 1851) and letter 0588 (7 or 14 December 1851). Francis, it seems, had not written to Tyndall since late November and apologises here for his 'long silence'. Enough time has elapsed between letter 0581 and this letter for Francis to receive that letter, pass on Tyndall's objections to Donovan and Phillips, and receive replies from them (n. 4 shows it dates after 11 December). Lastly, at the time of writing this letter, the review of the introductory lectures delivered at the School of Mines (n. 17), for which Tyndall corrected the proofs on 26 December and which was published in the January number of *Phil. Mag.*, was merely a proposal. Tyndall had yet to read the lectures, decide whether they warranted comment, and write the review. Thus it probably dates to mid (rather than late) December, and either Francis had only just received letter 0588 or 0588 dates to 7 December.

2. *the London fog*: London was notorious for its green-tinged winter fogs, called pea soupers, which were produced by a combination of the damp climate and smoke from coal fires.

3. *You talk of the difficulties of editorial life*: Tyndall in letter 0581.

4. *letter . . . return of Donovan and Phillips' MSS*: these had been sent by Tyndall to Francis in early December (probably with letter 0581), from Francis to the authors, and they had since returned the manuscripts with responses to the criticisms. The lengthy footnote in which Phillips responded to Tyndall (see letter 0581, n. 14) was dated 10 December, giving 11 December as the earliest possible date for this Francis letter.

5. *Donovan's paper*: see letter 0581, n. 5. Francis explains here why he published—against Tyndall's recommendation (in letter 0581).

6. *Woods will also be printed*: see letters 0550, 0567 and 0581.

7. *the objections . . . following Number*: earlier letters allude to a proposal to publish Tyndall's objections alongside Woods's paper; that proposal is here modified. Tyndall's criticisms were not published, although Woods published a correction to his paper ('On the Heat of Chemical Combination', *Phil. Mag.* 3, no. 18 (April 1852), pp. 299–303). This was dated February 1852 and did not refer to Tyndall.

8. *Mechanical Theory of Electrolysis 2nd paragraph*: W. Thomson, 'On the Mechanical Theory of Electrolysis', *Phil. Mag.* 3, no. 13 (December 1851), pp. 429–44 (second paragraph on pp. 429–30).

9. *the objection you made to Reuben Phillips' view*: in letter 0581 (see also letter 0550).

10. *send you his note . . . see what answer he makes*: Phillips's note to Francis was enclosed. Tyndall responded to both letter and footnote with an editorial note (that is, anonymous) rejecting Phillips's modification of electro-chemical theory. It was published immediately after the lengthy footnote in which Phillips responded to Tyndall's objections (cited letter 0581, n. 14).

11. *The review . . . pretty merciful*: Tyndall had expected Francis to find his review over-critical (letter 0588). It was published in 'Notices respecting New Books', *Phil. Mag.* 3, no. 15 (January 1852), pp. 57–66 (on 57–60).

12. *the Poetry of Science*: Francis alludes to Robert Hunt, *The Poetry of Science, or Studies of the Physical Phenomena of Nature* (London: Reeve, Benham, & Reeve, 1848), in which Hunt aimed to show that scientific facts 'have a value superior to their mere economic applications, in their power of exalting the mind to the contemplation of the Universe' (p. v).

13. *his introductory lecture*: see n. 17.

14. *the glass covering . . . miserable failure*: due to his work on the transmission of light through different media, Hunt had been consulted over the choice of glass for the Palm House at Kew when it was being designed in 1844–45. After testing various samples of green glass, he recommended one coloured with copper oxide which gave the building an unfortunate pea-green tint (Ray Desmond, *Kew: The History of the Royal Botanic Gardens* [London: The Harvill Press, 1995], pp. 162–63).

15. *proofs of your report on Magnus*: Tyndall had sent the report to Francis with letter 0581 (cited n. 20).

16. *portion of Schlagintweit . . . at least a part, this month*: Tyndall had translated the entire paper early in the year (see letter 0470). Hermann Schlagintweit's paper was published in

two parts, 'Observations in the Alps on the Optical Phænomena of the Atmosphere', *Phil. Mag.* 3, no. 15 (January 1852), pp. 1–16 and 3, no. 16, (February 1852), pp. 92–104.

17. *Introductory Lectures delivered at the School of Mines*: Royal School of Mines, *Records of the School of Mines and of Science Applied to the Arts, Vol. 1, Part 1.: Inaugural and Introductory Lectures to the Courses for the Session 1851–52* (London: Longman, Brown, Green, and Longmans, 1852).

18. *Forbes and Playfair's*: E. Forbes 'The Relations of Natural History to Geology and the Arts' (ibid, pp. 49–62), and L. Playfair, 'The Study of Abstract Science Essential to the Progress of Industry' (ibid, pp. 23–48). Tyndall noticed the lectures of Forbes and Playfair, plus those by Henry de la Beche and Robert Hunt, and overlooked those by Warrington W. Smyth and John Percy. His review was published (alongside the review of Hunt) in 'Notices respecting New Books' (cited, n. 11 above), pp. 61–6.

To William Francis   [14 or 21 December 1851][1]   0590

Queenwood Sunday

My Dear Francis,

What you have been pleased to call your 'miserable scrawl'[2] was a very pleasing scrawl to me—I write these reviews exactly as I feel concerning them and your note gratified me inasmuch as it proved that a unity of feeling subsists between us on these topics[3]—Might I beg of you to present my sincere thanks to Mr Huxley for his great kindness[4]—If any fellow ever had reason to congratulate himself upon kind treatment most assuredly I have.

I scarcely know what to make of Feilitsch,[5] it is hardly suited for an abstract, I will send you a summary of his results[6] by tomorrow's post and then you can exercise your discretion upon them. —I imagine from what you say regarding the 'bulletin'[7] that you would shoulder a rifle with *[great]* good will against Louis Napoleon[8]—It is wise to prepare for the gentleman.

believe me dear Francis | most truly yours | John Tyndall

StBPL T&F, Authors' letters

---

1. *[14 or 21 December 1851]*: the comments on reviews (n. 3) and the reference to Francis's 'miserable scrawl' (n. 2) suggest that this letter was written in reply to 0589 of mid December, and that it was thus written in mid-late December. However, the references to Huxley, Feilitzsch and Louis Napoleon are not in direct response to letter 0589, therefore there must be a missing letter from Francis written at about the same time as letter 0589. The relationship to letter 0589 makes a late December date likely, although we cannot rule out January 1852. Tyndall would not have written this letter on Sunday 28

December as he breakfasted with Francis in London that day before returning to Queenwood (JT/2/13b/556). That leaves 14 and 21 December. (There are reliably dated references to Feilitzsch in February 1852 (see n. 5 below), but we consider that this letter refers to a different, unidentified Feilitzsch manuscript.)

2. *your 'miserable scrawl'*: in letter 0589 Francis had apologised for his 'scrawl'. It is possible, however, that in the missing letter, Francis also apologised for a 'miserable scrawl'.

3. *I write . . . these topics*: Tyndall had been relieved by Francis's assurance that reviewers had a duty to be honest and that he shared Tyndall's estimation of Hunt (letter 0589).

4. *thanks . . . great kindness*: kindness in signing Tyndall's RS certificate. Tyndall had already thanked Huxley for being willing to sign his FRS nomination certificate in early December (letters 0581 and 0582). Here, he sends thanks for his actual signing.

5. *Feilitsch*: it is unclear to what paper Tyndall alludes. He wrote a short summation of a pamphlet by Feilitzsch for publication in the March 1852 number of the *Phil. Mag.* (see letters 0599, 0602, 0603, all February 1852), but the context here suggests that this letter is earlier and refers to an earlier paper; in which case Francis shared Tyndall's misgivings and chose not to publish. Perhaps, though, this letter dates from January 1852.

6. *send you . . . his results*: probably with letter 0591.

7. *the 'bulletin'*: voting papers associated with the plebiscite (see n. 8).

8. *against Louis Napoleon*: Louis Napoleon initiated a coup d'état on 2 December 1851. He seized power after failing in his campaign to have the Constitution, which barred the president for running for a second term, amended in his favour. He held a plebiscite on the new constitution on 20–1 December.

To William Francis [mid-late December 1851][1] 0591

My Dear Francis.

I send you Feilitsch[2]—do unto it what seemeth to thee good,[3] I shall probably request you to send a few memoirs to paris for me on the 1ˢᵗ—Lardner[4] is writing a work on Natural Philosophy[5] and I think he will be able to make use of them—if you advise me against it however I will take your advice

Sincerely yours | Tyndall

StBPL T&F, Authors' letters

---

1. *[mid-late December 1851]*: this letter is tentatively dated to mid-late December by the allusion to Feilitzsch, which suggests it was written only a day or two after letter 0590. The allusion to a regular posting to Paris on the 1st suggests that this is written late in the month (although does not identify a specific month).

2. *Feilitsch*: paper not identified (letter 0590, n. 5).

3.   *do unto it . . . good*: it seems it was not published (see letter 0590, n. 5).

4.   *memoirs to paris . . . Lardner*: Tyndall had made a similar request previously (see letters 0554 and 0557). This letter suggests that he had been conducting further correspondence (missing) with Lardner. His uncertainty over whether it was advisable to send memoirs to Lardner reflects the latter's reputation (letter 0557, n. 4).

5.   *Natural Philosophy*: Lardner's *Hand-book of Natural Philosophy and Astronomy. Second Course* (London: Taylor, Walton, and Maberly, 1852), was published in September 1852 (advertisement in *Morning Chronicle*, 29 September 1852). It included chapters on heat, electricity (both common and voltaic), and magnetism. After it was published, Tyndall asked Francis to send him a copy (see letter 0687).

To Thomas Archer Hirst   30 December 1851   0592

Queenwood Dec. 30th 1851

My Dear Tom

Christmas has past and we have not seen each other:[1] all things remain in the same uncertainty as heretofore, nothing yet decided—I am not rigorously decided myself, for a little weight thrown in would be sufficient to turn the scale and induce me to take my chances in England, though what these chances are likely to be is utterly unrevealed to me. Your request with regard to Booth has been attended to[2]—the two halves of a £5 note were posted to him on two successive days from London. On a bit of paper which enclosed one of them I wrote 'make use of this'—that was all—I shall be very glad to go halves with you on this act, so instead of giving me credit for 5 pounds set down only 2–10–0. I would have made an advance in this direction myself when you sent me Booth's letter[3] had I not been utterly cashless at the time.

I was left utterly alone here yesterday week—Debus went to London & I remained until Christmas morning when the sun beamed out gloriously and propositioned me to take a dash out: I had many little things to settle regarding the Phil. Mag.[4] this pointed my thoughts towards London,[5] and Debus being there comfortably situated was an additional inducement: started in the morning, reached Dunbridge[6] 7 minutes before train time according to my calculation but in reality 2 hours after it—the trains on x-mas day were similar to those on Sunday so I had to stroll through the country for nearly 7 hours waiting for the next train—part of the time I spent pleasantly in a church yard reading the tombstones—strange to say from my boyhood I have taken great pleasure in a stroll among graves, there is a quiet soothing sadness in the place like the falling of autumn leaves. I reached London however that night, and left it again on Sunday last. I have not got your last letter[7] by me; I have got a chaotic batch of letters from you here and I suppose the last must have got in among them from which to rescue it would be an hour's labour; this I feel no inclination to spend over the search—All I remember of it is certain impudent insinuations regarding the imperfections of my auricular organs,[8] and a quotation from Montaigne[9] the contents of which I wanted no Frenchman to tell me. Talking about letters—I have a mind to give you a specimen of a few which passed between myself and Mrs Edmondson some time ago. Mrs Edmondson is a proud woman, and I believe has been accustomed to rule in her own household; Mr Edmondson is an exceedingly active little man corporeally but weak and irresolute mentally and his lady has probably taken him as a specimen of manhood in general—This I think has spoiled the lady to some extent: she has been so long accustomed to attention that she now seems to demand it and his[10] best and kindest acts are in some way vitiated by the assumption of a certain dignity of demeanour which chills most people—It is a pity for at bottom she is a noble-hearted woman and all she wants is a man strong enough[11] to shake this artificial nonsense away from her. But I think as I have dipped so far into the story I must tell you the whole of it. —No I wont I will copy a note which she sent to me and my reply to the said note and you may pick what you can out of both.

[...][12]

<blotched postmarks render the final page partly illegible; the following transcription therefore relies on the LT Typescript at some points.>

Thus endeth the epistle—Having thus clearly defined my relations I came to a treaty and since that time the lady and I have been firmer and surer friends than before.

Tomorrow night the cry of 'Hail to the new year!' will arise in the Ketzerbach[13]—I hear it like a faint echo from the gone Eternities.[14] What a dream-image this life of ours is when we look back upon it! Like the landscape after sunset from the Summit of Amöneburg[15] it becomes more and more impalpable, twilight and mist e[n]velop it—but I look aloft and find that I am not alone; overhead and around me are the star-radiances of memory, of friendship, and of hope; gaily I cast my shuttle across the loom of time, and trust that when the summons comes to quit the loom I shall depart just as cheerfully—But where am I rambling to? Right forward into the Inane, as Tom would say. I'm not gone yet however, I shall make your knuckles crackle before the energy of my grasp ere I depart. Thou great Whale[16] good-bye— May the Gods nurture thee O thou Leviathan! John.

Just look to a Review of Elementary Physics in the January *[no.]* of the Phil. Mag. Do you know the writer?[17]

I intended to write a long letter to Prof. Knoblauch but will defer it till next time—greet him most kindly for me—if I can get astride my Pegasus[18] I will send him an <u>epithalamium</u>![19]

RI MS JT/1/T/547
RI MS JT/1/HTYP/178–181

1. *have not seen each other*: Tyndall had planned to make a farewell trip to Marburg before leaving for Toronto or Sydney (see letters 0568 and 0587).
2. *Your request . . . attended to*: in letter 0587.
3. *Booth's letter*: enclosed with letter 0516.
4. *many little . . . Phil Mag*: Tyndall had many reviews and translations under way, but whether any required his meeting with Francis is unclear. He was revising his review of Hunt's *Elementary Physics* (letter 0588 n. 3), reviewing the lectures delivered at the opening of School of Mines (see letter 0589), and a paper by Woods was in process (letter 0550, n. 3).
5. *towards London*: Tyndall's visit to London is recorded in his Journal (entries for 25–28 December, JT/2/13b/555-6).
6. *Dunbridge*: roughly 4 miles from Queenwood.
7. *your last letter*: Tyndall's comments refer to letter 0580 (30 November) which supports the suggestion made above that it was held back and posted with letter 0587 (7–8 December).
8. *impudent insinuations . . . auricular organs*: Hirst had dismissed Tyndall's claims to musical talent; see letter 0580.
9. *a quotation from Montaigne*: Hirst had cited a passage from Montaigne in accepting that Tyndall's critique of his translations was a sign of friendship (see letter 0580, especially n. 3).
10. *his*: 'her' would make sense, but Tyndall, perhaps in error, wrote 'his'.
11. *a man strong enough*: an expression of the view, which Hirst shared with Tyndall, that men should be strong and guide their wives. Tyndall and Hirst thought Mrs Edmondson was stronger than her husband.
12. *I will copy . . . both*: letters 0578 (short) and 0579 (very long) were written a month previously on 30 November 1851. Tyndall copied them out in full here.
13. *the Ketzerbach*: see letters 0565, n. 22 and 0587.
14. *the gone Eternities*: emphasizes the time passed since he lived in Marburg; to capitalize 'Eternities' was a Carlylean allusion.
15. *Amöneburg*: a castle and surrounding town built on a mountain (or prominent basalt cone) about 6 miles east of Marburg.
16. *Thou great Whale . . . Leviathan*: this may allude to Carlyle's 'Signs of the Times' (1829) in which, in commenting on disorientating effect of the repeal of anti-Catholic Test Act, Carlyle used a leviathan to symbolise intolerance. Perhaps, Tyndall teasingly accused Hirst of intolerance, but we have no suggestion as to why.
17. *Review . . . the writer*: Tyndall himself (see letters 0569 and 0588).
18. *get astride my Pegasus*: the mythical Pegasus was regarded as the horse of the Muses. By striking the ground with his hoof, he created the inspiring well Hippocrene, on Mount Helicon, home of the Muses (as well as similar wells in other parts of Greece). Tyndall hoped for inspiration.
19. *epithalamium*: a poem or song written to celebrate a marriage; Knoblauch was engaged.

From Elizabeth Steuart   31 December 1851   0593

Steuart's Lodge | December 31st 1851

Dear John,

I have been expecting and wishing to hear from you,[1] feeling anxious to know what your prospects are with respect to the Professorship at Toronto. Though I cannot help thinking that if anything were decided on, you would not delay writing to me—The idea of your leaving England causes me great regret, and puts an end (for the present at least) to my visions of seeing you in the <u>first</u> ranks of scientific men, however I should hope that distance will not preclude the possibility of your being recalled, whenever any suitable situation is vacant, as I make no doubt you have friends who will not lose sight of your interests.

I have a small request to make,[2] which I hope you will not consider troublesome: I am anxious to have '<u>Mistletoe</u>' growing here, on some of the old trees, and as the berries are <u>only</u> to be had in England, I thought if the plant should be in your neighbourhood, you would kindly have some gathered, and sent to me in a little box by post: I shall return stamps to whatever amount is necessary, and feeling very much obliged for your kind compliance with my wishes. The seeds cannot be put in until Spring, but I am told none are left by the birds at that time, and therefore should be secured now.

Mr Steuart has not been so well this last week,[3] but seems better to-day: he unites in every kind wish for you with

your very sincere friend | E. D. Steuart.

Doctor Tyndall | Queenwood College | Stockbridge | Hants.[4]

RI MS JT/1/TYP/10/3333
LT Transcript Only

---

1. *hear from you*: Tyndall had not replied to her last letter (0586, 5 December).

2. *I have a small request to make*: the request was not easily carried out; it took Tyndall two months (until letter 0608).

3. *Mr Steuart has not been so well this week*: he had been seriously ill (letter 0586) and died the following year.

4. *Doctor . . . Hants*: address from the envelope.

1852

From
John Frederick William Herschel   9 January 1852   0594

32 Harley Street. | 9th Jan /52

Sir J. Herschel presents his Compliments to Mr Tyndall and returns Prof. Faraday's letter[1] which he may require—the rest of the testimonial and works sent being printed matter Sir J. H. presumes need not be sent by post. Should Mr T. wish for them they will be delivered to any one whom Mr T. may depute to call for them

The choice of the Council[2] has fallen on

> Dr. Wooley[3]   Class.
>
> Mr Pell[4]   Math.
>
> Dr. J. Smith[5]   Chem.[6]

RI MS JT/1/TYP/2/496
LT Transcript Only

---

1. *Prof. Faraday's letter*: letter 0506.
2. *the Council*: the professors for the newly-founded university in Sydney were chosen by Herschel, Airy, and 'two others' (letter 0568); these were Professor Maldon and H. Denison (see Wilfrid Airy, ed., *Autobiography of Sir George Biddell Airy* [Cambridge: Cambridge University Press, 1896], p. 207).
3. *Dr. Wooley*: the Rev. John Woolley, foundation Professor of Classics.
4. *Mr Pell*: Morris Pell, foundation Professor of Mathematics.
5. *Dr. J Smith*: Dr. John Smith, foundation Professor of Chemistry and Experimental Physics. He was a medical doctor who had trained at Marischal College in Scotland.
6. *Chem.*: the position Tyndall applied for. In letter 0561 Tyndall described it formally as the Professorship of Experimental Philosophy and Chemistry.

To Elizabeth Steuart   [9–10][1] January 1852   0595

Queenwood | 10th. Jan. 1852

Dear Madam,

You were right in supposing that had anything as regards my future scene of labours been decided I should have written to you; as regards Toronto nothing I believe will be known for some time—Indeed I have sometimes misgivings as to the propriety of leaving England, for the contact with men of science affords a great stimulus, and from this I should be in a great measure withdrawn in Toronto—When the time comes I shall look clearly and calmly at the matter and make my choice accordingly. I have to return you my best thanks for the kind trouble you have taken in procuring the letters of introduction for me, and beg of you to thank Mr Singleton[2] also in my behalf. Should I go to Toronto it will be a source of great satisfaction to me to be permitted to make the acquaintance of Mr Magrath;[3] and I trust that each of us, in his particular sphere, may be able to do something towards upholding the credit of the land of our birth.

I send you by this post a quantity of misletoe berries[4]—and hope that the day will come when you will be able to point them out to me clustering round your fine old trees. With regard to the postage stamps I beg of you to forget them utterly[5]—I may probably want some shamrock roots to plant here in England and if so this little matter will be brought to a perfect equilibrium.

With sincerest wishes for the health and happiness of yourself and Mr Steuart.[6]

believe me dear Madam | most faithfully yours | John Tyndall.

Should any thing occur regarding Toronto I will write to you at once.

RI MS JT/1/TYP/10/3334
LT Transcript Only

---

1. *[9–10]*: as this letter mentions the Toronto application but not the Sydney one, we assume that Tyndall wrote it before receiving letter 0594 and writing (in reply) 0596. The date (10 January) is an LT date and may therefore be the postmark rather than the date of writing, which could have been 9 January.

2. *thank Mr Singleton*: for a letter of introduction (see letter 0586, where Steuart suggested that Tyndall send Singleton's letter, which is missing, to Magrath).

3. *Mr Magrath*: see letter 0586, n. 5.

4. *misletoe berries*: requested by Steuart in letter 0593. Tyndall habitually wrote 'misletoe'; Steuart (as transcribed by LT), wrote mistletoe (letter 0593).

5. *postage stamps . . . utterly*: Steuart had offered to pay for postage.

6. *Mr Steuart*: he had been ill (see letters 0586 and 0593).

To John Frederick William Herschel   10 January 1852   0596

Queenwood College 10 Jan. 1852

Sir,

It would gratify me much if you would do me the honour of accepting the memoirs already in your hands.[1] Long before I heard any thing of the University of Sidney I had thoughts of requesting permission to lay them before you.

I remain, Sir, | Your obedient Servant | John Tyndall
Sir John Herschell Bart. | &. &.—

RS HS 17.383

---

1.   *the memoirs already in your hands*: Tyndall had sent five memoirs with his application (letter 0561).

To Thomas Archer Hirst   1 February 1852   0597

Queenwood Sunday February 1st 1852

My Dear Tom.

I lay in bed this morning until nearly 8 after I had been awake from 6, but lay there thinking or endeavouring to think; sometimes a bright idea would flash in upon me, but I usually relapsed very soon into that slumberous state which, though not actual sleep, is just as sterile as regards true thought as sleep itself—My old admonitor nudged me at times with the exhortation to get up—another of my mental occupants would then present a picture to my mind's eye of what I was to expect if I did get up—on the picture the tedium of shaving and washing were very prominently brought forward— this repeatedly turned the balance of judgement in favour of repose, and there I lay until it was close upon 8. For a quarter of an hour previous, however, I felt a certain anger rising within me, a force at first feeble, but which increased as I lay, finally overcame the inertia of my body, and placed me steadfastly upon the floor—In such experiences I think one may obscure the isolated operations of what we are accustomed to call intellect—it is simply a <u>seeing</u> faculty. My mind was clear during part of the time I lay there—I could see my duty plainly enough, I could see that the mental occupant which invited me to repose, was a cajoler and a liar, and as he suggested that I might enjoy myself at least that morning and rise early the next I knew very well that it was arrant humbug and that the scoundrel was willing to play me the same trick next morning if I permitted him—all this I saw clearly nevertheless I lay there. Now it struck me that I had made a mistake in deeming the act of washing and shaving a nuisance—could this gap have been removed and could I have at once stepped into the harness and commenced my work I should have done so. But what right had I to estimate one portion of my necessary work more highly than another? The act of washing and cleansing was as necessary as any I could perform; it formed, or ought to form, a fractional part of a true day's work, but in my eagerness to engage in my other occupations, in my exorbitant estimate of these occupations as compared with washing and cleansing, I am sometimes led to ignore the claims of the latter altogether. This is not as it should be. Were the object of existence the performance of a certain work, the neglect of every thing else might be perhaps justifiable; but if the end of existence be not the performance of work, if work itself have an end and object, namely the cultivation of human powers and the enriching and expanding of human experiences—then I say that washing and cleansing if devoutly done and at its proper time, is as important as any other portion of our labour.[1] Again, our forces are aroused by our contemplations. Man feeds his ambition by picturing to himself the excellences of power, and can work himself into phrenzy by the contemplation of his wrongs. The lover feeds his flame by the image of his mistress, the light of her eye at a certain moment, a certain smile, a particular glance, a word, all form nuclei round which the beauty and odours of thought congregate and thus is his passion deepened. Now here I conceive the human will comes into play; it is in a mans power to present objects to the intellect for contemplation which shall stir noble aspirations within him, and by exercise in this way the will may become very powerful. That it is in his power to present ignoble objects is sufficient illustrated by what occurred to me this morning in bed. But a truce to theories; I got up with a certain calm humiliation of thought and washed myself, taking a certain pleasure in thus beautifying (!) the temple of my mind. I felt a kind of selfrenouncement as I proceeded and looked upon the world with a loving and forgiving eye. 'Why should I ever be hasty', I said, or feel any thing approaching to anger? Why should egotism or self ever lift its head within me? Why should I be envious at the success of others, or why oh why should the gratifications of sense, the love of repose or comfort, invade the triumphant mental calm which I now enjoy I will always be thus' I exclaimed 'I will renounce all I will feed my soul on high contemplations, I will make it the study of my life to enoble myself; my work shall be the instrument of my culture and not the object of my existence. On this high ground body and soul will shake hands together, and the reasonable demands of both be cheerfully granted'.

In this mood Oh, Tom did I purpose to write to thee this morning and the twilight of this mood still glimmers in my soul. It is difficult however to preserve that serenity of mind amid the clouds and vapours which encircle daily transactions. The highest peak of the Andes, above which is the clear heaven, and below which the clouds usually sail is sometimes visited by a storm. There is a certain fallacy in this figure but I will let it go. Shortly after breakfast this morning a complaint reached me that two of the farmers had been guilty of an act considerably disgusting. I went to consult with Mr Edmondson about it and found his little brain exasperated in a variety of ways against them; they had been letting of squibs of powder and making a noise the night before which vastly excited his ire—These little breaches of discipline are the things upon which his eye chiefly rests; his opposition to them belongs to what Tennyson calls 'the falsehood of extremes',[2] and hence I rarely sympathize with him. Well I made proposals and he made proposals, and under all the circumstances I found it difficult to preserve that sunclear frame of mind which I enjoyed while dressing. Sufficient of it however remained through all to demonstrate what might be done by culture in this way. These very sources of irritation the contact with narrow, weak, and defective men, are perhaps the very purchases by which one is exalted into this cloudless region and without which it perhaps could not exist.

Surely I have given you enough of speculation this morning Tom. I will now dismiss it, and talk of something which the majority of men would call more real, though I doubt its claim to the title. The Sydney matter is finished, decided against me—I received a note from Sir John Herschel[3] a few days ago apprising me of the fact.[4] to the name of the person chosen for the post[5] for which I applied was attached the term 'chemistry'—I am perfectly contented with the result. With regard to the Toronto affair it is still undecided. The matter has to undergo a variety of siftings from a variety of gentlemen before the candidate is chosen.

And now I should like to know a little about yourself and your future movements—I suppose you are preparing for an examination in Marburg. What afterwards? Would you not like to spend a little time at Göttingen with Weber and Gauss, or at Berlin, or at both? If you fix on Berlin first let me know, so that I may write to some of the people there about you—If you have no particular objection to it I should recommend you to try Göttingen for a semester; had I had the means I should undoubtedly have done so. But I trust your next letter will contain a full account of your intentions, and leaning on this trust I will for the present refrain from asking any more questions.

And now my dear brother good bye. This letter was commenced early this morning—it is now deep in the shades of evening,[6] and I have met many little matters to test my philosophy during the day—across them all however I look to thee at this closing hour of the day, and surely it is a blessing to have one little visual speck in that strange land, one incarnated thought, towards which amid all these petty janglings my heart can turn an unclouded side.

John.

JT/1/T/548

1. *washing and cleasing if devoutly done . . . labour*: this is a Carlylean allusion, as is much of
    the rest of the letter. The phrase is reminiscent of a passage in *Past and Present* (book 3,
    chap. 15): 'What Worship, for example, is there not in mere Washing! Perhaps one of the
    most moral things a man, in common cases, has it in his power to do . . . thou wilt step out
    again a purer and a better man'.
2. *what Tennyson calls the 'falsehood of extremes'*: Tyndall quotes the final line of Tennyson's
    poem 'Freedom': 'Of Old Sat Freedom on the Heights', first published in *Poems* (cited let-
    ter 0568, n. 17), vol. 1, pp. 221–2. It is usually read as a political poem, espousing a liberal
    conservative compromise between democracy and oppression.
3. *a note from Sir John Herschel*: letter 0594.
4. *fact.*: there is a long deletion here which is heavily scrawled out. Perhaps Tyndall expressed
    disappointment, but then decided not to reveal to Hirst his initial feelings about the Syd-
    ney position.
5. *the person chosen for the post*: Dr. John Smith (letter 0594, n. 5).
6. *commenced . . . evening*: although finished and signed, Tyndall did not post this letter until
    weeks later (see letter 0604).

From Edward Sabine   5 February 1852   0598

11. Old Burlington S^t | Feb. 5^t 1852

My dear Sir—

As you desired, your certificate for the R.S. was sent (with Dr Faradays
signature affixed) to Mr Sylvester, six weeks ago,[1] from the RS., *[day]* post,
addressed according to the direction in the RS.^ies list furnished by himself—
After waiting a long while, a second letter was written to him to the same
address, to which he has replied saying he knows nothing of the former letter—Had there been a mis-direction the letter would have been returned at
the P.O. to the R.S. from which the certificate was dated. I conclude therefore
that the letter was duly delivered but has been mislaid.[2]

I must plead this as a cause for troubling you once more for a statement
of your works, as I have destroyed the one you formerly sent me[3]—I will then
obtain Dr Faradays signature again, together with Mr Huxley's, Mr Grove's, &
Mr Gassiot's (the 2 last named gentleman having expressed to me their wish
to sign) and will send it to you to forward to Mr Sylvester[4] & to return to
me—It must be all done in the present month[5]—

Sincerely yours | Edward Sabine.

Dr Tyndall;[6]

RI MS JT/1/S/8

1. *your certificate . . . six weeks ago*: in letter 0585 (5 December 1851) Sabine informed Tyndall that he was about to send the certificate around to be signed.

2. *I conclude . . . mislaid*: Sabine holds Sylvester responsible for the loss of the letter.

3. *statement . . . sent me*: the list requested in letter 0571. Tyndall sent the information to Sabine on 3 December 1851; Sabine acknowledged its receipt in letter 0585.

4. *obtain Dr Faradays signature . . . Sylvester*: the certificate was signed (from personal knowledge) by Faraday, Grove, Huxley, Sylvester, and also John Phillips, in that order. Gassiot did not sign the certificate.

5. *It must be all done in the present month*: the signatures were collected by 26 February.

6. *Dr Tyndall*: we place Tyndall's name, as addressee, at the bottom of the letter, although Sabine wrote it at the bottom of his first page.

To William Francis [6 February 1852][1] 0599

Queenwood Friday

My Dear Francis.

Your letter[2] reached me yesterday morning and it was my righteous intention to reply to it by last night's post. But being pushed to the last extremity in preparing apparatus and experiments for the two lectures which are to come off at Warminster and Southampton on Monday and Wednesday next[3] I forgot my duty till it was too late. I rose early this morning and my first job is to write to you while every body else is asleep. Brewster[4] has reached me and I am exceedingly glad to have secured it at the price. The mag.[5] has also come and Feilitsch.[6] Feilitsch[7] is a representative now to some extent—I think I have a very clear picture of him in my mind's eye—He is evidently the 1st man of a little circle there at Greifswald;—but I wont run off into a biographical sketch—His pamphlet is written in a clear and satisfactory manner, It is not like a scientific memoir as the laws of interference which he explains at some length would in the latter case be assumed as known. I had a portion of a review of it written, but I will look into it again and if any thing can be *[squeezed]* out of it for the Magazine it shall be done.

Your treatment of[8] the extra figure in Magnus's paper[9] meets the difficulty concisely and completely.

I did not at all expect you to reinsert that advertisement,[10] for the mistake was certainly as much ours as yours. I can only say that I am exceedingly obliged to you.

The accounts[11] I have looked over and found all correct. I now return them.

These lectures have made a tremendous draw upon my time for I have had to make a considerable portion of my apparatus—Had it not been for *[this]* unlucky interference I should now have a paper in the hands of Sabine,[12] and Kopp[13] and Feilitsch done besides—I shall have both of the latter most certainly ready in time for the next number of the Magazine. —Will not the 16th be sufficiently early?[14] You know they wont be long.

Could you tell me in your next letter how is the misletoe propogated[15]— that is how are the seeds sown.

most faithfully yours, | J Tyndall

StBPL T&F, Authors' letters

1. *[6 February 1852]*: the reference to the lectures at Warminster and Southampton 'on Monday and Wednesday next' dates this letter to early February 1852 (n. 3 below). The many allusions to papers in the *Phil. Mag.* also suggest early February. The only possible Friday is 6 February.

2. *Your letter*: missing.

3. *two lectures . . . Wednesday next*: Tyndall lectured at Warminster on Monday, 9 February and at Southampton on Wednesday, 11 February (Journal, 20 January and 20 February 1852, JT/2/13b/558). The *Hampshire Advertiser and Salisbury Guarding* reported that Tyndall lectured on 'Light and Electro-Magnetism', and proferred on Tyndall 'great credit for the manner in which he conducted his numerous delicate experiments under circumstances of great difficulty' ('The Southampton Polytechnic Institution', *Hampshire Advertiser*, 14 February 1852, p. 4).

4. *Brewster*: a book by David Brewster, possibly his *Treatise on Magnetism* (Edinburgh: Adam and Charles Black, 1837) or *A Treatise on Optics* (Edinburgh: Adam and Charles Black, 1837).

5. *The mag.*: i.e. the *Philosophical Magazine*. Allusion suggests that this letter was written early in the month, soon after the publication of the latest issue.

6. *Feilitsch*: F. Feilitzsch, *Optical Untersuchungen veranlasst durch die total Sonnenfinsternis 28. Juli 1851* (Greifswald: C. A. Koch, 1852), a 75–page pamphlet. A short summary was published in the *Phil. Mag.* (see letter 0603, n. 2).

7. *Feilitsch*: Fabian Ottokar Carl von Feilitzsch (1817–85) was a German physicist who worked on magnetism and galvanic currents. He had been professor 'extraordinary' of physics at the University of Greifswald since 1848.

8. *treatment of*: Tyndall first wrote 'correction of', which might imply a fault in the version he submitted.

9. *Magnus's paper*: an abridged translation of Magnus's 'Ueber thermoelektrische Ströme' (cited letter 0556, n. 10) was published in the February issue of the *Phil. Mag.* (see letter 0581, n. 20). The published paper included one plate containing six figures.

10. *that advertisement*: not identified.

11. *The accounts*: perhaps accounts for Tyndall's translation work for the *Phil. Mag.*

12. *a paper in the hands of Sabine*: this was sent in late February (see letter 0606).
13. *Kopp*: Hermann Kopp, 'Ueber die Ausdehnung einiger fester Körper durch die Wärme', *Poggend. Annal.* 81, no. 1 (1852), pp. 1–67. Tyndall sent an abridged version to Francis with letter 0602.
14. *in time . . . sufficiently early?*: in time for the next issue of the *Phil. Mag.* Tyndall sent the summation of Kopp to Francis in mid-February (with letter 0602); it was published in the April issue. The summation of Feilitzsch was sent slightly later (letter 0603) but published earlier, in the March issue.
15. *How . . . propagated*: Tyndall was asking on behalf of Mrs Steuart (see letter 0593).

To William Francis [7 February 1852]¹ 0600

Queenwood Saturday night

My Dear Francis.

How would you advise me to act with regard to the enclosed?² Shall I send the document³ when I receive it from Col. Sabine to Mr Sylvester by post, or would it not be surer to send it first to you and you might oblige me by sending a boy with it to Lincoln's Inn Fields⁴—I dont know what to think of it, the obliging readiness with which Mr Sylvester consented to sign forbids the idea that he has any objection to do so. Had I known of the matter while in London⁵ I should have procured an introduction or two by which the difficulty might have been obviated—but I thought the matter ended.

very sincerely yours | John Tyndall

The letter reached me this morning—I should be glad if you would return it to me.

Be kind enough to send me gassiot's address.⁶

StBPL T&F, Authors' letters

---

1. *[7 February 1852]*: this letter was written after Tyndall received letter 0598 (of 5 February) from Sabine; the next Saturday was 7 February.
2. *the enclosed*: letter 0598.
3. *the document*: the nomination certificate for his election to the RS.
4. *Lincoln's Inn Fields*: Tyndall was mistaken. Sylvester, a recently qualified barrister, was a member of one of the inns of court, but of the Inner Temple, not Lincoln's Inn.
5. *while in London*: Tyndall had been in London over Christmas (see letter 0592, esp. n. 5).
6. *gassiot's address*: Sabine had told Tyndall (letter 0598) that Gassiot had expressed a desire to sign the certificate.

From Thomas Archer Hirst   8 February 1852   0601

Marburg | <u>Feb 8th/ 52</u>

My dear John

We have not yet I am glad to say, seen each other and I am beginning to hope that our next interview will be in England where you will remain.[1] I delayed writing sooner, on account of the probability that I might have a chance of speaking what I had to say I scarcely know where to begin however for I have forgotten where I left off. I will start with the most important topic however, my health, and tell you that it would be difficult to find a stronger and healthier student; to pills and Senna[2] thank God I am a complete stranger, and although my bowels have in their time undergone what would have ruined 99 out of a hundred I'll back them now against your own. Brown bread I can digest when necessary and as for veal, though it is as insipid to me as ever I am no longer frightened at it. Impatient as I was when suffering, the chapter on food which can be seen in my life history has been an instructive one. I started with a radically false theory of what, and how much, to eat, drink, and avoid, with a theory of abstinence that defeated itself and engendered the very evil it would avoid, namely the encumbrance of sensual appetite on mental activity. At present I venture to say I eat twice as much and think twice as little about eating as formerly. Formerly I thought it a duty daily to walk my health out, and in grudging it the trouble robbed not only myself of the enjoyment of walking but deprived my health of the very object of the walks, for both are mysteriously connected. —Now my health makes less claim on me in this respect and has learned to be satisfied with the exercise that other circumstances and objects require.

But I will cry halt or I shall waste my paper and time on a subject you have studied enough yourself. —Work goes on with me very well and unless I mistake very much my future path in science lies open before me; namely mathematics and the mathematical part of physics. I find I can throw my interest more completely into such work, and therefore I will stick to it; for strange as it may appear I take far more interest in a beautiful theory and calculation than in a beautiful experiment. It gratifies me to see the latter confirm the former, but I am willing to let others undertake it. Another fact also suggests to me that the former is my task, namely it is far more suggestive to me. In Physics proper I can march straight ahead on the beaten track well enough but I notice few or no by-paths worth following. In mathematics on the contrary I long to turn aside, and come in contact oftener with thickets that promise to repay looking into. I have promised as you may recollect to be in England next April, but if possible I will obtain longer leave of absence and in the next vacation sit down and attempt a dissertation. I would have done it before but I have a notion my time is better occupied as it is, in traversing new fields, to me, and gathering materials for future use, than if I were to sit down and employ those I have—I know well enough this latter is by far the most important object, but at the same time, it is also true, that a certain position is necessary to be first attained in order that you may not employ yourself altogether on things already investigated, and well known. One subject Stegmann has suggested to me and which I shall very probably choose it inasmuch as having a little time last Christmas to look <u>round</u> the matter, I found it promising. Namely to investigate the analogy between certain theorems in Conic Sections, and the corresponding Solid Bodies. To give an example, to see how far the beautiful relations existing between conjugate and principal diameters in the Ellipse are true for the Ellipsoid, when instead of two, three dimensions are present. One Sunday a few weeks ago I read for the first time your Dissertation[3] and was all the more pleased with it when I remembered that exactly a year before it had appeared to me a mere assemblage of incomprehensible hieroglyphics. It was a very interesting little dissertation indeed. As far as I know at present, I shall, after remaining in England a month, return to Paris for a time and there hear Cauchy and Regnault. I would have liked to have spent a Session in Göttingen or Berlin, but I do not exactly see my way clear thereto at present. Perhaps I may do that after the Paris residence in order that the latest impression before returning to England more permanently may be a German one. The French language will be necessary for me but in the German I take far more pleasure. Wrightson[4] is still working away at the Laboratory but I believe not nearer any result than at the beginning of last Semester. He has attained an immense notoriety for explosions, *[scarcly]* a week passes without a sand bath or some other chemical utensil being shattered in a thousand pieces, and the foundations of the old laboratory built by the Holy St Elizabeth[5] being shaken. The fellow is actually to be pitied he is industrious, and anxious to <u>do</u> something but positively cannot; apart too from actual incapacity I believe the secret of his failure is to be traced to his temper. He has the will to learn and the perseverance after a fashion, but it is an ill tempered dogged perseverance, he is never on good terms with his subject, instead of approaching it as a friend who can and will reveal much to him if he earn and deserve it, he regards it as an enemy who out of spite holds the result above his head exactly out of his reach. He considers himself as a Tantalus,[6] & snatches impatiently at the fruit above his head and curses the Gods that he cannot grasp it. What is worse for him however he is not exactly as others in similar circumstances often are, he is conscious of his incapacity, inwardly ashamed of it & consequently far from happy. He is so terribly afraid of being thought ignorant that he[7] gives his opinion on all topics, forgetting that he thereby but too often palpably manifests it. On a Sunday Evening I go through a part of the Theory of Equations with him, Davy and Dickinson, and have often an opportunity of seeing truly curious traits of character.

RI MS JT/1/H/165

1. *We have not yet . . . you will remain*: because Tyndall had not yet visited Marburg, prior to departing for a position in either Toronto or Sydney, Hirst was hopeful that Tyndall would remain in England. He had not yet heard that Tyndall had not received the Sydney post.
2. *Senna*: a laxative which Hirst had previously used (see letter 0553).
3. *your Dissertation*: Tyndall's dissertation was a mathematical analysis of screw surfaces.
4. *Wrightson*: Francis Wrightson. Hirst frequently made disparaging comments about him.
5. *the Holy S[t] Elizabeth*: Hirst alludes to St Elizabeth's Church (letter 0482, n. 8). Several buildings in the church complex were used by the university.
6. *a Tantalus*: Tantalus was a figure in Greek mythology who (as a punishment for killing his son and offering him as a sacrifice to the gods) was made to stand in a pool of water beneath a fruit tree, with the fruit ever eluding his grasp and the water always receding before he could take a drink.
7. *he . . . .*: the remainder of the letter was written vertically along the top margin of the first page of the letter. Although Hirst squeezed in these last sentences as if he were finishing the letter, he did not sign it and, it seems, did not post it until later.

## To William Francis	[mid-February 1852][1]	0602

Dear Francis,

I send you Kopp.[2] I should feel obliged if you would just look along the tablular statement[3] and see wheth[er] there are any technical names which might be introduced instead of those which I have used—Eisenpath have called carbonate of iron Eisenkies sulphide of iron[4]—I dont know wheth[r] there are any technical names for them in England. Bleiglanz[5] and Eisenglanz[6] come under the same head. I have adhered to your instruction as to giving a very short abstract of the paper,[7] the table indeed is the chief object of value.

I will next look into Feilitsch[8] but cannot yet say wheth[er] it will furnish any thing suited to the magazine

Sincerely yours | Tyndall

StBPL T&F, Authors' letters

1. *[mid-February 1852]*: this letter follows letter 0599 (6 February) when Tyndall hoped to complete the Kopp translation by 16 February, and precedes letter 0603 (18 February or, less likely, 25 February), when he sent the translation of Feilitzsch's pamphlet to Francis. Thus, it dates to 7–17 February and probably towards the later end of that range.
2. *Kopp*: published as Hermann Kopp, 'On the Expansion of some Solid Bodies by Heat', *Phil. Mag.* 3, no. 18 (April 1852), pp. 268–70 (see letter 0599, n. 13 for German original).
3. *tabular statement*: the tabulated results are on pp. 269–70 (ibid.); all the terms discussed are found in the table.
4. *Eisenkies suphide of iron*: in the published table this was changed to 'Iron pyrites'.
5. *Bleiglanz*: Tyndall translated this as 'galena', a widespread iron ore.
6. *Eisenglanz*: Tyndall translated this as 'oxide of iron'.
7. *a very short abstract of the paper*: the 67–page paper was condensed to under three pages.
8. *Feilitsch*: see letter 0599, n. 6.

To William Francis [18 February 1852][1] 0603

Wednesday

My Dear Francis,

I send you Feilitsch.[2] I am much obliged for the trouble you have taken with the certificate[3]—One word in your last letter I must take exception to, and that is the word 'prevail'. —I would not prevail on any body. I would sooner wait for 7 years[4] and gain the matter on the grounds of having worked up to it than accept it tomorrow through interest. Though I talk thus I have not the shadow of a doubt that you think as I do on the matter and that you will act accordingly. I should be sorry indeed if Mr Sylvester thought the accident alluded could have altered my feelings towards him—I shall not soon forget the kind promptitude with which he responded to my request.[5]

I will think of a report[6]—It struck me yesterday when looking over Poggendorff that the paper of Lamont N° 12. page 572 ought to be given fully in the Magazine[7]—The periodic law which he establishes is very remarkable and every thing bearing on this subject will be read with interest.

The little article from Ragona Scina page 590 might also appear.[8]

If you say 'yes' to Lamonts paper let me know is it not possible to print the numbers in the tables without copying them. This would reduce the translation to a trifling matter.

Tomorrow morning I shall begin to throw Kohlrausch[9] into a readable form and I shall probably have it ready by the 10th of next month.

Sincerely yours | Tyndall—

Suppose I want to sow a quantity of Misletoe can you tell me how I am to manage it?[10]

StBPL T&F, Authors' letters

1. *[18 February 1852]*: the allusions to Sylvester, the RS certificate and Feilitsch place this letter in February, some days after letters 0600 and 0602 (7 February and mid-February). Given that Tyndall had hoped to have Feilitzsch done by 16 February (0600), Wednesday 18 February seems more likely than 25 February. Also, the summary of Feilitzsch, enclosed in the letter, was published in the March issue of *Phil. Mag.* which makes 25 February (received 26th in London) unlikely, although possible.

2. *Feilitsch*: summary of a pamphlet by Feilitzsch (cited letter 0599, n. 6) which was published as '*Optical Investigations occasioned by the Total Eclipse of the Sun on the 28th of July 1851. By* Dr. v. FEILITZSCH. Greifswald, 1852: Th. Kunike' under 'Notices respecting New Books', *Phil. Mag.* 3, no. 17 (March 1852), pp. 232–33.

3. *the certificate*: the certificate for Tyndall's election to the RS. Francis (and Sabine) had advised Tyndall and assisted with the process of getting signatures (see letter 0600).

4. *wait for seven years*: biblical allusion to the patriarch Jacob, who served Laban seven years to gain his daughter Rachel as wife (*Genesis* 29:15–28).

5. *I should be sorry . . . my request*: Sylvester, who had agreed to be one of the signatories from 'personal knowledge' of Tyndall, had lost the first copy of the certificate.

6. *a report*: a topic for one of his 'Reports on the Progress of the Physical Sciences'.

7. *the paper of Lamont . . . in the Magazine*: Johann von Lamont, 'Ueber die zehnjährige Periode, welche sich in der Grösse der täglichen Bewegung der Magnetnadel darstellt', *Poggend. Annal.* 84, no. 12 (1851), pp. 572–82; translation published as Lamont, 'On the Ten-year Period which exhibits itself in the Diurnal Motion of the Magnetic Needle', *Phil. Mag.* 3, no. 20 (June 1852), pp. 428–35.

8. *the little article . . . also appear*: Dominico Ragona-Scinà, 'Ueber die Longitudinallinien des Sonnenspectrums. Schreiben an Hrn. Prof. Dove', *Poggend. Annal.* 84, no. 12 (1851), pp. 590–92; translation published as Ragona-Scinà, 'On the Longitudinal Lines of the Solar System. From a Letter to Professor Dove', *Phil. Mag.* 3, no. 19 (May 1852), pp. 347–49.

9. *Kohlrausch*: what Tyndall intended is unclear, but he made three papers by Kohlrausch the basis of his next 'Report' (n. 6 above): 'Die elektromotorische Kraft ist der elektroskopischen Spannung an den Polen der geöffneten Kette proportional', *Poggend. Annal.* 75, no. 10 (1848), pp. 220–28; 'Die elektroskopischen Eigenschaften der geschlossenen galvanischen Kette', ibid., 78, no. 9 (1849), pp. 1–21; and 'Versuch zur numerischen Bestimmung der Stellung einiger Metalle in der Spannungsreihe', ibid., 82, no. 1 (1852), pp. 1–18. Tyndall's discussion was published as 'Reports on the Progress of the Physical Sciences', *Phil. Mag.* 3, no. 19 (May 1852), pp. 321–30. He gave the title 'On the Electroscopic Properties of the Voltaic Current; being an experimental verification of the Theory of Ohm' to Kohlrausch's three papers.

10. *Suppose . . . manage it?*: Tyndall had previously asked Francis how mistletoe is propagated (letter 0599).

To Thomas Archer Hirst   23 February 1852   0604

23rd Feb. 1852

My[1] dear Tom

The enclosed[2] has been long written as you will see. Shortly after it was finished, I had to give two lectures which came upon me almost simultaneously at Warminster and Southampton[3] and the bustle and confusion of travelling hither and thither and of removing a large quantity of apparatus from town to town took away all my peaceful thoughts. Your letter[4] reached me yesterday and converted what has often been a source of sorrow and of darkness to me into joy and sunshine; I mean the state of your health. I know nothing that could have imparted more solid satisfaction to me than this piece of intelligence. Gradually, my brother, you will learn what Nature intends and you will ascertain that in many cases 'her ways are not our ways, neither are our ways her ways'.[5] Doubtless even after you have learned thus much you will often have to set bounds and limits to indulgence but you will do it with a clear eye and understanding heart; you will see where it is necessary and where it is not necessary. You will act the part of a king confident in his own power and in the love of his subjects, not that of a tyrant who fearing lest his subjects should abuse their liberty makes them all slaves. Many and many a time I have drunk the cup of disappointment in these matters for which I now thank the gods—how should I have ever known the false from the true, how should I have ever sounded the depths of my own nature and that of men generally without such experiments. Even in a physical investigation it is not[6] by fighting through defeat and disappointment that we arrive at the true end, and shall we complain if the same law hold[7] good in the sphere of mind? I think it good[8] that you have got thus far, and my prayer and hope is that each succeeding year will clarify your vision and increase your stability more and more.

Nobody has yet occurred to me as likely to suit you in Marburg.[9] I had a letter written to one young fellow but on reconsideration I thought you would consider him too young being not more than 16; I did not send the letter.

I dont think it would be wise in you to quit Germany so soon as next spring. I would strongly recommend you to spend some time in Gottingen and afterwards some time in Berlin. <u>What's the need of haste</u>? Were I in your place I should certainly feel no anxiety about getting away from Germany— Recollect you have laid your hand to the plough and must stick to it[10]—you have begun a work,[11] it must be creditably finished—I dont think you could have it upon a happier choice than that which you have made, mathematics and the mathematical part of optics. But remember in the land of England there is a University of Cambridge which piques[12] itself upon its knowledge of these matters—now you must not knuckle down to a Cambridge fellow, unless nature has stated the law to be so—you must prove that her gifts are superior to university drill—to sum up all you must simply do your work in the best manner possible and <u>by next April you will not have learned half your trade</u>. The French language will come in due season, it is all right and necessary that you should spend some time in Paris but no retreat as yet from Germany—that's my voice, And now again good bye—If you halt between two opinions.[13] then take the bravest,

Yours affectionately | John

RI MS JT/1/HTYP/185–187a
LT Transcript Only

---

1. *My*: as notes 5, 6 and 7 indicate, this letter has some conspicuous transcription errors by LT. Readers should allow that there may be other, less conspicuous, errors.

2. *The enclosed*: letter 0597.

3. *two lectures . . . Southampton*: delivered on 9 and 11 February (see letter 0599, n. 3).

4. *Your letter*: letter 0601.

5. *'her ways . . . her ways'*: Tyndall applies to Nature biblical language used for God: 'For my thoughts are not your thoughts, neither are your ways my ways, saith the LORD' (Isaiah 55:8).

6. *is it not*: our suggested correction for LT's 'it is not'.

7. *hold*: our suggested correction for LT's 'told'.

8. *thank the gods*: our suggested correction for LT's 'think the good'.

9. *Nobody . . . in Marburg*: allusion uncertain. Perhaps Hirst was looking for a tutor in French (see letter 0601). Tyndall's reassurance later in this letter that Hirst would learn French in due time supports this interpretation.

10. *you have laid . . . stick to it*: an allusion to Luke 9:62 where, when a would-be follower wanted to farewell his family first, 'Jesus said unto him, No man, having put his hand to the plough, and looking back, is fit for the kingdom of God'.

11. *you have begun a work*: possibly an allusion to Philippians 1:6: 'Being confident of this very thing, that he which hath begun a good work in you will perform it until the day of Jesus Christ'.

12. *piques*: takes pride in or congratulates oneself (*OED*).-

13. *If you . . . two opinions*: allusion to Elijah's challenge to the people of Israel: 'How long halt ye between two opinions? if the LORD be God, follow him: but if Baal, then follow him. And the people answered him not a word' (1 Kings 18:21).

From Hector Tyndale 23 February 1852 0605

Philadelphia 23ᵈ February 1852

Dear Sir

It is with pleasure that I acknowledge the receipt, in due course, of the two memoirs and also of your kind letter of the 15ᵗʰ November last,[1] which acknowledgement will serve to show my own unfitness for correspondence, inasmuch as I have so long delayed to answer one for whom, from all I have heard, I, in common with all my Fathers' family, entertain a high respect and esteem. I most sincerely reciprocate your kind regrets caused by our not having met during my recent visit to Europe; more especially as I have the mortification of believing, from your memoirs, that I was in Berlin at the time of your temporary residence in that city during the last summer. I had hoped to have met you at Leighlin Bridge, when I had the pleasure to meet and to form an acquaintance with your kind Mother and Sister to whom, in your correspondence with them, please present my best regards and the assurance that their kind attentions are strongly impressed upon my memory. All the more so as they were the only members of my Father's family I met with.

My visit to Leighlin Bridge and its vicinity was one of the most interesting and sad I have ever made. It is easy to understand why the birth place of a Father should be interesting to a Son. I heard, during my childhood, so much of the various precincts of Leighlin, both from my Father[2] and my Uncle William,[3] that I had become familiar, in thought, with every field, lane and hill which lie around the Barrow and the 'Dineen River', as my Father called it,[4] had worn a channel in my imagination almost as deep as that little streamlet, as we would call it, has worn upon the rocks below it. While the names of persons <u>then</u>, resident were living in my mind as familiarly as those of my own playmates. While looking upon these scenes I did not reflect that those from whom I had so often heard of them had left the country when young, a country where youth is more youthful and certainly longer lived than in our climate, and that what to them was grand or beautiful was so because painted with the vivid recollections of childhood colored by all the beauties that happy ignorance (or innocence) of childhood alone can know. Not remembering this you will readily understand the difficulty I had to realize the fact of my actual presence upon the birth-place of those so dear to me and long since passed away, and I confess to a feeling of disappointment, which I did not fail to reproach myself for.

You who had the advantage—for in one respect, at least, it is so—of being born in the land of your Father, where every locality is filled with your fore-Fathers memories; will perhaps fail to understand why I should dwell upon what, at best, partakes more of the feeling of the child than of the sentiment of the man. And it <u>was</u> exactly with the feeling of a child that I saw those places for I had never known them, whose names had filled so large a space in my memory, and therefore all the experiences of after life rested under the control of youthful feelings and childish recollections. It was the most complete 'overthrow of the present order of things and the most decided triumph of first principles'[5] I ever encountered.

It is simply because it was so interesting to me that I should trouble you with this, perhaps, tedious mention of it. That you are yourself a native of the same vicinity, bearing the same name and therefore will, I hope, pardon this long recital of personal feelings without attributing it to the improper cause of egotism. Will you however allow me to add to this weary relation[6] the cause of my disappointment in my visit to 'Augheroo'[7]—the place which my Grand Father and his Father held and upon which my Father and all his family were born—for I was so notwithstanding all my early impressions, or rather perhaps by the cause of their brilliancy. I saw nothing of the charms and beauties my Father saw a child but merely a desolate moor-land, hilly, dark and barren. Without even the chiefest Irish characteristic of gayity and vivacity.

I met with another disappointment (for I might as well—to use an old Northumbrian Saw—from which county by the way I believe we date our name, <u>Tyne</u>-dale[8] (?) '/be/ hung for an old sheep as a young lamb'[9] so long as I have already bored you so very deeply) in not being able to find any family records, at least so far as to amount to nothing: The clerk, or Sexton, of the church of Old Leighlin informed me that the Church Records, with the small exception remaining, had been destroyed by time, carelessness and the—butter dealer![10]

For these things I have, myself, but little, if any, regard but my eldest brother Sharon[11] (you see my Father thought his duty filled by the adoption of one scriptural name[12] and therefore felt himself at liberty to take choice from pagan traditions and accordingly he dragged poor Hector (Hector's name but not, I trust, his mores) once more from out the buried walls of his city and tied him <u>neck</u> and heels to me;[13] And I have humble trust that if my sponsorial title be a down drag that my brothers' scriptural one will help me upwards especially as I can say with great truth and affection he is a good brother and an honest and sincere man.) but my brother has a desire to climb the genealogical tree as high as possible. And if you can aid him in this laudable desire both he and myself will be indebted to you.

I thank you for your kind attention in sending me the two addresses, or scientific papers, delivered by yourself and though I have not the slightest claim to an acquaintance with any scientific subject this simple style enabled me to understand them clearly. I read them with pleasure and must ask of you the favour to forward at any time any additional papers from yourself as I must, notwithstanding our want of personal acquaintance, always feel an interest in you and your pursuits.

It would give all our family great pleasure to see you here and in the course of events it is likely we may do so, but permit me to say that while there is occupation in England or in Europe it is safer to remain there. Our country is so new, so much in need of mere physical labour that as yet there are only the poorest prospects for a person of scientific pursuits and professions, unless there is a positive call, or request, from some of our institutions. Yet I will hope that at some future day we shall have the pleasure of seeing and of forming a personal acquaintance with you. Should you visit the United States I shall be glad to see you and so will all the family of my Father. If you would wish to make any inquiries or to obtain any information relative to this country or in any manner in which I can aid, it will give me satisfaction to do so, and I hope you will make use of me should you desire it.

Notwithstanding our very limited acquaintance you may be assured that 'anything relating to my (your) personal life would be of interest to you (me)'[14] and I shall be happy, whenever your time permits, to hear from you in relation to yourself, to your profession and prospects as also concerning your Mother and Sister. And if anything relative to our family on this side the Ocean will interest you I will with pleasure give you any information.

I trust that the distinction of election as a Fellow of the Royal Society has been or will shortly be conferred upon you. I am glad, nay all our family are glad, that you have by your own exertions worked yourself to distinction, they trust that you will achieve still greater things and that your progress will not be limited by that honorary de[g]ree but that your name will add lustre even to that institution, rich as it is in names of eminence.

My brother Sharon, to whom I have alluded, and who, by the way, desires his respects to be sent to you, wishes to know whether you or ourselves spell our name correctly. Tynda<u>ll</u> or Tynda<u>le</u>? It has been spelt, so far as I would see by the records, in all sorts of ways, Tindal, Tindall, Tyndal, Tyndall and Tyndale and a long time ago, it is said, Tynedale. Have you any information on the subject?

You already know that my Father is deceased, 1842, he commenced business with nothing and though he built the business up to a large amount and had made quite a fortune yet at the time of his death he left but little property, owing to an unfortunate (Irish) habit of loaning—anglice[15] <u>giving</u>— money but I may say that he left an honest name and a good business, the importation and dealing in China and Glass ware,[16] to which myself and my brother in law, Mitchell, the husband of my eldest sister Elizabeth, succeeded. My Mother[17] is living in good health and is as good a woman as ever lived.

My eldest brother, to whom I have referred several times, found, at rather a late period, that his tastes did not lead him to traffic[18] and has within two years again betaken himself to study and is now at one of our first colleges preparing himself in Mathematics as a Civil Engineer. My youngest, living, brother—Harold[19] (so called not after the unlucky Saxon King but after a good Catholic—Roman—priest who formerly resided here and now I believe in Dublin) is in the land of golden-dreams, California, where I hope he will do well. All my Sisters are well married and the whole family have taken off their hats, Caps and overcoats to stay for life. My uncle William died, as I believe you know, in 1845 leaving a small business—though comfortable, which his widow manages—and five children.[20] They are comfortably situated and the family is a promising one.

Thus then, my dear Sir, I have given you a sketch of ourselves, or rather of our positions, and have nothing further to say save the begging of you to pardon this long, rambling, tedious tax upon your time and good humour and to subscribe myself very faithfully and truly Your friend

Hector Tyndale

John Tyndall Esq

RI MS JT/1/T/43

1. *the two memoirs . . . November last*: Tyndall's letter is missing. Hector mentions that the memoirs reveal that Tyndall had been in Berlin, which strongly suggests that one was the paper on diamagnetism published in the *Phil. Mag.* (cited letter 0498, n. 6) which was dated 'Berlin, June 1851' at the end. The other may have been Tyndall's recent paper on bismuth (cited letter 0525, n. 2).

2. *my Father*: Robinson Tyndale (c.1775–1842); a prominent Philadelphia businessman (see n. 16).

3. *my Uncle William*: William Tyndale (1783–1845); born in Carlow; died in Philadelphia.

4. *the Barrow . . . as my Father called it*: the Barrow is the river that runs through Leighlin; the Dineen River is possibly the River Dinin (or Dinnin) which is to the northwest of Leighlin Bridge and Old Leighlin on the Carlow-Kilkenny boundary, and adjacent to Agharue (see n. 7).

5. '*overthrow . . . principles*': not identified.

6. [illegible]

7. '*Augheroo*': also written Augharoe or Agharue, a village in Carlow located northwest of Old Leighlin.

8. *Tyne-dale*: referring to the river Tyne.

9. '*be hung . . . young lamb*': a well-known saying, usually used to justify committing a greater offence having already perpetrated a minor one.

10. *butter-dealer*: allusion not identified.

11. *my brother Sharon*: Sharon Tyndale (1816–71), was born in Philadelphia and pursued a commercial career in Philadelphia and Illinois before turning to engineering. He studied mathematics in Illinois and engineering at Harvard College, Cambridge, Mass., and graduated in engineering in 1851. He began his engineering career as a railway surveyor in Illinois. Later he became secretary of state for Illinois (1864–69). He was murdered in 1871.

12. *one scriptural name*: Sharon is a Biblical place name.

13. *poor Hector*: alludes to the prince of Troy who was killed by Achilles during the Trojan War and his body then dishonored, to which tied 'neck and heels' may be an allusion. The allusion to Hector's mores is unclear.

14. *'anything . . . you (me)'*: Tyndall may have (formulaically) expressed doubt that news of his 'personal life' would be of interest to Hector and Hector is quoting from Tyndall's missing letter.

15. *anglice*: 'in English' or 'as the English would say it' (*OED*).

16. *a good business . . . Glass ware*: initially Tyndale & Company; renamed Tyndale & Mitchell after Robinson Tyndale's death.

17. *My Mother*: Sarah Tyndale (1792–1859).

18. *traffic*: the buying and selling or exchange of goods for profit; bargaining; trade (*OED*).

19. *Harold*: Harold Tyndale (1826–?).

20. *his widow . . . five children*: Mary Tyndale, née Carter (1804–82) and William Emmett Tyndale (1834–64), Mary Tyndale (1837–1907), Jane Tyndale (1838–1927), John Tyndale (1840–1925), Tecumseh Thomas Tyndale (1845–1912). Their first child, Rosa Tyndale (1833–39) had died in childhood.

From Edward Sabine 24 February 1852 0606

February 24. '52 | 11. Old Burlington S[t]

Dear Sir—

I have read the paper which you sent me[1] with much interest. It is perfectly clear & intelligible at first reading. but I think that when read at the R.S. (which is I suppose your intention) it should have the diagrams exhibited in a larger scale. It would perhaps be difficult to follow the <u>description</u> of the apparatus; but it is most clearly understood <u>with</u> the diagrams.

On the first perusal I received the impression that you viewed the existence of forces as wholly abstract from matter—matter being merely an accident which made us acquainted with the presence of forces by being susceptible of being acted upon by them—If such be supposed in the case of Magnetic forces, it would not be an analogous force to Gravitation. A Body placed midway between Earth & Moon would make us <u>see</u> that a force of Gravitation existed there, & would be analogous to the piece of soft iron between the poles of a magnet—But the origin of the force of gravitation is in the attractive quality with which matter is endowed, without the existence of matter there would I suppose be no force of Gravitation; so without the existence of matter there would I suppose be no magnetic force.

On a second perusal I think I inferred more distinctly that the abstraction referred to was by no means to be implied—that, (still keeping in view the analogy of the forces) your view is that as Gravitation (independent of the distance) is dependant on density so is Magnetism on <u>Molecular arrangement;</u> and, subordinately, not on the particular form of the ultimate particles.

I have forwarded your paper to the R.S.—this is your wish I believe.

Sincerely yours | Edward Sabine

RI MS JT/1/S/9

---

1. *the paper which you sent me*: an early version of 'On Molecular Influences. Part I. Transmission of Heat through Organic Structures', *Phil. Trans.* 143 (1853), pp. 217–31. Tyndall reported that he completed the research for the first section of the paper on 20 January (Journal, JT/2/13b/558). See letter 0613 for Sabine's further advice.

From Lyon Playfair 5 March 1852 0607

33 Ladbroke Sq | Notting Hill | 5th March /52

Dear Sir,

Having been engaged in the preparation of some public lectures I regret to find that I have not answered your letter of the 24th Feb[y].[1] The Cavendish Soc[y] which published a translation of Bunsens first Memoir[2] does not intend to translate or publish any more Scientific Memoirs, confining its operations to self-contained works—I have therefore been unable to obtain its translation. I think it possible if you were to write to Francis of the Phil. Mag. that he would order its translation for Taylor's Scientific Memoirs.[3] I had much pleasure in signing your certificate for the Royal Soc[y] a few days since[4]

Truly yours | Lyon Playfair

J Tyndall Esq[r]

RI MS JT/1/P/113

---

1. *your letter of the 24th Feb[y]*: letter missing.
2. *Bunsens first Memoir*: cited letter 0520, n. 4.
3. *its translation for Taylor's Scientific Memoirs*: Tyndall had been trying since September 1851 to get this very long memoir by Bunsen (cited letter 0526, n. 2) translated and published. It was taken up by *Scientific Memoirs* in October (see letters 0666, n. 1 and 0676, n. 3).

4.   *signing your certificate . . . a few days since*: Playfair had not signed Tyndall's certificate
     in advance, but had signed from 'general knowledge' when the certificate was displayed,
     along with those of other candidates, in the Society's rooms.

To Elizabeth Steuart   10 March 1852   0608

Queenwood | 10th March 1852

Dear Madam,

I have much pleasure in sending you the enclosed directions regarding
the propagation of the Misletoe which have been written by one of the ablest
vegetable physiologists in London.[1] I shall be very glad indeed if they assist
you in any way in your endeavour to transplant the misletoe to the banks of
the Barrow.[2]

I remain dear Madam | most faithfully yours | John Tyndall

RI MS JT/1/TYP/10/3335
LT Transcript Only

---

1.   *written by . . . in London*: not identified. Enclosure missing.

2.   *your endeavour . . . the Barrow*: Steuart had asked Tyndall to send mistletoe seeds in letter
     0593. The Barrow is the river that runs through Leighlin Bridge.

To William Francis   [12 March 1852][1]   0609

Friday

My Dear Francis,

There is a memoir by Savart on the vibration of plates of wood crystal &c
in the Annales du Chemie et de Physique[2] which I am extremely anxious to
read—it stands in vol. *[40]*—you would confer a great favour on me if you
would send it to me by rail as soon as you can—dont pay the carriage—or if
you do be good enough to set it down against me—[3]

Sincerely yours | J Tyndall

I hope the books I sent you[4] arrived safely

StBPL T&F, Authors' letters

---

1.   *[12 March 1852]*: dated to March 1852 by an entry in Tyndall's Journal: 'I was most anx-
     ious for a memoir of Melloni's and for one of Savart's and Francis has sent me a number of
     the Scientific Memoirs unexpectedly containing both' (18 March 1852, JT/2/13b/561).

We assume that Francis sent the translations on receiving this letter. The Friday on which Tyndall made the request is therefore probably 12 March, with 5 March possible but unlikely.

2. *memoir by Savart . . . Physique*: Félix Savart, 'Recherches sur l'élasticité des corps qui cristallisent régulièrement', *Annal. Chim. et Phys.* 40 (1829), pp. 5–30. Tyndall's letter is hurried: he mixed French and German in identifying the journal and the handwriting is difficult to decipher. The translation, sent by Francis, was 'Researches on the Elasticity of Bodies which crystallize regularly', *Scientific Memoirs*, vol. 1 (1837), ed. Richard Taylor, pp. 139–52 and 255–68.

3. *set it down against me—*: against what Francis paid Tyndall for translations for the *Phil. Mag.*

4. *the books I sent you*: not identified.

From Andrew Ramsay　　　18 March 1852　　　0610

London 18 March 1852

Dear Sir,

I am much obliged for your offer of Bunsens Memoir.[1] I have no doubt we have it in the library here, but I am too much occupied to avail myself at present of the information your kindness has supplied me with.

I remain | *[dear]* Sir | your ob[t] st[2] | And[w] Ramsay[3]

RI MS JT/1/R/2

---

1. *your offer of Bunsens Memoir*: this could be an offprint from *Pogg. Annalen* which Bunsen had asked Tyndall (in letter 0520) to forward on his behalf (cited letter 0526, n. 2). If so, Tyndall was slow to act.

2. *ob[t] st*: obedient servant.

3. *And[w] Ramsay*: Andrew Crombie Ramsay (1814–1891), Local Director for Great Britain (that is excluding Ireland) of the Geological Survey and lecturer in geology in the recently established Government School of Mines. The two institutions were closely associated.

To Thomas Archer Hirst　　　24 March 1852　　　0611

24th March 1852

My Dear Tom.

Brutus under the eye of Caeser's Ghost[1] was hardly more bewildered than I was for 5 minutes this morning—even now I can hardly believe that you are so near me, and shall not be thoroughly convinced until I have had the pleasure of kicking you for treating me so scurvily—Come to me before Sunday if you can—nay if you could come by Friday evening and deliver a lecture for me here it would please me much[2]—I am condemned to give a lecture at Romsey on tuesday next and am now preparing for it so a substitute at Queenwood would relieve me. If you cant come before you will surely manage to witness my dexterity at 'Shooting dry rubbish'[3] in Romsey,[4] but if you are here on Sunday we shall wander out together—Easter is coming boy and if you are here at the time I will sacrifice a week's work to take a ramble with you—Now write to me and say when I may expect you[5]—I have got a cold too. But this glorious day and your arrival has thrown a flush of vigour into my veins which has completely quelled my infirmity and expanded me into a Hercules—

now boy write | John—

Kindest remembrances to your <u>brother</u>[6]

RI MS JT/1/T/549

1. *Brutus . . . Ghost*: an allusion to Shakespeare's play, 'Julius Caesar', in which Brutus is visited by Caesar's ghost, who tells him that they will meet again at Philippi (Act 4, Scene 3). In the final scene, Brutus commits suicide after losing the battle at Philippi, thus fulfilling the ghost's prediction (Act 5, Scene 5).

2. *Come to me . . . please me much*: Hirst did not arrive in time to deliver the Friday lecture (see n. 5) but he gave a lecture to the boys on the evening of Tuesday 6 April and, because Tyndall became ill, took his classes over 5–7 April (Hirst, 'Journals', 3–7 April).

3. *'Shooting dry rubbish'*: means, literally, to dump or tip dry rubbish, as distinct from wet rubbish. We cannot identify Tyndall's figurative meaning.

4. *witness . . . in Romsey*: Hirst did attend Tyndall's lecture at Romsey on Tuesday 30 March (Journal, 4 April 1852, JT/2/13b/562).

5. *when I may expect you*: Hirst replied that he would arrive in Queenwood on Sunday 28 or Monday 29 March (Journal, 28 March, JT/2/13b/561); he arrived on the Monday and left again on the evening of the 7 April (Hirst, 'Journals', 29 March–7 April).

6. *your brother*: Hirst was staying in Bristol with his brother John. (Hirst, 'Journals', 29 March)

From George Wynne   25 March 1852   0612

Ry.[1] Dept. | Board of Trade |
Whitehall | 25 March/52

My dear Tyndall,

It is a very long time since I have heard from you[2] and I am anxious to know what your successes have been with the many irons you have had in the fire, I hope some of them have heated properly and <u>profitably</u>, pray let me know as no one can feel more interested in your success than

yours very truly | Geo. Wynne

We are all well at home and my little girl who was a great invalid during the winter is daily improving in health—My boys continue to go on very satisfactorily at the school particularly my eldest boy who is a tiger for work—Tho' my wife is not <u>here</u> I may take it on myself to send you her kind remembrances G.W.

RI MS JT/1/TYP/5/1844
LT Transcript Only

1. *Ry.*: Railway
2. *heard from you*: the most recent letter of which we have evidence was in late 1851 (alluded to in Wynne's letter (0573) of 26 November 1851).

From Edward Sabine   29 March 1852   0613

11 Old Burlington S^t [1] | March 29. '52

My dear Sir

Are you likely to have the 2^d section of your paper finished in time to be read in the present Session of the R.S., which ends in June? If you are, it might be worth while to delay the reading of the 1st Section till both could be read together: otherwise I consider that the description of the apparatus you have devised, & the illustration of its use afforded by its application to the experiments on the conduction of heat form of themselves a quite sufficiently important paper—I am led to make the inquiry by a reperusal of your letter of Feby 29th, in which you have left to me to decide whether the first section should be read by itself, or should be reserved for the addition of the 2^d section—I scarcely feel able to judge on such a point myself.[2]

Are you restricted in your experiments by their cost? or if you had a grant of £50 from the Gov^t Grant of this year would it enable you to have more complete apparatus for them? If you were to state a definite object in your researches, & ask for a grant to the amount above stated, I am sure that it would be <u>very favourably</u> rec^d and would be very likely to be allotted.[3] Mr Thomson of Glasgow & Mr Joule are applicants for sums to that amount from the £1000 at the disposal of the R.S. this year, for experiments on very kindred points to yours.

Sincerely yours | Edward Sabine
Dr Tyndall.[4]

RI MS JT/1/S/10

1.   *S^t*: the superscript 't' is one of the many small superscript squiggles which Sabine used in abbrevating words.

2.   *Are you likely . . . of the 2^d section*: Tyndall seems to have followed this advice and held the first part of the paper (sent in February—see letter 0606) until the second part was ready. The completed paper (cited letter 0606, n. 1) was received by the RS on 20 October 1852 and read on 6 January 1853. It bore the original title, 'Part I', but no reference to the initial submission in February 1852.

3.   *likely to be allotted*: for a later stage of the process see letter 0625.

4.   *Dr Tyndall*: Sabine, as usual for him, identified the addressee at the bottom of his first page, not at the bottom of the letter.

To James David Forbes   5 April 1852   0614

Queenwood College | Stockbridge Hants |
5th April 1852

Dear Sir

Were I expert at making apologies this would be a fitting time for the exercise of the talent; but as I am not, and as I have too much faith in your kindness to believe that a lengthened apology for my present act is necessary, I proceed at once to the subject which induced me to write to you.

You have worked at the 'conduction of heat'—this I learned long ago from statements in the handbooks of Baumgartner, Ettingshausen and Müller;[1] but a complete account of your experiments I have not yet been able to lay my hands on. I have looked through the last 40 volumes of the Philosophical Magazine but to no purpose save that of finding out what I conceive to be the locality of the memoir which contains your results upon conduction. At page 27 of the 4th vol. of the Magazine[2] you incidentally remark that the result of certain enquiries was contained in a paper read to the Royal Society of Edinburg on the 7th of Jan, 1833[3]—this paper I imagine is the one which excites my present interest.

An enquiry in which I am at present engaged renders me anxious to make myself accurately acquainted with what has been hitherto done in determining the conductive power of bodies; my distance from London throws an obstacle in the way of obtaining the necessary books and thus the thought occurred to me that in one important case at least you might be able and willing to assist me. If you could kindly lend me a copy of your enquiry upon conduction you would oblige me very much—I will take great care of it and lend it back to you uninjured.

I remain | Dear Sir | Most faithfully yours | John Tyndall
Prof. Forbes | &c. &c &c

StA JDF Incoming letters 1852, no. 33

1.　*the handbooks of Baumgartner, Ettingshausen and Müller*: could refer to any of a number of handbooks by Andreas von Baumgartner, Andreas von Ettingshausen, and Johann Heinrich Müller.

2.　*At page 27 of the 4ᵗʰ vol. of the Magazine*: in Forbes, 'Experimental Researches regarding certain Vibrations which take place between Metallic Masses having different Temperatures', *Phil. Mag.* 4, no. 10 (January 1834), pp. 15–28.

3.　*a paper read . . . 1833*: see Forbes's reply (letter 0618).

From William Francis　　　　　[6–13 April 1852][1]　　　　0615

I have made inquiries respecting Shortredes Logarithms[2] and beg you to present my respects to your friend Mr Hirst and inform him that the

Logarithmic Tables complete royal 8ᵛᵒ costs £1.10.0

　　do[3]　　1 to 120.000 places decimals costs 12/

If he wishes for a copy I can easily procure it for him or a note from you to our clerk C. Gyde will be all that is necessary

Lastly my respects to friend Debus and tell him I am greatly obliged to him for his papers which shall find a place in one of the next Numbers[4]—I wish he had exerted his talents a little and put it into English—However I will tell him that when we meet—

Yours most sincerely | W Francis

Dʳ Tyndall | & &

I have just received a remarkable paper from Faraday on Lines for[5] Magnetic Force for the Mag—[6]

RI MS JT/1/T/550

1.　*[6–13 April 1852]*: this letter was enclosed to Hirst in letter 0617 (14 April), and was therefore written before 14 April. It has no address or heading which suggests that we have only the end section of a longer letter, the section which Tyndall forwarded to Hirst. Tyndall must have received the letter on or after 7 April because until that date Hirst was with him at Queenwood. Thus the earliest date of writing is 6 April.

2.　*Shortredes Logarithms*: Robert Shortrede (1800–68), *Logarithmic Tables: Containing Logarithms to Numbers from 1 to 120,000 . . . to Seven Places of Decimals, etc* (Edinburgh: Adam & Charles Black, 1849). We have not been able to identify the 'royal' edition.

3.　*do*: ditto.

4.　*his papers . . . next Numbers*: Francis seems to promise Debus a place in *Phil. Mag.*, but no paper by Debus appeared in this period.

5.　*for*: Francis meant lines 'of' magnetic force.

6. *remarkable paper . . . for the Mag*: Faraday, 'On the Physical Character of the Lines of Magnetic Force', *Phil. Mag.* 3, no. 20 (June 1852), pp. 401–28.

To Edward Frankland   [6–13] April 1852[1]   0616

Queenwood 13th April 1852

My Dear Frankland,

The arrival of our old father at his old habitation[2] has put me in mind of the long epistolary debt which you have against me—I'm a wretched Sinner in this respect I know—A few years ago I used to find an absolute pleasure in correspondence but a few years can work great changes as my present experience testifies—A short time ago I was full of life, and ready at any given moment to fall in love with the first pair of bright eyes which shot their glances into me—now I scarcely feel a vestige of the old fervour I am becoming daily more of a moral petrefaction; more and more a monk in habits if not in thought. I should begin to be very anxious if I thought this state likely to be permanent but well I know that there is still a balm in Gilead,[3] and that one of these fine mornings my heart and affections shall be found blossoming once more like the rose of Sharon[4]—For some weeks my health had been far from vigorous—I had an attack of influenza a short time ago and unluckily the very day it seized upon me I had to lecture at Romsey;[5] by an effort I stirred up fire enough to carry me through but the reaction came afterwards and I was weak and feeble for many days—the east wind is now yielding up its sting, the glorious sun is once more shining, and in sympathy with the revival of universal nature I feel a stirring in my dry bones[6] also. it is to be hoped that I shall be a vigorous man at the end of the half year.

Here are three mortal pages about myself—Well though I have heard nothing from you direct for some time I have heard something <u>about</u> you— Father tells me that you are about building a new house, that your laboratory is a beautiful one, and I have myself read that solid report of yours on the hydrocarbon gas[7]—a substantial thing which carries internal evidence to those who are able to appreciate it that the man who executed it knew what he was about. I shall take a run over to you sometime and see how the patriarch looks under his own vine and figtree[8]

Debus is now in Bristol having accompanied Tom Hirst thitherward—Tom made his appearance in England lately without letting any one know anything of his intention—He will return to Marburg in a couple of weeks—poor old Marburg—many odorous[9] memories cluster round the old spot. I suppose the apples are in bloom now at each side of the Professors Walk[10]—and the leaves are budding along the slope up to the castle walls[11]—although I have not a living, speaking memento of Marburg as you have[12] still I shall ever remember the old place with affection. The time I spent there is now becoming dreamy to me; but it is a dream which is here and there illuminated by Smiles and Sunshine—

Marburg naturally associates itself with the idea of M[rs] Frankland—I hope she likes England and that each succeeding year she will like it better and feel happier in the land of John Bull. Remember me most kindly to her. Little I thought *[4]* years ago before I had ever seen her face, and when you and I used to jest about her on the spot where I at present write[13] that the day would come when I should desire my remembrances she being a resident in Manchester—Wonderful world! is it not. If my paper were longer[14] I would write more—lebe wohl—Tyndall.

D[r] Frankland &c &c | Professor of Chemistry | Owen's College | Manchester[15]

JRL Frankland, 30

1. *[6–13] April 1852*: Tyndall probably began this letter on 6 April, because he initially wrote '6th', crossed it out, and wrote '13th' above.

2. *our old father at his old habitation*: this implies that Singleton, known affectionately as father to Tyndall and Frankland, had visited Queenwood, but we have found no independent reference to such a visit.

3. *a balm in Gilead*: a rare perfume produced in the region of Gilead that was used medicinally. The phrase, used occasionally in the First Testament (for example, Jeremiah 22:8), came in Christian tradition to represent a spiritual medicine, able to heal the soul.

4. *the rose of Sharon*: Song of Solomon 2:1. 'I *am* the rose of Sharon, *and* the lily of the valleys'.

5. *lecture at Romsey*: Tyndall lectured at Romsey on Tuesday 30 March (letter 0611, esp. n. 4).

6. *a stirring in my dry bones*: allusion to the vision of Ezekiel 37:1–10, in which old bones become living people.

7. *report of yours on the hydrocarbon gas*: possibly 'Contributions to the knowledge of the manufacture of gas', read 13 January 1852 and published in *Manchester Phil. Soc. Mem.* X (1852), pp. 71–120 and *Liebig Annal.* 82 (1852), pp. 1–49; abstract printed in *Chem. Soc. Journ.* 5 (1852), pp. 39–50.

8. *the patriarch looks under his own vine and figtree*: allusion, in biblical language, to Frankland having family and property.

9. *odorous*: sweet-scented (*OED*).

10. *the Professors walk*: not identified.

11. *the castle walls*: see letter 0479, n. 9.

12. *living speaking memento . . . as you have*: Frankland's wife (letter 0470, n. 9).

13. *[4] years ago . . . at present write*: Tyndall alludes to their shared time at Queenwood in 1847–8, when Frankland had made his first visit, of only a few months, to Marburg.

14. *were longer . . .* : these and the following words are written vertically in the left margin.

15. *D.r Frankland . . . Manchester* : address from envelope.

To Thomas Archer Hirst 14 April 1852 0617

Queenwood 14.th April 1852

My Dear Tom.

My intention was to write you yesterday—nay the day before, a half formed resolution to throw a bombshell into West Parade Huddersfield[1] where it should explode upon your arrival floated on the surface of my will. Yesterday I had risen nearly high enough to inscribe resolution in performance but I was fairly dragged down again by a species of magne-crystallic gravitation which works upon my brain and cripples its free action. Tom Carlyle mentions the case of Brindley[2] the engineer who when in a /net/ a difficulty which threatened to defeat him went to bed for 3 days and rolled about like one possessed by a devil, wrestling with his difficulty and at times omitting a groan or oath as if he were grappling with a lion. I am not a man of Brindleys intensity and would deem it irreverent to swear—I dont go to bed for the purpose of thinking, but when I go there for the righteous and peaceable purpose of sleeping I find I cant do it. The last image before my mind's eye at night is an attempt to give form and figure to the play of heat within a crystal—my first thought on waking is the same, as if the same ideas had flowed on all night like a subterranean stream and bubbled up to light again in the morning. I say to myself sometimes 'is it manly to be a slave of these things? are you doing your duty in this invading your health, drying up your heart, sacrificing the impulses of love and friendship and converting your religion into a scientific stalactite?' The question put in these strong terms startles me, and I halfresolve to smash my fetters and liberate myself. I am further encouraged to this by the ominous croak of every body around me—'why dont you go out? why will you waste yourself soul and body in this way? you are killing yourself, you are now an old man, and a premature grave is already opening its jaws to receive you!' I pause an instant and cogitate thus:— My brain is engaged in a battle and shall I give it up because the fight thickens and I am weary? Friendship, Love, religion may be invaded by this scientific demon but the question is, shall I run away from him or beat him down? In the thick of conflict a man forgets to pray, forgets every thing, except how to hit his foe to the best advantage. I do not think that it expresses the thing fairly to say that I am the slave of my pursuits. When two foes engage in fair and honorable contest neither can be said to be a slave to the other. A British tar grappling with a Turk at the battle of Navarino[3] and grappled by him in turn, cannot be said to be the turk's slave—he has it is true no time to woo his maiden or to pledge a bumper to his comrade while the struggle lasts; and the closer the grapple the more he will forget both mistress and friend, but fairly over and the Turk set at rest maiden and comrade will think all the more of poor Jack. I have a turk at my throat just now, but three months hence I hope to have him hewn to pieces and then by the gods I shall be as exuberant and joyful as any man among you.

The day we parted I accompanied Haas to Southampton—you know the difficulty regarding the London tickets—on good Friday however tickets were issued which were valid up to the following tuesday. Haas and I took a run across to Cowes[4] where we waited for a few hours; the sea breeze was a medicine to my bones, and that evening I found myself better than I had been for weeks before—next day I started for London and arrived in time to learn that Francis had just departed. In the afternoon I found myself flat and weak; this came on day after day, lessening however, for some time. I returned from London on Sunday; the east wind gradually yielded up its sting and the weather became warmer; under this influence I mended rapidly and am now greatly improved—We commence operations again today; I will take care of myself for some time and have no doubt I shall defeat all auguries regarding my dissolution and keep death and his marrow-bones at a respectful distance.

Surely this glorious weather is sufficient to put new life under the ribs of death—Would that I were at your side among those northern hills—I do not say it repiningly but I should enjoy it from my heart. When I look out upon the blue sky and the rich sunshine teemed down upon the world, and think of you, and January, and Jimmy,[5] and Beacon Hill, and Skircoat moor, and those billows of a granite sea which rear their crests over Todmorden;[6] by the lord I wish to be among you all! I think I could drink in a vigour which would put my former self to shame and with it as a matter of course all those (<u>you</u> included) which even the said poor former self used to crush, conquer, and bewilder!

If my paper were longer I would write more,[7] but here I pause—shake the hand of January for me, counsel Jimmy, kick Tom Perkinton [,] kiss my little sweetheart[8] if she lets you—spread out your arms to Beacon Hill and worship his bald crown. If you see Mr Larkin[9] present my respects—here my occupation ends—good bye[10]

John

RI MS JT/1/T/550

1. *West Parade Huddersfield*: an address where Hirst was going to meet up with old friends, most likely the address of Phillips (see letters 0619 and 0621).
2. *Tom Carlyle mentions . . . Brindley*: James Brindley (1716–72), the renowned builder of canals and aqueducts, was used by Carlyle as an exemplar of silent grappling with a problem (*Past and Present*, book 3, chapter 5 ['The English']). The groans and oaths are Tyndall's embroidering on Carlyle's account of his intellectual struggles.
3. *battle of Navarino*: fought on 20 October 1827 during the Greek War of Independence, on the west coast of the Peloponnese. An Ottoman fleet was destroyed by a force of British, French, and Russian ships.
4. *Cowes*: a seaside town on the Isle of Wight.
5. *January . . . Jimmy*: January Searle and Jimmy Craven.
6. *Beacon Hill . . . over Todmorden*: Beacon Hill was just outside Halifax, to the east; Skircoat Moor was on the SE periphery of Halifax (Booth lived nearby; see letter 0516encl and letter 0619, n. 9); Todmorden is a market town roughly 10 miles west of Halifax.
7. *would write more*: this last paragraph is squeezed onto the remaining paper.
8. *my little sweetheart*: probably Ada Piercy (see letter 0392, n. 12).
9. *Mʳ Larkin*: the Reverend Edmund Larken.
10. *good bye*: According to a note on the MS, letter 0615 was enclosed in this letter.

From James David Forbes   16 April 1852   0618

Clifton | Bristol | 16 April 1852

My Dear Sir

Your letter of the 5$^{\text{th}1}$ has been just forwarded from Edinburgh.

The experiments you refer to on the Conduction of Heat were never published in detail, because though carefully made, the difficulty of obtaining <u>numerical</u> results seemed at the time insuperable. The <u>order</u> of the metals obtained was published in the <u>Proceedings</u> of the R. Society of Edinburgh but without any details.[2] The only result particularly worth notice was the correction of the wrong position of Platinum given by Desprez, & which has been since adopted by him.

I have lately been engaged in experiments of a totally different kind made at the expense of the British Association, but an attack of serious illness which has obliged me to leave Edinburgh has of course interrupted them for the present.

After I wrote to you last, I read several of your later papers & was much pleased with the apparent consistency of the results obtained with the Rheostat & Tangent Galvanometer, & I had almost written to ask you what thickness & amount of wire your rheostat consists of—and whether those instruments are to be procured & at what cost in this country.

I saw a Rheostat at Watkins & Hills[3] but the wire appeared to me probably much finer than that you used.

I remain my dear Sir | Yours faithfully | James D. Forbes—
John Tyndall Esq | Queenwood College | Stockbridge

RI MS JT/1/F/29

---

1. *your letter of the 5th*: letter 0614.

2. *published in . . . any details*: J. D. Forbes, 'Researches on the Conducting Power of the Metals for Heat and Electricity, tending to establish a New Analogy between these principles', *Edinb. Roy. Soc. Proc.* 1 (December 1832–May 1844), pp. 5–7.

3. *Watkins & Hills*: optical, mathematical and philosophical instrument makers (see letter 0533, n. 8).

From Thomas Archer Hirst   16[–18] April 1852[1]   0619

Huddersfield, | April 16th 1852

My dear John,

About our Bristol visit Heinrich will have told you all, I have now a few moments to spare and will commence my letter by describing my route since then. On Wednesday evening arrived at Wovenden's Manchester.[2] I made the journey very short by reading your two memoirs on 'Diamagnetism' and the 'Polarity of Bismuth.'[3] It will be sufficient praise to you to tell you I never understood two memoirs as well. They possess one inestimable property which can result but from one thing, a thorough knowledge of what you are writing about in all its minutiae,—that property is a mathematical precision and clearness in reasoning, laying brick on brick with good mortar between until the building is complete. If I understand one part aright: Before the poles of a magnet the attraction and repulsion varies with the distance differently for different bodies; for instance soft iron and carbonate of iron; have you examined this further, or do you think any relation exists between the chemical constitution of the body and this variation? What suggested this to me, I was considering, in that little experiment where on your torsion balance you had a piece of zinc on one end and Bismuth on the other, if it could be mathematically shewn <u>where</u> by increasing the strength of the stream the balance would come to rest. You have proved clearly enough that it could not remain still.

If, with equal strength of current and equal distances of the Zinc and Bismuth from the ends of the magnet, the repulsion of the one varied differently to the attraction of the other at those several distances, it would bring another element into calculation which would render it a difficult perhaps but certainly an interesting mathematical problem.

But I will go on with my journey. I spent Thursday morning with Dr Frankland who shewed me round his beautifully arranged laboratory[4] and was as kind as ever. He told me, which you can retail to Debus, that Kolbe, after leaving Marburg for England, changed his mind in Braunschweig,[5] and went no further. I believe now he is not expected even. Huddersfield being the most convenient place to Manchester I went there next to surprise January. It was 4 p.m., he was at home, I was shewn into the sitting room and soon he came down to me; he stopped however in the middle of the room (I with my back to the fire very quietly looking at him) made one of the most comically astonished faces I ever saw, as he stood there for almost a minute speechless. Soon however his bowels began to shake. I knew what was coming. 'Why, Tom, what the devil are you doing here?' he shouted at last as he ran to me and pitched into me furiously with his fists. I could stand it no longer and, nearly killed with laughing, I ran away from him, he after me racing me round the table crying 'You damned rascal, what do you mean!' At last he caught me and with both hands nearly shook my arm off and then sat down in astonished exhaustion, to cool; he then got up once more steadily to shake my hand, and in his usual tone said 'Good Lord, Tom, but I'm glad to see you lad'. I knew then he was sane. When I was undressing that night I found to my surprise that a belt-purse I had carried the previous day round my waist, containing upwards of £40, was missing. I knew that the evening before I had pulled it off and put it under my pillow at Wovenden's, and consequently if not stolen it must still be there. There were two possibilities against my finding it again—a dishonest chambermaid, or a dishonest sleeper in the same bed this evening. There was nothing to be done that night, however, and I slept calmly till morning, dreaming that I had found it again, which however on waking I found to be a dream merely. I returned to Manchester, asked to see Mr Wovenden quietly, a sharp thin-faced woman looked at me very searchingly. 'Were you here on Wednesday evening, sir?' 'Yes, in No. 35', said I. The colour began to mount into her face. 'Well, sir, you deserve your head knocking off'. 'Thank you, Ma'am', I returned. 'I see you know my errand already and that all is safe'. She handed me the purse, correct to a penny. 'Yes, sir, but it would not have been if we had not had an honest chambermaid'. 'Well, tell the honest chambermaid I should like to see her'. A young handsome, timid girl appeared; she looked half frightened as if she was guilty in finding the purse. I put half-a-sovereign in her hand. 'There's something for your honesty my girl, and if there were not so many people about I'd give you a kiss in the bargain'. She smiled at me as if she would have preferred the kiss to the half sovereign, but I had not pluck enough to give her it, so I let her go. Now, there's an incident for you.

<u>Sunday morning</u>

I am at Brighouse[6] and have just risen from my knees, I will not say from praying for I merely listened to my orthodox uncle praying. I had told him previously I was going to consecrate this day by seeing some old friends at Halifax,[7] and he put in a significant clause about Sabbath-breaking into his prayer. I am now by myself, they are all gone to Chapel;[8] I wait until the train at 12 and join January to go to little Booth's at the Bird Cage.[9] They do not know that I am here yet. January has written to tell the old fellows to meet him there and I expect to give them no small surprise. I will leave the report until to-morrow. I received your letter[10] yesterday at Huddersfield, and shall have a word to say to you about it. With my lecture[11] which you will send to Halifax send me a prospectus of Queenwood College

The train will not be here yet for twenty minutes so I will give you a little translation of a song you will know; some verses are bad I know, but in the original there are some faulty I know, your musical ear is getting out of tune, however, and I don't care for you.

<u>Gretchen's Song at the Spinning Wheel.</u>[12]

My peace is gone
My heart is sore;
I shall find it never
And nevermore.[13]

Him not to have
'Tis but a grave,
The world to me
Is misery.

Distracted grows
My poor, poor head,
My poorer thoughts
Bewildered.

My peace is gone,
My heart is sore,
I shall find it never
And nevermore.

My looks but for him
From window they roam,
I go but for him
From my home.

His noble figure,
His stately pace,
His commanding eye,
The smile on his face,

And his flowing speech
Enchantment is,
His hand's soft pressure
And ah! his kiss!

My peace is gone,
My heart is sore,
I shall find it never
And nevermore.

My bosom heaves
Towards him, towards him,
Ah, could I but clasp
And cling to him!

And kiss him oft
As I could wish,
E'en on his kisses
I would perish!

RI MS JT/1/HTYP/191–192
LT Transcript Only

1. *16[–18] April 1852*: although begun on the 16th, the last part of the letter was written two days later, on Sunday the 18th.

2. *Wovenden's Manchester*: a lodging house run by the Wovendens (see letter 0442).

3. *your two memoirs . . . of Bismuth'*: both papers were published in the *Phil. Mag.* (see letters 0498, n. 6 and 0525, n. 2).

4. *his beautifully arranged laboratory*: Frankland was consulted on the design of his laboratory spaces at Owens College (Russell, *Edward Frankland*, pp. 149–50).

5. *Braunschweig*: or Brunswick, a city northeast of Marburg, en route to the port of Hamburg.

6. *Brighouse*: small town half way between Huddersfield and Halifax. Some of Hirst's Yorkshire relatives lived there.

7. *some old friends at Halifax*: the visit is recounted in the next Hirst letter to Tyndall (letter 0621).

8. *orthodox uncle . . . gone to Chapel*: the husband of his aunt Allatt (see letter 0406 about a previous visit), who had led family prayers. The family attended 'Chapel', that is, they were Non-conformists, not Anglicans; they considered that rail travel on Sunday was Sabbath-breaking.

9. *little Booth's at the Bird Cage*: Booth and his mother lived at Birdcage Cottage on Skircoat Moor (Brock and MacLeod, *Hirst Journals*, 21 December 1851, n. 273).

10. *your letter*: letter 0617.

11. *my lecture*: a MS copy, perhaps made for the school magazine, the *Queenwood Reporter*, of the lecture that Hirst had delivered, 'On Study and Students', on his last evening in Queenwood (Hirst, 'Journals', 6 April 1852). Hirst copied the lecture into his Journal (ff. 857–63) and it was later published (see letter 0624, n. 15).

12. *Gretchen's Song at the Spinning Wheel*: 'Gretchen am Spinnrade', a song written in 1814 by Franz Schubert and based on Goethe's *Faust*.

13. *nevermore*: in the version of the poem which Hirst entered in his Journal (f. 808), there is an extra verse at this point.

To James David Forbes   19 April 1852   0620

Queenwood College n^r Stockbridge | 19th April 1852

Dear Sir,

I beg to return you my best thanks for the intelligence with which you have so kindly favoured me.[1]

I send you a specimen of the wire of the rheostat which I use; its length is just sufficient to go once round the cylinder of the instrument—the latter is of serpentine stone and is a little over 4½ inches in length—a plate of brass $\frac{4}{10}$ of an inch in thickness is screwed on to each end of the cylinder and from the centers of these plates gudgeons protrude which fit into a pair of brass sockets and permit of the rotation of the cylinder. The wire coiled round the latter like a screw can be traversed from end to end by a small brass wheel with grooved rim, the point of contact of wheel and wire being the place where the current enters or emerges. The wheel moves along a brass rod and is pressed gently against the wire by means of springs to which the rod is attached; the rod is graduated to shew the number of revolutions and one of the brass plates at the end is graduated at its rim to shew the $\frac{1}{400}$ part of a revolution. It is a simplification of Jacobi's instrument, and owes, if I mistake not, its origin to Poggendorff[2]—I purchased mine in Berlin the price was somewhat under three pounds—I hardly think you will get any thing of the kind at present in England.

The following rough sketch may throw a little more light on the matter:—

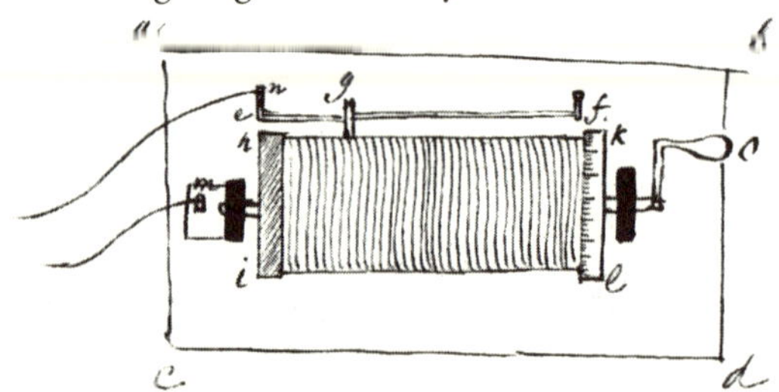

Abcd   the wooden base of the instrument

ef   the brass rod.

g   the brass wheel

hi,kl   brass plates at ends—the latter graduated

m   current enters—passes through gudgeon to the wire, and issues at
        g, thence to

n   and forward through the other wire.

o   the handle, by turning which g moves and any required length of
        wire is introduced.

But a sight of the instrument would be far more satisfactory than any description I could give—you will not I am sure deem me a forward fellow if I propose to you to take in Queenwood on your return to Edinburgh—There is 2 hours rail from Bristol[3] to Warminster—between 2 and 3 hours coach from Warminster to Salisbury and half an hours rail from Salisbury to Dunbridge Station. This establishment is 4 miles from Dunbridge and is reached by a fly.[4] London is three hours distant. If you could reconcile a visit to us[5] with your arrangements I can truly say that everybody here would consider it an honour and I for my part would rejoice to shew you whatever little instruments I possess. I know that in writing this I take a certain liberty but I absolve myself from any approach to impertinence nor do I think that you will charge me with it.

I remain Dear Sir | your obliged & faithful Servant | John Tyndall
Prof. Forbes | &c. &c. &c.

StA JDF Incoming letters 1852, no. 42

---

1. *the intelligence . . . favoured me*: in letter 0618.

2. *owes its origin to Poggendorff*: most histories attribute the rheostat to Charles Wheatstone and his partner Samuel Cooke. Tyndall here emphasizes the particular design of his rheostat.

3. *Bristol*: Forbes had left Edinburgh for Bristol, due to ill health (letter 0618).

4. *a fly*: a fast-travelling carriage.

5. *a visit to us*: Forbes did not take up Tyndall's invitation.

From Thomas Archer Hirst   22 April 1852   0621

<u>Brighouse</u>,[1] (3rd Jotting)[2] April 22nd 1852
(being my 23rd Birthday)

Well, I have been and appeared and disappeared like a vision at Halifax, if I had been a veritable ghost I could not have startled them more, though perhaps they would have received me less kindly.

Phillips and I on Sunday afternoon[3] found ourselves before Booth's garden gate which was locked; we jumped over, and found poor little Booth in the garden, much altered and more haggard-looking since I saw him; he did not manifest the least particle of surprise, but shook me warmly by the hand without speaking. When I asked him if he was not astonished to see me he replied 'I scarcely am, it looks so natural that you should walk in so'. In the house we found Roby sitting and smoking with old Mrs Booth. Phillips went in first; in speaking to him Roby had his back to me, Phillips however made him turn round quickly with the startling introduction, 'Mr Pridie allow me to introduce a friend of mine from Hell'. If it had been January's friend the Devil himself Roby could not have opened his eyes wider.

We passed the afternoon as we have done many before very pleasantly up there in little Booth's chamber. The old woman scarcely knew what to do with me, and many a time I caught Booth's eye fixed on me with a strange and yet significant expression. I have not much hope for him, I find a great alteration in him, he has a cough which will soon shake him to pieces, the only object for us is to prolong his life by lessening his labour and anxiety.

Yesterday he came to me when I was alone, I knew well enough what for, the poor fellow long hesitated and at last succeeded in spite of my efforts in turning the conversation on his difficulties and my letters. He rose to go, and as he held me by the hand he said, or sobbed, 'Your five pounds[4] I have used in the same spirit in which it was sent me, <u>no one but yourself could have guessed how it was needed nor have sent it so delicately</u>. One thing however you must promise, and that is that you will let me try to repay it'. 'With all my heart', I replied, 'if you will let me judge when that time comes; above all if I see you exerting yourself to do so, or denying yourself or your Mother any little comforts that you need I will never have I'. He almost ran away from me and all I heard more was 'God bless you'.

There is much in your letter that I wished to say a word about, but I will leave it now. The 8[VO] edition of Shortrede's Logarithmic tables,[5] price £1.10.0, is the one I want—will you be kind enough to write to Francis' clerk[6] to procure it and <u>I will call for it as I pass through London</u>. Did Debus hear anything of my stick or Swiss hat?[7] Write <u>directly</u> and send the lecture[8] to Perkinton and Cravens. You will hear from me again before I see you.

Yours affectionately, | T.A. Hirst.

Just enquire once more if the Log. Tables contain sines and tangents for intervals of 10 seconds.

RI MS JT/1/HTYP/193
LT Transcript Only

1. *Brighouse*: see letter 0619, n. 6.
2. *3rd Jotting*: this suggests that Hirst may have sent 1st and 2nd jottings to Tyndall, either in a missing letter, or in an earlier part of this letter. Tyndall's reply (letter 0624, for example, n. 6) seems to imply some missing communication.
3. *Sunday afternoon*: i.e., Sunday 18 April; the visit is described more fully in Hirst's Journal (18 April 1851).
4. *your five pounds*: Hirst and Tyndall had sent Booth £5 in late 1851 after hearing of his ill-health. They had tried to do so anonymously (see letters 0587 and 0592).
5. *The 8VO edition of Shortrede's Logarithmic tables*: Hirst chooses one of the editions listed by Francis in letter 0615.
6. *Francis's clerk*: Charles Gyde.
7. *my stick or Swiss hat*: Hirst often left things behind during this visit to England. The hat was a special purchase, bought in Lucerne, during his Swiss walking tour (see letter 0553).
8. *the lecture*: as in letter 0619, n. 11.

From Emil du Bois-Reymond   27 April 1852   0622

At Dr Bence Jones', | 30 Grosvenor str., |
Grosvenor Square | April 27, 1852

Dear[1] Dr Tyndall,

Your prophecy has become true! Not a year is elapsed since[2] and now I am in London. Last Christmass, Dr. Jones[3] went over to Berlin to see my experiments and he invited me to stay with his family in London as many weeks as I should like. So I am here and already introduced to the managers of the Royal Institution where I propose to show some experiments.[4] I do not know if you will be able to come to the town within these next three or four weeks. Should you be so then I hope I will be so happy as to see you again. At all events I was anxious to give you intelligence of my arrival in England. I am,

dear Sir, | yours, | very truly, | E. du Bois-Reymond

RI MS JT/1/M/145

1. *Dear*: du Bois-Reymond was writing in English, rather than his native German.
2. *since*: since Tyndall and du Bois-Reymond were together in Berlin.
3. *Dr. Jones*: Henry Bence Jones.
4. *the Royal Institution . . . some experiments*: du Bois-Reymond had arrived in England on 26 April. He gave private demonstrations of his experiments at the RI before delivering a lecture 'On Muscles' on Tuesday 25 May (see Gabriel Finkelstein, *Emil du Bois-Reymond: Neuroscience, Self, and Society in Nineteenth-Century Germany* [Cambridge, MA: MIT Press, 2013], pp. 132–33). Tyndall had performed these experiments in Berlin in 1851 (see letter 0480).

To Emil du Bois-Reymond 28 April 1852   0623

Queenwood College | 28th April 1852

Mein Verehrter Freund.

Sie haben mir English geschrieben und ich werde an Sie ein Paar Seilen Deutsch richten. Vor einer viertel Stunde habe ich Ihren Brief empfangen, ich sah ihn an—da standen die Englische Post-Stempel darauf, doch dachte ich 'kein Englander hat es geschrieben, diese Buchstaben sind mir bekannt, Du Bois, wenn er lebt hat sie gemacht'—Ich öffnete den Brief und fand meine Vermuthung bestätigt. Es macht mir tiefe Freude daß Sie dort in London sind und meine Freude wird zum dritten Potenz vermehrt Sie hier in Queenwood zu sehen. Dr Jones scheint ein hospitabler guter Mensch zu sein aber willkommener kann er Sie nicht machen als ich Sie machen werde. Erlauben Sie mir dann eine Einladung bei Seite derseinigen zu setzen und Ihnen sagen dass es mir die größte Freude machen werde wenn Sie nach Queenwood kommen, und je länger Sie hier blieben desto angenehmer wird es mir sein. Wir·sind zwar auf dem Lande, weit von allen Städten, und darum kann ich Ihnen kein Oper, kein Theater, kein Concert, keine schöne Dame, noch größe Männer— überhaupt keine Stadt Pracht darbieten; aber wir haben frische Luft, hübsche Bäume, große Kreide Massen worin Sie 'die kleine Verhältniße' untersuchen können gute Bette und gesunde Speisen—Die biete ich Ihnen dar so lang wie Sie uns mit Ihrer Gegenwart beehren wollen.

Wollen Sie mir ein Paar Seilen schicken um zu sagen wie lang Sie in London bleiben und <u>wenn</u> Sie in the Royal Institution Ihre Experimente machen wollen—Ich möchte sehr gern nach London gehen während Sie dort sind aber ich weiß nicht ob es möglich wird.

Eben habe ich Ihren Brief zum zweiten Mal gelesen—'three or four weeks' ist eine lange Zeit—höchst wahrscheinlich ist es—ja fast gewiß dass ich die

Umstande so einrichten kann um nach London gehen zu können innerhalb dieser Zeit—Sie werden Francis natürlich besuchen 'Red Lion Court Fleet Street'—Er wird sich freuen Sie zu sehen.

Ganz der Ihrige | J. Tyndall

Queenwood College | 28th April 1852

My esteemed friend.

You have written to me in English and I will address you a few lines in German. A quarter of an hour ago I received your letter,[1] I looked at it—there stood the English postmark on it, but I thought 'no Englishman has written it, I know these letters, Du Bois, if he is alive, has written them'—I opened the letter and found my suspicion confirmed. It gives me great pleasure that you are there in London and my joy would be increased to the third power to see you here in Queenwood.[2] D[r] Jones[3] seems to be a hospitable and good man, but he cannot make you more welcome than I will. Permit me then to add my invitation to his and to say to you that it would give me the greatest pleasure if you come to Queenwood, and the longer you stay the more pleasurable it will be for me. We are indeed in the country, far from all towns, and therefore I can offer you no opera, no theatre, no concert, no beautiful woman, nor great men—I can offer you absolutely nothing of the splendour of the city; but we have fresh air, beautiful trees, large chalk masses in which you can explore 'the small conditions', good beds and healthy meals—these I offer you as long as you want to honour us with your presence.

Do send me a few lines to say how long you will stay in London and <u>when</u> you want to carry out your experiments in the Royal Institution—I would very much like to go to London while you are there but do not know if it will be possible.

I have just read your letter for a second time—'three or four weeks' is a long time—it is highly possible—indeed almost certain that I can arrange things in order to come to London during this time—you will of course visit Francis[4] 'Red Lion Court. Fleet Street'. He will be glad to see you.

Entirely yours | J. Tyndall

RI MS JT/1/T/398

---

1. *your letter*: letter 0622.
2. *see you here in Queenwood*: there is no evidence that du Bois visited Queenwood.
3. *D[r] Jones*: Henry Bence Jones.
4. *Francis*: William Francis had, like Tyndall, studied in Germany and made many friends there.

To Thomas Archer Hirst   [25 April–3 May][1] 1852   0624

With this[2] M[rs] E.sent me 6 postage stamps [3]—I sent them back saying that I felt sure you would not permit me to take them. Your letter[4] was a pleasant one to me—I quite agree with the old Woman at Wovendens[5] as to the punishment you deserved. It was altogether a pretty incident and prettily described. —Poor January![6] —I had him before me almost as vividly as if his body were here—Remember me kindly to all, especially to Jimmy for although the little fellow appears to have forgotten me he is still very fresh in my memory. —But I wont tell sentimental lies, I don't believe that Jimmy has forgotten me. I will write to London for Shortrede and have it kept till you call for it[7]—This east wind is back again and I feel it—yesterday I strained my kidneys jumping up stairs and to-day I must walk with a stoop—I shall be better tomorrow. My muscles are not as strong and tensile as they used to be, and I am beginning to be able to sympathize with those weaklings who were a puzzle to me in my days of vigour. n'importe the trees are budding and I shall soon be in bloom and flower also—. I write this in the intervals between my experiments—I have not followed that subject further but I think the fact is a very remarkable one—I intend to look at it again when I have time—you might certainly make a pretty problem out of it,[8] but you shall have plenty of such problems at a future day. Your hat was sent from here,[9] Debus met the coachman to whom you gave directions, upon his return but he was on another coach and so could not speak to him. All these are little Wovenden matters on a small scale. —I hope you have seen Miss Carter[10] and that you remember well every word she has spoken to you—Blessed is the man that inclineth his heart unto wisdom![11] Now it is drawing near post time Tom so if I dont cease writing I shant get this off tonight—I shall therefore bid you good bye my boy. The prospectus which I send you is far from being after my heart—I will have a better one written in due time[12]—Our terms now are 40, 50, & 60 according to the age, and there are some extras done away with—I feel a sweet joy on looking at our boys sometimes they are so free, healthy and happy. Its a pity that I am not a father, surely nature intended me to be one— This is to me a blessed proof that my feelings are still un-withered[13]—but now good bye

John.

RI MS JT/1/T/894

[Enclosure] Thomas Morris
to Thomas Archer Hirst   24 April 1852   0624encl

22, Hardman St. Bedford St. |
Hulme, Manchester April 24 1852

Dear Sir,

Your note of the 16th inst. has been forwarded to me.[14] As the 'Lecturer'
is strictly confined to lectures, it would be unnecessary to preface with any
sketch of your visit, the lecture delivered by you at Queenwood College.

I shall be glad to receive your MS. at the above address, for perusal, prior
to my forwarding it to Mr Ward, Dewsbury, printer.[15]

Yours truly | Thos. Morris[16]

T.A. Hurst

RI MS JT/1/HTYP/194
LT Transcript Only

---

1. *[25 April–3 May]*: Hirst noted on this letter that he received it in May 1852. Tyndall wrote
    on or after 25 April (that is, after receiving the enclosed letter of 24 April). Hirst wrote to
    Tyndall on 4 May (Hirst Journal entry), a missing letter that was probably Hirst's reply to
    this Tyndall letter; if so, the latest date of this letter is 3 May.

2. *With this*: as the letter has no salutation, the beginning is missing. 'This' probably refers
    to the enclosed letter from Thomas Morris. Perhaps Mrs. Edmondson had forwarded it to
    Tyndall.

3. *Mrs E. . . . stamps*: Anne Edmondson. How the postage stamps might relate to the enclosed
    letter is unclear.

4. *Your letter*: Tyndall initially answered letter 0619, but this letter also replied to 0621 from
    Hirst.

5. *agree . . . you deserved*: see letter 0619.

6. *Poor January!*: Hirst had spent time with January Searle during his visit to Yorkshire. He
    recorded incidents in his Journal (for example, 14 and 19 April) that showed, in his mind,
    that January's wife was difficult. Perhaps he had communicated such matters to Tyndall.

7. *I will write . . . you call for it*: as he intended, Hirst paid a second visit to Queenwood (20–3
    May), just before departing England for Germany.

8. *that subject . . . out of it*: Hirst had suggested that Tyndall's experimental investigation
    might lead to an interesting mathematical problem (see letter 0619).

9. *your hat was sent from here*: Hirst had enquired after his Swiss hat in letter 0621. It would
    seem that he had left it behind at Queenwood, just as he had left his money belt behind in
    Manchester (see letter 0619).

10. *Miss Carter*: Sarah Carter.

11. *hope you have seen . . . remember well . . . his heart to wisdom*: Tyndall alluded, ironically, to Sarah Carter's orthodox theological views, which she often expressed dogmatically to Hirst and Tyndall. 'Blessed is the man . . . wisdom' has the cadences of a biblical quote, but is a Tyndall conflation of texts.

12. *The prospectus . . . due time*: Hirst had requested a prospectus for Queenwood in letter 0619. Tyndall did not send the final version, he explained in June, because it was too expensive to post to Germany (letter 0631).

13. *my feelings are still un-withered*: this may be an allusion to Hirst's accusation, made a year previously, that Tyndall's emotions were rusty from lack of use (see letters 0482 and 0484).

14. *Your note . . . forwarded to me*: Hirst had asked Tyndall (letter 0619, 16 April) to send a copy of his lecture to Halifax. Presumably the note that Hirst had written to accompany the lecture, which a friend in Halifax had forwarded to Thomas Morris.

15. *your MS . . . the printer*: the lecture was published as 'On Study and Students', *Monthly Literary and Scientific Lecturer*, 3 (June 1852), pp. 168–74 (from Brock and MacLeod, *Hirst Journals*, 6 April 1852, n. 296).

16. *Thos. Morris*: Thomas Morris was editor of the magazine (see n. 15).

From Edward Sabine   5 May 1852   0625

Woolwich May 5. 52

My dear Sir;

Before I rec^d yours of the 19th April,[1] I had sent your former letter of the 31^[t] of March[2] to Professor Miller of Cambridge,[3] who is one of the Council of the R.S. and a member of the Gov^t. Grant Com^[e], to know how far an application from you for an allotment of the grant towards the experimental researches named in that letter would be likely to receive his support;[4] as I am aware that on such subjects his opinion would be much thought of. His reply, which was decidedly favourable did not reach me till the day after your letter of the 19^[th] April, naming distinctly £50 as the amount you wish you to apply for. The grant committee meet tomorrow, when I will bring forward your applicati^[n], &. I have little or no doubt that you may look forward with confidence to the full amount being so allotted.[5] It is precisely such services to science which are most looked for.

in haste, | faithfully yours | Edward Sabine.

I am glad that your name is amongst the 15 recommended for election to the R.S. in the present year.

P.S. Professor Miller thought he might be useful to you in advising you where best to obtain apparatus: & will correspond with you with great pleasure if agreeable to you to do so. His address is. 7. Scroope Terrace, Cambridge.

Dr Tyndall.[6]

RI MS JT/1/S/11

1. *yours of the 19ᵗʰ April*: letter missing. Probably the letter recorded in his journal entry for 9 May 1852 (JT/2/13b/563), in which Tyndall told Sabine that he would apply for the grant if he 'knew the form of application'.
2. *your former letter of the 31ˡᵗˡ of March*: a missing letter, in which Tyndall (in reply to letter 0613 from Sabine) told Sabine about his research plans (Journal, 9 May 1852, JT/2/13b/563). Note that throughout this letter, which Sabine writes 'in haste', his super-script abbreviations are little squiggles which can be read only from context.
3. *Professor Miller of Cambridge*: William Hallowes Miller (1801–80) had been Professor of Mineralogy at Cambridge since 1832, a fellow of the Geological Society since 1830 and a FRS since 1838. He was on the Council of the RS in 1851–52 and Foreign Secretary of the Society from 1856–73.
4. *experimental researches . . . his support*: discussed in letter 0613.
5. *I have little doubt . . . allotted*: Tyndall was indeed awarded a grant for £50 (letter 0638 and Journal, 12–18 July, JT/2/13b/577). According to the journal entry, he had received a note from Weld (missing) informing him that the Council had voted for the grant.
6. *Dʳ Tyndall*: addressee name placed at foot of the first page.

To Emil du Bois-Reymond   7 May 1852   0626

7ᵗʰ May 1852

My Dear Dʳ duBois.

Should you visit the Isle of Wight[1] you can readily take a jump over to Queenwood. You arrive at the Bishopstoke station and this station is about half an hour's journey from Dunbridge station on the line to Salisbury. We live 4 miles distant from Dunbridge (weniger als eine Deutsche Meile[2]) If you decide upon coming and write to me I will meet you at Dunbridge and guide you in safety to Queenwood.[3]

I should rejoice to be able to witness your experiments in Albermarle Street[4] but I find it impossible to get away on Monday—I hope you will suc-ceed in making converts of all who hear you, but of this I have little doubt, eben weil 'die Wahrheit [gross] ist und Sie wird siegen'.[5]

[I am] still in hopes of [being able] to get up to London before [your] departure although I am extremely occupied at present. first I am striving to get a paper ready to put into the hands of Col. Sabine this month,[6] and sec-ondly we are preparing here for an examination. These both lay their claims upon me and hold me fast to my task.

On the last thursday of every month the 'Red Lions'[7] of whom you have heard Dove speak in Berlin meet and dine together; the meeting of the Royal

Society takes place also on thursday—I have written to Francis to say that I shall strive to get up to London on that day and remain perhaps until the following Sunday: if you remain so long in town—We might then go to the R.S. and afterwards accompany Francis to the Red Lions.[8] I hope we can so manage the matter

ever sincerely yours | John Tyndall

RI MS JT/1/T/399

1. *the Isle of Wight*: a holiday destination for the wealthy.
2. *weniger als eine Deutsche Meile*: less than one German mile (German).
3. *Should you visit . . . in Queenwood*: it appears that du Bois did not visit either the Isle of Wight or Queenwood. According to the account of his visit to England in Finkelstein (*Emil du Bois-Reymond*, cited letter 0622, n. 4) the only place outside London which he visited was the Lake District.
4. *your experiments at Albermarle Street*: du Bois was to conduct experiments at the RI; see letters 0622 and 0623.
5. *eben weil . . . siegen'*: precisely because 'truth is great and will prevail' (German). The famous modern version of this saying was Thomas Jefferson's 'Truth is great and will prevail if left to herself' (Virginia Statute for Religious Freedom, drafted in 1777); variants can be traced back to biblical and classical sources.
6. *a paper ready . . . this month*: probably the second part of his paper on molecular influences that Tyndall was writing for submission to the RS. Tyndall had sent the first part to Sabine in February (see letter 0606). Tyndall sent the 'papers' (presumably a revised version of the first memoir along with the second) to Sabine on 14 May (Journal, 15 May, JT/2/13b/565).
7. *the 'Red Lions'*: see letter 0569, n. 11.
8. *On the last Thursday . . . to the Red Lions*: it seems that Tyndall did not get to London to meet du Bois. According to Tyndall's Journal (13 May 1852, JT/2/13b/565) he had been persuaded, by a (missing) letter from Francis, to go to London 'next' Thursday (that is, 20 May), but it appears that Tyndall was at Queenwood from 20–3 May when Hirst visited (Hirst, 'Journals', 20 and 23 May 1852). Nor could they have attended a Red Lions dinner together on Thursday 27 May, as du Bois left London for the Lake District that day (Finkelstein, *Emil du Bois-Reymond*, p. 133).

From Thomas Henry Huxley   7 May 1852   0627

41 North Bank | Regent Park | May 7th 1852

My dear Tyndall[1]

Allow me to be one of the first to have the pleasure of congratulating you on your new honours—I had the satisfaction last night to hear your name read out as one of the selected of the Council of the Royal Society—for election to the Fellowship this year—and you are therefore as good as elected.[2]

I always made sure of your success, but I am not the less pleased that it is now a fait accompli—

I saw Francis two days ago—he was quite well and as jolly as usual

I am | My dear Tyndall | faithfully yours | T. H. Huxley

John Tyndall Esq[r]

P.S. | I have heard nothing of Toronto—and I begin to think that the whole affair University and all, is a myth[3]—

IC HP 9.3

---

1. *My dear Tyndall*: this letter was so significant to Tyndall that he copied both it and his reply into his Journal (9 May 1852, JT/2/13b/563–4). He described Huxley (who was a few years younger) as 'that able and learned young fellow'.
2. *as good as elected*: Tyndall was duly elected (see letter 0630).
3. *Toronto . . . is a myth*: no news had been received, although applications had been due by 19 November 1851 (letter 0504, n. 3); Tyndall's had been posted at the end of October.

To Thomas Henry Huxley   9 May 1852   0628

Queenwood 9th May 1852

My dear Huxley

It is said that every son of Adam has some spark of poetic sentiment in him, and that what distinguishes the poet proper from other men is the faculty of being able to tell you what he and all feel. Were I a poet (and I know not whether to upbraid or bless the gods for not making me one) I should sit down with delight to gather from birds and blossoms their prettiest imagery, and from the May its sunshine and odours into one sweet bouquet to present to you. There is a portion of human nature which is agreeably tickled when an honour is conferred—your letter[1] contributed to this result, but not to this alone—the pleasure which I derived from it was composite and the largest item in the mixture, the item which tinted all others with its own particular radiance, was the fact of your having written to me as you did. I will not further compare the pleasure imparted by the intelligence with that derived from its manner of communication, but I will say that the united effect of both was to awake in me a charivari[2] of as pleasant thoughts and feelings as it has ever been my lot to experience I hope the day may come when I shall be able to reciprocate your kindness—the gods know how joyfully I would do so if I could.

believe me dear Huxley | most faithfully & truly yours | John Tyndall
Henry Huxley[3] Esq[re] | &c. &c. &c.

IC HP 1.2

---

1. *your letter*: letter 0627.
2. *charivari*: usually, a confused, discordant medley of loud sounds (*OED*). Tyndall alluded
    to the confusion but implied the components were all pleasant.
3. *Henry Huxley*: at this stage of their acquaintance, Tyndall did not know that Huxley was
    usually 'T. H. Huxley'.

To Emil du Bois-Reymond   [12 or 19 May 1852][1]   0629

Queenwood, Stockbridge | Wednesday

My dear DuBois

Have you given up the Isle of Wight? If so I regret it. I am so deeply
plunged in business here that I cannot break loose. It would afford me great
pleasure to get a sight of you for there are many points over which I should
like to talk with you; but if the fates rule it otherwise I must be content
ever sincerely Yours | John Tyndall
Dr Dubois Reymond

RI MS JT/1/TYP/7/2427b
LT Transcript Only

---

1. *[12 or 19 May 1852]*: du Bois-Reymond arrived in England on 26 April and left London,
    for the Lake District and then Hamburg, on 27 May (letters 0622, n. 4 and 0626, n. 8).
    The first mention of the Isle of Wight in extant letters is letter 0626 (7 May). This let-
    ter reads as a successor rather than predecessor to 0626; Wednesday 12 May would allow
    more time than 19 May for du Bois-Reymond to travel to both the Isle of Wight and
    Queenwood.

From Charles Richard Weld   4 June 1852   0630

Sir,

I have the honour of acquainting you that you were on Thursday last[1]
elected a fellow of the <u>Royal Society</u>, in consequence of which the statute
requires your attendance for admission on or before the 4th Meeting from
the day of your election, or within such further time as shall be granted by

the Society or Council, upon cause shewed to either of them; otherwise your election will be void.

You will therefore be pleased to attend at eight of the clock in the evening on one of the following days,[2] viz:—

Thursday June 10th
Thursday _____ 17th   } 1852
Thursday Nov. 18th
Thursday _____ 25th

I am | Sir | your humble Servant | C R Weld[3] | assistant Secretary.
From the Apartments | of the Royal Society | Somerset Place Strand.
June 4th 1852
John Tyndall Esqre
The Fees must be paid previous to admission.

[Enclosure]
From President of the Royal Society[4]   [4 June 1852]   0630encl

The President of the Royal Society
requests the honor of
Mr Tyndall's
Company at the Soireés
at 13, Connaught Place, on
Saturday April 24th May 8th, 22nd & June 12th
at 9 oC.

RI MS JT/2/6/115–117
JT Transcript Only

---

1. *Thursday last*: Thursday, 3 June.

2. *On one of the following days*: Tyndall attended on 17 June and was introduced by Huxley (Journal, 17 June 1852, JT/2/13b/570).

3. *C R Weld*: Charles Richard Weld (1813–69) was assistant secretary and librarian of the Royal Society, 1843–61. He received an initial salary of £200 and accommodation on the premises.

4. *President of the Royal Society*: William Parsons, 3rd Earl of Rosse. After transcribing the letter from Weld into his journal, Tyndall further noted: 'Accompanying the secretary's note came the following card'.

To Thomas Archer Hirst   [late May–early June 1852][1]   0631

My Dear Tom.

Two letters have come to me from the Cathedral Hotel,[2] one from a M[r] Dressner[3] and the other from myself[4] which arrived too late to catch you. M[r] Dressner merely speaks of some things which he entrusted to your care and gives you advise as to the manner of stowing them so as to avoid the custom house harpies—saying at the same time that he did not know until informed of it that such matters were so sharply looked after. My own letter (which was the best of the two!) ran as follows.

My Dear Tom. My conscience pricks me for not having written yesterday[5] as this may arrive too late—not that I have any thing particular, or unparticular to say, but have merely a dumb unaccountable feeling that I ought to have written to you. Well boy may the gods beckon you onwards, and may you ever feel that 'your valours are your best gods'.[6]

Say to Wrightson that it was my intention to have written to him at length—say that I have now no time as the postman is pulling on his leggings to be off. Greet him from me kindly—greet Noll, greet Karl,[7] greet Schell, greet Prof. Stegmann—a man who of my German acquaintance ranks among those whom I esteem the most, greet the ladies great and small, blue eyes and brown—or better still engage Wrightson to do it, instead of you; from the specimens he has given me in this line I know he will do justice to my feelings towards them.[8] —I am speculating on a trip to Jersey; would that you were at my side, I would roar like the ocean for joy—I hope a long life of happiness stretches before Knoblauch and his bride. One last wish I waft to thee boy and that is that when you and I jostle at a future day—as jostle we will please the gods—the shock may be 'the shock of men!'[9]

John Tyndall.

Thus did I discharge myself and got rid of you to the infinite relief of my conscience—I received your letter to January—I don't know whether I should subscribe to every sentence in it, but I know that I do not like[10] a whit the less for this—It is a brave good letter and full of truth,[11] and as I read it I saw exactly the ground from which you had sketched—indeed the whole matter was as clear to me as the arrangement of the keys and the sound of the notes of your piano are to the ear of Schell. The more I reflect on this matter the more extraordinary it appears to me—I mean the extreme similarity of our ways of looking at things—this is a great mischief, for hang it we can get nothing out of each other—you are not able to tell me any thing that I don't know already and I am in the same unlucky predicament myself with regard to you—Well,

well, I suppose we must tolerate each other notwithstanding; perhaps, the secret of the matter is that we are both taught by the same master,[12] and that the voices which attend thee make themselves audible to me also.

I have been reading a translation of Plato's Dialogues lately. I like that old fellow Socrates almost as well as Jesus—his apology[13] is the best thing I ever read in my life, although he has said nothing [in] it that you could not conceive it possible for any man who had made up his mind to live and die for the truth to say. It is so solid! the words lie as thinly over the great thoughts as layers of moss upon a rock—This is what delights me, there is such a terrible reality under every sentence—Who can calculate a man's value?—/there/ this ugly old Athenian with the twilight of 2000 years between him and me is as fresh and vivid before me as if I had heard his tongue with my bodily ears in the halls of his native city—wonderful are the links wherewith the human family are bound—Socrates, though not so impetuous, is as free in his speech as Thomas Carlyle—he does not shrink from telling those in whose hands his life lies that their practices are scandalous—he treats the idea of death as a school boy would his teetum-totum,[14] plays with it, looks at it, tosses it from hand to hand and evidently cares nothing at all about it—Surely he hath attained that sublime elevation of which you spoke to January when he could afford to despise pain. When we hear the small-/beer/ philosophists[15] of the present day talk about 'progress' and imagine them set beside this mighty boulder stone of antiquity; how ridiculous their creed appears—or rather how ridiculous that such pismires[16] should undertake to be the preachers of such a creed—Where shall we look for such men as Socrates, Plato, Cato, Plutarch and others of the same stamp at present—one man alone in England I find fit at all to stand beside them and that is our friend Tom Carlyle.

Our half year is drawing to a close here; we shall break up next week—a day or two ago I received an official intimation from the secretary of the Royal Society of my being elected as fellow.[17] accompanying the letter was a card of invitation to the President's Soiree at Connaught place on Saturday next. I shall try and get up as opportunities of the kind may not often occur.[18]

Well I have managed to get out a prospectus[19] which pleases me at last. it is a good practical document and I would send you one did not the postage prevent me—I will send January[20] half a dozen. Our prospects for the coming half year are better than they have been for years—I have no doubt whatever that with a little care the place can be made to prosper.

I purchase the Leader every week and intend when I have time—perhaps in the vacation to write a reply to those articles of Lewes on the Positive philosophy[21]—At the commencement he excited a hope in me which his subsequent articles have by no means fulfilled—In many cases I think his position is untenable and if I have time I will tell him why I think so, meanwhile the paper appears to be making way and its talent if any thing is /in/ the increase—

I will say no more at present. I dare say you have found it difficult to make that well-fed body of yours bend to its work once more.—You cannot imagine what a miserable varlet I should be if I did not work—I deserve no credit for working for without it I should assuredly die of 'blue devils'![22]

your affectionate | John

RI MS JT/1/T/1011

1. *[late May–early June 1852]*: this letter was written over an extended period, beginning 29 or 30 May, and finished c. 8 June; in addition, an earlier (missing) letter was copied in (n. 4 below). Letter 0631 begins as a covering note to accompany letters for Hirst which were forwarded to Queenwood, after missing Hirst at his London hotel. Hirst left London for Marburg on the morning of 27 May ('Journals', 27 May 1852). Even if the letters arrived a few hours after Hirst left and were forwarded immediately to Queenwood they could not have arrived before 28 May. Tyndall probably began this letter on 29 or 30 May, with 30 May being likely because it was a Sunday, when he often set aside time to write. The remainder of the letter is later; it was not finished until a few days after Tyndall had received Weld's letter (0630) of 4 June (see n. 17). Moreover, Hirst had not received it on 13 June when he wrote letter 0632, so it was not posted before 8 June (see Note to Readers on post times).

2. *the Cathedral Hotel*: not identified. The hotel at which Hirst stayed in London on his way from Queenwood to Marburg. Although the name suggests it was near St Paul's Cathedral, Hirst mentioned he was near the 'new' Houses of Parliament, so perhaps it was near Westminster Abbey ('Journals', 23–24 May).

3. *Mr Dressner*: not identified.

4. *the other from myself*: original missing, but copied out in the following two paragraphs, presumably to transfer it to light-weight paper for foreign posting.

5. *not having written yesterday*: if Tyndall had posted his letter on 25 May it would have reached London on 26 May, before Hirst left. Therefore he is writing either late on 25 May or early on 26 May for his letter did not reach Hirst's hotel until after he had left on the morning of 27 May.

6. *'your valours are your best gods'*: 'His hidden meaning lies in our endeavours / Our valours are our best gods', Emerson, discussing the nature of prayer, in 'Self Reliance' (first published in *Essays* (cited letter 0393, n. 3), pp. 35–73, on p. 64). Emerson was quoting from a seventeenth century play about Boadicea, by John Fletcher.

7. *Karl*: probably a fellow student (see letter 0565, n. 16).

8. *engage Wrightson . . . feelings towards them*: one of many references to Wrightson's sociable ways with ladies.

9. *'the shock of men!'*: 'But 'midst the crowd, the hum, the shock of men, / To hear, to see, to feel, and to possess, / And roam along, the world's tired denizen, / With none who bless us, none whom we can bless' (Byron, *Childe Harold's Pilgrimage*, Canto II [1812], Stanza 26). Tyndall implies that he and Hirst will become equals, rather than mentor and student.

10. *like*: Tyndall probably meant to write 'like you'.

11. *brave letter ... truth*: letter missing. Tyndall's comments imply it included criticisms. Hints of what Hirst might have written can be drawn from the reflections on January Searle which he wrote in his Journal (26 May) while in London. Presumably he sent his letter to Tyndall to check before posting it on to January.

12. *the same master*: a Carlylean allusion, more likely to 'Nature', than to Carlyle himself.

13. *his apology*: Plato's version of a speech given by Socrates in 399 BC, in which he defended himself against charges of impiety and 'corrupting the young'.

14. *teetum-totum*: probably a teetotum, a form of spinning top, known from Roman times, often used for gambling. It has a polygonal body, with a letter or number inscribed on each side (*OED*).

15. *small-beer philosophists*: small beer was a widely drunk, low-alcohol or weak beer; a philosophist was 'an adherent or practitioner of what is held to be erroneous speculation or philosophy' (*OED*).

16. *pismires*: insignificant persons, literally, ants (*OED*).

17. *a day or two ago I received official intimation . . . the Royal Society*: letter 0630. There is a studied casualness in this allusion which suggests that Tyndall had not written to Hirst as quickly as Hirst would have wanted to hear of Tyndall's honour. Tyndall probably received the letter on 5 June, and we suggest, did not write to Hirst until at least 3 days later, giving a date of 8 or 9 June for finishing the letter.

18. *card of invitation . . . may not often occur*: the soirée referred to was on 12 June. Tyndall did not attend due to illness (letter 0633).

19. *prospectus*: the prospectus for Queenwood (see letters 0619 and 0624).

20. *January*: that is, Phillips, was deeply involved in education in Yorkshire and therefore could direct potential pupils to Queenwood.

21. *those articles of Lewes on the Positive philosophy*: George Henry Lewes (co-editor of the *Leader*) wrote a series of 18 articles espousing the philosophy of the French philosopher Auguste Comte, published from May to August 1852. The tenth ('On the Influence and Methods of Physics') of 5 June 1852, appeared shortly before this letter was completed. Tyndall's planned reply did not eventuate. The articles were later integrated into Lewes's *Comte's Philosophy of the Sciences* (London: Henry G. Bohn, 1853).

22. *'blue devils'*: usually 'the blue devils', feelings of depression or melancholy; low spirits, despondency (*OED*).

From Thomas Archer Hirst   13 June 1852   0632

Marburg, June 13th /52

My dear John,

I wrote to Debus last Sunday little thinking that I should be able to write to you to-day to say that I have knocked the dissertation on the head. Last Christmas, as I told you,[1] I looked round the matter and spent the best part of a week thereon; since I returned from England I have stuck closely to it for ten days and it is now done with the exception of the mere mechanical part. The thing is small, it is true, but I have Stegmann's and Schell's authority when I say it is a rather neat little investigation, and, what is better, it has opened the door to still more. Both the Professor and Schell kindly offered to give me any assistance they could, but as it happened I did not require it. Schell visited me one day in the middle of it, and appeared much interested. I had an opportunity of observing his abilities a little. Schell's is one of the quick, brilliant and impulsive intellects, ever rich in suggestions, though often in danger of <u>jumping</u> to immature conclusions. Such an intellect is, however, of great value if it be but accompanied with an ability for patient experiment and honest disinterestedness in, or contentment with, their conclusions. He made several such suggestions, all ingenious and pretty, with reference to a complicated expression on which hung most of my hopes. Many of these, however, were exploded there and then by maturer consideration, and although he thought he could solve the matter if I would let him take a copy home with him, he failed to do so as he afterwards told me. One morning at 5 a.m. in bed a thought struck me in reference to this identical expression, to interpret whose significance had baffled me for two days; acting on the hint I got up, washed myself and attacked it uninterruptedly until dinner time; it was one of the luckiest hints that ever entered my big head[2], obstacle after obstacle tumbled before it, and in the space of two days the whole problem, much to my surprise, as well as to that of Stegmann and Schell, was solved. The same day Stegmann came to sit an hour with me, not having seen me for a week. 'How do you get on with the dissertation?' he asked. 'I think I have finished it, Professor', I replied. He put on one of his significant smiles, in which perhaps one might have traced a little sarcastic scepticism, as he asked me to shew him it. I did so. He made no remark at all until it was finished except putting a few pertinent questions when I explained badly; at the end however he was pleased and said he could suggest no improvement, that I had nothing to do but set it in circulation among the faculty directly. Accordingly I have justly translated it into German and Schell is kindly correcting it for me.[3]

So much for the 'History of my Dissertation'. I shall remain here to the end of the Semester superintending it, printing, etc. and, as there are no lectures I care to hear, I will go on with more investigations connected with the subject as well as my mathematical readings with Schell. Knoblauch has brought a pleasant-looking wife back with him, I have not yet spoken to her, but shall do on Thursday for there is to be an English Kräntzchen[4] then. I have only seen <u>him</u> once since I returned and that was the other morning. I thought I would go quietly and hear him lecture. He entered the room and commenced his lecture in the usual style with 'Meine Herren' without seeing me, then his eye happening to catch mine he gave a slight start, then a broad grin and a nod at the end of it, and rather pothered at having interrupted his own lecture, thus he stammered on until he regained his equilibrium. He said afterwards he was astonished most with my looks, which were so much more cheerful and healthy than usual. So says all Marburg indeed, perhaps Johann[5] expressed it most graphically—he came to me with his hand extended (the same hand on which you may have noticed a peculiarly comical first finger). 'Nun Herr Hirst, wie gehts? Sie haben sich in der Zeit recht tüchtig gefuttert'.[6] Wrightson is in despair with his investigation.[7] Poor fellow, I am half sorry on his account that I have been so lucky with mine, it seems to have suggested uncomfortable thoughts in him. The Simpsons with their sister,[8] <u>intended for me</u> Wrightson says, are pleasant neighbours here. Fräulein Hille is I am glad to say somewhat better; for <u>thy</u> satisfaction I may add however that I have not yet seen her.[9]

Tom.

Dr. Tyndall, | Queenwood College, | Stockbridge, Hampshire, England.[10]

RI MS JT/1/HTYP/197–198
LT Transcript Only

---

1. *as I told you*: in letter 0601.

2. *my big head*: Hirst may have meant this literally: he was very tall and had a high forehead, its appearance reinforced by his balding quite young.

3. *translated it into German*: Hirst completed his dissertation, *Uber conjugierte Diameter im dreiaxigen Ellipsoid* (On conjugate Diameters of the Triaxial Ellipsoid), in July 1852, though he did not formally receive his doctorate until later in the summer and did not leave Marburg until 1 August (see letter 0650).

4. *English Kräntzchen*: see letter 0482, n. 6.

5. *Johann*: not identified; possibly Johann Hessel, though the tone of the comment is that of a fellow student rather than a professor of 25 years standing.

6. *'Nun Herr . . . tüchtig gefuttert'*: 'Well, Herr Hirst, how are you? You've really been scoffing yourself lately'.

7. *Wrightson is in despair with his investigation*: a recurring observation in Hirst's letters to Tyndall.

8. *their sister*: Mrs Simpson's younger sister was Anna Martin, whom Hirst later married. This is the first mention of her in Hirst's letters to Tyndall.

9. *Fräulein Hille . . . not yet seen her*: Hirst had been worried by her threat to throw herself into Lake Geneva if her health did not improve, hence his relief that she appeared better (see 'Journals', 29 June 1852). If Tyndall did not want Hirst to see her, then either he had a romantic interest in her himself or thought she would be a distraction to Hirst.

10. *Dr. . . . England*: address is probably from the envelope.

To [Thomas Archer Hirst][1]   15 June 1852   0633

June 15th 1852

I had every thing packed on Friday evening last—ready to start for London on the following morning—I rose on the following morning at 5 oC. but found that a cold which had been threatening me for a day or two previous had blossomed out into such exceeding ripeness that I was obliged to stagger into bed again—Thus I missed the Soirée[2]—I devoted the remnant of strength left me to getting up a kind of farewell experimental display[3]—We had a number of people from the neighbourhood here last night and it was my lot to entertain them for an hour or so with science. We fired off a number of small cannons & did sundry other scientific tricks to amuse the <u>péuple</u>.[4] I am better today after my effort than I anticipated—you must not blame me for this last illness it is the confounded weather and not I that is to blame—as a proof of this I may mention that Phelan[5] was the first to fall ill and his symptoms are exactly the same as mine, and Phelan, as you know, is not a man who overstrains his mental energies or permits his mind to invade the comfort of his body—I shall be soon well again—in a day or two I purpose going to London to be fully initiated[6] and I will write to you soon afterwards—

Have you been at your old tricks in the Cathedral Hotel?[7]—

Debus[8] won't go to Germany.

RI MS JT/1/T/551

1. *Hirst*: this letter is neither addressed to Hirst nor signed by Tyndall, but it was written in Tyndall's hand and was received by Hirst in Marburg on 20 June (see letter 0634).

2. *the Soirée*: Tyndall had planned on attending a soirée hosted by the president of the RS on Saturday 12 June (first mentioned in letter 0631).

3. *experimental display*: showy experiments, including the firing of cannon by electrical means, seem to have been part of an end of term celebration (Journal, 14 June 1852, JT/2/13b/569).

4. *péuple*: 'le peuple' is the people, or the common people. Tyndall's accent is incorrect.

5. *Phelan*: probably a fellow teacher at Queenwood. In his Journal (4 April 1852) Hirst described him as an Irishman.

6. *I purpose . . . fully initiated*: admitted into the the RS as described in letter 0630. Tyndall attended a meeting to complete the admission process on 17 June (Journal, 17 June 1852, JT/2/13b/570).

7. *old tricks in the Cathedral Hotel?*: Tyndall was accusing Hirst of leaving belongings behind at the Cathedral Hotel, as he had done at a number of lodgings during his summer travels in England. Hirst responded to the implied criticism in letter 0634.

8. *Debus*: this note was written in the left hand margin as the page was full.

From Thomas Archer Hirst   20 June 1852   0634

Marburg, | June 20th 1852

My dear John,

Thy letter[1] arrived to-day, which as bad luck would have it has crossed mine[2] on the way. I write now only a few words. Never mind sending Smith the £10, I have already arranged it otherwise and by this time he has it.[3] I did so because, from a letter received from him since, I judged that the sooner he had it the better it would be for him. Why is not Debus coming to Germany? Tell him to write and answer the question. If you have not called for my watch at the Cathedral Hotel don't. My brother John will probably be coming soon and will bring it me. Tell me whether you have called or not. I have not been at my old tricks at the Cathedral Hotel. I sent it to get cleaned and the watch cleaner played me the trick, forcing me either to miss the packet or leave my watch;[4] I chose the latter. The remark of thine, of our coincidence of thoughts, is true;[5] otherwise I might say something more than 'Hear, hear!' Ditto Ditto to the sensible remarks on the 'Ancient Philosophers' (the expression reminds me of our friend Sarah Carter).[6]

It is a pity you missed that Soirée;[7] the mere fact of catching cold is no fault of yours perhaps, but your ability to get quit of it again is smaller than it ought to be; and therein art thou, in the eyes of all the Gods and the best of men of thy own stamp and ideal, GUILTY. Do good penance during the Vacation. The Irish young lady here, (intended for me, Wrightson says) is a pleasant little girl indeed, full of fun and frolic (too full perhaps); she can dance all the German students—Wrightson into the bargain[8]—completely off their feet. That's the Irish blood in her. As for the rest she is a good-natured, bright-eyed Becky Do-Good. You may understand our relations to each other however when I tell you that she calls me 'Father,'[9] and that very prettily too sometimes. The mere fact of her not being able to whisk me round the room sets me on a steadier footing with respect to her. 'Father, will you play just one more waltz for me and Mr Gerland?' she asks; and the reply follows, 'No, my dear you have danced enough to-night. Bid Mr Gerland good night, and go quietly to bed'.

Write soon and tell Debus to do the same.

Thine affectionately, | T.A.Hirst.

Dr Tyndall, | Queenwood College, Stockbridge, Hampshire. | p^d, To be forwarded if from home.[10]

R1 MS JT/1/HTYP/200
LT Transcript Only

1. *Thy letter*: letter 0633. Hirst also replied to Tyndall's earlier letter 0631 (see n. 5 below).

2. *mine*: letter 0632.

3. *sending Smith the £10*: Hirst had reimbursed John Stores Smith for money he had given January Searle (Hirst, 'Journals', 6 and 21 May 1852).

4. *I have not been . . . or leave my watch*: Hirst defended himself from Tyndall's accusation of carelessness over his watch (letter 0633, n. 7).

5. *of our coincidence of thoughts*: Tyndall's agreement with Hirst's characterisation of January Searle had led him to recognise 'the extreme similarity of our ways of looking at things'; see letter 0631, n. 11.

6. *reminds me of . . . Sarah Carter*: the allusion is unclear, but perhaps Sarah Carter, in her commitment to a narrow theology, praised ancient authorities over modern, as had Tyndall in letter 0631. See letter 0624, nn. 10 and 11.

7. *that Soirée*: the soirée hosted by the president of the RS (letter 0633).

8. *Wrightson into the bargain*: another allusion to Wrightson's sociable ways with ladies.

9. *our relations . . . she calls me 'Father,'*: according to Brock and MacLeod (*Hirst Journal*, 6 June 1852, n. 345), 'Father-Son' relationships were frequently adopted among German students of different age-groups for 'purposes of friendship and moral and intellectual guidance', for example between Hirst and Dickinson. Whether this convention hid the romantic nature of Hirst's interest in Anna Martin from Tyndall or from himself is unclear.

10. *Dr. . . . home*: address probably from the envelope.

From William Thomson   21 June [1852][1]   0635

32 Duke Street, St James | Monday June 21—

My dear Sir

Mr Sylvester has asked me to communicate to you (as he did not know your address) an invitation to breakfast with him tomorrow at 9 a.m.

Do not trouble yourself answering this, but if you can accept Mr Sylvester's invitation, please wait for me at your hotel tomorrow morning and I shall

call about a quarter before 9, to go with you to Mʳ Sylvester's house which is
No 26 Lincoln's Inn Fields—
    Your's very truly | William Thomson
    John Tyndall Esqr

RI MS JT/1/T/31

---

1. *[1852]*: Tyndall's journal entry of 22 June 1852 (JT/2/13b/571–2) recorded breakfast
    with Thomson, Sylvester, Owen, and unnamed others. Tyndall was in London from 17
    June to 3 July, between school terms (JT/2/13b/569-73).

From Edward Sabine 29 June [1852]¹ 0636

June 29

My dear Sir,
    Your letter² has arrived just as I am setting off for Kew on appointment.
But I will write to Lord Dunraven³ by tonight's post for London
    Sincerely yours | Sabine
    Dr. Tyndall.

RI MS JT/1/TYP/4/1302
LT Transcript Only

---

1. *[1852]*: the year 1852 is supported by references to Dunraven in Tyndall's journal
    (JT/2/13b/572-3) and letter 0644.
2. *Your letter*: missing.
3. *Lord Dunraven*: Edwin Wyndham-Quin, 3rd earl of Dunraven and Mount Earl (1812–
    71). Sabine wrote to Dunraven, on Tyndall's behalf, regarding a position at the Queen's
    College of Galway. Tyndall, who learned of the vacancy through Francis, had been advised
    by Wynne to approach Sabine (Journal, 28 June 1852, JT/2/13b/572–3). For Dunraven's
    reply see letter 0644.

From Anne Wynne 3 July [1852]¹ 0637

Harrow Satʸ—3ᵈ July

my dear Sir.
    I wish you would send a copy of your testimonials² to the Earl of Clan-
carty | Garbally, | Ballinasloe | Ireland
    I have written to lady Clancarty, I did not say much, because I believed

my husband to be writing at the same time, but he does not think it necessary, and with the testimonials, my letter will do.

I am yours, dear Mr Tyndall | very truly A. Wynne

RI MS JT/1/W/88

1. *[1852]*: Tyndall received this letter on 6 July 1852 (Journal, JT/2/13b/573).
2. *your testimonials*: the testimonials that Tyndall had collected in 1851 for the Toronto application, plus some additions gathered for his application for the position in Queen's College, Galway. Among those Tyndall asked for testimonials was G. G. Stokes (see letter 0643). LT noted that Tyndall also received testimonials from Sylvester and Baden Powell (LT 'Biography', pp. 510–11).

To Thomas Archer Hirst   5 July 1852   0638

Queenwood 5th July 1852

My Dear Tom,

Had you told me how matters stood with regard to Smith[1] I should have managed the matter somehow; however the thing is now done and it is useless to talk further about it. I read your account of your successful attack upon your dissertation[2] with great interest; it is strange how light dawns in upon a man and your recent experience confirms a theory of mine which I elaborated while in Berlin—that the mind of man is a Daguerreotype plate which intellectual effort goes to <u>polish</u>. Truth is not hid in a well as the ancients thought, it exists like sunbeams around us but owing to the rust of our mental Daguerreotype plate we are unable to take its impressions—this rust as I have said is removed bit by bit by intellectual effort and finally we get a polished surface impressible to the rays of knowledge.

Debus has posted your watch to Frankland who will soon set out for Germany and will deliver the article into your hands.[3] I was very much interested by your account of Miss Simpson[4]—If she executes her other little duties as earnestly and effectually as she does her dancing she is a very good girl.

I came from London yesterday having spent the last fortnight or better there I went up for the purpose of going through the formalities of admission to the Royal Society;[5] and while there, learned that the chair of Natural Philosophy in Queen's College Galway[6] was vacant. I have already commenced operations as a candidate—the formal application is not yet made for I received private intelligence of the matter[7] before it was announced publically—I learn however that there is one of the ablest fellows in Trinity College Dublin[8] in the field; so we are likely to have a hard fight for it—I have already brought some very heavy artillery to bear and am prepared for a sturdy conflict—Sanguine I am not, nor am I ever so, but I will do what I can notwithstanding.

I have written to Prof. Knoblauch to procure me a thermo-saüle[9] and a number of cubes of crystals—these I should like to have as soon as possible and should therefore be glad if you would urge him a little—I do not know of course what the crystals will cost but should not mind going as far as 30 or 40 Thalers for them—Heinrich has written to you—he is working hard at his investigation which I calculate is much more rational than squandering his money in Germany.

Send me a copy of your Dissertation back by Frankland—I should very much like to see it. The R. S. has granted me £50[10] and it is on the strength of this that I am purchasing the apparatus—I have invented a <u>ducky</u> little instrument for experiments on conduction,[11] it is as pretty and as effectual a little affair as I have seen for some time. I look upon it with somewhat of the feelings of a father. Good bye boy—Wert thou near me I would kiss thee

affectionately | <u>John</u>

RI MS JT/1/T/552

---

1. *how matters . . . Smith*: concerning money for John Stores Smith; see letter 0634, n. 3.

2. *your dissertation*: see letter 0632.

3. *into your hands*: Hirst had left the watch in London (see letters 0633, n. 7 and 0634, n. 4).

4. *Miss Simpson*: Anna Martin, Hirst's future wife. Hirst had praised her dancing in letter 0634 (see also letter 0632).

5. *admission to the Royal Society*: Tyndall recalled in his journal for 17 June 1850, 'I met Huxley whom I arranged to meet at the Royal at 3,25. Both of us were punctual to the minute; sat together near the President and at the proper time was introduced by Huxley; The President repeated a certain formula which declared my admission as Fellow, at the end of which we shook hands and the matter was complete' (JT/2/13b/570).

6. *chair . . . Galway*: see letter 0636, n. 3 and 0637, n. 2.

7. *private intelligence of the matter*: very likely from Edmund Ronalds, via Francis. Ronalds had previously fed information on the Cork position (letter 0551); Tyndall later asked Francis to thank Ronalds in the context of discussing the Galway position (letter 0664); and Francis and Ronalds had known each other as far back as 1841–42 when both were at Giessen under Liebig.

8. *one of the ablest fellows in Trinity College Dublin*: George Johnstone Stoney, who eventually received the position.

9. *thermo-saüle*: spelled thermosäule, alternately called a thermopile: an electronic device that converts thermal energy into electrical energy. It was developed in 1823 by Hans Christian Ørstead (1777–1851) and Joseph Fourier (1768–1830). More than urge Knoblauch, Hirst liaised with Kleiner, the instrument maker, on Tyndall's behalf and, in December, organized the delivery of the instrument (see letters 0689 and 0695).

10. *The R. S. has granted me £50*: Tyndall had applied for the grant at the insistence of Sabine (see letters 0613 and 0625).

11. *ducky little instrument for experiments on conduction*: as Tyndall was at this time beginning his investigations into the conduction of heat by organic structures (particularly numerous cubes made of different types of wood), this instrument is certainly the one described in his report of these investigations in the *Phil. Trans.* (cited letter 0606, n. 1) and the *Phil. Mag.* (J. Tyndall, 'On Molecular Influences—Part 1. Transmission of Heat Through Organic Structures', *Phil. Mag.* 6, no. 37 [August, 1853], pp. 121–38). A figure of the instrument appears on p. 219 of the former and p. 124 of the latter.

To Edward Sabine 5 July 1852 0639

Queenwood College | 5th July 1852

Dear Sir,

During the last three or four months a thought, which you doubtless will deem a strange one, has repeatedly suggested itself to me. Had the fates ordained that I should be called away to Toronto it was my intention to reduce this thought into tangible shape and to place it in your hands. To be brief, it was my design to write a brief sketch of my life history for the purpose of furnishing you with some information regarding the private relations of one whom you had so befriended.

Circumstances have occurred which precipitate this resolution of mine. I met you a day or two ago[1] in the Strand and from that time the above thought has been more than ever present with me. I left London yesterday and avail myself of my first tranquil moment to put the thought into execution. Your kindness is now leading you to interest yourself in an especial manner on my account, with regard to this Galway matter;[2] but even were I successful I could not feel otherwise than unhappy if my success were associated with the thought that your support might have been modified, if you had been more fully informed regarding me.

Of course you have judged me on scientific grounds alone and taken for granted that my private character is unblameable; any thing I have now to communicate will not interfere with the opinions which you hold at present upon these points, for from my childhood to the present hour I have borne a fair character. But, in this country, other circumstances besides character and ability come into play. There is social position for instance, and it is on this point that I want to relieve my mind by giving you precise information.

I believe a couple of generations back the people from whom I am sprung[3] belonged to higher portion of the middle classes. But my father was a poor man,[4] who made a livelihood by selling leather and shoes, during a portion of his life he was a policeman. In these few words I sum up all his shortcomings. I have nothing more to say against him, for a man of more inflexible integrity and intrinsic truthfulness of heart I have never met. From his father he inherited a considerable amount of intelligence, and I well remember when sitting by his death bed tracing the veins upon his forehead and listening to his remarks, being deeply impressed with the thought that if his natural ability had had fair play in this world he would have proved a superior man. His love of intellectual culture induced him to keep me at school until I was upwards of 18 years of age. At this time I joined a division of the Ordnance Survey[5] which was under the superintendence of a gentleman whose friendship now forms one of the sweetest facts of my life—namely Captain Wynne of the Royal Engineers. I was soon afterwards sent to England and remained connected with the Survey for about four years and a half. During this time I made myself acquainted with all its practical details, as a surveyor and leveller without and as a draughtsman within. In 1843 I separated from the O.S.[6] and went to Ireland. During my stay there the Grand Jury of Carlow had sufficient confidence in me, though but a very young man, to vote me the entire applotment of my native county—a job which necessitated the outlay of some hundreds of pounds. They afterwards found that they had exceeded their authority in so doing, and the matter fell to the ground. I then obtained an appointment in Manchester[7] in connexion with railway matters; from Manchester I went to Halifax where I remained nearly three years. During the railway commotion I was constantly in the thickest of the struggle. The lines with which I was connected being the most hotly contested in England, and I being the principal assistant in my employer's office heavy and arduous duties fell to my lot. If I thought you required it I would shew you testimony to prove that I discharged these duties to the satisfaction of all concerned. Through all my pursuits the idea has been constantly before me that what I profess to do I ought to know how to do well, in fact from my boyhood I have been stimulated by a certain idea of self culture; and looking back upon my life I can call facts and circumstances to mind which marked my tendencies and often indeed make me smile. I was fond of Natural Philosophy, and remember, when quite a lad, borrowing a volume of the Encyclopaedia Britannica, which was almost as large as myself, and reading the chapters on Aerostation, Electricity and Phlogiston with intense pleasure. I made balloons at the time, and manufactured gas to the infinite annoyance of almost everybody but myself. My poor father used to inspire me sometimes by calling me Newton, and, with reference to my scriptural knowledge, Stillingfleet.[8] In 1847 I was induced to accept a situation in this establishment,[9] the opportunities of improvement which it promised mainly influencing my choice. In 1848 I found myself with two or three hundred pounds. I had long entertained the idea of going to Germany,[10] and read and studied with reference to this object. Well I did go, and remained there three years which was just as long as my funds lasted. As Natural Philosophy connected itself more immediately with my previous pursuits and as I had loved it and dabbled in it from a boy I devoted myself to it. In one respect, at least, I earned a certain reputation in Germany and that is for being a hard worker, a reputation which I hope to maintain as long as my fibre lasts and I have an opportunity to use it. Since my return from Germany I have not been idle. I lecture twice and sometimes three times a week on Natural Philosophy. I have mathematical and physical classes for some hours each day. I have a number of farm students to keep in order, and what spare time I can command I devote to my investigations.

It is, of course, impossible for me to predict what effect the above sketch will have upon your future bearing towards me. I believe it is my duty to make you acquainted with the facts of my position and I leave consequences in the hands of the author of them. I believe plain speaking upon my treated me. For the last ten years the tendency of my life as regards social position has been an upward one, but in no case have I forsaken a straightforward course nor purchased a single advantage by other than honourable means. I will not do so now, nor, even by my silence, permit you to make a mistake regarding me. If you support me[11] it must be with the fact of what I am clearly placed before you. I have endeavoured, with what success I know not, to render myself fit for the companionship of cultured men, believing it to be a duty which I owed both to myself and to society. For the rest, I have endeavoured to reap warning, courage, and instruction from the lives of men whose circumstances have been in some degree like my own. I am a debtor to both Burns and Socrates in this respect;[12] and it now remains for me to make the experiment whether a man with nothing but naked character to recommend him, may not in these kingdoms, find the doors to an honourable activity open to him.

I have now done, and standing before you with a clear conscience and unburdened mind

subscribe myself, dear Sir | with sentiments of deep respects | Your faithful Servant | John Tyndall

RI MS JT/1/TYP/3/1026–1027
LT Transcript Only

---

1.   *a day or two ago*: Tyndall noted in his journal for 2 July 1852: 'Met Col. Sabine in the street with a lady on his arm, . . . "ha!" said he "I learn that you are an Irishman, well I should never have found it out". Bade him good bye and before I had separated many paces from him had made up my mind to let him know something more regarding me' (JT/2/13b/573).

2.   *this Galway matter*: the open Chair of Natural Philosophy. G. Wynne had advised Tyndall to write to Sabine to garner his support for the Galway possition.

3.   *people from whom I am sprung*: Tyndall's immediate forbears descended from Gloucestershire emigrants who settled in southeast Ireland around 1670.

4.   *a poor man*: Tyndall's parents, in spite of being well-educated, remained poor. Tyndall's mother, Sarah McAssey, was disinherited for marrying against her father's wishes. Tyndall's father was a shoemaker who also served as a sargeant in the Royal Irish Constabulary.

5.   *Ordnance Survey*: Tyndall joined the Irish Ordinance Survey in 1839 as a civil assistant; he was transferred to the English Survey in August 1842.

6.   *I separated from the O.S.*: in November 1843, Tyndall was dismissed for protesting against the Survey's inefficient administration and poor treatment of Irish assistants.

7.   *an appointment in Manchester*: Tyndall worked as chief surveyor for a proposed railway line between Halifax and Keighley.

8.   *Stillingfleet*: Edward Stillingfleet (1635–99) was a British theologian and scholar known for his staunch defense of orthodoxy.

9.   *in this establishment*: in 1847, Tyndall began working as an instructor at Queenwood College.

10. *going to Germany*: Tyndall went to the University of Marburg beginning in 1848, where he completed a dissertation under the direction of Friedrich Stegmann.

11. *If you support me*: see n. 2.

12. *Burns and Socrates in this respect*: the poet, Robbie Burns, grew up in poverty and endured financial hardship before establishing his reputation as a poet, while Socrates was the son of a stone mason and sculptor, and learned his father's trade before devoting his life to philosophy.

From Edward Sabine   6 July 1852   0640

Woolwich July 6. '52

Dear Sir

I am much obliged to you for taking the trouble to write the letter[1] which I have received this morning. I imagine that you attach more importance to the bearing of the subject matter which it communicates on your reception by the world & prospects in it than really belongs to it the social state of this country. There are but two <u>real</u> points, moral uprightness and intellectual cultivation & attainment. With respect to scientific grounds I did make those enquiries from persons engaged in similar pursuits to yourself, which give me reason to believe that if placed in suitable circumstances you are likely to advance science, and lead up others to do so—and I therefore felt it a duty holding the situation I do in the R.S. to do what might be in my power towards aiding you in your reasonable desire to acquire a more independent position. I did not make any enquires regarding your private character, assuming it as you say, and not being in a position in which I was the person required to investigate more closely. I do not doubt that in both respects the more closely enquiry is made the more satisfactory will be the result. The circumstances which you relate shewing you to be the architect of your own fortunes, would, with most persons I should think entitle you to additional respect. I should greatly doubt there being a contrary effect with any one.

The enquiry I made regarding the part of Ireland you came from was a mere accident arising from the circumstance that I also like yourself am an Irishman,[2] as far as being born there gives that title.

Believe me, my dear Sir, | faithfully yours | Edward Sabine

RI MS JT/1/S/12

---

1. *the letter*: letter 0639.

2. *an Irishman*: Sabine was born in Dublin.

To Michael Faraday   11 July 1852   0641

Queenwood College | near Stockbridge Hants. | 11th, July 1852

Dear Sir,

I return, with many thanks, the memoir which you were kind enough to lend me.[1] I believe I must plead guilty to the charge of keeping it a day beyond the time permitted me, and the reason of my doing so is, that during my stay in London (which I quitted yesterday) I was so tossed about, in mind and body, that the tranquility necessary to the profitable reading of such a paper was denied me. If I dared I would keep it a day or two longer, for a man requires to brood over such a subject, to get as near as possible to the writers point of view, and to familiarize himself with the peculiar form under which the matters treated of present themselves to the writers mind. The constancy of the magnetic force regardless of distance, as indicated by the moving wire, is an exceedingly striking and significant result:—I am still a little in the dark as regards your remark in remark in 3090, that 'the system of power about the magnet must not be considered as revolving with the magnet'.[2] Now in your bar magnets the force emanating from one of the corners is, I imagine, more intense than that emanating from any other point; in the rotation of the magnet this corner moves, and it appears to me that as the line of force emanating from the corner will always preserve the same position with regard to the corner, when the latter revolves the line of force peculiar to it must revolve also. In the case supposed by yourself in 3110,[3] where the wire and the line of force are conceived to coincide, if the wire and the magnet revolves together I should imagine that the wire and the line of force would continue to coincide, that one would be perfectly motionless with regard to the other, and that hence no current could be formed.

Most heartily do I subscribe to the sentiment expressed in 3159.[4] The object of the analyst in many cases seems to be to draw boundaries and limitations which the advancing experimentalist ever tends to break through. Far be it from me to decry the labours of the analyst—they are the necessary complement to experimental knowledge; but our knowledge is a thing <u>flowing</u> and not a thing <u>fixed,</u> and as long as this is the case the widest generalization which man has yet made is in danger of being swallowed up by one still more comprehensive. The analyst is the exponent of the centripetal tendency of the human mind, while the experimentalist is <u>centrifugal,</u> and seeks continually for still wider expansions.

I am making preparations at present to enter a struggle for a post in one of the Queen's Colleges in Ireland. The chair of Natural Philosophy in Galway College has become vacant and I intend to become a candidate for the post. Some of my friends have already commenced operations on my behalf—Col Sabine has written to a very influential quarter and other agencies are also at work.[5] The election will probably not be made till towards October; in the mean time I will invoke all the aid I can, and if I fail, the consciousness of having done all it behoved me to do in the matter will be sufficient to render me content.

believe me dear Sir | most faithfully Yours | John Tyndall
Professor Faraday | &c &c &c

RI MS JT/TYP/12/4015
LT Transcript Only

---

1. *memoir . . . to lend me*: Faraday's 'ERE 28'. Faraday had personally loaned Tyndall a copy of the paper, as recounted in a playful entry in his journal for 19 June 1852: 'Talked with Faraday for some time regarding his new theory of lines of magnetic force. "Did you not read my last paper" he asked "No I replied I could not get at the Phil. Trans." "Did I not send you a copy?" "No" I replied "Ah then they must be all picked up. Well I must see whether any body is likely to die and would be generous enough to return me a copy for you, really Tyndall you are such a worker that if a copy be obtainable you must have one. In the mean time I will lend you my private copy, it is the only one I have or I would give it to you"' (JT/2/13b/570).

2. *3090, that 'the system . . . magnet'*: 'ERE 28', § 3090.

3. *3110*: 'ERE 28', § 3110.

4. *3159*: 'ERE 28', § 3159.

5. *other agencies are also at work*: see letter 0636, n. 3.

To William Francis   11 July 1852   0642

My Dear Francis—

I really cannot call to mind where Brewster's sky investigations are to be found[1]—I think in the Transactions of the Royal Society of Edinburgh—On them Wheatstones invention of the <u>Polar Uhr</u>,[2] for telling the hour of the day by the polarized light of the sky is founded and Wheatstone could inform you where they are to be found—I know Darker of Paradise Street Lambeth can also tell for I remember having some conversation with him on the subject

sincerely yours | J. Tyndall | 11 July 1852

RDS 27/8

1. *Brewster's sky investigations are to be found*: Brewster published numerous papers on the polarization of light in the atmosphere, including 'Observations on the Lines of the Solar Spectrum, and on those produced by the Earth's Atmosphere, and by the action of Nitrous Acid Gas', *Edinb. Roy. Soc. Trans.* 12 (1834), pp. 519–30 and, more recently, 'On the polarisation of the Atmosphere', *Phil. Mag.* 31 (1847), pp. 444–54.
2. *Polar Uhr*: Wheatstone presented his Polar Uhr or Polar Clock to the BAAS meeting in 1848. See letter 0417, n. 12.

To George Gabriel Stokes   11 July 1852   0643

Queenwood College | n^r Stockbridge | Hants 11th July 1852

Dear Sir

I cannot be otherwise than satisfied with the entire matter and tone of your letter[1] which reached me this morning, and I thank you for it just as heartily as if it were the document for which I became your petitioner.[2]

Did I not fear that you would deem me a forward fellow I would request a favour of another kind at your hands. you know I am examining the influence of molecular structure upon the conduction of heat—Professor Thomson informs me that you have treated the subject analytically and have published your researches in the Cambridge and Dublin mathematical journal.[3]

If you could spare me a copy of the investigation I should feel greatly indebted to you.—

I am anxious both to inform myself as to your results and also to refer to them in my paper, and I really do not know how I am to procure them otherwise than by direct application to yourself.

I remain | dear Sir | very faithfully yours | John Tyndall
Prof. Stokes &. &.

The feeling of the necessity of a discovery is generally anterior to its accomplishment and certainly the present state of electrical science does originate such a feeling. We want a generalization which shall embrace all the modifications of electricity which you have mentioned.

Cambridge University, Stokes Correspondence - add 7656

1. *your letter*: letter missing.
2. *the document . . . petitioner*: Tyndall probably requested a testimonial to submit with his application for the Galway professorship, but received a letter from Stokes instead, which he could then use. This was an option often taken by more eminent men of science when asked for a testimonial (see, for example, letters 0542 from Whewell and 0506 from Faraday).
3. *Cambridge . . . journal*: G. Stokes, 'On the Conduction of Heat in Crystals', *Cam. and Dubl. Math. Journ.* 6 (1851), pp. 213–38.

From Edward Sabine 12 July 1852 0644

Woolwich July 12 | '52

My dear Sir,

I return you Lord Dunraven's note,[1] as you may wish to have it again—I own that I was surprised at Capt Wynne's supposition[2] that Lord Dunraven had any voice in the matter—but I wrote nevertheless.

I thank you for the quotation from Poggendorff.[3] I had just heard to the same effect by a letter from M. de Humboldt.– and further that the Kremsmunster *[1 word illeg.]* Tower[4] observations show the decennial period also.

Sincerely yours | Edward Sabine

RI MS JT/1/S/13

1. *Lord Dunraven's note*: Dunraven's reply to Sabine (see letter 0636).
2. *Capt. Wynne's supposition*: Wynne advised Tyndall to write to Sabine, so Sabine could write to Lord Dunraven regarding Tyndall's interest in the position. See letter 0636, n. 3. Wynne also wrote to Dunraven, who replied that he had no influence and could not be of use (Journal, 12–18 July 1852, JT/2/13b/576).
3. *quotation from Poggendorff*: not identified, but probably a reference to the works by P. A. Reslhuber or J. Lamont published in *Poggend. Annal.* concerning the decennial period in magnetic phenomena. See letters 0647, n. 1 and 0649, n. 1.
4. *Kremsmunster . . . Tower*: the observatory (built 1749–56), also known as the 'Mathematical Tower', at Kremsmünster Abbey in the town of Kremsmünster, east of Salzburg in northern Austria (see also letter 0647, n. 1).

From George Gabriel Stokes   13 July 1852   0645

Pembroke College | Cambridge | July 13th. 1852

My dear Sir,

I have very great pleasure in sending you the accompanying paper,[1] and am sorry I was so stupid as not to think of it before.

I am dear Sir | Yours very truly | G.G. Stokes.

RI MS JT/1/TYP/4/1358
LT Transcript Only

---

1. *the accompanying paper*: requested by Tyndall in letter 0643 (cited n. 3).

From Ferdinand Reich   16 July 1852   0646

Hochgeehrtester Herr!

Mit vielem Vergnügen sende ich Ihnen umstehend das gewünschte Certificat. Ob dasselbe von einer so wenig bekannten Persönlichkeit als die meinige Ihnen von einigem Nutzen sein kann, muss ich dahingestellt sein lassen. Jedenfalls werde ich fernerhin mit noch erhöhterm Interesse alle Untersuchungen studiren, die die wissenschaftliche Welt von Ihnen erfahren wird, und es wird mir viel Freude machen, wenn Sie Gelegenheit finden, dieselben zu vervielfältigen.

Hochachtungsvoll | Ihr | ganz ergebenster | F. Reich

Freiberg | am 16ten Juli 1852.

Herr John Tyndall ist mir durch die von ihm in den letzten Jahren in dem Philosophical Magazine und in Poggendorffs Annalen veröffentlichten Arbeiten als ein Physiker bekannt worden, der eben so durch Geschicklichkeit und Genauigkeit der Versuche, als durch durchdachte Benutzung derselben sich auszeichnet, und ich würde es für einen grossen Gewinn für die Wissenschaft halten, wenn er sich in eine Lage versetzt sähe, die ihm Gelegenheit böte, seine physikalischen Forschungen weiter verfolgen zu koennen.

Freiberg in Sachsen, am 16ten Juli 1852.

Ferdinand Reich, ph. Dr. | Professor der Physik.

John Tyndall | Queenwood College | near Stockbridge | Hants | England

Most highly esteemed Sir!

It is with great pleasure that I send you overleaf the desired certificate.[1] Whether it can be of some use to you, coming from such a little known personality as myself, I must leave undecided. In any case, I shall in future study with still greater interest all the investigations which the scientific world will learn from you, and it will give me much pleasure if you find the opportunity to increase their number.

Yours faithfully | Your | most humble servant | F. Reich

Freiberg | 16th July 1852.

Herr John Tyndall has become known to me through the works published by him in recent years in the Philosophical Magazine and in Poggendorff's Annalen as a physicist who distinguishes himself as much through the skilfulness and precision of his experiments as through their well-considered application, and I would deem it a great gain for science if he were to find himself in a position which would offer him the opportunity of being able to pursue his research in physics further.

Freiberg in Saxony, 16th July 1852.

Ferdinand Reich, PhD | Professor of Physics.

John Tyndall | Queenwood College | near Stockbridge | Hants | England

RI MS JT/1/R/12

---

1. *the desired certificate*: Tyndall recorded in his journal writing a letter (missing) to Reich on 6 July 1852 requesting the testimonial (JT/2/13b/573). Tyndall also noted receiving this letter on 25 July 1852, when he wrote, 'Received a <u>not</u> enthusiastic testimonial from Reich' (JT/2/13b/580).

To William Francis 20 July 1852 0647

20th July 1852

Dear Francis

I send you Reslhuber[1] as I promised—I believe it will be easier to print the tables from the figures in the book[2] than from mine and I therefore have not copied them.

I have looked through Woodhead.[3] It is written by a man extremely ignorant of the subject he pretends to handle—It is in truth for the most part perfect nonsense—now I dont know whether you review such books—Who is Woodhead? do you know him?—Does he cut any figure in the world? if so it would perhaps be worth while to demolish him at all events it will be a kind of scientific curiosity if you think it worth while to devote space to it—In your next just say a word on this point. If he was a man of any mark—like our friend[4] zum Beispiel![5] the review might be made a spicy article.

Sincerely yours | Tyndall

RDS 27/11

1. *Reslhuber*: P. A. Reslhuber, 'On the Decennial Period Observed by Dr. Lamont in the Magnitude of the Diurnal Motion of the Magnetic Needle' *Phil. Mag.* 4, no. 24 (September 1852), pp. 219–23 (from *Poggend. Annal.* 85, no. 3 [1852], pp. 412–20). Reslhuber was the Director of the Observatory at Kremsmünster. See also letter 0649.

2. *the tables from the figures in the book*: the 'book' probably refers to the *Poggend. Annal.* (vol. 85 ran from January to April 1852). Reslhuber's paper contains five tables; all were reproduced in the translation.

3. *Woodhead*: George Woodhead, whose book, *Atmosphere: A Philosophical Work* (London: Hippolyte Bailliere, 1852), was reviewed under 'Notices respecting New Books' in *Phil. Mag.* 4, no. 24 (September 1852), pp. 228–30. 'With regard to the work,' the review reads, 'it is our duty to state that we have rarely seen so much nonsense crammed into so small a space' (p. 228).

4. *our friend*: not identified.

5. *zum Beispiel*: for example or for instance (German).

To William Francis   [21 July 1852][1]   0648

My Dear Francis,

My brain is hot and requires cooling, so I am off to the Isle of Wight for two or three days. I shall be back here on Saturday and will therefore have time to correct any proofs[2] that you may send to me—

Sincerely yours | J Tyndall
Queenwood Wednesday

RDS 27/10

1. *[21 July 1852]*: a trip to the Isle of Wight was recorded in Tyndall's Journal entries for 22–5 July 1852 (JT/2/13b/577–80). He spent three days on the island. He must have written this letter on Wednesday, 21 July, the day before he left for his excursion.

2. *proofs*: most likely translations for the *Phil. Mag.*, particularly Lamont (published in August; see letter 0649, n. 1).

To William Francis　　　July 25 [1852]　　　0649

Queenwood 25th July

My Dear Francis

I return Lamont:[1] the reason why I recommended its insertion is that I knew Sabine[2] & others will take great interest in the matter, particularly in the remark which he makes regarding temperature at the Conclusion. the other paper[3] is a natural accompaniment and therefore I recommended both to go together. Next month I imagine will do quite well.

I quite accord with your notion regarding Mr Gore—send him his paper[4] by all means—the sentiment contained in the line 'Men rush in where angels fear to tread' often occurs to me on reading such productions as that of Gore to you—'to shew a connexion between gravity, heat & electricity!' easily said Mr Gore; but I would recommend you to hood me the subject for the next half century (if you live so long) before you trouble the world with your notions of the matter.

In these magnetic papers[5] the tables appear to me to be the only things of permanent value I don't think it would be well to shorten them—

I got home from the Island[6] last night with a face as red as a Chippewaw savage—this morning the skin is all scaling away—I spent three days on the island to the manifest profit of my health—and shall begin immediately to apply the face which I gathered there. Sincerely yours J Tyndall

RDS 27/12

---

1. *return Lamont*: J. Lamont, 'Addenda to the Investigation on the Decennial Period in the Magnitude of the Daily Motion of the Magnetic Needle', *Phil. Mag.* 4, no. 23 (August 1852), pp. 145–46 (translated from *Poggend. Annal.* 86, no. 5 [1852], pp. 88–90).
2. *Sabine*: see letter 0644.
3. *other paper*: probably the paper by Reslhuber (see letter 0647, n. 1), though it did not appear until September.
4. *Gore . . . his paper*: not identified, but perhaps George Gore (1826–1908), elected FRS in 1865. From 1853 he published articles on electrolysis and heat in the *Phil. Mag.* and other journals.
5. *magnetic papers*: presumably the papers by Lamont and Reslhuber.
6. *the Island*: Isle of Wight (see letter 0648).

From Thomas Archer Hirst   7 August 1852   0650

Göttingen. | Aug. 7th /52

My dear John,

You see by the above that I have at last fulfilled my intention of visiting Göttingen.[1] A week ago I left Marburg with my new title 'Dr Hirst'[2] and a few letters of introduction from Gerling and Kohlrausch to Gauss and Weber. I must confess I do not like letters of introduction; indeed, visiting great guns of all species is at first a task to me, but if I must do it I would rather make my way slower and on my own merits than thank any body for giving me a certain start; in the one case the first impression you make has a chance of being gradually improved, in the latter it is not so sure, for the friend who introduces you may have made a mistake and introduced somebody better than you. However, they were offered me and I could not refuse them. Weber is very kind to me, he is an interesting fellow indeed. Do you remember Turner,[3] the Unitarian parson at Halifax? For he has great similarity in his personal appearance and painful stuttering utterance. In spite of this however he is a great talker; when I go I have nothing to do but give him a start and he talks away the whole interview without a veritable full stop. It is true his conversation goes scientifically often beyond my depth. But judging from what I do understand he has an extraordinarily clear head. I am making some experiments with him, on the determination of the Magnetic inclination by a new and beautiful method of his which in a future letter I will try and explain to you.[4] At present I have too many things on hand to spare the time. I am studying his Electrodynamics[5] which he has lent me and shall ask him to make some experiments therein also. In spite of his hesitating expression his lectures are splendid. He forgets everybody and everything in his absorption, uses his experiments and his blackboard as they ought to be used, viz. as a language—in this respect he equals Bunsen. He has also his eccentricities which are not without interest. Sometimes he stutters for several seconds, to the great pain of all his hearers, and cannot get one syllable out: he knows his liability to do this and enters the room with the first sentence on his lips so as to have a fair start, before all are seated and almost before he reaches the table he has commenced his lecture. Yesterday he introduced me to Gauss. We did not find him at home but at the Museum,[6] reading his newspaper. I said only a few words to him, however, for I did not want to disturb him, and there were too many people about. I shall go alone and meet the lion in his den. There is something about Gauss that fascinates me—a more venerable, gentlemanly looking old fellow I do not remember to have seen. I thought too in his face I could trace a peculiar expression, I hardly know what to call it—it seemed to suggest a certain repose after a nobly spent life. I must look at it again. There is another fellow here, Stern[7] the Mathematician, also an able fellow. I had no introduction to him but attended his lecture on Mechanics, and then went to ask his leave to do so again. In ten minutes I felt more at home with him than either Gauss or Weber, and as I came away he invited me to take a comfortable 'butter-brod'[8] with him when I wished. I took him at his word and went again, and was greatly pleased, for he made not the least stir or bustle, but let me see him as he was. We had the plainest of all possible butterbrods together and no lack of conversation thereto. I meet him too almost every evening <u>in the water</u>; for we both go regularly to a bathing place in the neighbourhood. All Göttingen bathes daily at a little spot about half a mile from the town. By the Lord, I am getting to be a magnificent swimmer, you used to make pretensions in that line but I shall cut you out very soon.

You will have commenced your Session once again. I have been in such a bustle since I received your last[9] that I have both mislaid your letter and forgotten what it was about, or whether there was something in it I ought to answer. Oh! I remember, about those Crystals and Apparatus.[10] I told Knoblauch once or twice about them before he wrote to order them—the poor little fellow was so absorbed with his wife (By the bye, one of the most interesting German ladies I have met). The crystals he says you will not get for some time yet, the 'Thermo Saule'[11] perhaps sooner; my brother[12] will <u>perhaps</u> come over in September and we shall go through Berlin together—if so, and if it is ready, as Knoblauch is going to try and arrange it, I will send it by him. The crystals are ordered in Frankfort, and there appears to be much uncertainty both as to when and how you are to get them. How about the Irish Situation?[13] If I understood Weber right you have written to him about it.[14] I gave him copies of your memoirs.[15] And to-morrow I shall give Gauss some also—Frankland is over (Tell Debus my watch arrived safe[16] and thank him for me, I will write to him soon)—confound my parentheses!! I am in a terrible hurry to get to bed and things crowd on me at the last moment. Well, as I was saying Frankland is over, I have met him several times in Marburg and Cassel. Two of his pupils[17] are in Marburg too for a short time (I cannot resist another parenthesis, for in it I have to record the before unheard-of fact that in Marburg at present are no less than <u>fifteen</u> British subjects, men, women and children). One of these pupils is a chip off the true block, a sturdy, common-sensible young fellow, who knows as well how to work as play; another monument of Carlyle's influence on Young-England.[18] He returns at the end of the month and I shall take advantage of the opportunity and send you a copy of the Dissertation.[19] Very neatly got-up, you will say, when you see it, that is, as far as the printing etc. goes. But now I <u>will</u> go to bed, for I have to be up at 5:30 a.m.

Good night and God bless thee. | Thine affectionately, | T.A. Hirst.

Write <u>by return to Marburg</u>. At the end of this month I shall be on my way to Berlin most likely.

R1 MS JT/1/HTYP/202-203
LT Transcript Only

1. *visiting Göttingen*: Tyndall had suggested that Hirst go to Göttingen in letters 0597 and 0604, and requested that he write back his intentions on this matter. Hirst responded in letter 0601.

2. *'Dr Hirst'*: Hirst had recently received his doctorate at the University of Marburg (see letter 0632, n. 3).

3. *Turner*: the Rev. William Turner (1788–1853), a tutor in mathematics, physics, and mental and moral philosophy at Manchester College from 1808–27, before becoming the Unitarian minister of Northgate End Chapel, in Halifax, in 1829. He was a long-time supporter of the town's literary and philosophical society, and its mechanics' institute. He is best-known for *Lives of Eminent Unitarians*, 2 vols, (1840–43) (*ODNB*).

4. *new and beautiful method . . . explain to you*: Hirst mentioned his experiments on magnetic inclination again in letter 0652.

5. *studying his Electrodynamics*: Hirst probably alludes to W. Weber, 'Electrodynamische Maassbestimmungen', *Leipzig, Abhandl. Jablon. Gesell.* (1846), pp. 209–378, part of which was republished in *Poggend. Annal.* 73, no. 2 (1848), pp. 193–240 and as W. Weber, 'On the Measurement of Electro-dynamic Forces', in R. Taylor (ed.), *Scientific Memoirs*, vol. V (1852), pp. 489–529.

6. *the Museum*: likely the Königliches Academisches Museum (Royal Academic Museum) at the University of Göttingen.

7. *Stern*: Moritz Abraham Stern (1807–94), a mathematician who studied and taught at Göttingen, where he attained his degree in 1829, became extraordinary professor in 1848, and full professor, succeeding Gauss, in 1859. He was the first non-baptized Jew to become a full professor at a German University (Birgit Bergmann, et. al. [eds.] *Transcending Tradition: Jewish Mathematicians in German-Speaking Academic Culture* [Berlin: Springer, 2012], p. 22).

8. *'butter-brod'*: a slice of buttered bread, traditionally eaten by itself or with one other topping, such as sliced meat, cheese, or jam.

9. *your last*: letter 0638.

10. *Crystals and Apparatus*: Tyndall was attempting to procure cube crystals and a thermosäule from Knoblauch, who responded in letter 0651.

11. *'Thermo Saule'*: see letter 0638, n. 9.

12. *my brother*: John (see letter 0652).

13. *the Irish Situation*: Tyndall's application to a position at Galway.

14. *written to him about it*: Tyndall wrote to Weber regarding his application to a position at Galway. Tyndall's letter is missing, but Weber's reply is letter 0655.

15. *copies of your memoirs*: see letter 0518, n. 2 for a list of possible memoirs and letter 0512, n. 5 for their German translations. To that list can be added his paper on bismuth (see letter 0525, n. 2), although the translation (J. Tyndall, 'Ueber die Polarität des Bismuths, nebst einer Untersuchung des magnetischen Feldes', *Poggend. Annal.* 87, no. 10 [October 1852], pp. 189–205) had not yet been published.

16. *watch arrived safe*: Hirst had left his watch in London; Debus had sent it to Frankland to take to Germany (see letter 0638, especially n. 3).

17. *Two of his pupils*: not identified.

18. *Young-England*: a group of young, chiefly aristocratic Tory politicians active, especially, in the 1840s advocating the preservation of the established Church and social hierarchy, whilst promoting religious toleration and a philanthropic paternalism towards industrial workers (*OED*).

19. *the Dissertation*: Hirst's dissertation (cited letter 0632, n. 3).

From Karl Hermann Knoblauch　　　10 August 1852　　0651

Marburg 10th August 1852

My dear friend,

I but now write to you to answer and send you my best thanks for your kind letter of the 5th of July.[1] But it is not now that I have performed your commissions[2] at least as far as I could do with them. A short time after the arrival of your letter I wrote to Kleiner[3] in Berlin to order the <u>Thermosaule</u>.[4] I have instructed him to give it the same size and arrangement with my own instrument, which I have found satisfactory in all my inquiries since a great number of years. As I had bespoken the matter with M[r] Hirst I ordered him to send the apparatus to Marburg directly or in the course of August, as I am here during this time. Or if it were impossible to get the instrument ready in this time, keep it in Berlin until M[r] Hirst comes to fetch it. As for the krystals, I have written to Albert[5] in <u>Frankfurt</u> on the Main the same from whom I get all my krystals myself as you are aware of. He at least is the best I know though I have not always been satisfied by his cutting of the crystals. Certainly he will send a number of cubes. Among them one or the other will do for the purpose. It was the same in my last inquiries and I confess that without Albert I indeed have not been able to perform the best part of them.

Kleiner has given me no answer. Albert writes that it will be impossible to get Agate (white or yellow) clear. It is always unhomogeneous and only imperfectly transparent so that it will scarcely possible to determine the optic axes. *[Dichroite[6]]* also is never perfectly translucid. The size will be 3/10 inch as near as possible but not exactly so in every case, some crystals will be smaller.

You wish to have some convenient means to fix the crystals in order to pass the heat through them. I have now ordered something of this kind with a mechanist to spare you the expense for it. I fixed the crystal in my own experiments on a square piece of wood, which I had prepared myself, able to be turned round its vertical axis for 90 degrees.

Of course I have hastened your commissions as much as possible but my own experience with German mechanics gives me no particular trust in their functionality. I *[must]* wait about half a year for my thermo-pile, and some crystals, which I had ordered in october last, didnot arrive here before July. —However I shall do what is in my power: I shall write from time to time to Berlin and Frankfurt and treat your affair like my own.

M[rs] K. and my father return their best compliments to you. I am very happy to have established my home in my house and wish that you might soon follow my example.

I was in this semester very near to leave Marburg as I had got a vocation to Halle. But when the Cassel government had offered to me the ordinary professorship with more than the double of my former income, I followed my inclination to stay further in Marburg a town which I like very much, as well as M[rs] K. who has made some parties in our beautiful neighbourhood. By my new situation I have become a member of the Examinators and M[r] Gerland, whom you no doubt remember, was the first who was examined by me in Physics.

Prof. Bunsen leaves Breslau in this vacation-time to go to Heidelberg where he has accepted a professorship[7] and the directory of the chemical laboratory.

I address you my best thanks, my dear friend, for the great pains you have undergone in translating my paper on the transmission of radiant heat through crystals.[8] You are indeed very kind! Believe me that I feel thankful to you. With great pleasure I remember the time you were working in my rooms and I sincerely wish that a similar time may come again.

Your communication about M[r] Powell's lecture on my experiments[9] was the first I heard of it. Where is the whole lecture printed?

I have been glad hearing your election as a member of the Royal Society and congratulate you with all my heart.

I expect my father this evening in Marburg. He comes from Reinerz,[10] a bathing place in Silesia. I am just going to the station to receive him.

With my best wishes for you, believe me most sincerely | ever yours | Herm. Knoblauch.

For the relation between the <u>deviation</u> of the needle (and the <u>force</u>) in the thermomultiplier look:

<u>Becquerel</u>. Traite de l'électricité IV, 3, 5.[11]—Ann. de Chim. et de Phys. XXXI, 371.[12] Pogg. Ann. IX, 345;[13] LVI, 325.[14]

Pogg. XVII, 543[15]

<u>Nobili</u> Ann. de Chim. et de Phys. XLIII, 146.[16] Pogg. Ann. XX, 226 sqq. 232 sqq.[17] LVI, 325[18]—

<u>Maced.</u> <u>Melloni</u> Biblioth. univ. LV, 19 sqq.[19] Mém. de l'Acad. XIV 445 sqq. (ref. Biot).[20] An. de Chim. et de Phys. LIII, [5] sqq[21] LV, 337 sqq.[22] Pogg. An. XXXV, 128-134,[23] 395 sqq.[24] XXXVIII, 11 sqq. 14, 23.[25] XXXIX, 7, 8,[26] LVI, 326.[27]

Mém. de L'Acad. XIII.[28]—An. de Chim. et Phys. LXXXV, 337 sqq.[29] L'Institut 89 p. 22.[30] Pogg. An. LII, 574, 575[31]

<u>Petrina</u>, v. Holgers Zeitschrift fur Phys. I, 171.[32] Pogg. An LVI, 328.[33]

<u>Poggendorff</u>. Pogg. An. LVI, 324, 329 sqq.[34]

<u>Melloni</u> has found for his instrument the following table: | (Pogg. XXXV p. 133).[35]

deviations	forces	deviations	Forces	deviations	forces
20°	20,0	31	37,4	41	65,5
21	21,1	32	39,6	42	69,3
22	22,3	33	41,8	43	73,2
23	23,7	34	44,1	44	78,0
24	25,1	35	46,7	45	83,2
25	26,6	36	49,5		
26	28,2	37	52,4		
27	29,9	38	55,4		
28	31,6	39	58,5		
29	33,4	40	61,9		
30	35,3				

Farewell! | yours | most truly | H.K.

RI MS JT/1/K/18

1. *letter of the 5ᵗʰ of July*: letter missing. Tyndall referred to this letter when he wrote to Hirst (letter 0638).

2. *your commissions*: Tyndall asked Knoblauch for a thermosäule and a number of cubes of crystals. See letter 0638. Knoblauch's English was not always perfect.

3. *Kleiner*: or Kleimer; an instrument-maker or dealer.

4. <u>*Thermosaule*</u>: see letter 0638, n. 9.

5. *Albert*: J. W. Albert, a mineral dealer.

6. *Dichroite*: alternate name for iolite, a blue crystal, though commonly showing various colours when viewed from different angles.

7. *accepted a professorship*: Bunsen succeeded Leopold Gmelin at the University of Heidelberg in 1852.

8. *translating my paper. . . crystals*: H. Knoblauch, 'On the Dependence of Radiant Heat in its passage through Crystals upon the direction of transmission', in J. Tyndall and W. Francis (eds.), *Scientific Memoirs* (London: Taylor and Francis, 1853), pp. 99–113. Originally published as H. Knoblauch, 'Ueber die Abhängigkeit des Durchgangs der Wärme durch Krystalle von ihrer Richtung in denselben', *Poggend. Annal.* 85, no. 2 (1852), pp. 169–88.

9. *Mr. Powell's lecture on my experiments*: Baden Powell lectured on Knoblauch's work in his 'On the Analogies of Light and Heat' at the RI on Friday, 23 April 1852.

10. *Reinerz*: a spa town in the Klodzko Valley on the Bystrzyca River in lower Silesian Voivodeship, Poland where there were iron springs.

11. *Traite de l'électricité IV, 3, 5*: A. C. Becquerel, *Traité d'électricité et du magnétisme*, 7 vols (1834–40).

12. *Ann. de Chim. et de Phys. XXXI, 371*: A. C. Becquerel, 'Recherches sur les effets électriques de contact produits dans les changements de temperature et application qu'on peut en faire à la determination des hautes températures', *Annal. Chim. et Phys.* 31 (1826), pp. 37–92.

13. *Pogg. Ann. IX, 345*: A. C. Becquerel, 'Untersuchung über die durch Temperaturdifferenzen erzeugte Contactelektricität und deren Anwendung zur Bestimmung hoher Temperaturen', *Poggend. Annal.* 9, no. 2 (1827), pp. 345–59.

14. *LVI, 325*: J. C. Poggendorrf, 'Von dem Gebrauch der Galvanometer als Messwerkzeuge', *Poggend. Annal.* 56, no. 6 (1842), pp. 324–44.

15. *Pogg. XVII, 543*: A. C. Becquerel, 'Vom thermo-elektrischen Vermögen der Metalle', *Poggend. Annal.* 17, no. 12 (1829), pp. 535–54. This was written in the left margin.

16. *Ann. de Chim. et de Phys. XLIII, 146*: L. Nobili, 'Sur la mesure des conrants électriques ou projet d'un galvanometer comparable', *Annal. Chim. et Phys.* 43 (1830), pp. 146–87

17. *Pogg. Ann. XX, 226 sqq. 232 sqq*: L. Nobili, 'Ueber die Messung elektrischer Ströme, oder Vorschlag zu einem vergleichbaren Galvanometer', *Poggend. Annal.* 20, no. 10 (1830), pp. 213–45.

18. *LVI, 325*: see n. 14.

19. *Biblioth. univ. LV, 19 sqq*: Knoblauch may have meant M. Melloni, 'Appréciation exacte

de l'intensité des courants électriques, au moyen du galvanomètre', *Archives des sciences physiques et naturelles* á la Bibliothèque Universelle, Genève 55 (1834), pp. 9–15.

20. *(ref. Biot)*: J. B. Biot, 'Rapport fait à l'Académie des Sciences sur les experiences de M. Melloni relatives à la chaleur rayonnante', *Comptes rendus hebdomadaires des séances de l'Académie des Sciences* 14 (1838), pp. 433–572.

21. *An. de Chim. et de Phys. LIII, [5] sqq*: M. Melloni, 'Mémoire sur la transmission libre de la chaleur rayonnante par différents corps solides et liquides', *Annal. Chim. et Phys.* 53 (1833), pp. 5–73.

22. *LV, 337 sqq*: M. Melloni, 'Nouvelles recherches sur la transmission immediate de la chaleur rayonnante par différents corps solides et liquides', *Annal. Chim. et Phys.* 55 (1833), pp. 337–97.

23. *Pogg. An. XXXV, 128-134*: M. Melloni, 'Ueber den freien Durchgang der strahlenden Wärme durch verschiedene starre und flüssige Körper', *Poggend. Annal.* 35, no. 5 (1835), pp. 112–39.

24. *395 sqq*: M. Melloni, 'Neue Untersuchungen über den unmittelbaren Durchgang der strahlenden Wärme durch verschiedene starre und flüssige Körper', *Poggend. Annal.* 35, no. 5 (1835), pp. 385–413.

25. *XXXVIII, 11 sqq. 14, 23*: J. B. Biot, 'Bericht an die Academie der Wissenschaften zu Paris über Hrn. Melloni's Versuche in Betreff der strahlenden Wärme', *Poggend. Annal.* 38, no. 5 (1838), pp. 1–50.

26. *XXXIX, 7, 8*: M. Melloni, 'Ueber die Polarisation der Wärme', *Poggend. Annal.* 39, no. 9 (1838), pp. 1–32.

27. *LVI, 326*: see n. 14.

28. *Mém. de L'Acad. XIII*: not identified.

29. *Annal. de Chim. et Phys. LXXXV, 337 sqq*: Knoblauch may have meant M. Melloni, 'Mémoire sur la constance de l'absorption calorifique exercée par le noir de fumée et par les métaux', *Annal. Chim. et Phys.* 75 (1840), pp. 337–88.

30. *L'Institut 89 p. 22*: not identified.

31. *Pogg. An. LII, 574, 575*: M. Melloni: 'Ueber die Beständigkeit der Wärme-Absorption des Kienrusses und der Metalle, und über das Daseyn eines Diffusionsvermögens, welches durch seine Veränderungen den Werth des Absorptionsvermögens bei den übrigen Körpern verändert', *Poggend. Annal.* 52, no. 4 (1841), pp. 573–85. This was written in the left margin.

32. *Holgers Zeitschrift fur Phys. I, 171*: F. A. Petrina, 'Beiträge zur Kenntniss elektischer Ströme', *Wiener Zeitschrift für Physik, Chemie und Mineralogie* 1, no. 3 (1840), pp. 165–80.

33. *Pogg. An LVI, 328*: see n. 14.

34. *Pogg. An. LVI, 324, 329 sqq*: see n. 14.

35. *(Pogg. XXXV p. 133)*: see n. 23.

# From Thomas Archer Hirst					[17 August 1852][1]					0652

Dear John.

You will see that in my last[2] I had anticipated the questions contained in your little note which reached me in Gottingen a day or two after I had posted mine.[3] I returned to Marburg yesterday the Semester there being closed and I having a few things yet to settle before I leave Marburg on a journey with my Brother John. I visited Gauss as I promised and the record of the interview you will some day read in my Journal,[4] it was to me a very interesting one. Weber introduced me to Prof. Poggendorf who came on a visit from Berlin, I saw very little of him but he seemed a very pleasant old gentleman and gave me a kind invitation to visit him in Berlin. I have not yet finished the calculations of some experiments on the Magnetic Inclination[5] made in Gottingen and shall defer them to another time. They have made me a member of a Mathematical Kräntzchen here in Marburg,[6] with Stegmann Schell Gerling, Hessel Kohlrausch & others. next Saturday I expect to lay before them the results of said experiments in Gottingen and am preparing for the same—Weber has just been making some new experiments on the polarity of Bismuth, which will interest you I have only once seen the apparatus[7] and can not explain its merits fully, but I will try, what I miss or mistake you will for yourself perhaps supply or correct—

on reconsideration I will wait a week and make a few experiments with Knoblauch on the subject in order to compare the result with your memoir[8] on the subject. By-the-bye send my Brother a few more copies of your memoirs if you can spare them I have used up those I brought

What steps have you taken towards trying for the Cork professorship?[9] Enclosed I send you a little note[10] I received the other day from Stegman which pleased me much. It was accompanied by two large volumes of a rather scarce and certainly valuable work.[11] Poor January Searle he wrote to me for £20[12] the other day, just before receiving yours. I could not spare it and had to tell him so. I would rather have given him £50 than have written the letter you posted.[13] but I sincerely believe I did right and that in spite of the present pain it will have cost him, that in the end it will be better. I hope so at any rate

Thine affectionately | T. A Hirst

R1 MS JT/1/H/493

1.	*[17 August 1852]*: Hirst returned to Marburg after two weeks in Göttingen on 16 August (Hirst, 'Journals', 15 August 1852).

2. *my last*: letter 0650.

3. *your little note*: letter missing.

4. *record of the interview*: see Hirst, 'Journals', 12 August 1852.

5. *experiments on the Magnetic Inclination*: Hirst worked with Weber on experiments 'on the determination of Magnetic inclination' while visiting Göttingen (see letter 0650).

6. *made me… Marburg*: correct spelling Kränzchen. This was a small, informal mathematical society in Marburg. In his journal, Hirst recorded that it met weekly 'in the open air' (3 July 1852). He was invited only after he completed his dissertation.

7. Weber … *apparatus*: Hirst discussed Weber's experiments and described the apparatus in letter 0658.

8. *your memoir*: Tyndall's 'On the Polarity of Bismuth' (cited letter 0525, n. 2). Hirst reported the results of his experiments in letter 0658.

9. *the Cork professorship*: Hirst likely meant Galway. Tyndall had told Hirst of his intention to apply for a professorship at Galway in early July (letter 0638). The possibility of applying for a professorship at Cork had arisen almost a year earlier, in October 1851 (see letter 0551).

10. *a little note*: the note appears in Hirst, 'Journals', 25 July 1852.

11. *two large volumes… valuable work*: a copy of Heinrich Wilhelm Brandes's (1777–1834), *Manual of Higher Geometry*. See Hirst, 'Journals', 25 July 1852. The enclosed note is missing.

12. *wrote to me for £20*: see Hirst, 'Journals', 8 August 1852, where he recounted the difficult decision not to provide a loan to January Searle.

13. *the letter you posted*: letter missing. Tyndall's growing disillusionment with January Scarle, who had asked him for a loan of £10, is recorded in his journal entry of 25 July 1852 (JT/2/13b/580): 'The anchors of my opinion regarding January are becoming very loose, not because he asks this but because of the manner in which he asks it. It seems very much an attempt on January's part to live upon his acquaintances'.

To William Francis [early–mid August 1852][1] 0653

If you intend to publish a prospectus besides this preface[2] then the latter hardly needs alteration. But if you do not intend to publish a prospectus then the preface ought to be more explicit.

When will the prospectus be needed? It must be a good one and therefore will require thought. Now I am up to my neck in work Struggling to get something ready for Belfast.[3] If it would do <u>afterwards</u> I am your man—But Huxley I doubt not would make the thing exactly what it ought to be. Had you told me earlier you should have had my version in your hand at present.

I have looked every where through my papers for Barter's letter[4] and regret to say that I have looked in vain. I do not remember ever to have seen the document but I will lay no stress on this as my memory cannot be trusted. Have you looked through the book—it may be hid somewhere between the leaves—What's to be done?

Sincerely yours | J Tyndall

I shall make an effort to get up to town[5] on Saturday.

RDS 27/40

1. *[early–mid August 1852]*: written after his decision to go to Belfast (made some time between his journal entries for 25 July and 26 August; see n. 3), but prior to letter 0656 (27 August), when he returned the prospectus and preface to Francis. At the latest this letter was written a few days before Saturday 21 August.
2. *prospectus beside this preface*: for the new series of *Scientific Memoirs*, of which Tyndall (along with Francis, Huxley, and Henfrey) was an editor (see letter 0656, n. 4).
3. *for Belfast*: Tyndall (after initially deciding against it) had decided to attend the meeting of the BAAS in Belfast after Sabine asked him to serve as a secretary of the Physical Section (see Journal, 26 August, JT/2/13b/580–1).
4. *Barter's letter*: not identified. Barter may be Richard Barter (1802–1870), an Irish physician and well-known hydropath.
5. *up to town*: London.

From George Wynne [26 July–mid August 1852][1] 0654

My dear Tyndall,

I have just seen my cousin Mr Wynne the Under Secretary,[2] he says he has not received your memorial[3] yet—how has this happened? if you have not received an official acknowledgement you had better write again. He says the proper channel is thro' the Lord Lieutenant's private Secretary, but had it come to him it would have done quite well—I am sure you will have his interest in the matter and I think unless good interest is brought to bear against you that you are pretty sure of the appointment, he said he was sure I would not have interested myself in the matter unless you were in every respect a proper person, I told him he was quite right then and that you would be a credit to any Government that appointed you and that I believed the professors of the College themselves were very anxious you should be appointed. I am writing in great haste—write to me

Sellerna | near Clifden | Co. Galway

Believe me very truly yours | Geo. Wynne

Mr Wynne's address is—

John Wynne Esq. | Under Secretary's office | Castle | Dublin

RI MS JT/1/TYP/5/1848
LT Transcript Only

1. *[26 July–mid August 1852]*: this letter clearly refers to Tyndall's application for the position at Galway in mid 1852. Wynne had offered to write to his cousin to secure his support for Tyndall's appointment (see Journal, 28 June, JT/2/13b/572–3). It is unclear when Tyndall sent in his application: in mid-July he sent his testimonials to Francis's contact, Dr. Edmund Ronalds (journal entry of 12–18 July, JT/2/13b/577); he probably submitted the full application soon after. There is a break in his journal between 25 July and 26 August, when he wrote, 'My application and testimonials are safe in Dublin. First in acknowledgement I received a note signed by Lord Naas, and to a second letter I received an acknowledgement from Mr Wynne the Under Secretary' (JT/2/13b/580). This letter seems to have led Tyndall to write a 'second letter', to which he received an answer before 26 August. Thus, this letter dates between late July and mid-August, 23 August at the latest. Tyndall was ultimately unsuccessful in his application (see letter 0679).

2. *Mr Wynne the Under Secretary*: John Arthur Wynne.

3. *your memorial*: Tyndall's application for the Professor of Natural Philosophy at Galway.

From Wilhelm Weber   26 August 1852   0655

Hochgeehrtester Herr Doctor!

Verzeihen Sie gütigst dass ich Ihr Schreiben vom 6ten Juli erst heute beantworte, woran ich bisher durch verschiedene Ursachen, hauptsächlich durch eine mit Prof. Poggendorff gemachte Reise, verhindert worden bin. Ich danke Ihnen verbindlichst für die 3 Abhandlungen aus dem Philosophical Magazine, welche ich durch die Güte des Herrn Dr. Hirst richtig erhalten und mit grossem Interesse gelesen habe, und werde auch Ihre Abhandlung über Molecular-Wirkungen sogleich studieren, sobald ich den Band der Philosphical Transact., in welchem sie enthalten ist, empfangen haben werde.—

Ich lege einige Zeilen diesem Briefe bei, mit dem Wunsche dass dieselben für den Zweck, dem sie dienen sollen, nicht zu spät kommen und dass es mir geglückt sei, ihnen meiner Absicht gemäss eine solche Fassung zu geben, wie sie für den Erfolg am günstigsten wäre.

Mit grösster Hochachtung | ganz | der Ihrige | Wilhelm Weber
Göttingen, 1852 Aug. 26.

Most highly esteemed Herr Doctor!

Please be so very kind as to excuse the fact that I am only today answering your letter of 6th July,[1] from which I have been prevented doing until now through various reasons, mainly a journey I took with Prof. Poggendorff. I thank you most kindly for the 3 papers from the Philosophical Magazine,[2] which I have received safely through the kindness of Dr Hirst and have read with great interest, and I shall also study your paper on molecular actions[3] straight away, as soon as I have received the volume of Philosophical Transact. in which it is contained.—

I am enclosing some lines in this letter, with the wish that they will not come too late for the purpose which they are supposed to serve and that I may manage to give you, in keeping with my intention, such a version as would be most congenial to success.[4]

With the greatest respect | entirely | Yours | Wilhelm Weber
Göttingen, 1852 Aug. 26.

R1 MS JT/1/W/12

---

1. *your letter of 6th July*: letter missing.

2. *3 papers from the Philosophical Magazine*: Hirst gave Tyndall's memoirs to Weber (see letter 0650, nn. 14 and 15).

3. *your paper on molecular actions*: Tyndall's paper in the *Phil. Trans.* (cited letter 0606, n. 1).

4. *enclosing some lines . . . success*: the lines—that is, the testimonial—are missing. It seems unlikely that it arrived in time to be sent with the Galway application (see letter 0654).

To William Francis   27 August [1852][1]   0656

Queenwood 27th Aug.

My dear Francis.

I send you my version of the prospectus[2] and shall <not> be at all dismayed if you should quietly thrust it in the fire, I am an impertinent devil I know for daring to deviate from Huxley[3]—In all probability his is the best of the two. I have simply spoken my own mind and written a thing which I imagine I should like to read myself as a prospectus. I think it due to the old Scientific Memoirs[4] to say what I have said in their favour—I can only partially judge as to the translation but the subjects selected in the physical department are most of them classical. Remember however I am not a man of the world and if Huxley thinks me too speculative by all means send me to the devil. A bushel of thanks for the time [-hill] it gives me a day and a half breathing time more than reckoned on—I shall be in London on Monday and shall then squeeze your knuckle bones to powder.[5] It will be time enough to set out for Belfast on Tuesday.[6] if Huxley has no better company and is willing to travel 2nd Class. I shall be happy to offer myself as his shadow from London to Belfast—I am almost [worn] to [nil] god knows.

Sincerely thine J. Tyndall
You may change [Nov.] if you like. I am ready for anything.[7]

StBPL T&F, Authors' letters

---

1. *27 August [1852]*: the letter is dated to 1852 by allusions to Belfast and the prospectus of the *Scientific Memoirs*. The Belfast meeting of the BAAS was held in early September 1852.

2. *the prospectus*: for the new series of *Scientific Memoirs*. Although the request is missing, letter 0653 and this letter suggest that Francis had asked Tyndall's advice on both a preface and a prospectus.

3. *deviate from Huxley*: Tyndall had suggested that Huxley write the prospectus (letter 0653), when he was concerned he would not have time to do it himself.

4. *the old Scientific Memoirs*: that is, the 5 volumes of *Scientific Memoirs, Selected from the Transactions of Foreign Academies of science and Learned Societies and from Foreign Journals*, edited by Richard Taylor. The new series divided into two parts, representing natural history (edited by Huxley and Henfrey) and natural philosophy (edited by Tyndall and Francis).

5. *squeeze your knuckle bones to powder*: firmly shake your hand.

6. *I shall be in London . . . on Tuesday*: Tyndall left Queenwood on Monday 30 August, left London for Dublin on Tuesday 31 August, and arrived in Belfast on Wednesday 1 September (JT/2/13b/581).

7. *You may change . . . anything*: this could refer to the first part of the *Scientific Memoirs*, which Tyndall believed would be published on 1 November (see letter 0669), or to the November issue of the *Phil. Mag.*

To William Francis [28–9] August 1852[1] 0657

Queenwood Aug. 1852

My dear Francis.

I send you back the preface[2] with the modifications which I would propose. —They are very slight as you will see and refer to the first and last paragraphs exclusively.

I also return the Prospectus—Huxley I doubt not will feel inclined to kick me when he is locally near enough to permit of his boot-toe reaching me. The remarks I have to make on the prospectus are on the margin. The title I would alter thus

Scientific Memoirs

New Series

Prospectus

You must remember I am *[invincibly]* ignorant on all these practical matters —I state my opinion bare & naked and you must take it for what it is worth.

486   *The Correspondence of John Tyndall, Volume 3*

I shall be with you shortly after you receive this letter as I intend to start from here in good time on monday morning. till then dear Francis goodbye

    J Tyndall

StBPL T&F, Authors' letters

1. *[28–9] August 1852*: written after letter 0656 (27 August) and prior to Tyndall travelling to London on Monday 30 August (see Journal, 1 September 1852, JT/2/13b/581).
2. *this preface*: for the *Scientific Memoirs* (see letter 0656, n. 4).

From Thomas Archer Hirst   29 August 1852   0658

Marburg. Aug[st] 29th / 52

My dear John

I have now taken three days to think of the proposal in your letter,[1] and the advantages are so equally balanced that I have had the greatest difficulty in coming to a conclusion, it is one of those cases that require rather prompt decision for the more it is looked into and the more difficult it is; one thing that would have helped me much you did not state in your letter and that is, <u>when</u>, supposing you get the Professorship at Galway,[2] you will be required to enter upon your duties there—If I could have had one Semester in Berlin I should have been considerably more reconciled, for I am not alone concerned therein, I had made a promise to see young Dickinson settled there, to whom I could have been of some service and who is worth it—I hope however that at soonest you will not be required to leave Queenwood before November the later the better for me—

Well, the prompt decision I make is that if M[r] Edmondson is willing I am also prepared to come to terms with him. I can make a year at Queenwood useful to me and I hope to him also but at the end of that time I <u>must</u> return to Berlin and Paris. That is the only consolation I have for giving up the plans I had set out for myself—And now about arrangements I am willing to undertake what you have done, 3 hours per day and lectures, and if required I should not hesitate to deal with the Farmers[3] either; it is no pleasant task I know and I should therefore be willing enough to be excused, I leave that to M[r] Edmondson As to the Salary, I am prepared to receive as much as I can get for I can use it though I shall not be obstinate about that matter, if I come I shall in any case throw my whole soul into the work and, between ourselves, considering the expense of my apprenticeship I think I might be worth £150 per annum to him. That however you will have most voice in deciding for you understand best my relation to pecuniary matters—

But now about my lodging that is the principle consideration with me. I <u>must</u> have a few square yards sacred to myself, it may be a weakness of mine but I am not inclined to forego it, your room there by no means suits me, it is not quiet enough, it has not the homeliness about it that my nature and custom demands, at proper times I am body and soul at Mr Edmondsons service but after that I claim my perfect freedom and independence. I must be out of the way of interruption and must not be overlooked. If this appears too much to demand in M^r E's eyes, who has been brought up under the eye of boys, and finds his happiness therein, he must remember it is inexpressibly disagreeable to me who have been accustomed to very different habits. In short on this point I am strict, but on the other hand I am willing to make some sacrifices if necessary to attain it.

Do you remember our once visiting Queenwood and in the Evening we paid Yeats a visit in the adjoining house by the farm yard—I have that little room in my eye whenever I think of Queenwood and my request is that it or a similar one apart from the main building be given up to me should I come—As I said I shall not grudge a pecuniary sacrifice to attain it but I think it ought not to be demanded—

I am expecting every day to leave Marburg on a journey with my Brother,[4] but I am anxious to hear again from you to know when in the event of success you will be leaving Queenwood Write therefore to me under cover to Herrn Pfarrer Schnackenberg, Braach bei Rotenburg Kurhessen. Dickinson will be there during the vacation and will know where to forward the letter to me—If by that time you have proposed me to Mr E. you can tell me how he takes it

<Handwritten letter ends; hereafter LT Transcript Only>

Since I last wrote[5] I have repeated your experiments with Bismuth[6] and corroborated every experiment. I succeeded after some trouble in getting a bar that set axial. I saw there that the Bismuth behaved exactly as a bar of a magnetic substance when in the latter the lines of greatest density were parallel; in the former however perpendicular to the length of the bar. This deportment you explained by shewing that the line of greatest density in the Bismuth strove to set itself in the weaker, and in the shale in the stronger quadrants of the Magnetic Field. You say however that a North pole in the Magnet induces a South pole in the Bismuth as in the shale. Now in the shale you consider the extremities of the line of greatest density as the poles, but in the Bismuth the lines perpendicular thereto. Now I am as yet not prepared to attempt any explanations but the thought struck me: Which must be considered as poles in a substance whose molecules are arranged symmetrically? In a bar of soft iron the molecules are homogeneous throughout and we take the ends as poles, the relation of the length to the breadth of the bar settles the question, but how when the molecules are differently arranged?

Then again, the repulsion of the whole mass of Bismuth by a magnet is after all somewhat difficult to reconcile with an induced polarity similar to a magnetic substance. If we send a current around a bar of soft iron we know what polarity is induced; if we replace the iron by a bar of bismuth with its particles arranged similarly (that is not crystallized) will it be similarly polarised to the Iron or not? Weber's experiment I almost think is capable of answering this question, though for want of apparatus I have not done it at all satisfactorily yet.[7] By means of a fine string attached to a wooden disc bearing a horse shoe magnet N R S, the latter is so suspended that it revolves in a horizontal plane and sets its poles N S in the magnetic meridian. This wooden disc bears a small plane Mirror whose plane is perpendicular to that of the magnet, faces R, and is parallel to the line joining the poles N S ab (Fig. 1) is the ground plan (c b) (Fig. 2) its elevation. A telescope T with a scale p q attached (as in magnetic observations) is placed at a little distance (6 feet) and the smallest movements of the Mirror noted. C P D is a copper tube of about an inch in diameter and two feet long. This is placed perfectly vertical to the plane of the Magnet and in the position shewn at P (Fig. 1) and projects one foot above and below that plane as shewn in (Fig. 2). Around this copper tube are two windings of isolated copper wire, and the ends are connected with one of Bunsen's cells. Inside this tube is a bar of Bismuth A B. at B is attached a string which bears a small weight G, at A another string is fastened which passes over the pulley H to the observer at the telescope where an arrangement is so made that he can raise or depress the Bismuth Bar at will. A continuous current is sent around the tube C D and after seeing that the magnet is not influenced by it, the bar of Bismuth is raised and depressed at proper intervals (so as to increase the deflection of the Magnet) so that alternately the ends A and B are brought into the same plane as that of the magnet. By repeated elevations and depressions Weber has produced a considerable deflection and measured the same. This deflection can only be caused by an induced polarity in the Bismuth Bar. As I said my experiments are by no means fine enough to say if the Bismuth behaves also here as a bar of soft iron would. I do not know either what kind of bar Weber used, whether it was crystallized or not and how long. To make one for myself I poured molten Bismuth into a hot glass tube of about 3/8 of an inch in diameter and 5 inches long; kept it in a state of fusion in a sand bath until all air bubbles were expelled and the bar perfectly solid; I let it cool then very slowly and on taking it out of the sand bath it, on cooling further, broke its glass covering and came out of its shell like a chrysalis—a piece of such a bar sets axial between the poles of the magnet, so that the line of greatest cleavage (for it is all crystallized very regularly) is in direction of the length of the bar—looking at the section of one end it appears crystallized in concentric polygons about the centre or axis of the bar. I could not as yet get a suitable theodolite and mirror so as to give you any certain results.

RI MS JT/1/H/171
RI MS JT/1/HTYP/205–207

1. *proposal in your letter*: Tyndall proposed in a letter to Hirst (missing, but received by Hirst on 26 August and thus written approximately 21 August) that he succeed Tyndall at Queenwood if Tyndall was offered and accepted a position at Galway (Hirst, 'Journals', 29 August 1852).
2. *Professorship at Galway*: see letter 0636, n. 3. Tyndall did not receive the appointment, which was granted to George Johnstone Stoney.
3. *deal with the Farmers*: there were a number of students at Queenwood who would become farmers. They were often a rowdy bunch. See letter 0639, where Tyndall told Sabine that he had 'a number of farm students to keep in order'.
4. *brother*: John (see letter 0652).
5. *I last wrote*: last extant letter is 0652.
6. *your experiments with Bismuth*: see letter 0652, esp. n. 8. Tyndall responded in letter 0661.
7. *yet*: LT added an editorial note, 'Diagram here', which she did not include.

From George Wynne   [August–early September 1852][1]   0659

Whitehall | Thursday

Dear Tyndall,

I hope to be myself in Dublin on Monday next—if I had one or two more of your testimonials[2] I might be able to dispose of them beneficially—if you send them here and write 'to be forwarded' on them they will be sent on to my address.

farewell | Most truly yours | Geo. Wynne

RI MS JT/1/TYP/5/1845
LT Transcript Only

1. *[August– early September 1852]*: date is uncertain. Wynne was involved in promoting Tyndall's application for the position at Galway. Wynne became involved at the very end of June (see Journal, 28 June 1852, JT/2/13b/572–3); the application was submitted (to Dublin) well before 26 August 1852 (letter 0654, n. 1); and a decision was made in late October (letter 0679). We believe the letter probably dates to late August or early September, although a date between late July and late September is possible.
2. *your testimonials*: the testimonials Tyndall submitted with his application. Many had been acquired for his earlier application for a position at the University of Toronto.

[To Thomas Andrews][1]   18 September 1852   0660

Queenwood College near Stockbridge |
Hants– 18th Sept. 1852

Dear Sir.

I think I heard you express a wish to see a good and convenient electro-magnet previous to getting one made for yourself. If the following sketch of the instrument I use, and which I find to be very serviceable, be of any service to you it will give me much pleasure.

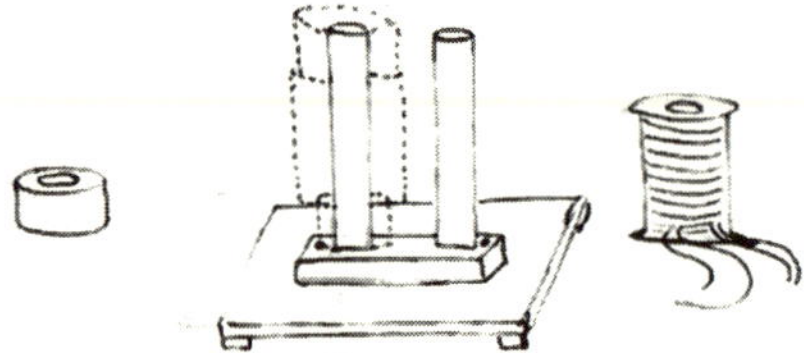

Instead of the common horseshoe form the magnet is composed of two upright soft iron pillars united at the bottom by a cross piece of the same material. The contact of the ends of the cylinders with the cross piece must be very good—otherwise a loss of power will ensue. The cross piece is fastened to the cylinders by a pair of strong screws which pass through the former and pierce the interior of each cylinder—the latter being fitted with a thread for the purpose

Height of the cylinders	8 ⅞ inches
diameter —————	1 ⅝ inches
Distance from centre to centre	4 ¼ inches
Breadth of cross piece	2 inches
Depth ————	1 ¼ —
Length ————	8 inches

The whole rests upon a slab of mahogany 12 ½ inches by 10 ½, being made fast to the wood by screws passing through the crosspiece

Round each cylinder is a coil of copper wire. The coil is made exactly on the principle of the common spool (I'm not sure whether this is the term ladies give the article round which thread is wound) the coil slips down along its cylinder being quite close to it—the aperture through its centre being not much more than sufficient to permit the cylinder to pass through.

Height of coil	5 ⅞ inches
diameter	3 ⅞ —

Each coil contains 6 lbs of copper wire—*[overspread]* with cotton—silk is expensive, and when the cotton is soaked in a solution of shell lack[2] it is equally good as an isolator.

In each coil I have caused <u>two</u> wires to be wound—you can either send the current through both at the same time as through as single unit of twice the thickness, or you can send it through the whole length of the single wire—

By reference to the dimensions you will see that the cylinders are higher than the coils—The latter I cause to embrace the centre of the cylinders by causing them to rest upon a ring of wood which raises them to the required height—A similar ring of wood is placed <u>above</u> each coil high enough to bring it to the same level as the upper end of the cylinders.

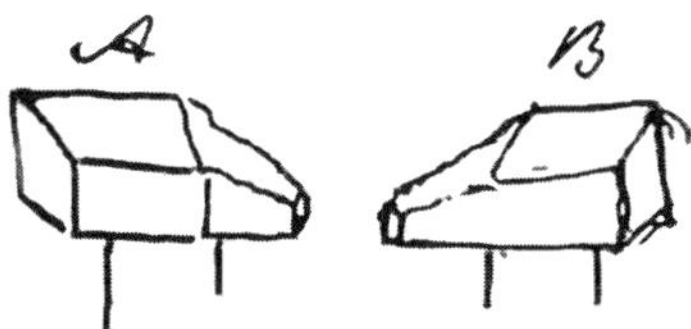

on the ends of the cylinders two moveable pieces of soft iron are placed which you can place with their flat faces opposite and as close as you please or with their points opposite as in the figure—between these masses the crystal or other body experimented with is hung—

Over my instrument is a glass case of this shape having a tube of glass fixed on the top and through which a silk fibre descends—the substance to be examined is attached to the fibre.

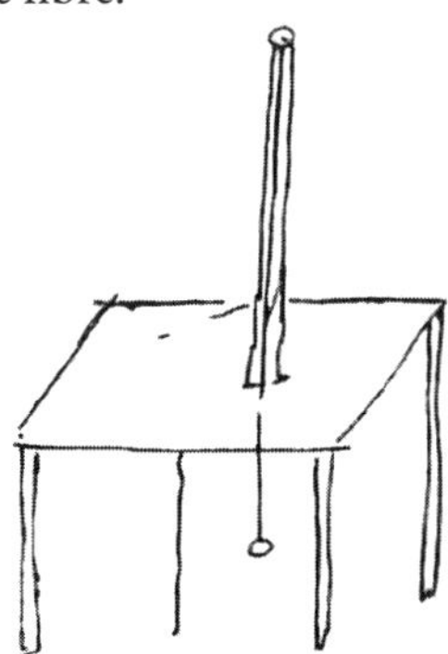

I hope you will understand me—
very sincerely yours | J Tyndall
Enclosed is a specimen of the wire used in the coils.

Science Museum, Wroughton MS 350/1/128

1.   *[To Thomas Andrews]*: British chemist, Thomas Andrews (1813–85). He had attended the Belfast meeting of the BAAS, where he presumably met Tyndall and requested information about an electro-magnet (see also letter 0668).

2.   *shell lack*: i.e., shellac.

To Thomas Archer Hirst   19 September 1852   0661

Queenwood Sunday 19th Sept. 1852

My dear Tom.

I have got through a mortal mass of work this day and now I bring my fagged energies to its completion in writing to you. It was not and is not in my power to give you the least notion as to when I should be required in Galway[1] in the event of my being called there—I know nothing about the matter and am therefore unable to enlighten you. I am glad to hear you speak of Dickinson as you have done and can most cordially sympathise with your wish of seeing him safely settled in Berlin. Those worth taking care of are so rare that they ought to be taken care when found. Dickinson has a little brother here who though quite a child promises well—it is amusing to see the grim earnestness with which the little fellow copes with his task. When the time comes I will lay the whole matter[2] before M[r] Edmondson and if possible secure you the quiet nook you desire—I don't know how far this is practicable as the place I believe has been converted into a place of meeting for the quaker boys and M[r] Edmondson's family on Sundays. We must never forget that all this is only conditionally spoken; I have antagonists in the field[3] who have the advantage of being upon the spot and what I have to depend upon is that I have more to shew in the way of work and testimonials than they have.

When I last wrote to you[4] I had no notion of going to Belfast where the British Association met this year. Col. Sabine was president and he made many kind enquiries of Francis as to whether I intended to go; asking at the same time whether I would like to be one of the secretaries of the physical Section as he thought it might be of service to me to be put in such a post. Well I resolved to go, and did go—We had a most pleasant meeting—the weather was good natured, the subjects brought forward interesting the speakers eloquent and cordial all went off capitally My old scientific antagonist Thomson[5] was President of the physical Section; once after he had made a communication which contained much that I dissented from I rose to express my dissent—I heard a bustle of making way behind me, when lo! the Lord Lieutenant[6]—his countess,[7] Lord Naas[8] his lady[9] and several others entered the room—They seated themselves and I was called upon to proceed—I stated my case as coolly as if his Excellency had been my schoolfellow.

An incident took place one morning in the Committee room from which I derived both pleasure and instruction, Sir David Brewster has usually appeared cold and indifferent towards me, an exception I must say among the scientific men of England. Well I contrived to feel indifferent about Sir David's indifference whispering to myself that a day might come when he perhaps could not afford to be so great as he was. I went therefore on my own quiet way and let Sir David pursue his—Well one morning in the Committee Room he came up to me asked me how I did in the most friendly manner possible, shook me cordially by the hand and apologized for not answering a letter which I wrote to him two years ago.

With regard to the bismuth[10] I must have expressed myself equivocally I never meant to say that its polarity is the same as that of iron—I believe it has the polarity which is attributed to it by Weber and my experiments go to prove this. In fact it was on this very point that Thomson and myself had the discussion before the Lord Lieutenant.

Of the progress of the Galway matter I can say nothing—I have sent in my testimonials which have been duly acknowledged and so the matter rests—I am told they are anxious to get the best men possible for such posts; and am also informed that I have pretty strong interest on the spot more I know not.

If you can conveniently do without it for a time don't draw upon me for the 2nd £10—I have paid the first—My going to Belfast has swamped all my loose cash and I am afraid Edmondson wont be in a condition to stand and deliver till after Christmas. If you have already done it its no matter—I will manage to meet it.

thy affectionate | <u>John</u>

RI MS JT/1/T/553

---

1. *required in Galway*: see letter 0658, nn. 1 and 2.
2. *the whole matter*: concerning a teaching position for Hirst at Queenwood. See letter 0658, n. 1.
3. *antagonists in the field*: probably a reference to G. Stoney.
4. *When I last wrote to you*: the last extant letter to Hirst is 0638, but letters from Hirst allude to a note of early August (0652, n. 3) and a letter written in mid-August (letter 0658, n. 1).
5. *scientific antagonist Thomson*: William Thomson, with whom Tyndall differed in opinions concerning magnetic induction in crystalline substances.
6. *the Lord Lieutenant*: Archibald William Montgomerie (1812–61).
7. *his countess*: Theresa Howe Cockerell, née Newcomen (d.1853).
8. *Lord Naas*: Richard Bourke, 6th Earl of Mayo (1822–77), from whom Tyndall had received acknowledgment of his application for the Galway position (Journal, 26 August 1852, JT/2/13b/580).

9.   *his lady*: Blanche Julia Wyndham (1826–1918).
10.   *with regard to bismuth*: see letter 0658.

From Thomas Archer Hirst   20 September 1852   0662

München, | Sept. 20th 1852

My dear John,

 You see from this heading that I am in the celebrated and beautiful city of Munich—I have been here four days doing nothing but looking at beautiful buildings pictures and statues until I am almost satiated. I am even neglecting my journal for want of time, so I must cut you rather short and let you see further particulars in the common receptacle of my thoughts sayings and doings. I only write in order that you may know my movements in the event you should require me in England. As I expected, I <u>shall be in Berlin early in October</u>. If you have occasion therefore to write after the receipt of this it will find me most speedily if directed Post Restante, Berlin. In any case indeed I should like to hear from you on my arrival in Berlin as <u>my arrangements there will depend entirely on your letter</u>. I cannot close my letter without telling you a little incident I encountered today. There are two Kings of Bavaria living, one the old King Ludwig who some time ago abdicated in favour of his son the reigning King Maximilian the 2nd.[1] The old King is further notorious from his connection with Lola Montes.[2] In spite of this piece of folly, however, he is a fine old fellow and greatly loved. He it is who has changed Munich from a dirty puddle hole to one of the most magnificent of European cities.[3] He is just now building out of his own private purse a new and splendid building called the Pinakothek[4] for a modern Picture Gallery. On the outside all round are cartoons now painting, illustrative of the development of art in Munich, which will immortalize all the architects, sculptors, painters, etc. that have worked for him, for they are all pictured there at their several occupations. The old fellow makes this his hobby, and spends all his fortune on it. He is mightily pleased also to let people see it. To-day we[5] went there and as we were in one of the rooms intent upon a cartoon in walked an old gentleman, a young one in Grecian costume, four ladies, and at their back an immense man in livery who immediately scowled on us and beckoned us to retire. It was his Majesty old King Ludwig, Otto King of Greece, the young reigning Queen of Bavaria, and 3 of the [six] princesses. I knew him immediately from his portrait and of course doffed my hat and was about to retire according to the scowling footman's orders, but the Queen stepped up to us and in a very graceful manner addressed me in German, telling me to remain, that the King was very glad indeed to shew his building. Upon this the old fellow stepped up to us also and began chatting to us in a quite familiar way, asking us what we were, if we were brothers, and from what country. On telling him we were English and my brother William an architect[6] he was still more gracious, and he and the queen continued the conversation in English (not very excellent I royalty I had ever the luck to meet, and I felt my heart somewhat expanded towards kings and queens in general.

Write soon. With love, | Thine affectionately, | T.A. Hirst.

R1 MS JT/1/HTYP/209
LT Transcript Only

---

1. *two Kings . . . King Ludwig . . . King Maximilian the 2nd*: Ludwig I was King of Bavaria from 1825 until 1848. Maximilian II served as King from 1848 until 1864.
2. *notorious from his connection with Lola Montes*: Ludwig I was notorious for having affairs, including with Lola Montez, an Irish dancer and courtesan. See B. Seymour, *Lola Montez: A Life* (Vail-Bailou Press: New York, 1996), p. 287.
3. *most magnificent of European cities*: Maximilian II was the main benefactor of the Maximilianstrasse and the Bavarian National Museum. His enduring legacy was one of cultural revival and support of German arts.
4. *Pinakothek*: an art museum in the Kunstareal in Munich, Germany.
5. *we*: Hirst and his brother, John. Hirst recorded the incident in his journal entry of 20 September.
6. *my brother William an architect*: this is incorrect. It is John, Hirst's companion, who was an architect. Their brother, William Aked Hirst, was a farmer. The error was most likely made by LT in transcribing the letter.

To the Editors of the *Phil. Mag.*   21 September 1852   0663

Gentlemen,

In writing his remarks on the 'supposed absorption of heat by a bismuth and antimony joint', which occur in your last Number,[1] Mr. Adie[2] appears to have forgotten the experiment of Lenz,[3] in which water was frozen at the point of junction, and the Centigrade thermometer sunk to 3.5 degrees below zero.

John Tyndall
Queenwood College,
Sept 21, 1852

From the published letter, 'On the absorption of heat by a bismuth and antimony joint', *Phil. Mag.* 4, no. 25 (October 1852), p. 318.

1. *last Number*: R. Adie, 'On the unequal Heating Effect of a Galvanic current while entering and emerging from a Conductor', *Phil. Mag.* 4, no. 24 (September 1852), pp. 224–25.

2. *Mr. Adie*: Adie replied to this letter in the November number of the *Phil. Mag.* (cited letter 0670, n. 1), which Tyndall responded to in the following number (letter 0685).

3. *experiments of Lenz*: E. Lenz, 'Einige Versuche im Gebiete des Galvanismus', *Poggend. Annal.* 44, no. 6 (1838), pp. 342–49. See letter 0685.

To William Francis   21 September 1852   0664

21ˢᵗ Sept. 1852

My dear Francis.

In all my life I never had such a mass of work before me[1]—Some days ago I dreaded breaking down but the Gods are good to me and my brain holds out amazingly—It appears as if a little vigour was lent me to cope with the task before me. I promised to have my own paper[2] sent in to the Royal early in October. My experiments are finished thank heaven but then the writing is to come. I have laid hold of Clausius[3] and shall have it with you by the end of the month.[4] by the 15ᵗʰ of Oct. you shall Helmholtz über die Erhatung der Kraft.[5] He and Clausius must go together.[6] And then there is the farmers and the boys and the lectures,[7] and Lovell Reeve[8] which I have just finished—I will say a word or two to you tomorrow about Reuben Phillips[9]—I <u>am</u> delighted to hear of Mʳ Taylor's improvement[10]—Now that you have him safe and sound you must take good care of him and keep all vexatious matters far away from him. It must be a bright and gladdening thought to you & Miss Taylor[11]—for indeed it is to me—to see him rescued from such a melancholy fate and set once more erect in the enjoyment of his faculties. I have examined Quetelet root and branch[12]—he is all right—Remember me most kindly to Ronalds when you see him. I owe him a great deal[13]—

Yours as ever | Tyndall

StBPL T&F, Authors' letters

1. *mass of work before me*: Tyndall was busy with translations for the *Scientific Memoirs* and preparing his own paper for the Royal Society in October. See Journal, 19 September 1852 (JT/2/13b/585).

2. *my own paper*: see letter 0606, n. 1. The manuscript was sent to Sabine before 6 October (Journal, JT/2/13b/585).

3. *Clausius*: two papers by Clausius were published in the first two parts of the *Scientific Memoirs*. The first, R. Clausius, 'Ueber das mechanische Aequivalent einer elektrischen Entladung und die dabei stattfindende Erwärmung des Leitungsdrahtes', *Poggend. Annal.* 86, no. 7 (1852), pp. 337–75, was translated as R. Clausius, 'On the Mechanical Equivalent

of an Electric Discharge, and the Heating of the Conducting Wire which accompanies it', in J. Tyndall and W. Francis (eds.), *Scientific Memoirs* (London: Taylor and Francis, 1853), pp. 1–32. This paper appeared in Part I. The second, R. Clausius, 'Ueber die bei einem stationären elektrischen Strome in dem Leiter gethane Arbeit und erzeugte Wärme', *Poggend. Annal.* 87, no. 11 (1852), pp. 415–26, was translated as R. Clausius, 'On the Work performed and the Heat generated in a Conductor by a Stationary Electric Current', in ibid, pp. 200–209. This paper was split over Parts II and III. It is unclear to which paper Tyndall alludes here (and in later letters), although the request that it be published alongside the paper by Helmholtz suggests that he alludes to the second paper.

4. *Clausius . . . end of the month*: Tyndall sent the translation to Francis before 6 October (Journal, JT/2/13b/585).

5. *Helmholtz . . . der Kraft*: H. Helmholtz, *Über die Erhaltung der Kraft*, read before the Physical Society of Berlin on 23 July 1847 and later published (Berlin: Reimer, 1847). The translation was published as H. Helmholtz, 'On the Conservation of Force', in J. Tyndall and W. Francis (eds.), *Scientific Memoirs* (London: Taylor and Francis, 1853), pp. 114–62.

6. *He and Clausius . . . together*: the first Clausius paper was published in the first number, Helmholtz was published in the second number along with the start of the second Clausius paper, the remainder of which appeared in the third number.

7. *farmers and the boys and the lectures*: his work at Queenwood.

8. *Lovell Reeve*: Reeve (1814–65) was an English conchologist and editor of the *Literary Gazette* from 1850–6. The text to which Tyndall alludes has not been identified.

9. *word or two to you tomorrow about Reuben Phillips*: letter, if written, is missing. Tyndall probably alludes to a paper by Phillips that appeared in the January number of the *Phil. Mag.* (see letters 0666 and 0684, n. 7).

10. *Mr Taylor's improvement*: as Taylor's health declined, he entered into partnership with William Francis, co-founding the Taylor & Francis Publishing House in May, 1852. At that time, Richard Taylor's brothers, John and Edward, attempted to prove Richard mad in order to have Francis and Richard's partnership declared null. In June, he had been committed to a private asylum in Sydenham. This is the first of several allusions to Taylor's ill health and family bickering. (See Brock and Meadows, *Lamp of Learning*, pp. 132–4.)

11. *Miss Taylor*: Sarah Taylor (1808–84), Richard Taylor's legitimate daughter and half-sister of William Francis.

12. *Quetelet root and branch*: not identified, but probably having to do with the French statistician Adolphe Quetelet's publications in the October issue of the *Phil. Mag.* ('On Atmospheric Electricity, according to the Observations at Munich and Brussels', and 'On the State of Static and of Dynamic Electricity during several heavy Showers observed at Brussels on the 14th of June 1852', *Phil. Mag.* 4, no. 25 [October 1852], pp. 249–52 and 253–56.)

13. *I owe him a great deal*: Ronalds's had passed on news to Francis about the Cork position in 1851 (see letter 0551 and 0654, n. 1 for indications of his involvement); he may also have been the source of the early news about the Galway position (see letter 0638).

To William Francis   [late September 1852][1]   0665

[x]A notice of the principle of a chromatic stereoscope was communicated by Sir David Brewster to the Physical Section of the British Association at Swansea in 1848. See Report of the Association[2] and Phil. Mag. January 1852. P31.[3]

My dear Francis,

You must manage to put the above as a foot note to the 1st page of the Report of Doves enquiries[4]—you may choose the place of the star—after the word 'ingenious inventor' it might appropriately be placed—or perhaps in the title—the place is of not much importance so as the note appears.

Sincerely yours | J Tyndall

RDS 27/24

---

1. *[late September 1852]*: dated by the allusion to the 'Report' of Dove's investigation.

2. *Report of the Association*: D. Brewster, 'On the Vision of Distance as given by Colour', *Brit. Assoc. Rep. 1848* (pt. 2), p. 48.

3. *Phil. Mag. January 1852. P31*: D. Brewster, 'Notice of a Chromatic Stereoscope', *Phil. Mag.* 3, no. 15 (January 1852), p. 31.

4. *1st page of the Report of Doves enquiries*: J. Tyndall, 'Reports on the Progress of the Physical Sciences', *Phil. Mag.* 4, no. 25 (October 1852), pp. 241–49. The footnote was not placed in the published article. The phrase 'ingenious inventor' appears in the first paragraph.

To William Francis   4 October 1852   0666

4th Oct. 1852

My Dear Francis.

The more I think on it the more advisable I deem it to have Bunsen[1] if possible in the 1st number. I shall have Helmholtz[2] ready by the middle of this month and as you have given notice of it in a note to Thomson's paper,[3] it, Bunsen & Knoblauch[4] may appear and Clausius[5] can remain over to the next number. I certainly think Helmholtz and Clausius will be too great a dose for ⟨a⟩ single number.[6]

Sincerely yours | Tyndall.

Dont be too hard on poor Reuben[7] as there is something in him though clouded by egotism and impaired by want of proper culture—I wish the marrow could be distilled from his experiments it might be worth recording.

RDS 27/14

1. *Bunsen*: R. Bunsen, 'On the Processes which have taken place during the formation of the Volcanic Rocks of Iceland', in J. Tyndall and W. Francis (eds.), *Scientific Memoirs* (London: Taylor and Francis, 1853), pp. 33–98. The original paper is cited in letter 0526, n. 2.

2. *Helmholtz*: see letter 0664, n. 5.

3. *Thomson's paper*: the note '*Ueber die Erhaltung der Kraft*, von Dr. H. Helmholtz. Berlin, 1847. [A translation of this essay will appear in the First Part of the New Series of the Scientific Memoirs.~ED.]' appeared in W. Thomson, 'On the Mechanical Action of Radiant Heat or Light: On the Power of Animated Creatures over Matter: On the Sources available to Man for the Production of Mechanical Effect', *Phil. Mag.* 4, no. 25 (October, 1852), p. 257.

4. *Knoblauch*: see letter 0651, n. 8.

5. *Clausius*: see letter 0664, n. 3. Tyndall's recommendation that Clausius's paper appear after those by Bunsen and Knoblauch suggests that he is alluding to the second of Clausius's papers included in the *Scientific Memoirs*.

6. *the next number*: in the end, the paper by Bunsen appeared in the first number, the paper by Knoblauch was split between the first and second numbers, the paper by Helmholtz appeared in the second number, and the second of Clausius's papers was split between the second and third numbers.

7. *Dont be too hard on poor Reuben*: Tyndall had mentioned in letter 0664 that he was going to write to Francis about Reuben Phillips, probably in relation to a paper that appeared in the January number of the *Phil. Mag.* (see letter 0684, n. 7). The nature of the conversation and what Francis said to prompt Tyndall to write this is unclear. A year previously, while discussing the inclusion of a paper by Phillips in the *Phil. Mag.*, Tyndall had disparaged his competence and the value of his investigations. On this occasion, it seems Francis was more dismissive than Tyndall.

From Thomas Archer Hirst   [5 October 1852][1]   0667

Berlin Oct. 7th 52.[2]

My dear John

I have received your letter[3] and on the strength of the uncertainty as to my coming suddenly to England I have taken lodgings and shall enter on a course of lectures with Derichlet, Steiner[4] and Eisenstein,[5] mathematicians and perhaps Dove on Physics. My brother has now returned[6] and I can enter unfettered into my work. God willing I will make up for the month's, lost time, I was going to say, at any rate superfluous holiday.[7] I was glad to hear you went to Belfast and were Secretary there,[8] it will no doubt be of service to you. The incident with Sir David Brewster[9] is very natural to a great class of people, at first he thought you a sort of bold adventurer perhaps, when you appeared in the Scientific lists as the antagonist of Plücker;[10] he is a chary fellow and naturally looked askance, now he begins to think it safe to patronize you. —Continue to go your own quiet way as before and Brewster his and then as Carlyle says.—

we shall see.—

You say in your letter 'well, I contrived to feel indifferent about Sir David's indifference, whispering to myself that a day might come when he perhaps could not afford to be so great as he was'. These few words John remind me again of something I often wish to say, but on account of its vagueness, for it is a mere intuitive feeling, I could never get it expressed. I hardly ever yet saw any thing in or about thee John that I could not from my soul sympathize with, and myself strive after, hardly any point in which I did not feel thee to be the stronger so that when I would have spoken or made attempt to advise, it seemed to me that silence was more becoming; it is therefore with real and no assumed diffidence that I venture to express a kind of instinctive misgiving I have had sometimes when thinking about thee lately, and as instinct is worth listening to whatever height we may have gained through Reason I will break my silence. It may be all fancy on my part to be laughed away merely. I hope so. It is that Science may be demanding too much of thee[;] it too thou wilt grant has allurements and temptations all the more dangerous because not recognized as such by the more intelligent part of the world. Its demands too are great, tending to swamp the man in the philosopher; all this you know far better than I do; are you sure however that you are keeping a sharp eye to it? I know that like a garment preoccupation can be sometimes cast aside when requisite; but a garment can be worn until it become a necessity. (Old Coton[11] used to assert that we came by our breeches in some such way). In short John it appears to me that few have made such sacrifices to science during the last few years as you, perhaps few obtained as brilliant and durable results, which but increase the danger; and lastly what diminishes this danger I allow, few as able to carry their brilliant results. I have said now what I wanted on the strength of our mutual friendship, the purest and best of God's gifts to me on earth, you will take this word of caution as from a brother,[12] pure instinct has suggested it to him and he gives it on no other ground.

<Tyndall's copy in his journal ends; hereafter LT Transcript of Hirst's letter>

I have called on Kleiner about your Thermo Saüle,[13] it is not ready or my brother would have brought it you, he promises to have it ready in 14 days, I must then contrive to send it. What about the payment—I have plenty of money for myself and had no intention of drawing on you for another £10; I did it before more as a precaution on starting on a long journey than because I was in want of it; I am not aware what the Thermo Saüle will cost and therefore cannot promise to pay for it. Write me instructions on the matter.

I received Debus's letter[14] also the other day, he will hear from me soon. Adieu.[15] | Thine affectionately, | Thomas A Hirst.

RI MS JT/2/6/161–162
RI MS JT/1/HTYP/210
JT Transcript and LT Transcript Only

1. *[5 October 1852]*: Hirst's journal entry for 5 October records that he wrote this letter to Tyndall; one version of parts of this letter is found in Hirst's journal.

2. Oct. 7th 52: probably inserted by JT and LT from a postmark. It is incorrect.

3. *your letter*: letter 0661.

4. *Steiner*: Jakob Steiner (1796–1863), a Swiss mathematician and professor of geometry at the University of Berlin from 1834 until his death.

5. *Eisenstein*: Ferdinand Gotthold Eisenstein (1823–52), a German mathematician. He taught at the University of Berlin from 1847 and had recently been elected a fellow of the Göttingen Academy of Sciences (1851) and the Berlin Academy (1852). See also letter 0678, nn. 8–11.

6. *My brother has now returned*: Hirst's brother, John, had arrived in Germany on 9 September, and the pair had spent the remainder of September travelling around Germany. See Hirst, 'Journals', for this period.

7. *superfluous holiday*: after his return from Göttingen, Hirst left Marburg and travelled around the countryside.

8. *went to Belfast and were Secretary there*: the meeting of the BAAS in Belfast (see letter 0661).

9. *incident with Sir David Brewster*: described in letter 0661.

10. *appeared in the Scientific lists as antagonist to Plücker*: Hirst was noting here that Tyndall quite consciously set himself up as an antagonist to Plücker.

11. *Old Coton*: not identified.

12. *word of caution as from a brother*: Tyndall certainly took Hirst's admonition to heart. He transcribed the pertinent paragraphs into his Journal on 16 October (the chief source for this letter). Over a week later, he wrote, 'Tom's letter has I think been of service to me; it has called out "Patience!" to my scientific haste, and induced me to withdraw from a time to higher and holier contemplation. I think that a good deal of Faraday's week day strength and persistency might be referred to his Sunday exercises. He drinks from a fount on Sunday which refreshes his soul for the week. I think I will try and do the same according to my own methods; for I believe the same source of power is substantially open to me and Faraday although we approach it by different routes' (24 October 1852; JT/2/13b/588).

13. *Thermo Saüle*: Tyndall had asked Knoblauch to obtain apparatus and crystals for him, and asked Hirst to ensure they were obtained in good time (see letters 0638 (especially n. 9), 0650, and 0651).

14. *Debus's letter*: in letter 0650, Hirst mentioned that he would write to Debus concerning Hirst's watch.

15.   *I have called on Kleiner . . . Adieu*: this last section of the letter (excluding the final saluta-
      tion) does not appear in the copy Tyndall made in his journal, nor in the copy in Hirst's
      journal (see n. 1). We believe JT and Hirst copied only the most pertinent part of the
      letter—concerning Hirst's admonishment of Tyndall—and not the business matters
      concerning the thermosäule. The LT transcript, made from the original Hirst letter (now
      missing) includes the two paragraphs.

[To Thomas Andrews]   10 October 1852   0668

Queenwood Stockbridge | 10th Oct. 1852

My Dear Sir

I hope you will pardon my quite unintentional delay in replying to your
last letter.[1] I have had such a mass of work pressing upon my time and attention of late that the matter was washed away from my memory. I am aware of
no single experiment, however fine, that cannot be made by a magnet such
as I have described to you—all Plückers experiments are shewn by it with
ease—one experiment however succeeds better with a larger magnet and that
is the experiment of Zantedeschi on the repulsion of the flame of a candle[2]—
It would appear as if <u>quantity</u> of magnetism was the principal thing here. You
saw the large magnet in Berlin,[3] with that magnet and a very weak excitation
a repulsion of a flame is obtained which, with a small magnet, would be quite
unattainable, no matter how highly it might be excited. But for all ordinary
purposes, and more especially for actual work the magnet I have described to
you is most convenient. You might make it a little larger preserving the same
proportions, but permitting the copper wire to surround the <u>entire</u> /legs/ from
top to bottom, and /thus/ dispensing with the wooden rings. The cross-piece
is I think more convenient than and equally powerful with the /hors<e>/.[4]
Although possessing seve<ral> magnets Prof. Poggendorff w<as> induced
to order one exactly similar to mine while I was in Berlin.[5] I am well aware
of /your/ having laboured at Physics for I have derived both pleasure and
instruction from the perusal of your papers.[6] Some of your results I have long
imagined to possess great significance and cannot but <thin>k that this will
eventually <con>tribute to the elaboration of a <mo>re satisfactory theory
of electric <a>ction than we at present possess.

believe me dear Sir | most faithfully yours | John Tyndall

Science Museum, Wroughton MS 350/1/129

———————

1.   *your last letter:* letter missing. It was perhaps a reply to letter 0660.

2. *repulsion of the flame of a candle*: refers to F. Zantedeschi's experiments showing the repulsion of a flame by a strong magnetic field ('On the motions presented by flame when under the electro-magnetic influence', *Phil. Mag.* 31, no. 210 [December 1847], pp. 421–24). See letter 0396, n. 10.

3. *large magnet in Berlin*: perhaps in Magnus's laboratory, which was exceptionally well-equipped.

4. *[hors]*: the edge of the paper is damaged and this could also be 'horse' or, if closely squeezed in, 'horseshoe'. The following word, possibly 'form', has been crossed out. Tyndall probably was referring to the horseshoe form of most electromagnets, which he eschewed (see letter 0660).

5. *Berlin:* this probably refers to the 1851 trip to Berlin where Tyndall worked with Magnus on diamagnetic research.

6. *your papers*: Andrews had published numerous papers on various aspects of physics, including electricity and heat, since the mid-1830s. See, for example, T. Andrews, 'On the thermo-electric Currents developed between Metals and fused salts', *Phil. Mag.* 10, no. 63 (June 1837), pp. 433–40 and 'On the Latent Heat of Vapours', *Journal of the Chemical Society* 1 (1849), pp. 27–41.

To Emil du Bois-Reymond   10 October 1852   0669

Queenwood 10th Oct. 1852

My dear du Bois.

Will you permit me to introduce to you my friend Mr Hirst[1] who intends to spend some time in Berlin. I dont know what he intends to do there—probably pursue mathematics, but he has a physical head and I am therefore anxious that he should cultivate the acquaintance of the physicists of Berlin. You will not perhaps find him 'funkelnd'[2] but if you give him a little time you will I doubt not learn that he possesses abilities of no ordinary kind. In fact he is a young fellow who has been bound to me through years by the closest ties of friendship and concerning whose future I entertain great hopes. I should like him to have an opportunity of visiting the physical society[3] and of looking in to the Academy[4] if it can be accomplished. I may remark that he has already passed his examination and received his Doctor diploma in Marburg.[5] It may be in your power to give him some information regarding the mathematicians whose lectures it would be desirable to attend. I do not think that he would like you to lose much time with him but I am very desirous that he should at least make your acquaintance.

I am hard at work translating the paper of your friend Helmholtz[6] which you gave me on leaving Berlin—it will probably be published on the 1st of

November. Die Sache ist nicht immer leicht zu verstehen,[7] but it is a valuable paper. Remember me kindly to D[r] Beetz[8] when you see him—there are many others also to whom I could desire to be remembered, but my sheet would not contain their names. Trusting that you feel yourself happy and vigorous after your Italian and Swiss journey

believe me dear du Bois, | most faithfully yours, | J Tyndall

Staatsbibliothek zu Berlin, Sammlung [Collection] Darmstaedter, F 1 e 1855 (2): Tyndall

1. *introduce to you my friend M[r] Hirst*: this is the only extant letter of introduction for Hirst of several written by Tyndall.
2. *'funkelnd'*: 'scintillating' (German).
3. *physical society*: The Deutsche Physikalische Gesellschaft (German Physical Society), an organization of scientists, of which Emil du Bois-Reymond was a founding member.
4. *the Academy*: likely the Königlich-Preußische Akademie der Wissenschaften (Royal Prussian Academy of Sciences), to which Emil du Bois-Reymond was admitted in 1851. It is usually referred to in English as the Academy of Sciences of Berlin.
5. *received his Doctor diploma in Marburg*: Hirst received his doctorate degree from the University of Marburg under F. L. Stegmann.
6. *translating the paper of your friend Helmholtz*: *Über die Erhaltung der Kraft* (see letter 0664, n. 5).
7. *Die Sache ist nicht immer leicht zu verstehen*: 'The thing is not always easy to understand' (German).
8. *Dr Beetz*: Wilhelm Beetz (see letter 0512, n. 9).

To William Francis   12 October 1852   0670

Queenwood 12th Oct. 1852

My dear Francis,

I return you Adies paper;[1] by all means give him a place—a <u>prominent</u> one and not in the miscellaneous if you can afford it. I will submit the matter to an experimentum crucis myself in a few days and hope to furnish a reply to this note which will set M[r] Adie at rest forever.[2]

Most cordially do I I[3] return M[r] Taylor's kind greeting. You must studiously avoid all the old topics of vexation[4] and still avoid them so as not to arouse his curiosity—he must be treated I imagine on these subjects with apparent frankness for if he gets the idea that you are keeping matters secret, his own imagination will do the rest—Your care and the quiet of Sydenham[5] will I doubt not eventually make his cure complete.

I will send you Clausius very soon and along with it the remainder of

Helmholtz.[6] Thanks for the little reference to Ronalds[7]—any thing you hear pro or con I should be glad to be made acquainted with.

Sincerely yours | <u>Tyndall</u>

RDS 27/15

1. *Adie's paper*: published as R. Adie, 'On the temperature of a bismuth and antimony joint during the passage of an electric current', *Phil. Mag.* 4, no. 26 (November 1852), pp. 380–81. See also letter 0672, n. 3.

2. *at rest forever*: see letter 0685 for Tyndall's reply to Adie, which was published as a letter to the editor in the December number of the *Phil. Mag.*

3. I: Tyndall repeated this 'I', which appeared as the first word on the second page of the letter.

4. *old topics of vexation*: see letter 0664, n. 10.

5. *Sydenham*: ibid.

6. *Clausius . . . Helmholtz*: for Clausius see letter 0664, n. 3 and 0666, n. 5; for Helmholtz see letter 0664, n. 5.

7. *Ronalds*: Edmund Ronalds.

From Henry Bence Jones 12 October 1852 0671

My dear Sir

I hope you will consider the subject which I am about to mention as a sufficient excuse for this note. Dr Hoffmann and I are about to continue the publication of the English translation of Liebigs Report[1] and it would be very satisfactory to us if we could obtain your services in making the translation of the Physics.

We shall be much obliged to you if you will let us know[2] whether this is possible on any terms.

The work is far from flourishing but if I hear from you that you can undertake this part, in my next note some more definite arrangement may be proposed. There will probably be from 12 to 14 sheets of the translation.

Believe me | yours very truly | H Bence Jones

30 Grosvenor Street | Oct. 12. 1852

RI MS JT/2/6/159–160
JT Transcript Only

1. *Liebigs Report*: J. Liebig and H. Kopp, eds., Jahresbericht über die Fortschritte der Chemie, Physik, Mineralogie und Geologie. Für 1850 (Giessen: J. Ricker, 1851). Translated as J. Liebig and H. Kopp, *Annual Report of the Progress of Chemistry and the Allied Sciences,*

Physics, Mineralogy, and Geology, Vol. 4, eds. A. W. Hofmann and H. Bence Jones (London: Walton and Maberly, 1853).

2.	*let us know*: Tyndall's reply is missing but is summarised in his Journal for 16 October 1852: 'Wrote back saying that I had many heavy claims upon my time and would gladly avoid incurring further responsibility but that nevertheless if he desired I should be reluctant to decline the translation' (JT/2/13b/586). Bence Jones sent Tyndall the parts of the *Report* he was to translate on 19 October (letter 0674).

# To William Francis	17 October 1852	0672

Queenwood 17th Oct. 1852

My dear Francis.

I want you to do two favours for me:

No 1. To send me Ann. du chim. et du Phys. 56. At page 371 there is an account of <u>Peltier's</u> experiment[1] which I want to see

No 2. To send me Pogg. 44. At page 342 there is an account of a repetition of Peltiers experiment by Lenz.[2]

I will return these to you without delay. I have already a good many nos. beside me but am just waiting to get done with the subject to which they refer; when this is accomplished I will send them to you at once,

I intended yesterday to enquire whether you would think it well to publish a reply to Adie[3] side by side with his remarks in the next no. but as it is full the reply may remain over.

I have already sent one paper in to the Royal Society & hope that it will appear in the next vol. of the Transactions.[4]

Bence Jones has written to me to enquire whether I could be induced to undertake the translation of the Physics of Liebigs Bericht.[5] I shall probably do a small portion of it. He also has kindly mentioned me as likely to deliver a good course of lectures at the Royal Institution; and asks me[6] whether I would not undertake to deliver one some Friday evening before or after Easter—I have sent *[in]* an affirmative reply. So that after Christmas I hope to hold forth some evening.

I will get Helmholtz ready, but the 9th number has not reached me—I will draw the papers into the form of a Report[7] and perhaps say a word or two introductory.

most sincerely yours | John Tyndall
<u>Success to the Memoirs!</u>[8]

StBPL T&F, Authors' letters

1. *At page 371 there is an account of <u>Peltier's</u> experiment*: C. Peltier, 'Nouvelles expériences sur la caloricité des courants électrique', *Annal. Chim. et Phys.* 56 (1834), pp. 371–86.
2. *a repetition of Peltiers experiment by Lenz*: H. Lenz, 'Einige Versuche im Gebiete des Galvanismus' *Poggend. Annal.* 44, no. 6 (1838), pp. 342–49.
3. *reply to Adie*: concerning the ongoing disagreement about the absorption of heat at a bismuth antimony joint. Adie's paper was published in November (cited letter 0670, n. 1) and Tyndall's reply in December (see letter 0685).
4. *next vol. of the Transactions*: Tyndall's first publication in the *Phil. Trans.* appeared in the 1853 volume (cited letter 0606, n. 1).
5. *written to me . . . Liebig's Bericht*: see letter 0671.
6. *asks me*: letter missing. Tyndall received it on 16 October (see letter 0673, n. 11 and letter 0675).
7. *Helmholtz . . . form of a Report*: as the following letter makes clear, Tyndall planned to translate papers by Helmholtz from numbers 8 and 9 of *Poggend. Annal.* (H. Helmholtz, 'Ueber Hrn. D. Brewster's neue Analyse des Sonnenlichts', *Poggend. Annal.* 86, no. 8 [1852], pp. 501–23 and H. Helmholtz, 'Ueber die Theorie der zusammengesetzten Farben', *Poggend. Annal.* 87, no. 9 [1852], pp. 45–66). Tyndall reported his plans differently to Francis and, in the following letter, to Hirst. Here he proposes to summarise Helmholtz's articles in a 'Report on the State of the Physical Sciences'; there he plans two full translations. The latter plan was adopted (see letter 0673, n. 15).
8. *Memoirs!*: see letter 0656, n. 4.

To Thomas Archer Hirst   17 October 1852   0673

<u>Queenwood</u> Sunday morning | Oct. 17th 1852

My dear Tom,

I rose this morning with the intention of devoting this day to the gods, to an endeavour to refresh myself in some way by a communion with the higher powers; it was a raw cold morning, I sponged myself and drew on my clothes—too light perhaps for the season, for I wear no flannel. I sat for a time in a room without a fire and felt my chest and throat begin to tickle; the breakfast hour arrived, I had two flowing cups of good hot tea, meanwhile the maid constructed a fire for me and there it is bubbling and flirting and sparkling before me, while I, sitting at a respectful distance, can pursue my primitive thought in comfort. Little incidents of this kind have always a tendency to degrade this body of mine in my eyes to the position of an instrument merely, a kind of flute, harpsichord, bass fiddle, or what you will, through which a performer behind plays sweetly or indifferently as the case may be. How inextricably woven both are though! I cannot conceive of what the spirit would be without its instrument—a Paganini[1] perhaps deprived of his string—silent! But still less can I confound spirit and instrument, both are necessary to the outward and visible result which we behold, but still they are not identical. Here before me is a little engraving of the Duke of Wellington which I cut out of last week's Punch[2] and have nailed over my chimney piece—I like to hold communion with men like the Duke. Silently looking at his picture, with the British Lion lying beneath him, I am more instructed in fortitude and courage than by the perusal of a volume upon these subjects. If I had money I would surround myself with the portraits of great men—I admire strength and like to feel strong and still I do not admire this alone. I would not confine myself to the portraits of great men which should make my atmosphere one of heroism, but if I had a little more money still, I would add a few portraits of beautiful women. I am very sensitive to such influences. I have kissed pictures of beautiful maidens when a boy and I believe I could do so still. A girl is such a transparent thing and the soul's beauty is deduced by such sweet inferences from the play of mind within the eyes and upon the cheek—feeling or imagination would perhaps be a better word than 'mind'. O, I should like the communion of a girl of fine feelings which should diffuse their aroma through the sterner stuff of which I am myself composed. But are such to be found? Is it not far better for me to be content with the portraits [wherein] I can separate the ideal from the real and convert the former into spiritual nutriment. In the portrait the mind beholds the essence of loveliness—it does not dwell upon the defects which doubtless appertain to the original; but dwells, like memory, upon the sunshine and flowers of the idea before it and forgets the shadows and the weeds. But where am I wandering? I scarcely know where.– I was going to say something with regard to Wellington, was I not? I dwell upon his career, and picture those masses of men which he moved upon the continent. Here was an instance of the degradation before alluded to on the grand scale. What moved those serried columns? what lifted those tons of human flesh over the red walls of Badajoz.[3] Look at the terrible amount of mechanical energy developed suddenly by the few words 'up guards and at them!' at Waterloo.[4] what did it all? the soul of one man. This soul caused columns to crash, walls to tumble and the earth to quake beneath the tramp of thundering squadrons. The soul I represent to my mind as something like a pure force, now a force is modified by the medium through which it passes a solar ray for instance in its passage through a crystal,—and thus the force of the soul is modified in passing through the human understanding, the understanding controls and applies a force of which it is neither the creator nor generator; hence the model man is he who has soul in abundance and understanding commensurate. Subtract from Wellington his cool understanding and his force will dissipate itself like steam under no pressure; subtract from him his soul and you reduce him to the condition of an eye which sees how things ought to be done but is impotent to enact what it sees. In life it must be acknowledged that the soul's force is the thing most needed. *[The]* material wants of men are sufficient to keep the understanding active; but the understanding *[deriving]* its activity from such sources tends to become degenerate if not diabolical. Even in my own little experience it is the want of spiritual force which I have oftenest to deplore; and yet I believe it to be perennially accessible, that god has provided sources of strength and exaltation for the sons of man if they only seek them *[fait[h]fully]*. But the gauds and trumpery, nay even the reputed excellences of the world must not stand in our way here. To be the recipient of this bounty, the possessor if this force of the spirit, necessitates this willing sacrifice of what the world pronounces desirable—yea the sacrifice of scientific reputation if need be, although this seems to rest upon a spiritual basis. When I see great men squabbling about questions of priority I know that they have not the spirit; that their object is fame and the admiration of men; whereas a communion with the spirit is so satisfying, so exalting, so expanding in its nature as to degrade the differences of scientific men into the squabbles of children quarrelling about a toy.

A sweet silence reigns around me where I sit Tom. I hear nothing save the cawing of some rooks, the song of a cockrobin and an occasional spurt from my fire. The boys are all gone to church and their absence is a relief to me. I wished last night to have a cottage somewhere in the midst of a wood of pines where I might see the trees

> tossing their cones
> to the day song of their waterfall tones[5]

for I was wearied of Science and the clatter of boys, and lo! this Sunday morning has descended gently and gives me the repose I wished for.

I feel a kind of degradation in stepping down from the platform from which I have hitherto addressed thee to say something of secular matters. I am glad you have taken lodgings—just proceed on your way as if this Galway matter[6] had no existence. I wrote to you a few days ago[7] inclosing notes of introduction to Magnus, Dove and duBois Reymond.[8] I did not know that you had arrived in Berlin[9] so therefore addressed them <u>poste restante.</u> Send me word what the cost of the Thermo saüle[10] is and I will contrive to send the money somehow. I had a letter from Dr Bence Jones[11] a man of some eminence in London, asking me to undertake a translation of the physical portion of the Giessen Jahres Bericht; a portion of which I will probably undertake. In a note received from the same gentleman yesterday[12] he tells me that he has been speaking of me to the Revd. Mr Barlow the Secretary of the Royal Institution as likely to deliver a good course of lectures and asks me if I would engage to deliver a lecture on some Friday evening either before or after Easter.[13] I will see what can be done. You are aware I suppose that the 'scientific memoirs' are now edited by four persons of which I am one. Francis has the chemistry, Henfrey the Botany, Huxley the zoology and Tyndall Natural Philosophy.[14] Tell DuBois Reymond when you see him that I have already marked the two papers of his friend Helmholtz which appear in Pogg. 8 and 9 for the Phil. Mag. both will appear.[15]

God grant that I may be always ready to listen with reverent attention to the admonition of a friend who is competent to give it — I received thy letter boy[16] — never shrink from telling me your mind, for your counsel is ever sweet to me.

ever thine John.

You have already[17] made the acquaintance of Prof. Poggendorf. Greet him from me kindly when you see him, and should you ever meet M[rs] Poggendorf present my respects.

RI MS JT/1/T/554

1. *Paganini*: famous Italian violinist Niccolo Paganini (1782–1840).
2. *Duke of Wellington . . . Punch*: Arthur Wellesley, 1st Duke of Wellington, died on 14 September 1852. He was, from the British perspective, the hero of the battle of Waterloo, where Napoleon was defeated after his escape from Elba. *Punch* Magazine published a portrait of Wellington above a lion on 2 October 1852.
3. *human flesh over the red walls of Badajoz*: the Siege of Badajoz, from 16 March–6 April 1812. Wellesley besieged Badajoz, Spain, and forced French surrender, one of the bloodiest events of the Napoleonic Wars.
4. *'up guards and at them!' at Waterloo*: Wellington allegedly said these words at the Battle of Waterloo.
5. *tossing . . . waterfall tones*: Emerson, 'Woodnotes', lines 7–8 (in *Poems* p. 51). The original reads 'When the pine tosses its cones | To the song of its waterfall tones'.
6. *Galway matter*: Tyndall unsuccessfully applied for the chair of Natural Philosophy at Queen's College, Galway.
7. *I wrote to you a few days ago*: letter missing, but written on or around 10 October, when he wrote one of the letters of introduction sent with it. See n. 8 below and letter 0669.
8. *duBois Reymond*: see letter 0669. The other letters of introduction are missing.
9. *arrived in Berlin*: Tyndall had advised Hirst to visit other German universities after he completed his doctorate from Marburg.
10. *Thermo saüle*: see letter 0638, n. 9. For Hirst's query regarding payment, see letter 0667.
11. *I had a letter from D[r]. Bence Jones*: letter 0671.
12. *a note . . . yesterday*: letter missing.
13. *Royal Institution . . . after Easter*: see letter 0672.

14. *'Scientific Memoirs'* . . . *Tyndall Natural Philosophy*: see letter 0656, n. 4. Arthur Henfrey (1819–59), botanist and surgeon, was elected a member of the Royal College of Surgeons in 1843, a fellow of the Royal Society in 1852, and succeeded Edward Forbes as Professor of Botany at King's College, London in 1853.

15. *Pogg. 8 and 9 for the Phil. Mag. both will appear*: original papers cited in letter 0672, n. 7. The arrangement described here—which differs from that proposed in letter 0672 to Francis (see n. 7 of that letter)—was followed. The two papers were published separately, as H. Helmholtz, 'On Sir David Brewster's New Analysis of Solar Light', *Phil. Mag.* 4, no. 27 (December 1852), pp. 401–16, and 'On the Theory of Compound Colours', *Phil. Mag.* 4, no. 28 ([suppl.] 1852), pp. 519–34.

16. *received thy letter boy*: letter 0667, in which Hirst admonished Tyndall.

17. *You have already*: the entire postscript is written in the left hand margin of the fifth page (of 8) of the letter.

From Henry Bence Jones   19 October 1852   0674

30 Grosvenor St. | Oct. 19. 1852.[1]

My dear Sir,

I have ventured in addition to the Magnetism and Electricity to send you the first sheet on Molecular actions.[2] No one will do it more justice; not to say more. Will you put the figures in the text and foot notes as they stand in the German—If when you have finished each sheet you will send it with the text we shall go to press at once. We shall not trouble you with proofs and reviews unless you wish to have them. I have written to Mr Barlow.[3] You can have the great electro magnet[4] without doubt.

The lecture[5] should end as the hour strikes. Unless the subject is most attractive when five or ten minutes is not grumbled at.

I will write to you again when I know what day is fixed.

If any question arises regarding the report I shall be glad to answer it.

Asking your forgiveness for intruding the first sheet upon you

I am | Yours most truly | H. Bence Jones.

RI MS JT/1/TYP/2/669
LT Transcript Only

---

1. *30 Grosvenor . . . 1850*: LT has added 'Addressed to Dr Tyndall. F. R. S. | Queenwood College | Stockbridge' on the top of the transcript, most certainly from the envelope.

2. *first sheet on molecular actions*: a portion of Liebig's 'Annual Report' for Tyndall to translate (see letters 0671 and 0683).

3. *I have written to Mr Barlow*: Barlow also wrote to Tyndall directly (letter 0675).

4. *great electro magnet*: the large electromagnet housed at the RI that Faraday used to discover diamagnetism (see letter 0675).

5. *The lecture*: the Friday Evening Discourse at the RI on 11 February 1853, where Tyndall gave a lecture entitled, 'On the Influence of Material Aggregation Upon the Manifestation of Force'.

From John Barlow   21 October 1852   0675

Royal Institution | Albemarle S^t | oct^r 21/52

My dear Sir

Our friend Dr Bence Jones has given us hopes that you may be prevailed on to be one of those to whose free kindness we are indebted for our Friday evenings.[1]

The subject which you are likely to take is an excellent one for our audience and we have the same electro-magnet with which Dr Faraday rotated a ray of light & established diamagnetism. —It is half the 'mooring' of an Indiaman[2]

Perhaps you would tell me at your convenience whether a Friday in Feby or March would suit you—

Ever truly yours | John Barlow
J. Tyndall E^[sq] [3]

RI MS JT/1/B/21

---

1. *Friday evenings*: see letter 0674, n. 5.

2. *Indiaman*: a ship engaged in trade with India, or the East or West Indies; *spec.* a large trading ship belonging to the East India Company (*OED*).

3. *E^[sq]*: this semi-illegible squiggle probably stands for "Esquire".

To William Francis   [22–23 October 1852][1]   0676

My Dear Francis.

Thanks for the Pogg*[s]* &c.[2] which reached me last night.

I now return the proofs of Bunsen;[3] the translation appears to read well—I have not meddled with your queries but have added one or two of my own—If I remember aright there is a long list of errata in the next no. of Pogg. probably these have been attended to

I send you the 1st of Helmholz's papers.[4]

I send you the translation of Clausius's note.[5]

Your arrangement is the best that could be devised—

If the Government of the Heavens[6] be not something you are particularly anxious about then pray dont send it to me for by Heavens my hands are very full as it is.

I wish you would be kind enough to send me 10 or 12 prospectuses of the Memoirs[7] when you next write.

Could you manage to strike off a copy of Bunsen for me—you send the magazine to Heidelberg[8]—could you not enclose a copy of Bunsen's paper to him? It is a splendid one.

most sincerely yours | J Tyndall

Saturday

If it would not be too much trouble I think it would have a good effect if you enclosed a copy of each translation to the author of the paper—it might be sent unstitched

RDS 27/18

1. *[22–23 October 1852]*: dated by Tyndall's Journal entry of 24 October 1852 (covering 18–24 October) where he wrote, 'read proofs Bunsen on Iceland' (JT/2/13b/588). The postscript, written on a Saturday, was probably written on 23 October, while the first part was likely written on Friday 22 October. This date is consistent with the allusions to papers by Helmholtz and Clausius.

2. *Pogg[s] &c*: requested in letter 0672.

3. *the proofs of Bunsen*: Bunsen's long memoir on the volcanic rocks of Iceland (cited letter 0666, n. 1).

4. *the 1ˢᵗ of Helmholz's papers*: probably the article on Brewster cited in letter 0673, n. 15.

5. *I send you the translation of Clausius's note*: probably R. Clausius, 'On the Colours of a Jet of Steam and of the Atmosphere', *Phil. Mag.* 4, no. 27 (December 1852), pp. 416–17.

6. *the Government of the Heavens*: a reference to J. Purslo, *The Government of the Heavens* (Edinburgh: Adam & Charles Black, 1852). Tyndall eventually sent a review of the work to Francis in December (see letters 0692 and 0693).

7. *the Memoirs*: see letter 0656, n. 4.

8. *Heidelberg*: Bunsen became a professor at Heidelberg in late 1852.

To William Francis [c. 23–4 October 1852][1] 0677

My dear Francis

This translation[2] smacks strongly of its German origin—it reads well nevertheless, and the more I consider the paper the more I am struck with its *[beauties]*.

I hope you have a good many Geologists on your list. This paper must be circulated among them.

Appropos. Set down the name of M[r] Geo. Edmondson *[as]* a subscriber to the Memoirs.

I have not heard a *[word]* regarding Galway—perhaps they intend it to assume the mythical character ascribed by Huxley to Toronto.[3] —Remember me kindly to Ronalds when you write—

I had a kind note yesterday from M[r] Barlow Sec. R.I.[4] requesting me to name a day in February for my lecture—I will take the 1st Friday.

I trust M[r] Taylor is getting on well.

very Sincerely yours | J Tyndall

What do you intend to do with Knoblauch?[5] defer him to the second part?

RDS 27/17

---

1. *[c. 23–4 October 1852]*: based on reference to Barlow's letter (letter 0675), which Tyndall would have received on either 22 or 23 October.

2. *This translation*: Tyndall was translating Helmholtz, Brewster and Bunsen on Iceland at this time (Journal, 24 October, JT/2/13b/588). The reference to 'Geologists' suggests it may have been Bunsen's paper on Iceland (cited letter 0666, n. 1).

3. *by Huxley to Toronto*: neither Tyndall nor Huxley received professorships at Toronto. The process took so long that Huxley began 'to think that the whole affair University and all, is a myth' (letter 0627).

4. *note from M[r] Barlow Sec. R.I.*: letter 0675.

5. *Knoblauch*: probably the translation for *Scientific Memoirs* of the paper by Knoblauch mentioned in letter 0651 (cited n. 8).

From Thomas Archer Hirst   24 October 1852   0678

Mittel Strasse N° 5.- | Berlin- | Oct. 24th / 52

My dear John

Owing to some mistake I received the letters of Introduction[1] only yesterday and immediately made use of two Dove and Magnus Du Bois is not at home, and Reiss I have posponed. You know how much I value letters of introduction, it would be superfluous to thank you, I should as soon think of thanking myself, though I felt far from indifferent towards this new proof that our interests are common—As you seem to be in doubt what my precise object is here[2] I will say a word about it. It is almost entirely mathematical and Physico-Mathematical Lejeune Dirichlet and Steiner[3] here almost alone give me stuff enough to work over, they are two such men as I shall find no where else and Mathematics as ever remains steadily my object, and I find it all satisfying Physics is of no value much unless I can experiment myself; this I could do here I know, but not without sacrificing Mathematics, which I will not—If I have any 'pure force' in me I suppose in its own good time it will manifest itself, sufficient for the day is the duty thereof, If I am not some day a Mathematician it shall be because said force is not there—

Lejeune Dirichlet is a fine fellow, I made his acquaintance immediately on arriving in Berlin. He is an able and at the same time a hearty, good natured man, we suit each other well, I often go smoke a cigar with him and some-times spend an evening with he and Mrs Dirichlet;[4] she by no means suits me so well, she is (or has been) a handsome woman, as intelligent perhaps as I have met, speaks remarkably good English, and has a good <u>acquired</u> taste for literature Nevertheless she is too <u>prickly</u>, too intellectual, for an evening's 'tounge-fence'[5] she is as interesting a woman as one can readily meet, but one could never fall in love with her, all that gentleness, that maternity: that—you know all about it—well all <u>that</u> fails—She is accustomed to being praised, she assumes a kind of dignity, and as she feels that she cannot get it from me, but that she must rub against me; she I fancy takes some interest in me, I am a kind of experiment for her—you understand me? How different is Dirichlet himself, his honest, warm hearted treatment of me has done my heart good. I had scarcely been with him half an hour before I felt what kind of a rare man I had to deal with; he never meets me without crying, 'Nun wann sehen wir uns wieder?'[6] He is lecturing on 'Die Theorie der Kräfte, welche im umgekehrten Verhältniss des Quadrates der Entfernung wirken, mit Anwendungen auf Electricität, Magnetismus und die Bestimmung der Gestalt der Erde'[7] Moreover I have attacked Latin once more very fiercely, for my study it is absolutely necessary, I can already with difficulty read Newtons Principia for myself. Dirichlet sent the son of the late celebrated Mathema-tician Jacobi to give me lessons in Latin, he himself I regret to say is a jurist for he is a fine young fellow: of Steiner I must speak another time. Another fine mathematician here was Eisenstein,[8] his works have principally appeared in Crelle's Journal[9] and had been praised by all, I expected great satisfaction from his acquaintance One morning I set off to see him took Dickinson with me thinking he would be the likeliest to give him private lessons. I asked the maid who opened us the door if Dr Eisenstein was at home when she gave me the following startling answer 'Dr Eisenstein liegt jetzt als Leiche'[10] It was indeed true, consumption[11] had carried him suddenly off, at the very opening of a brilliant career he had to die.

So careful of the type she seems,
So careless of the single life:[12]

A letter[13] from poor little Booth yesterday completely upset me, I have written three or four letters[14] to day making all arrangements in my power for him and since then it seems as if a weight had fallen from my shoulders. (see next page)[15]

<*Handwritten page missing; next portion of letter is LT Transcript Only*>

He has been continually growing weaker and weaker, yet with a heroism I have never before seen he has denied all help and gone regularly to his daily work to earn his and his poor mother's bread, until now though he still continues he can hardly crawl home. He has written me a heart-wringing letter,[16] there is not one attempt at effect in it—nay, he has hidden much; it is a tale plainly and gently told, yet with a terrible earnestness in every line. You shall judge for yourself. 'I can measure its (his disease) progress[17] from week to week by my increasing difficulty of breathing, and general physical weakness. I am still able to go to my work at Dean Clough,[18] but I feel that the time is coming,—<u>and soon</u>—when I shall be compelled to give it up. I know what you will say, my dear Friend; that I ought to have given it up before now. But I think when you consider it rightly, that you will not condemn the course I have taken. Did I see a chance of recovering my health by so doing, I would stay at home from this moment, and apply to you for the assistance you so freely and generously offer: for I could then live on in the hope that at a future time I might be able in some shape or other to return your kindness, or at least to <u>imitate</u> it when the opportunity came, and thus repay what I might consider as a debt I owed to humanity. But I have given up all hope of getting better, and <u>such</u> being the case I consider it my duty to work on <u>as long as ever I can</u>, and then if it be my fate at last to live a parasite on my friends, I shall at least have the consolation of knowing that so long as my power to do so lasted, I continued to earn my right to a place in the world, by contributing to it my share of labour. But as I said, I feel now and then depressed in spirits, and my resolution wavers a little. Sometimes too I indulge in delusive hopes of recovery, especially when I look at my dear mother with her broken health and increasing infirmities which—poor soul!—she vainly strives to conceal from me. It is frightful to think I shall have to leave her—her, whose existence has been enfolded with mine till I have become to her as a part of herself. . . . . . Roby, from whom I could not, if I wished, conceal my circumstances and state of health, has repeatedly urged me to stay at home, and has kindly offered me assistance as you have done. I told him how I had promised to apply to you if ever I really needed it, but that in such case of necessity I should not refuse it from either. I confess to you, Tom, that I should feel less hesitation in receiving it from <u>you</u> than him. But I cannot give my reason here, for I am only a bungler in giving expression to my thoughts, and on this point above all others <u>I would not be misunderstood</u>. It is not however that I should have anything to say depreciatory of Roby. That his friendship for me through a long series of years has ever been of the warmest and most disinterested kind I have had ample proof, and I hope that I know how to value it—Instead of being gloomy and depressed I ought rather to feel joyously thankful that God has given me two <u>such</u> friends as you and he. May His blessing rest upon you both!'

Such a letter does not leave one word for me to tell you; I have written back as I said, and made arrangements that what he requires in the way of money be regularly sent him. Roby tells me how on his road home he has had to rest on the wet grass for fatigue. I have told him to stop at home from the moment of receiving my letter and I know he will obey me. As far as little personal attentions are concerned I know Roby will look after him well. Brave little fellow! As long as there is a breath in him he shall not want, and if there be yet hope for him he shall try it and to do that shall have every external care removed for him. For the present I have put it into the hands of an uncle of mine[19] in the neighbourhood who I know will willingly advance me any sum. I have given him instructions simply to enclose £5 when I order him in an envelope. Thus it will cripple nobody at present, and it will be something for me to work for when the time comes. I have tried to realize his condition, staring death there so calmly, so bravely in the face, and crawling with unflinching determination into its very jaws. 'I will work till the last'. I declare, John, nature has made of me such a spoilt child that I found it almost impossible to realize such a grim reality.

And your letter[20] too, John, was a pure pleasure to me, it was long since I had had such a one from you, and as I was reading it the thought struck me 'Is it possible that a few practical letters from him had made me uneasy?[21] Silly fellow, I don't know what it was, nor I do not care. John is as strong and healthy as ever, and as long as he lives will be so'. Bravo, John, devote a Sunday now and then to the Gods and _______ [22] | Tom.

Grüsse Heinrich herzlich! Bald schreibe Ich ihm.[23]

<Handwritten letter resumes>

Dickinson lives with me in a little palace of a lodging he is a good warm hearted lad, honest as daylight and surprises me sometimes with unconscious, healthy deep-remarks. He will be no sham. —He desires to be remembered kindly—Write him a line some time.—

Dʳ Tyndall- | Queenwood College- | Stockbridge | <u>Hampshire</u>[24]

RI MS JT/1/H/174
RI MS JT/1/HTYP/214-216

1. *letters of Introduction*: see letter 0673.

2. *in doubt what my precise object is here*: see letter 0669, where Tyndall wrote of Hirst in his letter of introduction to du Bois Reymond that 'I dont know what he intends to do there'.

3. *Steiner*: see letter 0667, n. 4.

4. *Mrs Dirichlet*: wife of Peter Gustav Lejuene Dirichlet; otherwise not identified.

5. *'tounge-fence'*: an argument or debate (*OED*).

6. *'Nun wann sehen wir uns wieder?*: 'Now when shall we see each other again?' (German).

7. *'Die Theorie der Kräfte. . . der Erde'*: 'The theory of the forces which act in inverse ratio of the square of the distance with applications to electricity, magnetism, and the determination of the shape of the Earth'.

8. *his works*: Ferdinand Gotthold Eisenstein (1823–52) (see letter 0667, n. 5).

9. *Crelle's Journal*: Eisenstein published numerous mathematical papers in *Crelle's* Journal, including 23 in the year 1844 alone.

10. *'Dr Eisenstein liegt jetzt als Liche'*: 'Dr Eisenstein is now a corpse' (German).

11. *consumption*: Eisenstein died of consumption (tuberculosis) on 11 October 1852 in Berlin.

12. *So careful of the type she seems, So careless of the single life*: from Alfred Lord Tennyson's 'In Memoriam' LV, lines 7–8.

13. *A letter*: Hirst received a letter from Booth on 23 October (Hirst, 'Journals', 24 October 1852).

14. *three or four letters*: see ibid.

15. *(see next page)*: Tyndall's note to Hirst to turn to the next page of the manuscript, as Tyndall was stopping here to leave room for the address.

16. *letter*: see n. 13.

17. *its (his disease) progress*: Booth had been ill with consumption (tuberculosis).

18. *Dean Clough*: a group of large factory buildings built in the 1840s–60s for Crossley's Carpets in Halifax.

19. *an uncle of mine*: probably his uncle Allatt, who lived in Brighouse, 4 miles from Halifax (see letter 0619, nn. 6 and 8).

20. *letter*: probably letter 0673.

21. *made me uneasy?*: Hirst had expressed his disquiet in letter 0667.

22. ___: as Tyndall often wrote to Hirst on Sundays, this is probably a humorous request for Tyndall to write to Hirst. It could also be a purposeful omission by LT.

23. *Grüsse Heinrich herzlich! Bald schreibe Ich ihm:* 'Warm regards to Heinrich! I will write to him soon' (German).

24. *D^r . . . Hampshire*: the address is written on the back of the last extant page of the handwritten letter.

From George Wynne 29 October 1852 0679

Board of Trade | 29th. Oct. 1852

My dear Tyndall

I was sorry to see today in the paper[1] that the Professorship in the Galway College has been filled up, M[r] Stoney appears to have had Lord Rosse's[2] interest and that was too powerful to contend against, otherwise I think there would have been little doubt of your success. Your application with the testimonials brought forward will pave the way in the event of another vacancy occurring. M[rs] Wynne will be greatly disappointed when I tell her this evening on my return home of your having failed in Galway,

most sincerely yours | Geo Wynne

RI MS JT/1/T/555
RDS 27/33[3]

———

1. *the paper*: we can find no such news item, either in the *Times* (the Digital Archive) or any other London newspaper of this date.

2. *Lord Rosse*: William Parsons, 3rd Earl of Rosse (1800–67), Irish astronomer, who served as President of the RS from 1848–54. He built a large observatory and supporting workshop on the family estate.

3. *RDS 27/33*: original missing; transcription is from the copy Tyndall wrote out and sent to Hirst in letter 0681 and Francis in letter 0682.

From Robert Bunsen 30 October 1852 0680

Heidelberg, den 30 Okt. 1852

Herzlichen Dank, mein theuerster Freund, für Ihren freundlichen Brief, und die angenehmen Nachrichten die er mir bringt. Es macht mir die grösste Freude dass Ihnen meine Isländische Arbeit gefallen hat. Wenn etwas Verdienstliches daran ist, so ist es wohl nur das gewissenhafte Streben, darin nicht über das unmittelbare Ergebniss der Beobachtungen und Versuche hinauszugehen, was leider in der Geologie nur zu wenig befolgt wird. Dr. Streng, mein Assistent, hat eben eine kleine Arbeit drucken lassen über die Isländischen, /Faröischen/ und Ungarischen vulkanischen Gebirgsarten, und auch für diese das für Island aufgestellte Gesetz bestätigt gefunden. Ganz anders verhält es sich mit den erloschenen Vulkanen des linken Rheinufers, deren Untersuchung sehr interessant zu werden verspricht. Ich sammle in dem chemischen Practicum noch immer Materialien zu Untersuchungen in der angedeuteten Richtung, obgleich ich selbst gegenwärtig mit einer ganz andern Arbeit beschäftigt bin, deren ersten Theil ich eben zum Druck redigire. Ich habe nehmlich ein sehr merkwürdiges Verwandtschaftsgesetz aufgefunden, das ein weites Feld für neue Experimentaluntersuchungen eröffnet und das mich gewiss noch über Jahresfrist beschäftigen wird. Doch <dazu> später einmal ausführlicher. Vor Allem muss ich Ihnen sagen, wie sehr ich mich gefreut habe, dass Sie entschlossen sind, England nicht zu verlassen, und welchen herzlichen Antheil ich an der Auszeichnung genommen habe, welche Ihnen von der Royal Society, als so wohl verdient, zu theil geworden ist. Möchte ich nicht bald die Freude haben, Ihnen das hier mündlich in unserm schönen Heidelberg wiederholen zu können!

In aufrichtiger Freundschaft | Ihr | R.W. Bunsen.

Heidelberg,[1] 30 Oct. 1852

Many thanks, my dearest friend, for your kind letter[2] and the pleasant news which it brings me. It gives me the greatest pleasure that you liked my Icelandic work.[3] If there is anything meritorious about it, then it is probably just the conscientious striving not to go in it beyond the immediate result of observations and experiments, which unfortunately is done in geology only too rarely. Dr Streng,[4] my assistant, has just had a small work printed on the Icelandic, /Faroese/, and Hungarian volcanic mountain types,[5] and has found the rule proposed for Iceland to be confirmed for these too. The situation is quite different with the extinct volcanoes of the left bank of the Rhine, the investigation of which promises to become very interesting. I am still collecting material in my chemistry practical course for investigations in the direction I have hinted at, although I myself am currently busy with a quite different work, the first part of which I am now editing for print. I have discovered, namely, a very peculiar relationship rule which opens a wide field for new experimental investigations and which will surely occupy me for over a year. But more details <about this> later. Above all, I must tell you how very happy I was that you have decided not to leave England, and what heartfelt interest I have taken in the honour, so well deserved, which has been bestowed upon you by the Royal Society.[6] Might I not have the pleasure soon of being able to repeat all this here to you in person in our lovely Heidelberg!

In sincere friendship | Your | R.W. Bunsen.

RI MS JT/1/B/143
LT Transcript Only

<hr>

1. *Heidelberg*: in 1852, Bunsen succeeded Leopold Gmelin at the University of Heidelberg.

2. *kind letter*: letter missing, but referred to in Tyndall's journal entry for 24 October 1852 (JT/2/13b/588).

3. *Icelandic work*: see letter 0526, n. 2 for the original paper and letter 0666, n. 1 for the translation.

4. Dr Streng: Johann August Streng (1830–97), a German mineralogist who began his career working with Bunsen at the University of Breslau before becoming a teacher at Clausthal in 1853, and then Professor of Mineralogy at Giessen from 1867 to 1895.

5. *a small work . . . mountain types*: not identified.

6. *the honour . . . the Royal Society*: Tyndall was elected FRS in June 1852.

To Thomas Archer Hirst   [30–31 October 1852][1]   0681

My dear Tom

The following has just reached me.

[. . .][2]

Thus this effort has exploded; immediately after the intelligence of my failure came your letter.[3] It seemed to me worth half a dozen such professorships. Poor Booth, he must not want. It would be a sad thing if two or three hearty fellows like you Roby and myself are not able to keep such a brave little fellow aloft. I will write to Booth myself, and would have extended my hand to him long ago were it not held back by the feeling that I ought first to clear off my own scores. However I shall be able to do both by God's help. Friendship and community of feeling would be all vapid hypocrisy and worthless winds if it could not step in in a case like the present.

A gentleman[4] has just dropped in from London and wants me to go to Salisbury with him—this smashes my arrangements for the day, for I will go with him. I met him but once before and was so pleased with him that he is now able to draw me from my writing table and convert me into a tourist. I am glad to hear your account of Dickinson and will undoubtedly write to him as you desire. Your account of Dirichlet[5] pleased me mightily—I met him once at Riess's—Riess is also a good and kindly fellow as you will find. I must now stop for I have two or three letters[6] to write before starting.

as ever | John

RI MS JT/1/T/555

1. *[30–31 October 1852]*: based on the date of letter 0679 from Wynne (n. 2), Hirst's letter (n. 3) and Tyndall's trip to Salisbury (n. 4), this was probably begun on the 30th and finished on the 31st.

2. *[. . .]*: Tyndall copied letter 0679 in full here.

3. *your letter*: letter 0678.

4. *A gentleman*: James Bevington. Tyndall described his trip to Salisbury (on Sunday, 31 October), and his affection for Bevington in his journal entry for 19 November (his first entry since 24 October) (JT/2/13b/589).

5. *account of Dickinson . . . Dirichlet*: see letter 0678.

6. *two or three letters*: one of these is probably letter 0682 to Francis.

To William Francis   [31 October 1852][1]   0682

My dear Francis,

Once more defeated but yet not dead. I received the following yesterday. [. . .][2]

Thus the vision has exploded! well I think I shall be able to bear up against it.

Sincerely yours | J Tyndall

RDS 27/33

---

1. *[31 October 1853]*: based on the date of the enclosed letter (0679) from Wynne, which Tyndall would have received on the 30th.

2. *[. . .]*: Tyndall copied letter 0679 in full here.

From Henry Bence Jones   6 November 1852   0683

My dear Sir

Many thanks for the MS[1] which arrived safely this morning The degrees are left centigrade & a table for transforming them has been given. The references you have rightly given as to numbers, but it is unnecessary to add the foot notes. The printer takes these from the German text which I shall be glad if you will return me <at> your earliest convenience.

The proofs shall be sent to you.[2]

I shewed D[r] Hofmann your manuscript he had never seen any thing like it. The delight of the printer will not be less than that of D[r] Hofmann.

With many thanks for so soon sending us copy

I am | Yours very truly | H Bence Jones

Nov 6. 1852 | 30 Grosvenor Street

RI MS JT/1/J/35

1. *MS*: Manuscript. Tyndall translated and edited the physics portion of Liebig and Kopp's *Annual Report* (cited letter 0671, n. 1).

2. *The proofs . . . to you*: in letter 0674 Bence Jones had told Tyndall that he did not need to check the proofs unless he wished to do so.

To William Francis   10 November 1852   0684

Queenwood 10th Nov. 1852

My dear Francis.

I am a post late and the reason is that your letter[1] found me in bed to which I have been confined all day.[2] With regard to Galway[3] the matter makes very little impression on me. Such an event is no new thing and therefore one can contemplate it without surprise—I rejoice to learn that Huxley has been so distinguished[4]—never mind both of us will yet compel the world to fund our bread and butter.[5] You did not send me his letter as you intended.

I return Helmholtz[6]—It is as you say an exceedingly interesting paper and will be read in England with avidity—My first idea was to preface it by a few words of introduction but this would compel you to alter your present arrangement and it is not worth while to do so. I am working at the second paper.

I send you Reuben's abstract.[7] My opinion is that if he permitted his theories of thunder clouds & so forth to take less hold of him and confined himself more to the facts of the case it would be much better, I feel myself compelled to state once more my belief that the new view expressed in the paragraph page 2 to which I have attached the x is at variance with the fundamental laws of electric action. I do not think the abstract in its present state worth publishing—Is it possible that Reuben cannot be got to state clearly what his results are with oil & soap and water?

I return you also my own paper[8] where the letters occur in pair such as mm' bb' &c. &c. a space or a comma ought to be placed between them—If you cast your eye on the figure you will see that they refer to distinct things, whereas bb' &c would appear as intended to point out a single thing—is it not so? b is one thing b' is another. I also return the paper of Clausius.[9] I have tacked my name to the paper of Helmholtz[10] It will explain the 'J.T.' attached to one of the notes.[11]

I wish I had something better than sympathy to offer for poor M[r] Taylor—it is a sad affair[12]—But you must support yourself in the conviction that you have left no means untried which seemed likely to promote his cure

ever yours | J Tyndall

StBPL T&F, Authors' letters

---

1. *your letter*: letter missing.
2. *found me in bed . . . confined all day*: Tyndall wrote in his journal on 19 November 1852: '. . . my health has not been strong for some time, 10 days ago I was suddenly struck down by sciatica and confined to bed for an entire day, and even at the present moment I feel residual twinges of the pain' (JT/2/13b/589).
3. *Galway*: Tyndall told Francis he had been unsuccessful at attaining the professorship at Galway in letter 0682. In his missing response Francis must have responded to this news.
4. *so distinguished*: probably refers to Huxley's receipt of the Royal Society's Royal Medal in 1852.
5. *fund our bread and butter*: a reference to Huxley's and Tyndall's unsuccessful attempts to gain professorships.
6. *interesting paper*: probably Helmholtz's 'On Sir David Brewster's New Analysis of Solar Light' (see letter 0672; cited 0673, n. 15). The second paper is 'On the Theory of Compound Colors' (cited ibid.).
7. *Reuben's abstract*: R. Phillips, 'On the Colours of a Jet of Steam and of the Atmosphere', *Phil. Mag.* 5, no. 29 (January, 1853), pp. 28–30. See also letter 0666, n. 7.
8. *my own paper*: see letter 0686. The figures referred to appear between pp. 420 and 421.
9. *paper of Clausius*: perhaps for the *Scientific Memoirs* (cited letter 0664, n. 3). Tyndall also translated a 'note' by Clausius (cited letter 0676, n. 5) for the *Phil. Mag.* during this time period.
10. *paper of Helmholtz*: Tyndall placed 'Communicated by Dr. Tyndall' as the first footnote (after Helmholtz's name) to both 'On Sir David Brewster's *New Analysis of Solar Light*' and 'On the Theory of Compound Colors'.
11. *the 'J.T.' attached to one of the notes*: Tyndall attached a footnote concerning experiments he had performed with coloured wafers to both of Helmholtz's papers: on p. 412 of 'On Sir David Brewster's *New Analysis of Solar Light*', and p. 531 of 'On the Theory of Compound Colors'.
12. *a sad affair*: see letter 0664, n. 10.

To the Editors of the *Phil. Mag.*   [10] November 1852[1]   0685

GENTLEMEN,

IN an abstract of Professor William Thomson's Mechanical Theory of Thermo-electric Currents, given in your Supplementary Number for July, reference is made to the well-known experiment of Peltier on the absorption of heat at a bismuth and antimony joint. This has drawn from Mr. Adie a brief communication, published in your Number for September, from which is appears that the writer has never been able to obtain Peltier's result; he virtually denies its existence, and affirms the true state of the case to be that *less* heat is developed at some junctions than at others, but that *cold* is never generated. An objection precisely similar to that now urged by Mr. Adie induced Lenz to repeat the experiment fifteen years ago.* To the experiment of Lenz I took the liberty of drawing Mr. Adie's attention in your October Number; I did so because Mr. Adie had never mentioned it in his remarks, and it seemed to me to offer a proof of the absorption of heat so obvious as to be immediately appreciated. It does not however appear so to Mr. Adie, for in your last Number I find that he suggests a hygrometric action as the probable cause of the diminution of temperature observed by Lenz. I shall ill occupy your space were I to dwell upon conjectures where the 'law and testimony' of experiment are so near at hand, and fact so readily attainable. If the following results do not convince Mr. Adie, they will perhaps be the means of clearing away whatever doubt his remarks may have created in the minds of others.

Experiment No. 1.-In Plate IV.[2] Fig. 1, A is a bar of antimony, B a bar of bismuth, both bars being brought into close contact at J. To the free ends of the bars the wires $w\ w'$ are soldered, and dip into the little pools of mercury $m\ m'$; c is a piece of cork through which the wires pass, and by taking which in the fingers the wires $w\ w'$ may be easily moved from the pools $m\ m'$ to $m\ m''$, the warming of the wires being prevented by the cork. From $m\ m''$ wires proceed to a galvanometer, G, whose needles prove themselves to be perfectly astatic by setting at right angles to the magnetic meridian.† B is a single cell of Bunsen, from which, when matters stand as in the figure, a current can be sent through the bismuth and antimony pair.

The voltaic circuit having been established, the current-a very feeble one- was permitted to circulate for two minutes, its direction being from antimony to bismuth across the junction ; at the end of the time specified the wires $w\ w'$ were moved from $m\ m'$ to $m\ m''$, a thermo-circuit being thus formed in which the galvanometer was included ; the index of the instrument was at once deflected, and the extreme limit of its first impulsion was noted ; it amount to

75°.

The deflection in this case was similar in direction to that produced when the warm finger was placed upon the junction.

The wires $w\ w'$ were moved back to their former position, and the apparatus was suffered to cool; by crossing the wires $b\ b'$, causing the former to dip into m and the latter into m', the voltaic current was reserved, its direction across the junction being now from bismuth to antimony; the same time

---

*   Poggendorff's Annalen, vol. xliv, p. 342
†   For an explanation of this, see an abstract of Du Bois Reymond's Researches on Animal Electricity, edited by Dr. Bence Jones.

of circulation being allowed, on establishing the thermo-circuit, as before, a deflection of

68°

was observed. The deflection was the same as that produced when a small glass containing a *freezing mixture* was placed upon the junction.

But Mr. Adie will probably urge, that it is not the cold developed at J, but the heat developed at some of the other points, which caused the deflection here. I will not pause to discuss the objection, but will proceed to an experiment which deprives it of all force.

Experiment No. 2.-AA′ is a bar of antimony, BB′ is a bar of bismuth east as in fig. 2, and in contact at the centre. From the cell B a current was sent through the system, and during its circulation the ends gg' were unconnected; neither heating nor cooling of these ends by the current was therefore possible. The direction of the current across the junction was first from antimony to bismuth. After a short period of circulation the current was interrupted, and the ends of the wires $w\,w'$ were dipped into the mercury cups gg', which were also in contact with A′B′; the index was driven through an arc of

40°.

The sense of the deflection in this case showed that the junction had been *heated*.

The current was reversed, its direction across the junction being now from bismuth to antimony; proceeding as before, the deflection was

30°.

The sense of this deflection was the same as that produced when the temperature of the junction was *lowered* by a freezing mixture.

I see no escape here from the conclusion that heat has been absorbed; for the ends gg', exposed as they are to the atmosphere, must have its temperature, while the ends $m\,m'$, on which suspicion might reasonably rest, the current having passed through them, are wholly excluded from the thermo-circuit. The reader will observe that this is merely a modification of Lenz's experiment with the metallic cross.

But Mr. Adie has tried the cross, and it does not satisfy him; very well, we will discard it, and proceed at once to an *experimentum crucis*. If the arms A′ B′ are not actually included in the voltaic circuit, they may seem to be in suspicious connexion with it. We must remove this source of doubt.

Experiment No. 3.-A and B, fig. 3, represent, as before, the bismuth and antimony couple, united at one end. M is a small chamber, hollowed out in a piece of cork and filled with mercury. A′ B′ is a second delicate thermo-electric pair, connected with the galvanometer, but wholly unconnected with A B. The wires $w\,w'$ are sufficiently strong to support A′ B′, so that the junction stands vertically over M, a slight pressure being sufficient to cause the wedge-shaped end of the pair to descend into the chamber of mercury. The whole arrangement was permitted to remain in a room until the temperature of the surrounding atmosphere was attained. Matters being in this state, when the pair A′ B′, which I will call the *test-pair*, was dipped into the mercury M, no effect was produced on the galvanometer. Now the mercury must partake of the changes of temperature of the junction with which it is in contact, and the nature of these changes will be ascertained with great precision by examining the mercury at proper intervals by means of the test-pair.

The voltaic circuit was closed, and the current allowed to circulate for three minutes, passing in the first place from bismuth to antimony. The current was then interrupted, and the test-pair was immediately dipped into the pool of mercury; the index of the galvanometer was driven through an arc of

40°.

The deflection was similar to that produced by immersing the end of the test-pair in a freezing mixture. Hence in this case heat was undoubtedly *abstracted from the mercury* during the passage of the current.

The apparatus being permitted to resume its equilibrium, the voltaic current was caused to traverse AB in an opposite direction. At the end of three minutes the test-pair was again immersed, and a deflection of

45°

was the consequence. The deflection was opposed to the former one, and demonstrated the *generation of heat* at the junction.

I am at present unable to see what possible objection can be brought against this last experiment. A hygrometric effect is out of the question; and the test-pair A′ B′ being wholly unconnected with the voltaic current, cannot in any way be influenced by the latter. The results observed are evidently pure effects of the heating and cooling of the junction.

It will perhaps be permitted me to cite a single additional experiment, which exhibits all the necessary evidence without the reversion of the voltaic current.

Experiment No. 4.-B, fig. 4, is a curved bar of bismuth, with each end of which a bar of antimony, A, is brought into close contact. In front of the two junctions are chambers, hollowed out in cork and filled with mercury as before. A current was sent from the cell B in the direction indicated by the arrow; at M it massed from antimony to bismuth, and at M′ from bismuth to antimony. Now if Peltier's observation be correct, we ought to have the mercury at M warmed, and that at M′ cooled by the passage of the current. After three minutes' circulation the voltaic circuit was broken, and the test-pair dipped into M; the consequent deflection was

38°,

and the sense of the deflection proved that at M′ head had been *absorbed*.

The needles were brought quickly to rest at zero, and the test-pair was dipped into M; the consequent deflection was

60°;

the sense of the deflection proved that a M heat had been *generated*.

The system of bars represented in fig. 4, being imbedded in wood, the junction at M cooled slowly, and would have taken a quarter of an hour at least to assume the temperature of the atmosphere. The voltaic current was reversed, and three minutes' action not only absorbed all the head at M, but generated cold sufficient to drive the needle through an arc of 20° on the negative side of zero.

These experiments, Gentlemen, corroborate a result which to my mind is sufficiently well established without them. Nevertheless I would say, that the conclusions of Mr. Adie are such as a restricted examination of the subject will most probably lead to. I have no doubt as to the correctness of his results described in the September Number of the Magazine; but I have just as little doubt, that had Mr. Adie varied the strength of his current sufficiently, he would have spared himself the statement, that 'in his experiments he had never met a fact which in the least encourages the view that electricity reduces temperatures'.

I remain, Gentlemen, | Your obedient Servant, | JOHN TYNDALL. Queenwood College, | November 1852.

From the published letter, Article LXVI. 'On the Reduction of Temperatures by Electricity [with a plate]', Phil. Mag., 4:27 (1852), pp. 419–23.

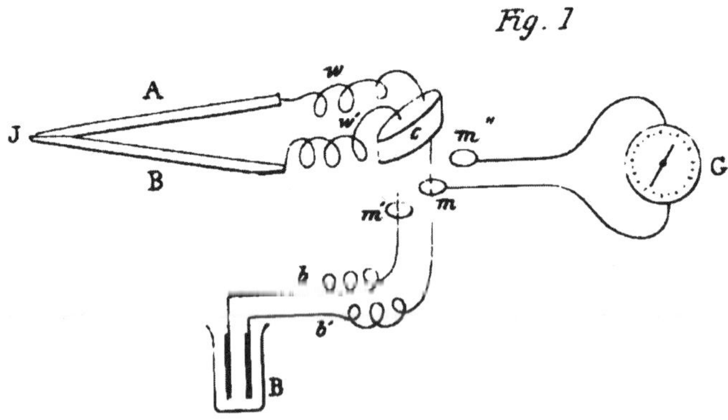

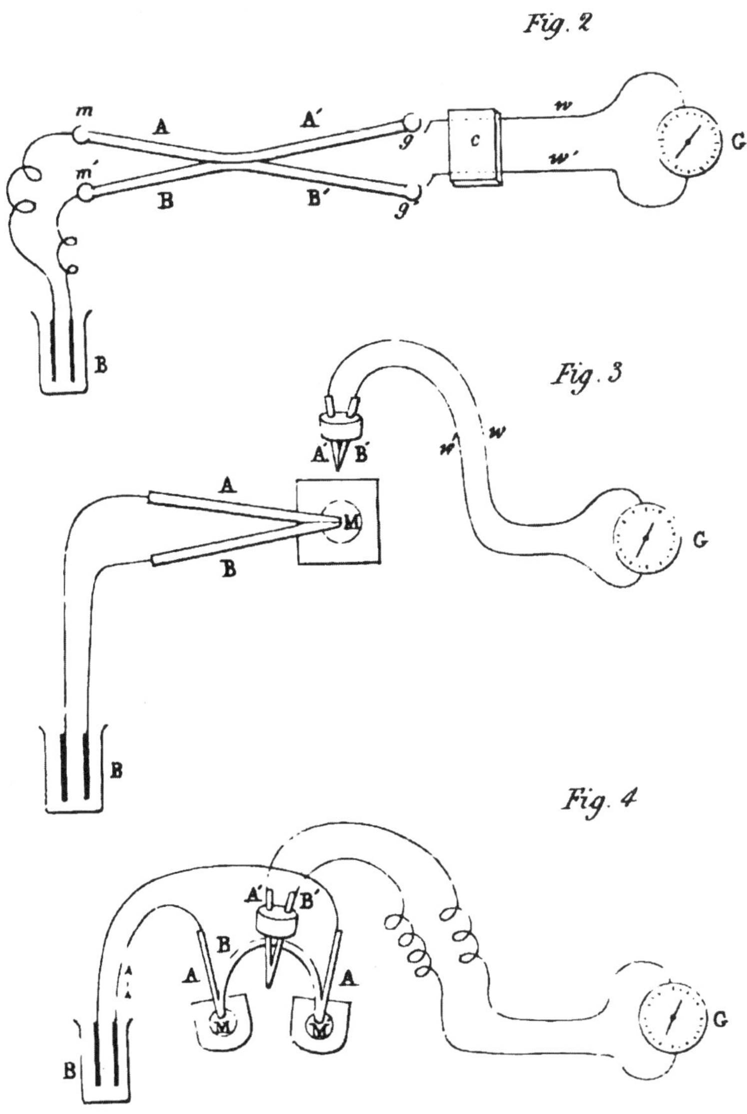

---

1. *[10] November 1852*: the initial draft of the letter was finished in the week before 24 October (Journal, JT/MS/2/13b/588); the final copy was returned to Francis on 10 November (letter 0684).

2. *plate IV*: a plate (no. 4) with four figures appeared in the published version of the letter, which we have placed at the end of the letter.

From Thomas Archer Hirst   14 November 1852   0686

Berlin | Mittel Strasse, N° 5 | <u>Nov 14th 1852</u>

My dear John

In my last letter[1] little Booths circumstances had so preoccupied me that I forgot what was my original object in writing which I will therefore now mention first. The 'Thermo-Saule'[2] will cost 20 thalers, the moment it is ready (Mr Kleiner is no man of his word)[3] I shall send it by post unless you can tell me a better way. I received notice of your failure[4] and on your account was disappointed on my own I must confess not so much so, for I have now advantages before me that I shall not readily find again,[5] and above all I am treated by all with the greatest kindness. I have seen all the great guns now except Bois Reymond who is not yet returned. I have heard from Booth, as I <u>felt</u> he has accepted my proposal[6] and remains at home; as far as distance will allow I will nurse him henceforth as if he were my brother. Some day you shall read the letter he has sent me. Meanwhile from the enclosed one[7] from Roby you will learn how the matter stands with respect to giving him money. I cannot accept Roby's proposal however If you have not offered him money or sent him it, <u>do not do so</u>.[8] you will know why. One thing I will accept from you however, namely if owing to squandering so much money in travelling I should be for a time straitened and thus the regularity of Booths receipts endangered you will help me out with a few pounds I do not fear this however, at any rate for the next quarter of a year all is already provided for and I am yet in full pocket at the end of that time—we shall see. One ought always to consider extreme cases I was meditating the other day on looking out for a situation, but I threw up the notion very soon, it is clearly my duty to stop here some time yet, and after that to spend a short time in Paris. Now suppose John, that by some accident (you understand, at present it is not in the least visible) my income should be cut short, I should not hesitate to ask thee to help me to complete my studies, here and in Paris. The possibility of such an accident I think I once told you. My Brother William is farming with £400 of mine for which he pays me interest as far as I can hear he is 'getting on' pretty well just now; he <u>might</u> however be unsuccessful and then I should be a little more cramped. Keep thy self in readiness therefore oh John to help thy big brother[9] when he becomes poor. And at present leave little Booth to me, for it is a source of joy to me that he will receive it from me. I have seen Riess and can endorse your opinion of him. He has opened his doors most hospitably to me at all times & I shall occasionally take advantage of his kindness. Meanwhile Dirichlet and I pull well together. I have written today such a mass of letters that I am used up so Good-bye.

Yours affectionately | <u>TA Hirst</u>
Dr Tyndall. | Queenwood College | Stockbridge | <u>Hampshire</u>

RI MS JT/1/H/175

---

1. *last letter*: letter 0678.
2. *Thermo-Saule*: see letter 0638, n. 9.
3. *Mr Kleiner . . . his word*: in early October, Kleiner had promised that the instrument would be finished in two weeks (see letter 0667). Hirst reported it ready in late November (letter 0689).
4. *your failure*: to acquire the position at Queen's College in Galway.
5. *on my own account . . . find again*: Hirst had considered returning from the Continent and taking Tyndall's place at Queenwood should Tyndall acquire the position in Galway; see letter 0658.
6. *my proposal*: the proposal concerned money for Booth, who was becoming increasingly ill. See n. 7 below, and letters 0678 and 0681.
7. *enclosed one*: enclosed letter missing.
8. *If you have not . . . <u>do not do so</u>*: Tyndall had written to Booth (letter missing) offering financial assistance the day before receiving this letter. In his journal entry for 23 November he wrote: 'Devoted the whole of last Sunday to working for poor Booth and found the product of my labour worth two pounds' (JT/2/13b/589-90). Booth rebuffed his offer (see letter 0691).
9. *big brother*: an allusion to Hirst's height (see 0398, n. 31).

To William Francis   [mid-October–mid-November 1852][1]   0687

My dear Francis.

I have 3 little commissions which I should feel greatly obliged to you if you would execute for me.

> To send me the <u>2nd</u> course of Lardner's Handbook of Nat. Phil.[2] published by Taylor & Walton—containing—Heat. Electricity Magnetism. Astronomy.

> 2nd. To purchase or borrow for me a prism of crown or plate glass—a right angled <u>isosceles</u> prism—having its hypotenuse about 2 inches in length and its thickness about ¾ of an inch—

> To send me a piece of crystallised carbonate of iron. Say a quarter of a pound weight or more if you can get it.

as ever yours | J Tyndall

Monday
Sketch of prism

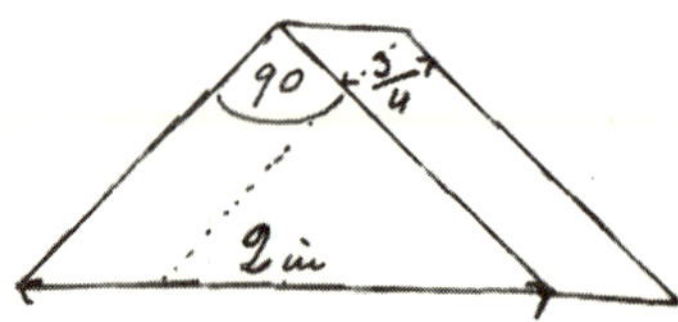

The thin vertical surfaces hypotenuse surface and two side surfaces must be polished—I dont care whether the bases are polished or not | JT

A single fully-developed crystal of carbonate of iron would be a great prize to me if I could procure it—The large mass for which I send is intended to be pulverized.

RDS 27/36

---

1. *[mid-October—mid-November 1852]*: date is uncertain; any time from the publication of Lardner (n. 2) in late September to mid-November is possible. We presume that Tyndall would have asked Francis to obtain a crystal in London prior to asking Hirst to send one from Berlin (in letter 0691 of 28 November). The latest possible Monday (before letter 0691) is 22 November.

2. *Lardner's Handbook of Nat. Phil.*: D. Lardner, *Handbook of Natural Philosophy and Astronomy, Second Course: Heat, Common Electricity, Magnetism, Voltaic Electricy* (London: Taylor, Walton, and Maberly, 1852).

To William Francis   [mid-November 1852][1]   0688

Friday,

My dear Francis.

I have adopted your plan for the first time and got a boy to write for me.[2] I hope the printer will be able to manage it—I have gone over every paper[3] carefully and made all necessary corrections—It is I think perfectly legible although possessing a harum scarum appearance. Were Helmholz never to do anything else his name will be remembered in connexion with these two papers.[4]

as ever yours | J Tyndall

I will now turn my thoughts on a paper for the Memoirs—Thanks for Huxley's letter[5] My 'Pluck' has not sunk the fraction of a degree—I will soon return the letter. I fear he stands a perilous chance of not being found among the 'private favorites' at the last day!

RDS 27/19

1. *[mid-November 1852]*: date is quite uncertain. Allusions to translations of Helmholtz papers (which were published in December) support a mid-November date, as does the reference to Tyndall's 'Pluck' (n. 5). Possible Fridays are 5, 12, 19, or 26 November.

2. *boy to write for me*: not identified, presumably a pupil at Queenwood. The boy had perhaps helped Tyndall make fair copies of his draft translations; their combined work could have given the translations a 'harum scarum' appearance.

3. *every paper*: probably the papers for inclusion in the December number of the *Phil. Mag.*

4. *these two papers*: although Helmholtz would forever be known for his 'On the Conservation of Force' published in the *Scientific Memoirs*, Tyndall is here likely referring to the two Helmholtz papers for the *Phil. Mag.* (cited letter 0673, n. 15). Tyndall referred to these as the 'first' and 'second' papers of Helmholtz in subsequent letters.

5. *Huxley's letter*: letter missing, but the allusions to Tyndall's 'Pluck' and to Huxley's 'perilous chance' of not being 'among the private favorites' suggest it dealt with Huxley's and Tyndall's unsuccessful bids for employment.

From Thomas Archer Hirst[1] 21 November 1852 0689

Mittel Strasse N° 5. | Berlin | Nov. 21<u>st</u> 1852

My dear John,

This morning I called upon du Bois who has only just returned. I liked the looks of him much he seems an open hearted energetic fellow. It seems he was with you in Belfast[2] he tells me you look pale and care worn, <u>I marked that</u> I know you seldom look fresh and handsome even when you are likeliest to do so nevertheless the remark has significance in it for both you and me—I am working hard here too and have to leave much undone in spite of it, but I am as strong as an ox, and can grapple with it somewhat better than I once could. My new acquaintances[3] here in Berlin will not allow me to grow a recluse and my companion Dickinson with his hearty <u>freshness</u> keeps me from all hypochondria. Thus all things considered I look upon the life here as healthy—About once a fortnight too on a Sunday evening I attend our only place of worship the theatre, and get my bowels shaken into order with a hearty laugh if I can—

What I write mainly for this time is to tell thee that the 'Thermo Saule' is ready.[4] Kleiner says it is safe to send it in a small box 8 inches square by Post addressed to you in Queenwood. I had however rather have your orders for doing so first 'Um es völlstandig zu machen sind drei Schirme *[dazu]*[5] You can either take these altogether or not; complete it will cost 25 thalers; Thermo Saule alone 20 thalers. Write immediately and give me orders. And now John stand aside and let me say a word to Heinrich.

Fortunately Heinrich I have mislaid your last letter and am consequently ignorant of the amount of my neglect towards you, my conscience however accuses me of its being considerable Luckily however we do not keep so strict a debtor and creditor account. I suppose and hope you are busy at your investigation I shall keep a look out in the Annalen[6] for a notice from you. John will have told you all I am doing here so there will be little need to repeat it. I may say however that Berlin fulfills my best anticipations. A great city when properly used has immense advantages, you can be more solitary there than in a small village and you have the satisfaction of knowing at the same time, that at a moments notice you can be in the busy world with all its bustle, its pleasures and its business. One greatest advantage however is the men you can meet. To look at a clever able fellow is somewhat contagious. To hear them converse on topics that require you to lift yourself up on to a certain platform in order to comprehend, is healthy. To sit at home and tackle hand to hand with your study is of course the greatest necessity; but from contact with other men our private difficulties seem to be made less, not actually but virtually so, by convincing us that daily greater are overcome. You are drawing near the close of another half year, how are you going to spend your Christmas? At your work as usual? Well if so I wish you a happy successful one

Yours most sincerely | T.A. Hirst.

RI MS JT/1/H/176

1. *Thomas Archer Hirst*: this letter was also for Heinrich Debus.
2. *Belfast*: at the BAAS meeting in early September.
3. *My new acquaintances*: after completing his dissertation at Marburg, Hirst visited Berlin, where he befriended several mathematicians (see letter 0678).
4. '*Thermo Saule*' *is ready*: see letter 0638, n. 9.
5. *Um es völlstandig zu machen sind drei Schirme* [*dazu*]: 'In order to make it complete, there are three screens [as well]' (German).
6. *in the Annalen*: Debus's next published paper was H. Debus, 'Ueber chemische Verwandtschaft', *Liebig, Annal.* 85 ([January] 1853), pp. 103–34.

From James B. Bevington   26 November 1852   0690

Neckinger Mills Bermondsey.[1] | London 26 Nov^br 1852

Dear Tyndall

I had a happy evening yesterday after the business of the day was over. You can fancy me coming out of the dark night into a well lit yet snug room; first the wife's welcome, then Geoffreys[2] bright eye lighting up at the sight of a little work on Photography[3] that I had bought for him; there was a pleasant sound from the kettle on the hearth; the good provender & the news of the day discussed, I had placed in my hands a packet with well known post mark & already familiar subscription. here I found material for thought & sympathy & many pleasant pictures were conjured up while with foot on the fender I perused more than once the various enclosures.[4] First the big printed document; here was professor Doctor & eminent scholar in full tilt, praising my friends by which I received a slight shade of reflected glory, there was a dash of kindliness & freedom about the set which gave them more than usual value, but then thought I what have I in common with so learned a set. Well for me that I know nothing of these honors when I shook you by the hand for I have a natural antipathy to Doctors of all sorts whether Clerical, Medical or Legal. I love them for their knowledge but I detest them for assumption & in my ignorance I fancy they have more of the latter than the former. What I like better is the wild freedom of the eratic Nazarene[5] who loved to dispute with the Doctors & to confute their systematized laws... Truly there are some Doctors of a nondescript genus. Ah! these I can look at ascance for a time & if they dont repel by their learned looks can perhaps fraternize with them. Thou my good friend hast no hard look for me therefore I will not disclaim the kindred which near thoughts entwine with jasamine flowers.[6] & let the flowers fade as they will, to the end of time they retain some fragrance.

It was a kind & generous thought of you to send me those confidential letters[7] which I have read over more than once & now return to you.

It is a vulgar thought to be ever offering money help; but little Booth has often occupied my thought during the last month & if my brotherhood with such spirits be sufficient I should be indeed glad to contribute £20—towards his comfort in any way most consonant with his feeling or rather with the feelings of his friends—For one year I shall hold this sum at your disposal for this object & at the end of the period I shall consider it sacred for some similar purpose if in the mean time you do not dispose of it. Mind you dont mention this subject to me—except it be in this phrase—Bevington I want £20—for our young friend, send me a cheque for it.

I was pleased to hear that with the stump of cigar in your mouth under shelter of some old tree you could still find pleasure in the wild moaning of the wind, I also have had pleasure in being wind rocked but unfortunately once upon a time I ventured my little wealth on a stormy sea, night after night the wind blew; it was too late to effect insurance & altho I had no loss I had the torture of 20 wrecks & the picture of property worked to worthlessness on the rocky shore. Since this event my love of the wild wind has abated but now & then the old passion revives.

Geoffrey I am glad to say is cheerful & nearly as well as usual except a paleness of complection & thinness of limb which tells the past illness—

The Doctors say that from the sounds of the chest they infer that the heart has a chronic disease which will require much care to mitigate they dont give much hope of any cure & if this be confirmed by others whom I intend to consult I must bring him up accordingly.

I cant bear to have him far away from me: last week when I left him at Bournemouth[8] & had rather unfavourable accounts: my bright thoughts ran so far into the dismal abodes, that fancying him already dead, I arranged how his little property was to be distributed among his playmates & where his little tomb was to be placed. But thank God he lives & gives us daily pleasure.

I have some thought of asking his school fellow—Le Quesne[9] to spend a fortnight with him during the Xmas holidays if the said youth does not go home. Geoffrey & he appear to have had some pursuits in common—would there be any thing inappropriate think you?

I have scarcely left room to say that I shall be glad to see you if time permits when you come to London.[10]

believe me affectionately yours | James B Bevington

RI MS JT/1/B/93

---

1. *Neckinger Mills Bermondsey*: the Bevingtons opened a leather-manufacturing firm on the Neckinger River, London, in 1800.

2. *Geoffreys*: Geoffrey Bevington (1838–72), the son of James; he died at the age of 34.

3. *little work on Photography*: not identified. Geoffrey became a keen amateur photographer. He was awarded the Silver Cup of the Amateur Photographic Society for a series of photographs, depicting his family's Neckinger Mills, displayed at the International Exhibition of 1862.

4. *enclosures*: Tyndall's letter and all of the enclosures are missing (see n. 7 below). Tyndall probably wrote to Bevington on either 23 November, when he received a letter from Bevington (Journal, 23 November, JT/2/13b/590), or the following day.

5. *the eratic Nazarene*: Jesus, sometimes known as a Nazarene because he came from the town of Nazareth, disputed with the religious authorities of his day.

6. *jasamine flowers*: jasmine flowers were often regarded as symbols of warm sentiment and attachment.

7. *confidential letters*: probably concerned with the ailing health of Booth (see letter 0686, nn. 7 and 8).

8. *Bournemouth*: a coastal resort town that during the 19th century became a place of recuperation.

9. *Le Quesne*: not identified, but perhaps the same student later involved in what for Hirst was a difficult incident at Queenwood, leading to the student's expulsion (see Hirst, 'Journals', 4 May 1856).

10. *when you come to London*: perhaps referring to Tyndall's visit to London on 30 November to attend the anniversary meeting of the RS. Tyndall does not mention seeing Bevington in his account of the trip (Journal, 12 December 1852, JT/2/13b/591).

To Thomas Archer Hirst   28 November 1852   0691

Queenwood 28th Nov. 1852

My dear Tom.

This is sloppy weather; cold wet, gray, and miserable—no cessation for the last 6 weeks so that the wet is beginning to ooze in upon us at all corners. I suppose moles and bats dont mind dirty weather and if men were moles and bats and not men they would by equally unaffected by it. But I wont complain, there is sunshine still aloft and it will reach us again in the natural rotation of things—

I see no better way of meeting the case of the thermo-säule than by telling Bang to draw on me at 3 months for [100] Thalers—I shall probably want you to send me some more things—Bang will I doubt not send you the money at once—With the thermo-säule I wish you would tell [Kleiner] to send me a dozen pairs of bars antimony and Bismuth[1] soldered together thus with about an inch and a half of

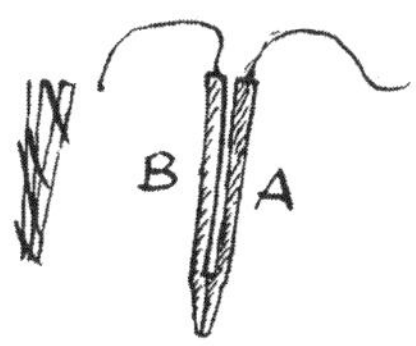

copper wire soldered to the free end of each bar. I want them in <u>separate pairs</u>. Between these two bars a bit of pasteboard or of wood must be introduced to keep them firm. They must be <u>very slight</u>, an inch, or an inch and a quarter in length and tapered down at the soldered junction to a small square end.

I wish you would further procure for me for love or money a complete crystal of carbonate of iron—you know it is [isomorphous] with carbonate of lime. A single crystal as large as possible is what I want—

I wrote to Booth before your last letter[2] reached me and received a reply from him[3] this morning—He ought to have accepted my offer—I will write to him no further about it, but can only say that I shall be perfectly glad to contribute any proportion whatever of the funds necessary for his sustenance.

That was certainly a strange thought of yours—very nearly indeed an absurd one—looking out for a situation!—nevertheless it is well to summon up the darkest possibilities at times & to make one's mind familiar with the worst. It has been my own habit all through—But you are in Berlin and must remain there—you must see Paris too, but that is [far] in the perspective. And if need be I will act the part of that old Greek whom we read of in Plutarch and give up every thing for the sake of making money for a time[4]—I think I could make money if I were to set my wits about it.

I send you a letter[5] received this morning from him who I told you had taken me away to Salisbury[6]—He is one of the choice ones of the world as you will see, He and I were sitting together at a fireside some weeks ago—'Tyndall' he said 'how long do you remain at Queenwood? Can I do any thing for you? One way only is open to me—I have money and I offer you it if you want it either as a loan or as this gift of a brother', very pleasant, is it not, Tom—But one such brother is enough for me at present.

thine affectionately | John.

Remember me to Dickinson. I hope to be able to let you have 20 or 30 pounds before the end of January.

RI MS JT/1/T/556

1. *antimony and Bismuth*: antimony bars and bismuth were connected in parallel series in the thermo-saüle.
2. *your last letter*: Tyndall replies here to Hirst's letter (0686 of 14 November), although a more recent letter from Hirst (letter 0689 of 21 November) should have reached him.
3. *I wrote to Booth . . . reply from him*: both Tyndall's letter and Booth's reply are missing, but see 0686, n. 8.
4. *that old Greek . . . for a time*: possibly Cato the Younger, as Hirst and Tyndall had read Plutarch's *Life of Cato* together early in 1851 (Journal, 13 January 1851, JT/2/13b/520).
5. *a letter*: letter 0690.
6. *him who . . . Salisbury*: James Bevington (see letter 0681).

To William Francis   15 December [1852][1]   0692

My dear Francis,

I send you Reich's paper[2]—Dealing as it does with experiments which occur very rarely and which interest every body it would not be well to omit it.

I send you also a condensation of Wartmann[3]—I would let it all go in one article merely drawing a line as I have done between the two *[portions]*.[4] One word I cannot translate it is acerdèse[5] at page 6 of the translation.

I shall probably take a run up to your soirée

ever yours | Tyndall | 15th Dec.

I send you a review of Purslo's work.[6] Such books I think ought to be permitted to die without notice

RDS 27/22

1. *[1852]*: the year is based on the allusions to Reich (n. 2), Wartman (n. 3) and Purslo (n. 6).

2. *Reich's paper*: probably F. Reich, 'New experiments on the mean density of the earth', *Phil. Mag.* 5, no. 31 (March 1853), pp. 153–59 (originally published as F. Reich, 'Neue Versuche über die mittlere Dichtigkeit der Erde', *Poggend. Annal.* 85, no. 2 [1852], pp. 189–98).

3. *condensation of Wartmann*: Tyndall translated and condensed É. Wartmann, 'Recherches sur la conductibilité des minéreaux pour 'électricité voltaïque', a paper read at the general meeting of the Society of Physics and Natural History of Geneva on 20 November 1851 (*Memoires de la société de physique et d'histoire naturelle de genève: Tome Treizième* [Genève: Librairie de Joel Cherbuliez, 1854], pp. 199–210) and É. Wartmann, 'Note sur quelques Expériences faites avec le Fixateur Electrique', *Archives des sciences physiques et naturelles* á la Bibliothéque Universelle, Genéve 19 (August 1852), pp. 282–87.

4. *one article . . . two [portions]*: the two papers were published together as Wartmann, 'Researches on the Conductibility of Minerals for Voltaic Electricity; and on the Electric Light', *Phil. Mag.* 5, no. 29 (January 1853), pp. 12–16. The line of separation appears on p. 15.

5. *acerdèse*: another name for manganite (p. 14 of Wartmann [n. 4 above]).

6. *Purslo's work*: see letter 0676, n. 6 and letter 0693.

To William Francis 19 December [1852][1] 0693

Sunday 19ᵗʰ Dec.

My dear Francis.

I return you the papers of Adie[2] and Phillips.[3] I do not at all think that the former has successfully met the case, nevertheless I should be reluctant to advise the suppression of his paper. I at first intended to propose a short reply in the answers to correspondents but on reconsideration it is, I am inclined to think, better to print it as it is. If you thought well of it I would write to Adie and point out where his reasoning fails; I shall *[either]* do this *[or]* reserve my remarks for the next no. of the magazine,[4] which ever you like.

Reuben's paper gives here and there evidence of singular *[acuteness]* and here and there also to my mind evidence of a rather extravagant tendency to theorize—I dont think you can do better than print it as it stands.[5]

I cannot obtain any distinct information regarding the supernumerary bow[6]—Mʳ *[Pratt's]* paper[7] I think may be printed.

I agree with your decision regarding Purslo—as the review is written it may be printed.[8] But I am inclined to think that it is a misapplication of criticism to bestow it on such books—When a man in high places says anything absurd let us have a shot at him but the legion of unknown speculators it would be better to pass in silence.

I am rejoiced to hear of Mʳ Taylor's improvement and wish him and you a happy christmas.

I may possibly take a run to Manchester at the end of the *[week]* and return to London a few days afterwards.

I send you a short tit-bit to make up weight. It is *[pretty]* and will be read with interest. It is cut off from an article in No. 5 of Poggendorff.[9]

I shall be glad to look over De la Rive[10] and trust he will give me an opportunity of praising him for I am almost tired finding fault.

ever yours sincerely | John <u>Tyndall</u>

StBPL T&F, Authors' letters

1. *[1852]*: the year is confirmed by allusions to papers by Adie and Purslo.
2. *the papers of Adie*: R. Adie, 'On the Temperatures of Conductors of Electrical Currents,' *Phil. Mag.* 5, no. 29 (January, 1853), pp. 46–49.
3. *and Phillips*: cited letter 0684, n. 7.
4. *next no. of the magazine*: Tyndall replied with one derogatory paragraph in 'On the Temperature of Conductors of Electrical Currents', *Phil. Mag.* 5, no. 30 (February 1853), p. 147. This was part of an ongoing exchange between Adie and Tyndall (see letters 0670, n. 1 and 0685).
5. *Reuben's paper . . . as it stands*: Tyndall had previously been more disparaging of Phillips.
6. *supernumerary bows*: extra bands, usually pale pink or green in color, that are often seen on the inside of the primary rainbow.
7. *[Pratt's] paper*: Tyndall probably alludes to J. H. Pratt, 'The supernumerary bows in the rainbow arise from interference', *Phil. Mag.* 5, no. 30 (February 1853), pp. 78–86 (a letter to the editor sent from India and dated 18 September 1852).
8. *regarding Purslo . . . printed*: see letters 0676, n. 6, and 0692. It seems that Francis changed his mind and did not publish the review.
9. *tit-bit . . . No. 5 of Poggendorff*: probably W. Haidinger 'On the Direction of the Vibrations of the Luminiferous Æther in Plane-polarized Light', *Phil. Mag.* 5, no. 29 (January 1853), pp. 49–51, which was extracted and translated from 'Ueber die Richtung der Schwingungen des Lichtäthers im geradlinig polarisirten Lichte', *Poggend. Annal.* 86, no. 5 (1852), pp. 131–44.
10. *look over De la Rive*: not identified. No translation, summary, or review of a paper or volume by de la Rive appeared in the forthcoming numbers of the *Phil. Mag.*

To William Francis   [23 December 1852][1]   0694

Thursday

My dear Francis;

I send you the proof of Wartmann[2]—The 1ˢᵗ paper was read at the Society of Physics & Natural History—I don't know whence it is taken[3]—Acerdèse must only be let stand[4]—Debus started this morning to Manchester—I will send the proof after him—I am on the point of starting to [Andover] where I must lecture tonight.[5] —I shall probably be off to Manchester tomorrow[6]— Sometime next week I shall make my appearance before you.[7]

Yours most truly | John <u>Tyn</u>dall

If you have anything to say to me Post office Manchester will find me.

RDS 27/25

1. *[23 December 1852]*: dated by the allusion to Tyndall's lecture near Andover (see n. 5).
2. *proof of Wartmann*: see letter 0692, n. 3.
3. *from whence it is taken*: see letter 0692, n. 3.
4. *Acerdèse. . . stand*: see letter 0692, n. 5.
5. *Andover . . . tonight*: the lecture took place in Abbott's Ann, a village 2 miles southwest of Andover (a town in Hampshire), on 23 December 1852 (Journal, JT/13b/592–3).
6. *to Manchester tomorrow*: Tyndall travelled to Manchester on 24 December and stayed until 31 December (Journal, JT/13b/593).
7. *my appearance before you*: Tyndall saw Francis while visiting London, 2–8 January 1853 (Journal, JT/13b/594–5).

To Thomas Archer Hirst   31 December 1852   0695

Manchester 31st Dec. 1852

My dear Tom.

I will not delay letting you know that the thermo-Säule has reached Queenwood safely, not in company with but after your letter[1]—I am so split up in the hubble bubble of Manchester[2] that I can scarcely say anything else. I will write to you at length in a few days—one thing presses upon me and that is that you may be hampering yourself as regards money matters. —I shall endeavour before I leave Manchester to arrange some way of transmitting funds to you and within the coming three weeks hope to make a large breach in your account against me.[3] Have patience with me until my thoughts tranquilize a little and then I will write you a letter full of all the melodies I can collect.[4]

Your affectionate Brother, | John

RI MS JT/1/T/557

1. *your letter*: letter missing.
2. *Manchester*: see letter 0694, n. 6.

3. *your account against me*: a reference to the money (£10 of a £20 loan) that he still owed
    Hirst.
4. *letter full of all the melodies I can collect*: see letter to Hirst, 9 January 1853.

To William Francis   [November–December 1852][1]   0696

My dear Francis.

Can you inform me how many pages of the Scientific Memoirs will be
filled by the manuscript[2] I have already sent you.

Will you be able to let me have the printed copies of my paper in the Phil
Trans[3] before the end of the month? I should like to send a few copies off with
the magazine[4] if it were possible.

Since[re]ly yours | John Tyndall

RDS 27/21

---

1. *[November–December 1852]*: date is quite uncertain. The paper in the *Phil. Trans.* alluded
    to (n. 3) was received by the RS on 20 October 1852 and read 6 January 1853. Tyndall was
    translating papers for *Scientific Memoirs* from September 1852 to January 1853.
2. *the manuscript*: Tyndall had translated several papers—by Bunsen (letter 0666, n. 1), Kno-
    blauch (letter 0651, n. 8), Clausius (letter 0664, n. 3), and Helmholtz (letter 0664, n.
    5)—for the *Scientific Memoirs*. It is unclear which he alludes to here, or whether he means
    all those sent to Francis thus far.
3. *my paper in the Phil Trans*: see 0606, n. 1. The *Phil. Trans.* was printed by Taylor & Francis,
    so Francis would have had early copies, though we are not certain when those copies (of
    perhaps only the first part) would have been available.
4. *the magazine*: the *Phil. Mag.*

BIOGRAPHICAL REGISTER

This register contains the names and biographical details of people who are mentioned three or more times in the letters included in this volume. Biographical information about people mentioned once or twice is contained in the notes appended to the appropriate letter(s).

Conventions

Tyndall and some of his correspondents were often unaware of the correct spelling of the names of people whom they met; in particular, it is clear that they spelled many German names phonetically—and incorrectly. In transcribing the letters the editors have generally retained the spellings of family names given in the letters or LT's transcripts. In this Register persons for whom we have a preferred spelling—for example, as used in biographical dictionaries—are entered under that spelling and any alternative usages within the letters are placed in "( )" brackets. Thus, Johann Friedrich Hessel, whose name was often mis-spelled "Hessell," is entered below as "Hessel (Hessell), Johann Friedrich Christian." Where we do not know the preferred spelling, the alternatives are given in alphabetical order, separated by a "/": thus, "Dickenson/Dickinson, Carl." Nicknames and pseudonyms were widely used by Tyndall and his friends; when they occur in the letters, the person they refer to is often given in an accompanying note. Nicknames are given in the Register and are placed in quotes inside parentheses: thus, Hirst, Thomas Archer ("Nep"). Pseudonyms are also given in the Register, placed inside parentheses and denoted by "pseud."; thus, Sterling, John (pseud. Caliban). Where known, the dates of birth and death are given in parentheses to the right of the nicknames and pseudonyms; thus, Phillips (Philips), George Searle ("January," pseud. January Searle) (1816–89).

Along with material from the Tyndall correspondence and from the journals of Tyndall and T. A. Hirst, the sources for the biographical entries are standard biographical dictionaries, such as the *Dictionary of Scientific Biography,* the *Oxford Dictionary of National Biography,* Poggendorff's *Biographisch-literarisches Handwörterbuch der exacten Wissenschaften,* and the *Dictionary of Nineteenth Century British Scientists.* Where additional sources add important material, they are given in "( )" at the end of the entry, often in abbreviated form.

Adie, Richard (1810–81), maker of scientific instruments. The third son of instrument maker Alexander Adie (1775–1858), who set up a shop in Edinburgh, which Adie took over after the death of his eldest brother. He also published in scientific journals, including the *Phil. Mag.* and the *Edinburgh New Philosophical Journal.* He and Tyndall had a running argument in the *Phil. Mag.* in 1852–3 concerning the temperature produced at the joint between antimony and bismuth.

Airy, George Biddell (1801–92), English astronomer and mathematician. He became the Plumian Professor of Astronomy at Cambridge in 1828–35, Astronomer Royal in 1835–81, and Fellow of the Royal Society in 1836. He equipped the Greenwich Observatory with a redesigned transit circle in 1850, setting a new standard for precision measurement, and in 1852 he pioneered the use of the telegraph to transmit Greenwich Mean Time at home and abroad.

Allatt, Mary (1806–?), Aunt Allatt, sister of Thomas Hirst's father, lived in Brighouse, West Yorkshire. The family were Dissenters, perhaps Methodist, as Methodism was strong in the region.

Bang, a Marburg banker with whom Tyndall and Hirst did business.

Barlow, John (1798–1869), Anglican clergyman and the Secretary of the Royal Institution between 1843–60. Together with Michael Faraday, he undertook the day-to-day running of the Institution. He took over many of Faraday's duties once his health declined in the 1840s. As the person responsible for organizing the Friday evening discourses, in October 1852 he invited Tyndall to give one of the lectures. In 1851, Barlow became minister of the Duke Street Chapel in London.

Battersby, one of the farmer students—that is, one of the older pupils, who studied the theory and practice of farming   at Queenwood. He was at the center of a dispute between Tyndall and Mrs. Edmondson.

Battersby, Eliza (c. 1806–?), Aunt Battersby, sister of Thomas Hirst's mother, was married to an Anglican clergyman. They lived in Bristol.

Baum, Weissbinder, and Frau Baum, Hirst lodged with Weissbinder Baum and his wife in Marburg, perhaps as part of a student boarding house.

Bevington, James Buckingham (1804–92), a wealthy and benevolent manufacturer of leather goods, later described as a leather merchant, of Neckinger-Mills in Bermondsey, south London. He enjoyed wide ranging religious discussion with Tyndall. They were good friends in the 1850s and remained in contact throughout their lives. ("Wills and Bequests," *Standard,* 13 October 1892.)

Booth, Francis (d. 1853), a mutual friend of Tyndall and Hirst. He was employed in an important position in a wool mill near Halifax. Tyndall had hoped Booth might be able to advance himself through school mastering. In 1851, he became ill with tuberculosis and his health deteriorated quickly. Both Tyndall and Hirst helped him monetarily during his illness.

Brewster, David (1781–1868), Scottish natural philosopher, editor, and educator. He is known for his scientific work in optics, especially the optical properties of crystals and the polarization of light, and for his editorial work as a constant contributor to numerous weeklies and monthlies, by which he earned his living. He was one of three founding editors of the *Edinburgh Magazine.* He was appointed Principal of the United College of St. Salvator and St. Leonard at St. Andrews in 1838, and Principal of the University of Edinburgh in 1859. He was an early supporter of the BAAS, serving as president in 1850.

Bunsen, Robert Wilhelm (1811–99), German chemist. He received his PhD from the University of Göttingen in 1830. He taught chemistry first at Göttingen in 1833, then in the Polytechnic School in Kassel in 1836, before accepting a professorship at the University of Marburg (extraordinary in 1838 and ordinary in 1842). In 1846, he also traveled to Iceland as part of a scientific expedition sponsored by the Danish government. At Marburg, he was an extremely popular teacher, mentoring both Tyndall and Frankland, and a renowned laboratory researcher. Tyndall's initial experiments in diamagnetism relied heavily on crystals that he secured from Bunsen's laboratory. He moved to Breslau in 1851, where he taught for three semesters, and then to Heidelberg in 1852, where he collaborated with Gustav Kirchhoff, making major contributions to spectral analysis and radiant heat.

Caliban, see Sterling, John.

Carlyle, Thomas (1795–1881), Scottish philosopher, essayist and historian. Tyndall and Hirst wrote long letters about their admiration—verging on obsession—for him, weighing their own merits and deficiencies against Carlylean ideals. They were particularly influenced by Carlyle's works on history, including *On Heroes, Hero-Worship, and the Heroic in History* (1841), *Past and Present* (1843), and *Latter-day Pamphlets* (1850), which first appeared as a series of pamphlets published from February through August, 1850.

Carter, Richard (1818–95), a surveyor in Halifax, and former employer of Tyndall (1844–47) and Hirst (1845–50). Later in life, Carter purchased significant property in the South Yorkshire town of Barnsley, where he was eventually elected mayor. In the 1880s he retired to Harrogate. (J. Forrest, ed., *Minutes of the Proceedings of the Institution of Civil Engineers,* vol. 123 [London: Institution of Civil Engineers, 1896], pp. 445–48.)

Carter, Sarah (c. 1812–?), Richard Carter's elder sister and housekeeper. She was theologically orthodox and had often disputed theological matters with Tyndall in his Halifax days.

Clausius, Rudolf (1822–88), renowned German scientist and one of the founders of modern thermodynamics. His famous paper on the theory of heat "Ueber die bewegende Kraft der Warme," was published in the *Annalen der Physik* in 1850. He taught physics at the Royal Artillery and Engineering School in Berlin in 1850 before moving to a professorship in mathematical physics at the Polytechnicum in Zurich in 1855. Tyndall translated many of his major works into English for publication in the *Phil. Mag.* and *Scientific Memoirs* in the early 1850s.

Craven, James ("Jemmy," "Jimmy") (1829–86), a young man who had trained as a surveyor in Halifax, very likely under Richard Carter, and, like Hirst, a young protégé of Tyndall. Craven and Hirst were good friends and supported each other's educational pursuits. Under Tyndall's guidance, Craven studied philosophy, geology, and chemistry; however, Tyndall did not feel that Craven was as dedicated to the work as Hirst. Craven set up independently as an "agent and surveyor" in Halifax late in 1850. He had a brother, possibly named John.

Darker, William H., optician and optical lapidary, and highly regarded maker of scientific instruments in Lambeth, London. He was a member of the Royal Microscopical Society.

Davy, an English student who came to Marburg with Wrightson, to whom he was articled in "Chemical Analysis" (letter 0516). Davy was in an anomalous position because he was not himself a Marburg student. He is perhaps the same Edmund William Davy (1826–c.1900), who in 1870 became a professor of medicine at the Royal College, Dublin.

Debus, Heinrich (1824–1915), chemical assistant in Bunsen's laboratory in Marburg until early 1851 and a friend to Tyndall, Frankland and Hirst at Marburg. Debus left Marburg when Kolbe replaced Bunsen as professor of chemistry, and took up a position teaching chemistry at Queenwood College, where he succeeded Frankland. He was an excellent teacher. After leaving Queenwood in 1867, he spent three years at Clifton College, Bristol, then returned to London as lecturer in chemistry at Guy's Hospital Medical School (1870–88), adding to this the position of professor of chemistry at the Royal Naval College, Greenwich (1873–88) under Hirst, who was founding Director of Studies at the College.

Dickenson/Dickinson, Carl (1833/1834–?), an English student who arrived in Marburg in mid-1851. He had probably been a student of Schnackenberg, who accompanied him on his journey to Marburg, and was therefore probably from Lancashire (possibly Preston or Blackburn). Hirst mentored him, and at the beginning of the 1852–3 academic year, he and Hirst moved to Berlin where they lodged together. He had a younger brother at Queenwood. (Brock and MacLeod, *Hirst Journals.*)

Dickson, Samuel, a London physician whom Tyndall strongly recommended that Hirst consult over his bowel problems. Probably the same Dr. Dickson that Tyndall had consulted on his own and his father's behalf in 1846 (for example, Journal, July 1846, JT/1/13a/135–7).

Dirichlet, Peter Gustav Lejeune (1805–59), German mathematician, who attended the University of Göttingen where he developed a close relationship with Gauss. Lejuene Dirichlet's most notable contributions to mathematics came in the fields of number theory, analysis and mechanics. He taught at the University of Breslau in 1827, and at the University of Berlin between 1828 and 1855 before succeeding his mentor Gauss at the University of Göttingen.

Dove, Heinrich Wilhelm (1803–79), Prussian meteorologist and experimental physicist. In 1829, Dove moved to Berlin where he became an ordinary professor in 1844 and Director of the Prussian Institute of Meteorology beginning in 1849. He also worked to expand the coordination between meteorologists internationally. Tyndall met Dove in Berlin in 1851.

du Bois-Reymond, Emil (1818–96), German physiologist. He began his studies at the University of Berlin, where he attained his degree in 1843 and began lecturing in 1846. He is known for his pioneering work in electrophysiology, particularly his ability to measure the electrical currents produced by nerves. He was working in Magnus's laboratory in Berlin when Tyndall visited in the summer of 1851. With du Bois, Tyndall was able to repeat Reymond's "celebrated experiment . . . of exhibiting an electric current by the action of the muscles of my arm" (letter 0480).

Edmondson, Anne (née Singleton) (d. 1863), wife of George Edmondson. She was active in the management of Queenwood College and like her husband, a committed Quaker. Tyndall and Hirst regarded her as the stronger character of the two.

Edmondson, George (1798–1863), Quaker, educationist, and headmaster of Queenwood College from 1847 until his death. His Quaker beliefs contributed to his innovative educational practices, for example the inclusion of manual pursuits in the curriculum. (For his earlier life see Volume 2.) The letters in this volume mention a daughter (Jane, c. 1833–?) and a son, George William. Jane gave lessons (in shorthand); George William was a young pupil at Queenwood. (Brock, "Queenwood.")

Emerson, Ralph Waldo (1803–82), American poet and essayist, and a leader of the Transcendentalist movement. Tyndall greatly admired Emerson's writings and urged his protégés Hirst and Craven to read Emerson's work. Tyndall and Hirst discussed Emerson at length, as shown in this volume, especially his poetry and the essay collection, *Representative Men* (1850).

Faraday, Michael (1791–1867), renowned English natural philosopher, best known for his discoveries in electricity, magnetism, and (in this volume) diamagnetism. Tyndall was inspired by Faraday's work and reputation. From their first meeting in June 1850, Tyndall kept Faraday informed about the progress of his research and sent him samples of crystals to use in experiments. Faraday became a friend and mentor, helping with Tyndall's appointment as Professor of Natural Philosophy at the Royal Institution in 1853.

Fichte, Johann Gottlieb (1762–1814), a German philosopher who is considered one of the founding figures of post-Kantian German Idealism. Fichte termed his Idealism *Wissenschaftslehre,* which translates to "Doctrine of Scientific Knowledge." Tyndall was interested in his notions of individual identity and vocation.

Fick, Sophie (1821–74), of Cassel, near Marburg in the state of Hesse-Cassel. Frankland met her in 1847, on his first visit to Marburg, and they became engaged in 1849. She came from a well-educated middle class family: she spoke English, two of her brothers were professors in the University of Marburg, and her father, Friedrich W. Fick (1783–1861), was chief engineer of the state. (Russell, *Edward Frankland.*)

Forbes, James David (1809–68), Scottish physicist and glaciologist and Professor of Natural Philosophy at Edinburgh from 1833. The well-connected Forbes had obtained the position over David Brewster, who had better scientific credentials. Forbes received the Royal Society's prestigious Rumford Medal in 1838 for his experimental investigations on the polarization of heat, which he was the first researcher to demonstrate in November 1834. He was also a pioneering mountaineer. While in the Alps, he studied the motion of glaciers, which brought him into controversy with Tyndall and Huxley in the late 1850s. Letters in this volume indicate both tensions and collegial relationships.

Foucault, Jean Bernard Léon (1819–68), French physicist known for the invention of the Foucault pendulum, an instrument he used to prove the rotation of the Earth in 1851. He also collaborated with Hippolyte Fizeau in experimental optics. While Fizeau made the first precise measurement of the speed of light, Foucault first demonstrated that light travels more quickly through air than water, which was used as verification for the wave theory of light. Tyndall translated and abridged Foucault's work for the *Phil. Mag.*

Francis, William (1817–1904), chemist, translator, and editor. In the mid-1830s he spent two years in Gera, an industrial town south of Leipzig, and a center of German printing, learning the printing trade and becoming fluent in German. He was apprentice to, and natural son of, Richard Taylor, who sent him to the University of Berlin in 1839 to study chemistry. In 1841 he moved to attain his degree under Justus Liebig and Gustav Rose in Giessen, all the while sending translations of German science back to Taylor. Francis began working for Richard Taylor's publishing firm in 1837, became editor in 1851, and partner in 1852 when Taylor & Francis publishing firm was established. Francis was Tyndall's most significant counselor and closest confidant as he maneuvered his way through the Victorian scientific scene. They edited *Scientific Memoirs* together, which was published in early 1853. Francis confided his paternity to Tyndall in 1851. (Brock and Meadows, *Lamp of Learning.*)

Frankland, Edward ("Trismegistus," "Tris") (1825–99), began his scientific career as an apprentice apothecary. While working in Lyon Playfair's London laboratory, Frankland was persuaded by his colleague Kolbe to visit Marburg in the summer of 1847. It was through friendship with Frankland at Queenwood College in 1847–8 that Tyndall became interested in studying at Marburg. Frankland left Marburg early in 1850 to take a position under Playfair at the Putney College of Civil Engineering. At the beginning of 1851 he left Putney for the chair of chemistry (half time) at the new Owens College, Manchester. He married Sophie Fick, whom he had met while at Marburg, in February 1851, just before moving to Manchester. (Russell, *Edward Frankland*.)

Gauss, Carl Friedrich (1777–1855), widely considered one of the greatest mathematicians of the nineteenth century. After earning his PhD at the University of Helmstedt in 1799, he focused his early research on astronomy, becoming professor of mathematics at Göttingen and the director of the Göttingen Observatory, positions he held until his death.

Gerland, B. Wilhelm (c. 1829–1905), was Hermann Kolbe's first PhD student at the University of Marburg, where Gerland also met Hirst. He then moved to England in the mid-1850s to study under Edward Frankland at Manchester, where he later worked in the chemical industry (Alan J. Rocke, *The Quiet Revolution: Hermann Kolbe and the Science of Organic Chemistry* [University of California Press, 1993]).

Gerling, Christian Ludwig (1788–1864), became professor of mathematics, astronomy, and physics at Marburg in 1817. He attained his PhD under Gauss, was the doctoral advisor for Julius Plücker, and became known for his work in geodetics. Hirst and Tyndall spoke despairingly about his teaching style.

Ginty, Margaret ("Peb"), née Roberts (c. 1826–?), married Tyndall's friend William Ginty in Ireland in 1846. Tyndall maintained a warm correspondence with both Margaret and her husband. At the 1851 census they had three children, Kathleen, William Roberts, and Fanny; George and Fred were born later. In 1855 the family followed William Ginty to Rio de Janeiro but the sons were educated in England (at Queenwood until 1863 when Edmondson died). Margaret Ginty returned to England within a few years of her husband's death.

Ginty, William Gilbert (c. 1820–66), close friend of Tyndall from Irish Ordnance Survey days and ally in the campaign for better treatment of the civil assistants employed in the English Ordnance Survey (see Volumes 1 and 2). He married Margaret Roberts, and the couple relocated to England, first in or near Wakefield then to Manchester in 1846. During his visit to England in the summer of 1850, Tyndall often visited the Gintys. Ginty became a prominent Manchester businessman, and was involved in various civil engineering projects, including the construction of the Liverpool-Bolton railway. He served briefly as honorary secretary of the Manchester Athenaeum (elected late 1853) before moving to Rio de Janiero in April 1854 to manage the city gas works. He was awarded a Brazilian knighthood—Knight of the Imperial Order of the Rose—in 1857 for his contributions to major construction projects. He died suddenly from unspecified causes in mid-1866.

Gladstone, John Hall (1827–1902), British chemist, who attained his PhD in 1847 under Liebig at the University of Giessen. He became a lecturer at St. Thomas's Hospital in 1850 and was elected FRS in 1853. Tyndall's correspondence with him on questions of religion and spirituality highlights the nature of Tyndall's reinterpretation of Christian belief and his predilection to Carlyle and Emerson.

Goëthe, Johann Wolfgang (1749–1832), preeminent German Romantic writer, best-known for his *Götz von Berlichingen, The Sorrows of Young Werther,* and *Faust.* Tyndall and Hirst frequently included quotes and poetry from Goëthe in their correspondence. Tyndall later wrote an essay on Goëthe's theory of color.

Grove, William Robert (1811–96), Welsh judge and electrochemist. He held a professorship of experimental philosophy at the London Institution from 1841–7, when financial constraints forced him to return to practicing law. He invented a nitric acid voltaic cell, known as "Grove's cell," that caught the attention of Faraday and Tyndall. He signed Tyndall's FRS nomination certificate.

Gyde, Charles, though described in these letters as a clerk, he was effectively the office manager of Richard Taylor's publishing firm. He was originally the company bookbinder, and continued in that position when he became clerk in charge in the 1820s. By the 1850s he was a close family confidant·to Taylor and William Francis (Brock and Meadows, *Lamp of Learning*).

Haas, John, had been a fellow teacher at Queenwood with Tyndall. He taught at Fellenberg's Hofwyl School in Switzerland.

Herschel, John Frederick William (1792–1871), mathematician and astronomer. He was the son of the famous astronomer Frederick William Herschel and himself became one of the most eminent scientists in Britain in the early Victorian era. His *Preliminary Discourse on the Study of Natural Philosophy* (1830) was widely read and greatly admired. He became Sir John in 1838, when, on his return from mapping the southern skies from Capetown, he was created a baronet. In 1849, he published his *Outlines of Astronomy,* a popular presentation of astronomy that went through eleven editions by 1871. He also edited the *Admiralty Manual* in 1849, a field guide in science for those traveling throughout the British empire.

Hessel (Hessell), Johann Friedrich Christian (1796–1872), German crystallographer and mineralogist, who taught at the University of Marburg (extraordinary professor in 1821 and ordinary professor in 1825), staying there for his entire career. He determined mathematically that only thirty-two classes of crystal were possible.

Higginson, Thomas Charles (c. 1821–98), an early companion of Tyndall on the Irish Ordnance Survey. Higginson had joined the Royal Artillery in 1842 and served in India (see Volume 1). In 1851 he was stationed at Shorncliffe and, as indicated in this volume, recently married. In the next few years he travelled on army business in the United States before being appointed to superintend recruiting in Ireland. Tyndall maintained contact over the years, writing letters and meeting on at least one occasion.

Hirst, Hannah Oates (1804–49), mother of Thomas Hirst, widowed since 1842 (see Volume 2).

Hirst, John Henry (1826–82), eldest brother of Thomas Hirst. He studied at the Royal Military Academy Woolwich, but instead of following a military career became an architect in Bristol. His wife (Harriet) left him in 1876.

Hirst, Thomas Archer ("Nep") (1830–92), English mathematician and Tyndall's closest friend in the years covered by this volume. In 1845 he took a place as apprentice surveyor in the office of Richard Carter, where Tyndall was senior surveyor. As the railway boom died down, he was one of a number of young men whom Tyndall encouraged in self-improvement. In this volume Hirst has become an admiring disciple of Tyndall, and through him of Thomas Carlyle, the master whom they both followed. In October 1850 he entered the University of Marburg to begin his PhD studies, which he completed in 1852. He could afford to remain in Germany because, since his mother's death in 1849, he had an annuity of approximately £150.

Hirst, William ("Bill"), (1828–?), elder brother of Thomas Hirst. He became a Methodist and took up farming near Badsworth in the early 1850s, at about the time that he married Rachel Ann Burkitt (c.1821–?) of Badsworth. Their first child was born in 1853.

Hoffmann, August Wilhelm (1818–92), German chemist and educator who studied under Liebig at the University of Giessen, receiving his PhD in 1841 and becoming one of Liebig's assistants in 1843. In 1845 he became the Director of the Royal College of Chemistry in London, bringing Liebig's laboratory-based teaching style to England. He received the Royal Medal in 1854 for his research in organic chemistry. While in London, he collaborated with Henry Bence Jones to edit Liebig and Kopp's *Annual Report of the Progress of Chemistry and the Allied Sciences, Physics, Mineralogy, and Geology* (1853), and they approached Tyndall to translate the sections on physics.

Humboldt, Alexander (1769–1859), the world renowned German naturalist and explorer, and by the 1850s one of the greatest scientific figures in Europe. He made a journey through South America with French botanist, Aimé Bonpland, from 1799 to 1804, publishing thirty volumes upon his return. His wide-ranging *Cosmos* appeared in five volumes between 1845–62, with volume three appearing in 1850. Financial constraints stemming from the publishing of his works forced Humboldt to spend the last thirty years of his life in Berlin working for the Prussian government. Tyndall met him at his residence in Berlin, but left the meeting disappointed as Tyndall was unable to get a word in.

Huxley, Thomas Henry (1825–95), anatomist and zoologist. He began his career as assistant surgeon on the *HMS Rattlesnake* in 1846. He returned to England in 1850, intent on establishing himself in a scientific career. He and Tyndall shared the disappointments of applying unsuccessfully for positions in Britain and abroad. He rose quickly in the scientific scene in England, elected FRS in 1851, winning its Royal Medal in 1852, and serving on its Council in 1853. He was one of the signatories for Tyndall's election to the Royal Society.

Hutchinson, John, a friend with whom Hirst discussed religious matters. Hutchinson had questioned his own beliefs, but by 1849 had returned to orthodoxy.

Jacobi, Moritz Hermann (1801–74), German architect, engineer and physicist. After studying architecture at Göttingen, he moved first to Königsberg in 1833, where he constructed one of the first electric motors, then in 1835 to teach civil engineering at the University of Dorpat, and finally to St. Petersburg as a member of the Imperial Academy of Sciences (privatdozent in 1839, extraordinary professor in 1842, and ordinary professor in 1847). While there he helped to advance the development of the telegraph.

January, see Phillips, George Searle.

Jesus (c. 4 BC–29/33 AD), was the founding figure of Christianity, known in Christian tradition as Jesus Christ. He was a Jewish teacher, prophet, and healer, who lived in the Roman province of Palestine; some Jews considered him to be the promised messiah or "Christ." According to early Christian sources, some Jewish leaders considered Jesus a threat and persuaded the Roman authorities to crucify him. Disciples, who saw him again after his crucifixion, believed him to have been resurrected. The Christian movement grew from these disciples, but under the cover of Judaism. Over centuries of doctrinal elaboration, it was claimed that Jesus was divine, and that his death was for the salvation for humankind. Tyndall, and many others in the nineteenth century, discounted the claims of miraculous powers, divine status and salvation, but praised Jesus as a great moral teacher.

Jones, Henry Bence (1813–73), English chemist, wealthy physician, and a manager (later Secretary) of the Royal Institution. He traveled to Giessen in 1841 to work in Liebig's laboratory to learn chemistry and animal physiology. In 1852, he asked Tyndall to help with the translation of Liebig's *Annual Report of the Progress of Chemistry and the Allied Sciences, Physics, Mineralogy, and Geology* (1853) and invited him to lecture at the Royal Institution that same year.

Joule, James Prescott (1818–89), the son of a brewer, he studied for a brief time under John Dalton before focusing his own work on electromagnetism. In 1838 Joule published his first paper on electromagnetic engines, and spent the next decade studying their efficiency. This research led to his formulation of the mechanical equivalent of heat, which he presented to the BAAS in 1843. The Royal Society awarded him a Royal Medal for his work in 1852. In the early 1850s, he corresponded with Tyndall concerning a testimonial for Tyndall and the publication of his own work in the *Phil. Mag.*

Kane, Robert John (1809–90), Irish chemist, editor, and educator. In 1834 he was appointed lecturer in natural philosophy at the Royal Dublin Society, and traveled to Giessen to study chemistry under Liebig in 1836. He contributed articles on chemistry to the *Phil. Mag.* throughout the 1830s, and in 1840 Taylor invited him to join the editorial board. From 1834 to 1849 he was professor of natural philosophy at the Royal Dublin Society; in 1845 he was appointed President of Queen's College in Cork (it did not open until 1849); and in 1846 he was appointed the first director of the Irish Museum of Irish Industry in Dublin (at which time he was knighted). (Brock and Meadows, *Lamp of Learning.*)

Kleimer/Kleiner, instrument maker and/or dealer in Berlin. Tyndall and Hirst corresponded about purchasing bars of antimony and bismuth from Kleimer, as well as a "thermosäule" for Tyndall.

Knoblauch, (probably Carl) (d. 1859), Hermann Knoblauch's father, a wealthy silk and ribbon manufacturer (perhaps also a merchant) of Frankfurt. He was an hospitable friend to Tyndall and Hirst.

Knoblauch, Hermann (1820–95), German physicist and chemist known for his contributions to the studies of radiant heat. He studied at the University of Berlin under Magnus, attaining his degree in 1847. From 1849–53, he was a teacher (extraordinary professor in 1849 and ordinary professor in 1852) at the University of Marburg. He co-authored two papers with Tyndall in 1850, Tyndall's first scientific publications. Each paper was published in both German (in Poggendorff's *Annalen*) and English (in the *Phil. Mag.*). Tyndall and Knoblauch continued to correspond for the next twenty-five years. Knoblauch married in 1852; his bride is not identified.

Kohlrausch, Rudolf Herrmann Arndt (1809–58), physics and mathematics professor, first at the University of Marburg (ordinary professor in 1853) then at the University of Erlangen in 1857. He is best known for his work in electricity and electromagnetism, including work with Weber showing the ratio of the electrostatic unit of charge over the electromagnetic unit of charge.

Kolbe, Adolf Wilhelm Hermann (1818–84), considered one of the founders of modern organic chemistry. He studied at the University of Marburg under Bunsen, receiving his doctorate in 1843. He traveled to England in 1845 to work in the organic chemistry laboratory attached to the Museum of Practical Geology, where he met Frankland. Together, they traveled back to Marburg in the summer of 1847. Thus, it was indirectly through Kolbe that Tyndall and then Hirst also went to Marburg. Kolbe succeeded Bunsen as Professor of Chemistry at Marburg in 1851 and then accepted a professorship at Leipzig in 1865.

Lardner, Dionysius (1793–1859), Irish astronomer, editor, and popularizer of science. He taught natural philosophy and astronomy at University College, London between 1827 and 1831, but his reputation came primarily from his popular lectures and writings, and through his editorship of scientific journals, including the *Cabinet Cyclopedia*. A scandal involving a married woman ended his scientific career in London in 1840, though he continued to be a successful popular lecturer, including in the United States between 1841 and 1844.

La Rive, Auguste or Auguste-Arthur de (1801–73), Swiss physicist who was one of the founders of the electrochemical theory of batteries. He was appointed professor of general physics (1823) and experimental physics (1825) at the Academy of Geneva. His research focused on the specific heat of gases and electrical studies, especially the theory of the voltaic cell.

Larken, Edmund Roberts (1809–95), rector of Burton-by-Lincoln in Lincolnshire, and the chief financial supporter of the *Leader*. His reforming, even democratic, views were far-removed from those of the typical Anglican clergyman.

Lees, Frederic Richard (1815–97), a promoter of teetotalism and Owenite socialism, a Carlylean journalist and fellow editor with George Searle Phillips of the *Truth Seeker*, which made him a subject of correspondence between Tyndall and Hirst.

Lenz, Heinrich Friedrich Emil (1804–65), Russian physicist specializing in electromagnetism, he accepted an appointment at the St. Petersburg Academy of Sciences in 1828. He rose through the ranks, holding the position of dean of mathematics and physics from 1840 to 1863, and then rector until his death.

Lewes, George Henry (1817–78), literary critic and editor. Along with Thornton Leigh Hunt, he established the *Leader* in London in 1850, with financial backing from the noted liberal clergyman Edmund Larken. He wrote a series of articles espousing the philosophy of the French philosopher Auguste Comte, published initially in the *Leader* in 1852, and subsequently brought together in *Comte's Philosophy of the Sciences* (1853). He had an open marriage with Agnes Jervis, who also had a relationship, including several children, with Hunt. Lewes met Mary Ann Evans (George Eliot) in 1851, and they lived together beginning in 1854.

Liebig, Justus (1803–73), German chemist and educator, he taught chemistry at the University of Giessen (extraordinary professor in 1824 and ordinary professor in 1825), where he significantly advanced the fields of organic and agricultural chemistry, advised hundreds of students, and pioneered laboratory-based methods of teaching. In 1840, he founded one of the leading journals of chemistry of the time, *Annalen der Chemie.* In 1852 he moved to the University of Munich but was plagued with poor health, which limited his teaching and research activity.

Mackay, Charles (1814–89), Scottish author, journalist and editor, best known for his book entitled *Extraordinary Popular Delusions and the Madness of Crowds* (1841). He began working for the *Illustrated London News* in 1848 and became its editor in 1852. Mackay had written encouragingly to Tyndall in 1848, after Tyndall thanked him for his writings and told him of his own ambitions (see letter 0362).

Magnus, Heinrich Gustav (1802–70), German chemist and physicist, who received his degree from the University of Berlin in 1827 under Mitscherlich. Magnus spent brief periods in Stockholm and Paris, but returned to the University of Berlin (privatdozent in 1831, extraordinary professor in 1833, and ordinary professor in 1845). He was an extremely popular teacher and established a world-renowned laboratory in Berlin. Tyndall worked in Magnus's laboratory for several months in 1851.

Martin, Anna (1831–57), sister of Mary Simpson. She met Hirst in Marburg in 1852 when she came to stay with the Simpson family. She was a sociable and playful young woman and Hirst tried to hide his interest in her by adopting a "fatherly" role (for example, letter 0634); his letters seldom refer to her by name. They were married in December 1854. Soon after her marriage, she showed signs of consumption. Hirst resigned from his position at Queenwood to take her to the warmer parts of Europe. She died in Paris in July 1857. (Brock, "Queenwood.")

Mitscherlich, Eilhard (1794–1863), German chemist and crystallographer, who taught chemistry at the University of Berlin (extraordinary professor in 1822 and ordinary professor in 1825). His research was wide ranging, including organic and inorganic chemistry and isomorphism in crystallography. He also wrote a successful textbook, *Lehrbuch der Chemie* (1829), which went through numerous editions. Tyndall met him in Berlin in 1851.

Montaigne, Michel de (1533–92), influential writer of the French Renaissance. Montaigne's famous *Essais* (1590) were reflections upon his self and his society, relativist and sometimes sceptical in tone. They were still read in the early 1850s, as they are today.

Müller, Johann Heinrich Jacob (1809–75), German physicist and mathematician. He studied first at the University of Bonn under Julius Plücker, then at the University of Giessen, where he received his PhD in 1833, focusing on crystals and polarized light. In 1844 he rose to professor of physics and technology at the University of Freiberg where he conducted research primarily on optics and radiant heat.

Noll (Knoll) (possibly Friedrich Noll), medical doctor and friend of Tyndall during his time at Marburg. Both Hirst and Tyndall were fond of Noll, but privately they were also condescending toward him, owing to his seeming lack of ambition. Friedrich Noll studied under Carl Ludwig, Professor of Physiology at Marburg, in the late 1840s, and published an article on the lymphatic system in 1850.

Perkinton, Thomas ("Tom"), friend of Tyndall and Hirst dating back to their surveying days in Halifax. The 1850 *Directory of Halifax, Huddersfield, Holmfirth, and Surrounding Villages* lists Perkinton as a surveyor and land agent. He also went into business with George Searle Phillips, though the nature of the business is not known.

Phillips (Philips), George Searle ("January," pseud. January Searle) (1816–89), a journalist, ardent Carlylean, and social reformer. He was secretary of the very successful Huddersfield Mechanics' Institute. From 1844 he was editor of the *Leeds Times;* he also edited the Transcendentalist publication the *Truth Seeker* along with F. R. Lees, and was a lecturer at the People's College in Huddersfield. Through Carlyle, he befriended Hirst in 1849. January Searle was one of his many pen names, by which name Hirst and Tyndall often referred to him.

Phillips, John (1800–74), English geologist and paleontologist, he was Assistant General Secretary of the BAAS from 1832 to 1862. He served as Keeper of the museum of the Yorkshire Philosophical Society from 1825 to 1840, and a reader (1853) then professor (1860) of geology at Oxford.

Phillips, Reuben, a chemist, who began publishing on electricity and then magnetism in the mid-1840s. His contributions to the *Phil. Mag.* were often reviewed negatively by Tyndall.

Piercy, Ada/Adah, a particularly beautiful young woman whom Tyndall knew in Halifax (see Journal, January–February 1847, JT/2/13a/173–8). The journal entries are ambiguous, perhaps a deliberate strategy to hide his romantic interest. Tyndall described her as a "sweet little girl," implying that she was a child: "she promises to become mine if I wait eight years and at the end of that time am not an old man with grey hair" (ibid., 24 January). She was not a child, but six or seven years younger than Tyndall (ibid., 22 January).

Playfair, Lyon (1818–98), British chemist, who from 1839 through 1841 worked in Liebig's laboratory in Giessen. In 1842 he became professor of chemistry at the Royal Manchester Institution. In 1845, he became chemist to the Geological Survey and professor of chemistry at the School of Mines, and in 1848 was elected a FRS. In October of 1851, Tyndall sent Playfair a copy of Robert Bunsen's article on the composition of the Plutonic rocks of Iceland in the hopes of it being published in the reports of the Chemical Society.

Plücker, Julius (1801–68), German mathematician and physicist, he worked primarily on the behavior of crystals in a magnetic field in his laboratory at the University of Bonn, where he taught physics and mathematics (privatdozent in 1825, extraordinary professor in 1828, and ordinary professor in 1833). His early work focused on mathematics, but in 1846 he turned to magnetism and diamagnetism. Weaknesses in Plücker's experimental work provided the chief stimulus to the joint research of Tyndall and Knoblauch in 1849–50.

Poggendorff, Johann Christian (1796–1877), German physicist at the University of Berlin. His major contribution to science was as editor for fifty-two years (1824–76) of *Annalen der Physik und Chemie,* which became known as *Poggendorff's Annalen* during his editorship. He used his broad knowledge of science and scientific men in producing a *Biographisch-literarisches Handworterbuch zur Geschicte der exacten Wissenschaften,* the first two volumes of which were published in 1863.

Poisson, Siméon Denis (1781–1840), French mathematician who applied mathematics to many branches of physics, including electricity and magnetism. He joined the faculty of the École Polytechnique in 1802, replacing Fourier as professor in 1806. After Thomson had mentioned Poisson's work during discussions at the Edinburgh BAAS meeting in 1850, Tyndall asked Thomson for references, which Thomson supplied, but Tyndall admitted that he struggled to understand the difficult mathematics involved.

Pridie, Roby, a friend of Hirst in Halifax, and son of the Rev. James Pridie, an Independent Minister in Halifax.

Riess, Peter Theophil (1805–83), German physicist who received his PhD in 1831 from the University of Berlin and whose work focused on electricity and magnetism. He was the first Jew admitted into the Royal Prussian Academy of Sciences.

Ronalds, Edmund (1819–89), English chemist and, in the period of this volume, founding Professor of Chemistry (1849–56) at Queen's College, Galway. He had been an assistant to Justus Liebig at Giessen and completed a PhD under him in 1842, before travelling to other European universities. Before moving to Galway he was a lecturer at Middlesex Hospital in London, served as Secretary of the Chemical Society (1848–50) and edited the first two volumes of its *Quarterly Journal* in 1849 and 1850. When he resigned from Galway, he took up a position in industry in Edinburgh.

Rose, Heinrich (1795–1864), German analytical chemist, and brother of the scientist Gustav Rose (1798–83). Heinrich taught at the University of Berlin (privatdozent in 1822; extraordinary professor in 1823, and ordinary professor in 1835), where Tyndall visited him in 1851.

Russell, John Scott (1808–82), a Scottish trained civil engineer and naval architect, who was active in the BAAS and was elected FRS in 1849. Russell was interested in all forms of transport and in improving technical education. In 1845 he became editor of the *Railway Chronicle,* and at the time of the Great Exhibition was secretary of the Royal Society of Arts. From 1852 he worked with Isambard Brunel on the design of the *Great Eastern.* In 1856 he was in severe financial trouble. Russell was brother-in-law to George Wynne, and thus a source of indirect advice and assistance to Tyndall.

Sabine, Edward (1788–1883), Irish astronomer and geophysicist, military officer, and science administrator. He was the astronomer for the Ross (1818) and Parry (1819–20) Expeditions to the Arctic in search of the Northwest

Passage. These expeditions led Sabine to his studies of the earth's magnetic field, and to his campaign to gather global magnetic observations. He was an outstanding lobbyist for science and effective administrator, and a valuable patron to Tyndall from 1851. In the early 1850s he was Treasurer of the Royal Society, General Secretary of the BAAS, and held the rank of Lieutenant-Colonel in the army. As President-elect of the BAAS in 1852, Sabine offered Tyndall one of the secretaryships at the Belfast meeting, in charge of several of the sessions and the order of papers. He proposed Tyndall for election to the Royal Society, using his influence to help garner support and gather signatures.

Schell, Wilhelm Joseph Friedrich Nikolaus (1826–1904), a student at Marburg, who obtained his PhD in mathematics in 1851 and immediately became a privatdozent in mathematics. He read mathematics with Hirst, and assisted with his thesis in mid-1852. He later held professorships at Marburg (from 1856) and the Polytechnic of Karlsruhe (from 1861).

Schiller, Friedrich (1759–1805), German philosopher and poet, whose works focused on ethics, aesthetics and notions of the sublime. His works were often discussed by Hirst and Tyndall.

Schmidt/Schmitt, Carl/Karl, a fellow student of Tyndall and Hirst at the University of Marburg. Hirst recorded in his Journal (1 February 1852) that Carl Schmitt was a poet, a cartoonist, and a member of the French Kränzchen.

Schnackenberg, Friedrich (later Frederic), and his father, a Marburg student who went to England in 1851 and initially worked for Singleton (Journal, 2 April 1851, JT/2/13b/524). He later set up in Blackburn, Lancashire, as a teacher of piano, German, French, mathematics and chemistry. In the later 1850s he became music master at a local grammar school, where he remained for decades. His father was a protestant pastor, "Pfarrer," in the village of Braach, about 80 miles ENE of Marburg. (Advertisements and articles in the *Blackburn Standard* and *Preston Chronicle*.)

Searle, January. See Phillips, George Searle

Simpson, Maxwell (1815–1902), Irish chemist, who traveled in 1851 to study chemistry at the University of Marburg. He had previously studied at Trinity College, Dublin (from 1832) and at University College, London. In 1847 he became lecturer in chemistry at Park St Medical School in Dublin and later at the Peter Street School of Medicine. He studied in Germany (1851–4) and Paris (c.1858–60). On return to Dublin he built a private laboratory where he conducted research in chemistry. With his family, he lived briefly in London, c.1870, and in 1872 he was appointed Professor of Chemistry at Queen's College, Cork, from where he retired in 1892. His wife was Mary Martin, daughter of Samuel Martin of Loughorne, County Down. They had three children, John, Lucy and Mary, when they arrived in Marburg. Hirst later married Mary Martin's sister, Anna. (For a Martin family tree, see Brock and MacLeod, *Hirst Journals,* opp. f. 1509.)

Singleton, Josiah ("Father"), younger brother of Anne Edmondson, he was previously a teacher at Queenwood, where he was a friend and ally of Tyndall and Frankland. They considered him unfairly treated when he was forced to leave Queenwood by the Edmondsons. He opened his own school at Spring Bank, in Over Darwen, Lancashire, and remained there until at least 1881. Tyndall helped teach there in the summer of 1850. (Brock, "Queenwood.")

Smith, John Stores (1829–92), a young literary man and disciple of Carlyle. He was introduced to Hirst by George Searle Phillips, and hence became known to Tyndall. Smith was born and educated in Manchester, set up a business in Halifax while pursuing literature as an avocation, and then sought to follow a literary life in London, where he financed and wrote for the short lived *Leigh Hunt's Journal* (7 Dec. 1850–29 March 1851). He suffered an unidentified marital crisis early in 1851, which may have affected his financial circumstances, for he attempted unsuccessfully to make a living in journalism and requested money from both Hirst and Tyndall. He published *Maribeau: A Life History* (1848), and *Social Aspects* (1850), both Carlylean in flavor. He returned to Manchester and commercial life, and in later life was politically active, for example, in the free trade movement.

Socrates (c. 470/69–399 BC), ancient Athenian philosopher and, especially in later life, a moralist. He is reputed to have taught through dialogue or cross-questioning. Condemned to death by the city of Athens for impiety and "corrupting the minds" of its youth, later generations have turned Socrates into a symbol of freedom of conscience for his choice to drink hemlock rather than escape from prison. He is known to us only through the reports of a later generation, most notably of the philosopher, Plato.

Stegmann, Friedrich Ludwig (1813–91), German mathematician, professor of mathematics at the University of Marburg since 1845, and PhD adviser to Tyndall and Hirst. He was the author of *Lehrbuch Der Variationsrechnung Und Ihrer Anwendung Bei Untersuchungen Über Das Maximum Und Minimum* (1854), an instructional calculus book.

Sterling, John (pseud. Caliban), Welsh surgeon and friend of Hirst, who probably met him through George Searle Phillips. Under the pseudonym "Caliban," Sterling published articles in the *Leader* and *Truth Seeker* on Carlyle, which were much discussed by Tyndall and Hirst.

Steuart, Elizabeth Dawson (née Duckett) (1802–93), was, like her husband, born into a wealthy family. She married William Richard Steuart on 18 August 1820, and they lived at Steuart's Lodge, Leighlin Bridge. She took a keen interest in Tyndall's career and was probably his most consistent and long-lived patron. She continued to support Tyndall after her husband died.

Steuart, William Richard (1798–1852), descended from an established family of Scottish ancestry that became wealthy landowners after settling in Ireland in the early eighteenth century. In 1820 he married Elizabeth Dawson Duckett, from another prosperous Protestant land-owning family. He held the rank of Captain, served as a local magistrate, and was appointed High Sherriff of Carlow in 1821. He owned many of the properties in Leighlin Bridge, including the house where the Tyndalls lived. In Ireland Tyndall and his father looked to the Steuarts for patronage. Steuart continued to look out for Tyndall's interests after he left Ireland, as is illustrated in this volume.

Stokes, George Gabriel (1819–1903), English physicist and mathematician, he held many distinguished positions throughout his life, including the Lucasian Chair of Mathematics at Cambridge in 1849, Secretary (1854–83) and President (1885–90) of the Royal Society. Stokes's earliest research focused on hydrodynamics, including a significant paper in 1850 where he applied his work on the friction of fluids to the behavior of pendulums. In the early 1850s he turned to questions of optics, and in 1852 published on fluorescence, a term he coined. Perhaps his major contribution to contemporary science was his oversight, for decades, of the refereeing for the *Philosophical Transactions.*

Stoney, George Johnstone (1826–1911), an Irish physicist and astronomer. He received his degree from Trinity College, Dublin, in 1852, the same year in which he accepted a professorship in natural philosophy at Queen's College, Galway, a position for which Tyndall also applied.

Sylvester, James Joseph (1814–97), English mathematician of Jewish descent. He was appointed to the professorship of natural philosophy at University College, London in 1838, elected FRS in 1839, and accepted a professorship at the University of Virginia in 1841, though he stayed there only six months. After returning to England the following year, he met Arthur Cayley, a fellow mathematician, and together they made significant advances in the theory of invariants. Through the 1850s Sylvester continued to be a prolific mathematical researcher, often publishing in the *Phil. Mag.* He signed Tyndall's nomination certificate for the FRS in 1852. In 1855, he was appointed professor of mathematics at the Royal Military Academy in Woolwich.

Taylor, Richard (1781–1858), prominent English publisher, especially of scientific journals, including *Annals of Science, Annals of Natural History,* and the *Phil. Mag.*, where Tyndall submitted his first scientific papers for publication. He also founded and edited seven volumes of *Scientific Memoirs* between 1837 and 1852, a collection of translations of major foreign language scientific works. Taylor married Hannah Corke in 1807 and with whom he had one daughter, Sarah. After Hannah was struck down by severe psychiatric illness, he had a long-term relationship with Frances Francis, his daughter's governess, with whom he had two children including his only son, William Francis. Taylor suffered a nervous breakdown in 1852 and was sent to an asylum at Sydenham, where Tyndall often visited him. He went into partnership with his son William at that time, founding the "Taylor & Francis" publishing firm, an imprint that first appeared on the cover of the *Phil. Mag.* in June 1852. Francis and Sarah moved to Richmond the following year to look after their father, and Tyndall moved into Francis's former lodging in London. (Brock and Meadows, *Lamp of Learning.*)

Thomson, William (1824–1907), brilliant young mathematician and physicist, he accepted the professorship of natural philosophy at Glasgow University in 1846, a post he held for fifty-three years. He published significant papers on the dynamic theory of heat in the late 1840s and early 1850s. Tyndall referred to Thomson as his "scientific antagonist" (letters 0418 and 0661) and for much of his career regarded Thomson as a scientific rival.

Tidmarsh, John C. ("Jack") (1824–1906), a close colleague and friend of Tyndall during his days on the Irish Ordnance Survey. In the early 1840s, he worked closely with Tyndall in Youghal, where he lodged with him and William Ginty, and again in Kinsale and Cork, where Tyndall enjoyed the hospitality of Tidmarsh's family. In 1845, he was employed, alongside Tyndall, making surveys for the West Riding Union Railway, and the two again shared lodgings together in Halifax. In March 1849 Tidmarsh emigrated to Australia, settling in Adelaide, where he initially worked as a surveyor and valuator for the city's first building society and later prospected during the Australian gold rush in 1852. He established a successful candle and soap manufacturing company, although overwork impaired his health. In November 1906 Tidmarsh ended his own life by firing a revolver into his head ("Death of Mr. John Tidmarsh," *Register* (Adelaide), 12 November 1906, p. 6).

Wartmann, Élie-François (1817–86), Swiss natural philosopher; professor of physics at the Academy of Geneva in 1851, when he wrote a testimonial for Tyndall. Tyndall translated Wartmann's paper on the polarization of atmospheric heat for inclusion in the *Phil. Mag.* in 1852.

Weber, Wilhelm Eduard (1804–91), German physicist and a professor of physics at Göttingen from 1831, and then Leipzig from 1843, before returning to his former position in Göttingen in 1849. Along with his colleague Gauss, he established the Göttingen Magnetische Verein to coordinate magnetic observatories, first in the Germanic states and then across Europe. He collaborated with Kohlrausch, establishing the ratio between the electrodynamic unit of charge and the electrostatic unit of charge. He also became involved in the experimental and theoretical development in diamagnetism. Hirst visited Weber in Göttingen, and he also wrote a testimonial for Tyndall.

Wheatstone, Charles (1802–75), an English experimental physicist and prolific inventor. In 1834 he accepted a professorship of experimental physics at King's College, London. In the same year he used a revolving mirror to estimate the speed of electricity moving through a wire, a similar experimental design that he later used to estimate the speed of light. Tyndall was especially interested in Wheatstone's invention of electro-magnetic clocks, which could be used to measure small intervals of time.

Whewell, William (1794–1866), English tidologist and historian and philosopher of science. He held professorships first in Mineralogy (1828), then in Moral Philosophy (1838), and ultimately he accepted the Mastership of Trinity College, Cambridge (1841). Whewell is best known for his multi-volume *History of the Inductive Sciences* (1837) and *Philosophy of the Inductive Sciences* (1840). His own research in physical astronomy focused on what he termed "tidology." He wrote fourteen major papers on the study of the tides, the last appearing in 1850.

Woods, Thomas (1817–1906), Irish physician, who by 1854 was physician to the Killyon and Balybritt Dispensaries in Parsonstown. In 1857 he was put in charge of the Parsonstown Workhouse Infirmary and Fever Hospital. He wrote several articles in the *Phil. Mag.* throughout the early 1850s on the heat of chemical combinations, which Tyndall reviewed and with which he often disagreed.

Wrightson, Francis, son of a Birmingham manufacturer, he was a student who struggled in his studies in Marburg—yet he had an articled pupil, Davy, with him. Hirst gave him advice and tutored both him and his pupil. Wrightson obtained a PhD in 1853 under the direction of Kolbe. It is possible that he was not as incompetent as Hirst represents him in these letters. He was later involved in agricultural chemistry in Warwickshire. (Brock and MacLeod, *Hirst Journals,* n. 193.).

Wynne, Anne (née Osborne) (d. 1864), daughter of Sir Daniel Toler Osborne, Bart. She married George Wynne in 1834. The letters in this volume show that she often took independent initiatives to promote Tyndall's interests. The Wynne's had three sons (Henry, Edward and Frank), to whom Tyndall sometimes sent letters, and one daughter, Lucy. In 1850–52 the Wynnes lived at Harrow.

Wynne George (1804–90), a former boss, and a friend, supporter and patron of Tyndall. Wynne was from a military family, although his father was a clergyman in Killucan. From 1825 he was a Lieutenant in the Royal Engineers and from 1835–40 employed in the Irish Ordnance Survey. When he and Tyndall met, Wynne was in charge of the Division in which Tyndall worked (see Volumes 1–2). He was a military officer until 1877, when he retired with the rank of General. Over the period of this volume he had the rank of Captain, and the position of Inspector of Railways (August 1847 to April 1858) under the Board of Trade. He married Anne Osborne in 1834.

Wynne, John Arthur (1801–65), a politician from Sligo, was undersecretary to the Lord Lieutenant of Ireland from February–December 1852. He was the cousin of George Wynne.

Yeats, John, a tutor at Queenwood during Tyndall's first period of teaching there (1847–8). He was rumored to be a spy for the Edmondsons. Frankland and Tyndall came to dislike him especially heartily when he undermined the position of fellow teacher, Jane Edmondson's brother, Josiah Singleton. Frankland and Tyndall sided with Singleton against Yeats and the Edmondsons. Yeats had left Queenwood before Tyndall returned there to teach in 1851.